AF684626

# Advances in Biochemistry in Health and Disease

Volume 26

**Series Editor**

Naranjan S. Dhalla, Institute of Cardiovascular Sciences, St. Boniface Hospital, Winnipeg, MB, Canada

**Editorial Board**

Roberto Bolli, Department of Medicine and Cardiology, University of Louisville, Louisville, KY, USA

Ramesh Goyal, Delhi Pharmaceutical Sciences and Research, New Delhi, India

Chandrasekharan Kartha, Cardiovascular Diseases and Diabetes Biology, Kerala Institute of Medical Sciences, Thiruvananthapuram, Kerala, India

Lorrie Kirshenbaum, St. Boniface General Hospital, Winnipeg, MB, Canada

Naoki Makino, Kyushu University, Fukuoka, Japan

Jawahar L. L. Mehta, Division of Cardiology, University of Arkansas for Medical Sciences, Little Rock, AR, USA

Bohuslav Ostadal, Institute of Physiology, Czech Academy of Sciences, Prague, Czech Republic

Grant N. Pierce, St. Boniface General Hospital, Winnipeg, MB, Canada

Jan Slezak, Institute for Heart Research, Slovak Academy of Sciences, Karlova Ves, Slovakia

Andras Varro, Department of Pharmacology and Pharmacotherapy, University of Szeged, Szeged, Hungary

Karl Werdan, Martin Luther University Halle-Wittenber, Halle (Saale), Sachsen-Anhalt, Germany

William B. Weglicki, School of Medicine and Health Sciences, George Washington University, Washington, USA

Advances in Biochemistry in Health and Disease focus on the latest developments in biochemical research with implications for health and disease. This book series consists of original edited volumes and monographs, presented by leading experts in the field and provides an up to date and unique source of information for all those interested in the fundamental, biochemical processes of the latest and emerging topics and techniques.

Covering a wide variety of topics, this book series is a valuable source of information from those at the lab bench through to the Health Care workers.

Lorrie Kirshenbaum · Inna Rabinovich-Nikitin
Editors

# Biology of Women's Heart Health

Springer

*Editors*
Lorrie Kirshenbaum
The Institute of Cardiovascular Sciences
St. Boniface Hospital Albrechtsen Research Centre
Winnipeg, Manitoba, Canada

Department of Physiology and Pathophysiology, Rady College of Medicine
Max Rady Faculty of Health Sciences
University of Manitoba
Winnipeg, Manitoba, Canada

Department of Pharmacology and Therapeutics, Rady College of Medicine
Max Rady Faculty of Health Sciences
University of Manitoba
Winnipeg, Manitoba, Canada

Inna Rabinovich-Nikitin
The Institute of Cardiovascular Sciences
St. Boniface Hospital Albrechtsen Research Centre
Winnipeg, Manitoba, Canada

Department of Physiology and Pathophysiology, Rady College of Medicine
Max Rady Faculty of Health Sciences
University of Manitoba
Winnipeg, Manitoba, Canada

ISSN 2512-2142 ISSN 2512-2150 (electronic)
Advances in Biochemistry in Health and Disease
ISBN 978-3-031-39927-5 ISBN 978-3-031-39928-2 (eBook)
https://doi.org/10.1007/978-3-031-39928-2

© The Editor(s) (if applicable) and The Author(s), under exclusive license to Springer Nature Switzerland AG 2023

This work is subject to copyright. All rights are solely and exclusively licensed by the Publisher, whether the whole or part of the material is concerned, specifically the rights of translation, reprinting, reuse of illustrations, recitation, broadcasting, reproduction on microfilms or in any other physical way, and transmission or information storage and retrieval, electronic adaptation, computer software, or by similar or dissimilar methodology now known or hereafter developed.
The use of general descriptive names, registered names, trademarks, service marks, etc. in this publication does not imply, even in the absence of a specific statement, that such names are exempt from the relevant protective laws and regulations and therefore free for general use.
The publisher, the authors, and the editors are safe to assume that the advice and information in this book are believed to be true and accurate at the date of publication. Neither the publisher nor the authors or the editors give a warranty, expressed or implied, with respect to the material contained herein or for any errors or omissions that may have been made. The publisher remains neutral with regard to jurisdictional claims in published maps and institutional affiliations.

This Springer imprint is published by the registered company Springer Nature Switzerland AG
The registered company address is: Gewerbestrasse 11, 6330 Cham, Switzerland

Paper in this product is recyclable.

# Contents

# Chapter 1
# The Emerging Need for Research on Women's Heart Health

**Huong Nguyen, Lorrie A. Kirshenbaum, and Inna Rabinovich-Nikitin**

**Abstract** The neglect of women's heart health in biomedical research has led to a lack of understanding about the unique physiological differences between men and women and how these differences impact the prevention, diagnosis, and treatment of heart disease. This lack of research has been, in part, due to the exclusion of women from clinical studies and clinical trials, as well as social and cultural factors such as the gender pay gap and the underrepresentation of women in leadership roles in the scientific community. Despite the fact that cardiovascular disease is the leading cause of death in U.S.A, there is still a lack of understanding about the prevention, diagnosis, treatment, and therapy of heart diseases in women. In addition, there is a lack of research on the unique physiological differences between men and women and how these differences impact the effectiveness and side effects of drugs. This lack of research is not only detrimental to women's health, but also to society as a whole, as women play a vital role in the health and well-being of their families and communities. Therefore, it is crucial that more research is done to address the neglect of women's heart health and to improve outcomes for women with heart disease. This research should focus on the development of gender-specific prevention, diagnosis and treatment strategies and on participation of women in clinical studies and drug trials. By addressing these issues, we can make significant progress in improving the outcomes for women with heart disease and ensure that women's health is no longer excluded from biomedical research.

**Keywords** Risk factors · Women physiology · Ethnicity · Awareness · Symptoms

---

H. Nguyen · L. A. Kirshenbaum · I. Rabinovich-Nikitin (✉)
Institute of Cardiovascular Sciences, St. Boniface Hospital Albrechtsen Research Centre, Department of Physiology and Pathophysiology, Rady Faculty of Health Sciences, University of Manitoba, Winnipeg, MB, Canada
e-mail: irabinovich-nikitin@sbrc.ca

L. A. Kirshenbaum
Department of Pharmacology and Therapeutics, Rady Faculty of Health Sciences, University of Manitoba, Winnipeg, MB, Canada

© The Author(s), under exclusive license to Springer Nature Switzerland AG 2023

L. Kirshenbaum and I. Rabinovich-Nikitin (eds.), *Biology of Women's Heart Health*, Advances in Biochemistry in Health and Disease 26,
https://doi.org/10.1007/978-3-031-39928-2_1

## The History of woman's Health Research

Historically, women have been excluded from most biomedical research and drug trials, including cardiovascular disease-related ones. This was mainly explained by safety considerations, such as women being in childbearing age or, alternatively, women having coexisting illnesses when included in research in older age [1, 2]. This biased approach led to a default assumption that what works for males would also work for females, resulting in an unfilled gap in the understanding of the effect of drugs on women's bodies. However, epidemiological reports of clinical studies have shown that men often respond differently to drugs than women and that the outcomes between men and women may vary, especially in cardiovascular diseases [3, 4]. As a result, in 1993, the U.S. National Institutes of Health (NIH) issued guidelines that required federally funded studies to include women and minorities in clinical trials to ensure that the safety and efficacy of drugs would be adequately investigated in the full range of patients who could benefit from the drug [5]. Despite the fact that a large amount of money was allocated towards funding gender-based studies, many of those who enrolled both sexes in their studies failed to report gender-specific analysis [6]. Consequently, eight out of ten prescription drugs were withdrawn from the U.S. market in the late 90 s because they proposed a greater health risk for women than men [7]. Consequently, one of the main reasons why scientists are hesitant to include women in their studies is because of how little is known about the physiology of women's bodies compared to men. For example, women's hormonal cycles are known to be a major factor in regulating women's health; however, the cycling nature of female hormones may affect the homogeneity of the study. Furthermore, another important consideration that prevented the reassessment of experimental data performed in men's models into women's models is the high financial cost and effort of repeating the experiments [8, 9]. For this reason, despite the increasing development of science, knowledge and understanding of women's physiology remains poor.

## The Need for Woman's Heart Health Research

According to the Centers for Disease Control and Prevention (CDC), cardiovascular disease is the number one cause of death in women in the U.S.A. Surprisingly, however, 44% of women are not aware of this fact [10]. With the long history of being excluded from biomedical research, we still do not have much understanding of prevention, diagnosis, treatment and therapy for women when it comes to heart diseases. As mentioned, many disease research and drug development were solely based on the fact that men represent the human species, which was proven ineffective due to the differences in physiological traits in both sexes. Hence, women might experience the efficiency and side effects of certain drugs differently than men; they even experience different symptoms and severity for the same diseases [11].

One of the most known examples in recent years is the difference in heart attack symptoms between men and women. While both men and women can experience chest pain or discomfort, shortness of breath, and pain or discomfort in the jaw, neck, back, arm, or shoulder, women can also experience signs of nauseous, light-headedness, or unusual tiredness (Fig. 1.1). These unique symptoms lead to misdiagnosis of women, which may cause late treatment and result in worse damage to the heart [12]. Moreover, women who report chest pain or slight chest discomfort are often overlooked as having anxiety, panic disorder, stress or heartburn. Hence, 41% of women wait for more than 12 h at the first sign of heart attack before seeking medical care and miss the treatment window [13]. Furthermore, since women are historically considered to be protected from heart disease during pre-menopausal age, many times, their chest pain is being ignored and attributed to non-cardiac conditions due to their age [14]. As a result, studies have shown that women have a lower chance of recovering after their first heart attack and are more likely than men to die within one year after an acute MI [15, 16].

Awareness of the increased risk for cardiac disease in women is slowly rising. Although only 53% of women report that they would reach for medical help if they are experiencing symptoms of a heart attack, the knowledge that heart disease is the leading cause of death for women has doubled since 1997 [17, 18]. Nevertheless, the gap in knowledge and awareness is also ethnically based, with reports showing that African American and Hispanic women are significantly less aware than white women of the risk for cardiac disease in women [18, 19]. Furthermore, mortality rates can also be affected by reduced awareness, limited access to healthcare, and inequalities in social and economic status [20]. This is best exemplified by the observation that mortality rates from cardiovascular disease among indigenous women in Canada are 53% higher compared to non-indigenous women [21]. Based on that, more efforts

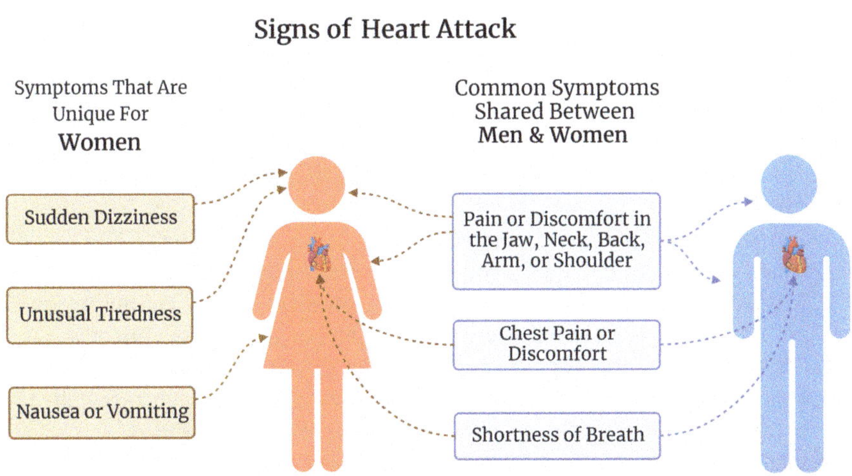


**Fig. 1.1** The similarities and differences in symptoms of a heart attack in women and men

to raise awareness and reduce gender disparities in research and clinical care should also focus on ethnic and cultural differences, in addition to sex differences.

Another important aspect that should be considered when aiming toward increasing awareness among women understands why women delay reaching medical help compared to men. Data derived from the standard survey have shown that 51% of women postpone reaching for medical help due to family/caretaking responsibilities and 42% due to confusion in the media [18]. These findings highlight once again the need to educate women on the different symptoms of heart attack healthcare providers and through public knowledge translation initiatives.

## Understanding the Risk Factors for Cardiovascular Disease in Women

Population studies have shown that 96% of myocardial infarction (MI) cases in women are attributed to traditional risk factors, such as smoking, alcohol consumption, hypertension, diabetes mellitus, obesity, unhealthy diet, and sedentary behaviour [22, 23]. However, despite the major effects of traditional risk factors on heart disease in women, another crucial risk factor that contributes to the incidence and risk of cardiovascular disease is the hormonal stage. Estrogen is a steroid hormone that is present in high levels in females from adolescence to menopause but drops significantly post-menopause. Women undergo hormonal fluctuations throughout life, making their susceptibility to cardiovascular diseases vary at different stages. For example, early age at menarche, irregular menstrual cycles, such as seen in polycystic ovary syndrome, and oral contraceptive therapy have all been shown to increase the risk of cardiovascular disease and predispose women to MI and/or stroke at an early age [24–26].

Furthermore, another dramatic risk factor for heart disease in women is pregnancy. Pregnancy presents major hormonal and metabolic stress to the body, which may contribute to the worsening or development of cardiac disease. In fact, studies have shown that cardiac disease mays present in 1–4% of all pregnancies [27]. Moreover, hypertensive disorders during pregnancy, such as preeclampsia, increase the risk for coronary heart disease, heart failure and stroke later in life [28, 29]. Another common pathology during pregnancy is gestational diabetes mellitus (GDM). 6–9% of pregnant women develop GDM [30], which increases the risk for type II diabetes (T2DM) up to 60% later in life [31], and consequently increases susceptibility to heart disease [32]. Therefore, it is important to highlight that although historically, much research has been done on women's health during pregnancy, not much progress has been done in the area of female-related risk factors during pregnancy, especially the risk for cardiovascular disease and non-communicable diseases [33].

Another important risk factor for cardiovascular disease in women is based on demographic differences. For example, lower socioeconomic status has been associated with reduced access to treatment, such as cardiac catheterization, resulting in

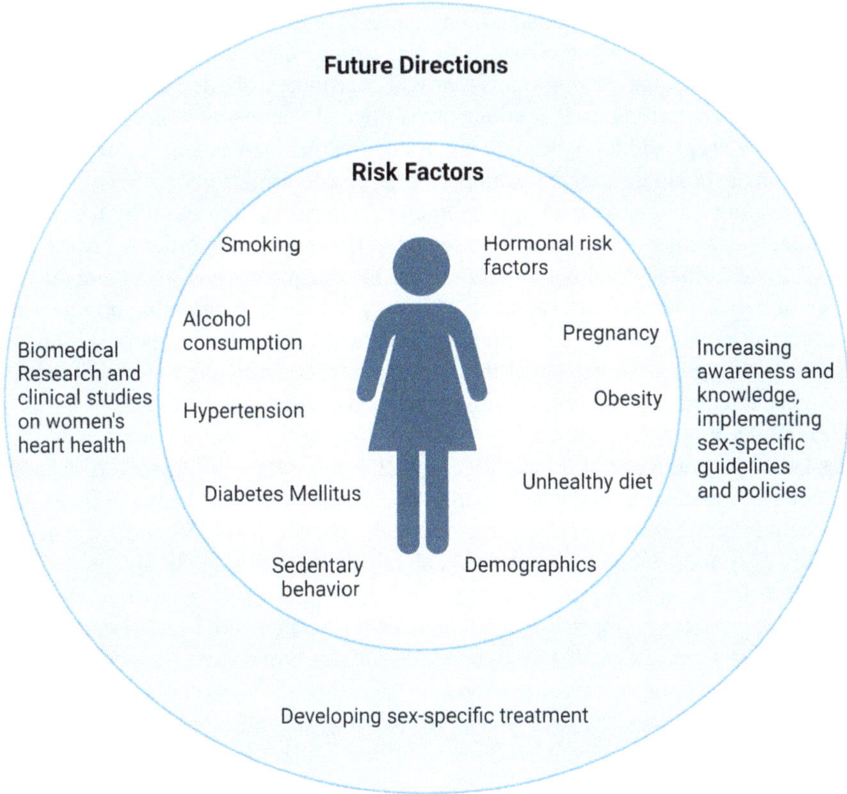


**Fig. 1.2** Schematic representation of the risk factors of cardiovascular disease in women and the future direction that need to be taken in order to transform women's heart health globally

higher mortality rates [34]. This link is especially prominent when comparing men and women from similar socioeconomic statuses. Therefore, understanding women's unique risk factors (Fig. 1.2) and promoting primary prevention is a crucial step in reducing the mortality rate due to cardiovascular disease in women.

## Future Directions

It is now clear that there is an emerging need for more research on women's heart health that will address the unique physiological differences between men and women and contribute to improved outcomes following cardiac disease among women. Such research should focus on several key areas in order to achieve significant progress in the field. One important area of research is the development of gender-specific prevention and treatment strategies for heart disease. This will be possible by increasing research that will focus on understanding the unique differences in physiology and risk factors of women while considering sex, ethnicity, culture and socioeconomic differences. Consequently, this knowledge will contribute to the development of interventions that are tailored to the unique physiological features of women and should involve new drugs and therapies that are more effective for women. Another important area that should be further developed when considering women's heart health is the inclusion of and participation of women in clinical studies and drug trials. It is essential that women are equally represented in these studies in order to ensure that the safety and efficacy of drugs and treatments are adequately investigated in the full range of patients who could use them. This may require implementing policies and funding initiatives that will increase the participation of women in these studies, as well as ensuring that the results of these studies are analyzed and reported in a way that takes into account the unique differences between men and women.

Finally, a major focus should be made on increasing training and knowledge translation for both general population, as well as healthcare providers. This should include the implementation of sex-specific guidelines and public health policies that will help increase awareness, remove physical, social and socioeconomic barriers and will encourage women to seek medical help at an early stage of symptoms.

With a focus on these multilevel approaches (Fig. 1.2), it is possible to make significant progress in improving the outcomes for women with heart disease and in ensuring that women's health is no longer neglected in biomedical research. Furthermore, the motivation to invest in women's heart health is important not only for women but also for society as a whole. Women play a central role in the economy, society and family life. Therefore improving the outcomes for women with heart disease can have a significant positive impact in many areas.

## Conclusions

Cardiovascular disease continues to be the leading cause of death for women, and it is now well-known that it can affect women at any age. However, there is still a major gap in knowledge that needs to be filled in order to explain the sex-specific characteristics of different heart diseases. Notably, sex-specific differences in heart disease are evident not only in research but also in education, prevention, diagnosis and treatment. Some barriers that may contribute to this gap have been discussed in this chapter and include the underrepresentation of women in therapeutic and biomedical research on cardiovascular disease, lack of public awareness, and insufficient resources invested in promoting women's heart health. Moreover, the sex-dependant differences in heart disease should be further focused on socioeconomic, demographic, cultural, racial, and ethnic differences that contribute to additional increased risk for specific groups of women. In order to be able to fill the gap in knowledge on women's heart health, it is important to enhance research and clinical trials by introducing more funding and study initiatives that would be the foundation to this emerging area of research and will eventually help transform women's heart health globally.

## References

1. Gurwitz JH, Col NF, Avorn J (1992) The exclusion of the elderly and women from clinical trials in acute myocardial infarction. JAMA J Am Med Assoc 268(11):1417–1422
2. Wenger NK (1992) Exclusion of the elderly and women from coronary trials: is their quality of care compromised? JAMA J Am Med Assoc 268(11):1460–1461
3. Soldin OP, Mattison DR (2009) Sex differences in pharmacokinetics and pharmacodynamics. Clin Pharmacokinet 48:143–57. NIH Public Access
4. Berger JS, Elliott L, Gallup D, Roe M, Granger CB, Armstrong PW et al (2009) Sex differences in mortality following acute coronary syndromes. JAMA 302(8):874–882
5. NIH (2000) NIH guidelines on the inclusion of women and minorities as subjects in clinical research—Updated 2 Aug 2000. NIH Guide
6. Hayes SN, Redberg RF (2008) Dispelling the myths: calling for sex-specific reporting of trial results. Mayo Clinic Proceedings, vol 83. Elsevier Ltd., pp 523–525
7. Simon V (2005) Wanted: women in clinical trials. Science 308:1517. American Association for the Advancement of Science
8. Medicine I of (2001) Exploring the biological contributions to human health. Exploring the biological contributions to human health. National Academies Press
9. Söderström M (2001) Why researchers excluded women from their trial populations. Lakartidningen 98(13):1524–1528
10. Gordon D (2021) Heart disease is the number one cause of death for women—Healthy Women
11. Holdcroft A (2007) Gender bias in research: how does it affect evidence based medicine? J Royal Soc Med 100:2–3. Royal Society of Medicine Press
12. Cleveland Clinic (2022) Heart disease in women: risk factors, symptoms and prevention
13. Benjamin EJ, Blaha MJ, Chiuve SE, Cushman M, Das SR, Deo R, de Ferranti SD, Floyd J, Fornage M, Gillespie C, Isasi CR, Jiménez MC, Jordan LC, Judd SE, Lackland D, Lichtman JH, Lisabeth L, Liu S, Longenecker CT, Mackey RH, Matsushita K, Mozaffarian D, Mussolino ME, Nasir K, Neumar RW, Palaniappan L, Pandey DK, Thiagarajan RR, Reeves MJ, Ritchey M, Rodriguez CJ, Roth GA, Rosamond WD, Sasson C, Towfighi A, Tsao CW, Turner MB, Virani SS, Voeks JH, Willey JZ, Wilkins JT, Wu JH, Alger HM, Wong SS, Muntner P (2017) American Heart Association Statistics Committee and Stroke Statistics Subcommittee. Heart disease and stroke statistics-2017 update: a report from the American Heart Association. Circ 135(10):e146–e603
14. Prabhavathi K, Tamarai Selvi K, Poornima KN, Sarvanan A (2014) Role of biological sex in normal cardiac function and in its disease outcome—a review. J Clin Diagnostic Res 8(8):BE01
15. Izadnegahdar M, Singer J, Lee MK, Gao M, Thompson CR, Kopec J et al (2014) Do younger women fare worse? Sex differences in acute myocardial infarction hospitalization and early mortality rates over ten years. J Women's Heal 23(1):10–17
16. Garcia M, Mulvagh SL, Merz CNB, Buring JE, Manson JAE (2016) Cardiovascular disease in women: clinical perspectives. Circ Res 118:1273–93. Lippincott Williams & Wilkins Hagerstown, MD
17. Roger VL, Go AS, Lloyd-Jones DM, Adams RJ, Berry JD, Brown TM et al (2011) Heart disease and stroke statistics-2011 update: a report from the American Heart Association. Circ 123(4):e18
18. Mosca L, Mochari-Greenberger H, Dolor RJ, Newby LK, Robb KJ (2010) Twelve-year follow-up of American women's awareness of cardiovascular disease risk and barriers to heart health. Circ Cardiovasc Qual Outcomes 3(2):120–127
19. Wenger NK (2010) Editorial: the female heart is vulnerable to cardiovascular disease emerging prevention evidence for women must inform emerging prevention strategies for women. Circ: Cardiovasc Qual Outcomes 3:118–9. Lippincott Williams & Wilkins
20. Anand SS, Razak F, Davis AD, Jacobs R, Vuksan V, Teo K et al (2006) Social disadvantage and cardiovascular disease: development of an index and analysis of age, sex, and ethnicity effects. Int J Epidemiol 35(5):1239–1245

21. Tjepkema M, Wilkins R, Goedhuis N, Pennock J (2012) Cardiovascular disease mortality among first nations people in Canada, 1991–2001. Chronic Dis Inj Can 32(4):200–207
22. Mehta LS, Beckie TM, DeVon HA, Grines CL, Krumholz HM, Johnson MN et al (2016) Acute myocardial infarction in women: a scientific statement from the American Heart Association. Circ 133:916–947. Lippincott Williams & Wilkins Hagerstown, MD
23. Norris CM, Yip CYY, Nerenberg KA, Clavel MA, Pacheco C, Foulds HJA et al (2020) State of the science in women's cardiovascular disease: a canadian perspective on the influence of sex and gender. J Am Heart Assoc 9. American Heart Association Inc.
24. Luijken J, van der Schouw YT, Mensink D, Onland-Moret NC (2017) Association between age at menarche and cardiovascular disease: a systematic review on risk and potential mechanisms. Maturitas 104:96–116
25. Kiconco S, Teede HJ, Earnest A, Loxton D, Joham AE (2021) Menstrual cycle regularity as a predictor for heart disease and diabetes: findings from a large population-based longitudinal cohort study. Clin Endocrinol (Oxf) 96(4):605–616
26. Kaminski P, Szpotanska-Sikorska M, Wielgos M (2013) Cardiovascular risk and the use of oral contraceptives. Neuro Endocrinol Lett 34(7):587–589
27. Elkayam U, Goland S, Pieper PG, Silverside CK (2016) High-risk cardiac disease in pregnancy: part I. J Am Coll Cardiol 68(4):396–410
28. Wu P, Haththotuwa R, Kwok CS, Babu A, Kotronias RA, Rushton C, Zaman A, Fryer AA, Kadam U, Chew-Graham CA, Mamas MA (2017) Preeclampsia and future cardiovascular health: a systematic review and meta-Analysis. Circ Cardiovasc Qual Outcomes 10(2):e003497
29. Alsnes IV, Janszky I, Forman MR, Vatten LJ, Økland I (2014) A population-based study of associations between preeclampsia and later cardiovascular risk factors. Am J Obstet Gynecol 211(6):657.e1–657.e7
30. CDC (2018) Diabetes during pregnancy|Maternal infant health|Reproductive health|CDC
31. Noctor E (2015) Type 2 diabetes after gestational diabetes: the influence of changing diagnostic criteria. World J Diabetes 6(2):234
32. Gu K, Cowie CC, Harris MI (1999) Diabetes and decline in heart disease mortality in US adults. JAMA 281(14):1291–1297
33. Woodward M (2019) Cardiovascular disease and the female disadvantage. Int J Environ Res Public Health 16(7):1165
34. Fabreau GE, Leung AA, Southern DA, Knudtson ML, McWilliams JM, Ayanian JZ et al (2014) Sex, socioeconomic status, access to cardiac catheterization, and outcomes for acute coronary syndromes in the context of universal healthcare coverage. Circ Cardiovasc Qual Outcomes 7(4):540–549

# Chapter 2
# Risk Factors for Ischemic Stroke in Women

**Amy Y. X. Yu, Tracy E. Madsen, and Moira K. Kapral**

**Abstract** Stroke is the second leading cause of death worldwide. Stroke incidence is higher in young women compared to men. Among older individuals where the sex difference in stroke incidence is small or absent, women still bear a higher burden of disease compared to men. The absolute number of strokes in women is higher because women have longer life expectancy and women who survive stroke have worse outcomes: higher disability, lower quality of life, and more need for long-term care. In this chapter, we review sex differences in risk factors for ischemic stroke. A better understanding of these differences may allow for a more precise estimate of stroke risk by sex, enable more accurate diagnosis, and promote sex-specific stroke prevention strategies.

**Keywords** Ischemic stroke · Hypertensive disorders of pregnancy (HDP) · Vascular health · Heart-brain interaction · Patent foramen ovale (PFO)

## Epidemiology

Stroke is the second leading cause of death worldwide [1]. In 2016, the global lifetime risk of stroke was estimated to be 24.7% in men and 25.1% in women, with variations by region [2]. The association between sex and risk of incident ischemic stroke across the age continuum is U-shaped with the risk being higher in women than men under

A. Y. X. Yu (✉)
Department of Medicine (Neurology), University of Toronto, Sunnybrook Health Sciences Centre, Toronto, ON, Canada
e-mail: amyyx.yu@utoronto.ca

T. E. Madsen
Department of Emergency Medicine, Warren Alpert Medical School, and the Department of Epidemiology, School of Public Health, Brown University, Providence, RI, US

M. K. Kapral
Department of Medicine (General Internal Medicine), University of Toronto-University Health Network, Toronto, ON, Canada

© The Author(s), under exclusive license to Springer Nature Switzerland AG 2023 
L. Kirshenbaum and I. Rabinovich-Nikitin (eds.), *Biology of Women's Heart Health*, Advances in Biochemistry in Health and Disease 26,
https://doi.org/10.1007/978-3-031-39928-2_2

the age of 30 years, lower in women between the ages of 40–80 years, and equal between the two sexes after age 80 years [3].

The higher stroke incidence in young women compared to men has been consistently observed in different populations [4, 5]. A recent systematic review and meta-analysis found that ischemic stroke incidence was 44% higher in women than men among individuals aged 35 years or less (incidence rate ratio 1.44, 95% confidence interval (CI) [1.18, 1.76]) [6]. In addition, ischemic stroke incidence is increasing among young adults, and this trend is disproportionately impacting young women [7]. Among older individuals where the sex difference in stroke incidence is small or absent, women still bear a higher burden of disease compared to men. The absolute number of strokes in women is higher because women have longer life expectancy and women who survive stroke have worse outcomes: higher disability, lower quality of life, and more need for long-term care [8–10]. Sex differences in the risk of ischemic stroke by age suggest important differences in stroke etiology and risk factors by sex, and there may be opportunities for sex-specific approaches to disease prevention across the lifespan [11] (Fig. 2.1).

In this chapter, we review sex differences in risk factors for ischemic stroke. A better understanding of these differences may allow for a more precise estimate of stroke risk by sex, enable more accurate diagnosis, and promote sex-specific stroke prevention strategies. We focus on ischemic stroke as this is the most common stroke type [1]. We recognize that stroke risk may be influenced by both biological sex as

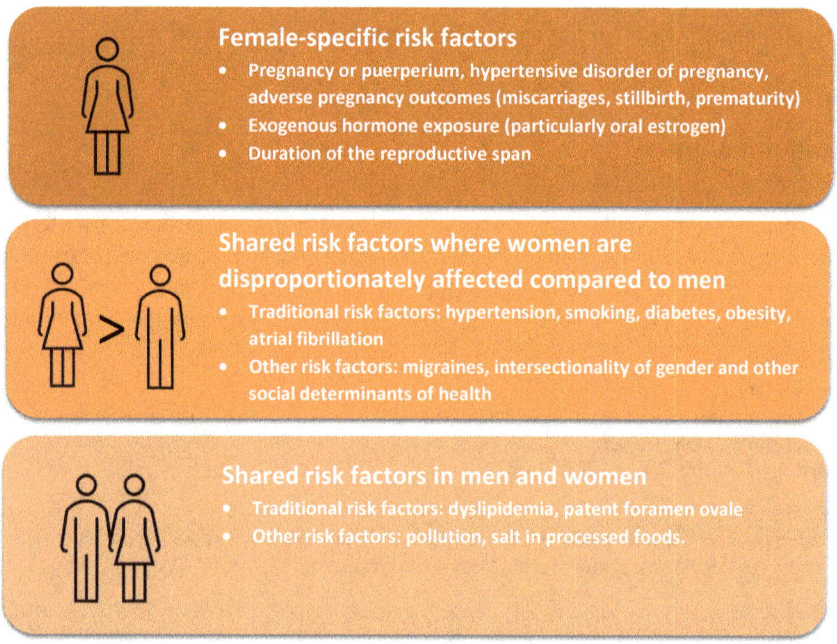


**Fig. 2.1** Sex differences in risk factors for ischemic stroke

well as the psychosocial constructs of gender and gender identity, but few studies have separately evaluated the effects of biological sex versus gender on stroke. Thus, we use the term "women" to refer to people of female sex throughout this chapter.

## Sex-Specific Risk Factors for Ischemic Stroke

### *Pregnancy or the Puerperium Period*

Throughout the lifespan, women are exposed to ischemic stroke risk factors that do not affect men. Pregnancy is a prime example of a period of important physiological and hormonal changes that uniquely affect women. A systematic review and meta-analysis of studies on pregnancy-related stroke incidence between 1990 and 2017 found an overall crude incidence rate of 30.0 per 100,000 pregnancies, 95% CI [18.8, 47.9] [12]. While pregnancy-related stroke is rare, its incidence is rising according to data from the Public Health Agency of Canada (2003–2016) and the US Nationwide Inpatient Sample (1994–2011) [13, 14]. The types of stroke that occur during pregnancy or the postpartum period differ from those occurring outside of pregnancy. In the general population, 60–70% of strokes are ischemic, [1] but during pregnancy, as little as 20% of events are ischemic [12]. Instead, patients with stroke who are pregnant are at higher risk of intracerebral hemorrhage or cerebral venous sinus thrombosis compared to the general population.

Various risk factors have been implicated in pregnancy-related stroke, including hypertensive disorders of pregnancy (HDP), maternal congenital heart disease, cervical artery dissection, connective tissue disorders, a hypercoagulable state, to name a few [13, 15]. Although some of these risk factors may appear to be specific to the pregnancy state, for example, HDP, gestational diabetes, or adverse pregnancy outcomes, there is now growing evidence showing that these risk factors may be associated with long-term adverse effects on cardiovascular health.

Indeed, HDP complicate 3–8% of all pregnancies and are associated with pregnancy-related stroke as well as long-term risk of stroke [16–19]. A recent population-based study from Taiwan showed that the association between HDP increased stroke risk was present even after 17 years of follow-up [20]. The authors found that compared to women without HDP, those with this condition had more than double the risk of ischemic stroke in the first three years after childbirth and more than quadruple the risk of hemorrhagic stroke after 10–15 years. Also, this relationship appears to be dose dependent. A Canadian study showed that patients with recurrent HDP had higher risk of subsequent cardio- and cerebrovascular events than those without recurrent HDP, and the time to first event was shorter in the former group [21].

The mechanism through which HDP increase the risk of future stroke is unclear. Hypotheses include shared risk factors, unmasking of underlying metabolic or vascular disease during the pregnancy "stress test," or induction of cardiovascular abnormalities [22]. Current guidelines recommend the use of low-dose acetylsalicylic acid to prevent pre-eclampsia after 12 weeks of gestations in individuals at high risk for pre-eclampsia [23]. Furthermore, an observational study of over 83,000 women in the California Teachers Study found that acetylsalicylic acid is associated with reduced risk of stroke in women with HDP [24]. More research is needed to understand if acetylsalicylic acid can be routinely used in patients with HDP to prevent stroke.

Adverse pregnancy outcomes have also been reported to increase risk of stroke. In a recent systematic review and meta-analysis of 18 studies and over 7.8 million women, the authors found a higher risk of stroke in women who had experienced miscarriage (7% increased hazard) or stillbirth (38% increased hazard) compared to women who did not experience miscarriage or stillbirth, and the risk of stroke increased with repeated exposure [25]. The authors did not identify any definitive association between infertility and stroke. In a population-based cohort study in Sweden, preterm delivery was associated with higher long-term risk of stroke, and the risk was highest among patients with recurrent preterm delivery (compared to a single exposure) and those with extreme preterm delivery at 22–27 weeks (compared to late preterm at 34–36 weeks or early term 37–38 weeks).[26] In addition, the authors found that children who survived preterm birth were at higher risk of stroke as adults [27].

On the other hand, in the Nurses' Health Study II, gestational diabetes was associated with subsequent cardiovascular disease, but this seemed to be mainly driven by cardiac events, rather than stroke [28]. Breastfeeding has been associated with reduced long-term maternal cardiovascular risk, including an approximate 10% reduction in stroke risk in observational studies with 8–12 years of follow-up time [29, 30]. Breastfeeding has also been found to be associated with reduced risk of stroke in postmenopausal women [31]. However, whether this association is modified or mediated by other psychosocial factors, such as gender identity, education, socioeconomic status, is not well understood.

## *Sex-Specific Risk Factors Outside of Pregnancy or Puerperium*

There are additional sex-specific stroke risk factors that are not associated with pregnancy. Exogenous hormones increase the risk of stroke, and the risk varies by formulation, dosage, and route. The authors of a Cochrane review found that users of combined oral contraceptives were at 70% increased relative risk of ischemic stroke compared to non-users (relative risk 1.7 and 95% CI [1.5, 1.9]) and the risk increased with increasing estrogen dose, with the highest risk associated with formulations including > 50 mcg of estrogen [32]. Similarly, post-menopausal exposure to exogenous estrogen is associated with increased risk of ischemic stroke, but not hemorrhagic stroke [33, 34]. In a nested case–control study in French women, there was an increased odds of ischemic stroke with post-menopausal exposure to oral estrogen (odds ratio 1.58 and 95% CI [1.01, 2.49]), but this was not the case with progesterone (0.78 [0.49, 1.26]) or transdermal estrogen (0.83 [0.56, 1.24]) [35].

The duration of the reproductive span is also associated with stroke risk. In the Million Women Study, a prospective study of 1.2 million women in the UK followed for an average of 11.6 years per woman, investigators found a U-shaped relationship between age at menarche and subsequent stroke risk, with the risk being higher among people with early ($\leq$ 10 years) or late ($\geq$ 17 years) menarche compared to women with menarche at age 13 years [36]. Later natural menopause timing has been associated with reduced all-cause mortality and cardiovascular disease, but there is emerging evidence suggesting that the association between early menopause and cardiovascular disease may be bi-directional, where women with poor pre-menopausal cardiovascular health may be at higher risk of early menopause [37]. A pooled analysis of individual data from > 170,000 women from 9 observational studies showed that a first cardiovascular disease event before age 35 years is associated with doubled risk of early menopause [38].

## Sex Differences in Risk Factors Shared by Men and Women

### *Traditional Vascular Risk Factors*

Although traditional vascular risk factors increase the risk of stroke in both sexes, some of these risk factors may affect women and men differently [11, 39–42]. In a population-based study from the UK Biobank where over 470,000 adults were followed prospectively for 9 years, investigators found that several vascular risk factors led to greater hazard of stroke in women compared to men, including hypertension (30% excess risk in women compared to men), smoking (20% excess risk), diabetes (25% excess risk), and obesity (30% excess risk) [41]. The authors did not find any sex differences in risk of stroke with hyperlipidemia. In a US-based study of over 20,000 participants from the REGARDS study (REasons for Geographic and Racial Differences in Stroke), fasting blood glucose was associated with a 70 to 100% increase in hazard of ischemic stroke in women compared to a 20–40% increase in men [42].

### *Sex and Heart-Brain Interactions*

Atrial fibrillation is present in approximately 15–20% of patients with ischemic stroke [43]. Women with ischemic stroke are more likely to have comorbid atrial fibrillation than men, and women with atrial fibrillation are at higher risk of ischemic stroke than men [44–46]. Female sex confers an extra risk point to the overall score in the $CHA_2DS_2$-VASc risk stratification tool [47]. Diagnosing atrial fibrillation in patients with ischemic stroke is highly relevant because they benefit from oral anticoagulation for stroke prevention, rather than antiplatelet treatment. However, despite the increased risk of stroke in women with atrial fibrillation, women are less likely to be treated with oral anticoagulation or more likely to receive subtherapeutic dosing of these drugs in routine clinical practice compared to men [48, 49]. Furthermore, women were under-represented in large clinical trials on the efficacy of direct oral anticoagulants for stroke prevention in patients with atrial fibrillation, where fewer than 40% of participants were female, limiting sex disaggregated analyses on the efficacy and safety of these medications [50–53]. Women with atrial fibrillation are more likely to experience symptoms compared to men, but less likely to receive antiarrhythmic medications, electric cardioversion, or catheter ablation [54]. Finally, atrial cardiopathy, a novel concept referring to abnormal atrial tissue substrate, with or without overt atrial fibrillation is proposed as a new risk factor for stroke [55]. There is currently little data on sex differences in the investigation or management of atrial cardiopathy, but this is a rapidly developing area of new research.

Patent foramen ovale (PFO) is another source of stroke-causing paradoxical emboli. PFO closure in selected patients has been shown to reduce long-term risk of stroke recurrence [56–58]. There is still little data on sex differences in effectiveness and safety of PFO closure for stroke prevention, but in a recent pooled individual patient data analysis of six randomized clinical trials that compared PFO closure versus medical treatment only (45% women), there was no signal of heterogeneity of effect by sex in subgroup analyses [59]. In a single-center prospective study of consecutive patients with cryptogenic stroke or transient ischemic attack imaged with trans-esophageal echocardiogram, the prevalence of PFO was found to be higher in men than in women (38% versus 28%), but there was no difference in PFO grade, prevalence of right-to-left shunt at rest, or coexistence of atrial septal aneurysm [60].

## *Beyond Traditional Vascular Risk Factors*

Migraine is a primary headache disorder, affecting up to 20% of the general population with about three times higher prevalence in women compared to men, and particularly young women because the majority of people with migraines will have their first episode before the age of 50 years [61]. Up to a third of patients with migraine experience “aura,” consisting of reversible neurological symptoms with their episode. In hemiplegic migraine, which can be a familial or sporadic condition, the migraine aura includes motor symptoms [62]. Thus, migraine is a common condition mimicking cerebral ischemia [63, 64]. However, patients with migraine with aura are also at increased risk of ischemic stroke. A large meta-analysis showed that compared to controls without migraine, migraine with aura is associated with double the relative risk of ischemic stroke, and this risk increases with concurrent exposure to oral contraceptives (sevenfold increase in relative risk) and concurrent exposure to both oral contraceptives and smoking (tenfold increase in relative risk) [65]. In addition, the association between migraine with aura and ischemic stroke is stronger in young individuals aged < 45 years [61]. No definite association between migraine without aura and ischemic stroke has been found. Thus, ischemic stroke should be considered in certain patients with migraines with aura (first-ever aura, complex or atypical aura, or changes in aura). Finally, a new diagnosis of migraine with aura should be seen as an opportunity to discuss stroke prevention, particularly when other vascular risk factors, such as smoking or oral contraceptive use are also present.

There is increasing recognition of the effects of healthy living environments (air quality, walkability, access to healthy food choices), psychosocial stress, medication access and adherence, and other social determinants of health on cardiovascular health [11]. In the UK Biobank study, the authors reported that women experiencing low socioeconomic status had a 20% excess hazard of ischemic stroke compared to men with low socioeconomic status [41]. In addition, researchers studied data from 176 countries reported to the World Health Organization and found that stroke mortality was higher in women than men in countries where women experienced overall higher inequalities, measured using the Gender Inequality Index [66]. More research is needed on the intersectionality between sex and other social determinants of health, including race and racism, socioeconomic status, education, and the factors that mediate these associations.

## Clinical Implications and Future Research

In 2014, the American Heart Association/American Stroke Association published guidelines for the prevention of stroke in women [22]. The Canadian Stroke Best Practice advisory committee published in 2018 a two-part review and consensus statement on acute stroke management during pregnancy and secondary stroke prevention during pregnancy [67, 68]. Finally, the recently published European Stroke Organisation guidelines committee followed the Grading of Recommendations and Assessment, Development and Evaluation (GRADE) approach to offer guidance on the management of stroke during menopause (including hormone replacement therapy), pregnancy, and the postpartum periods [69]. An important and consistent finding from these consensus statements and guidelines is the lack of high-quality evidence in this area. Women are under-represented in stroke clinical trials and sex disaggregated analyses are often under-powered or lacking [70, 71]. There is an urgent need for more population-based observational studies and clinical trials on sex specific considerations in stroke.

## Summary

In summary, it is important to recognize that many risk factors for ischemic stroke are sex-specific, while other risk factors shared by both sexes still may disproportionately affect women compared to men. Public messaging about sex-specific stroke risk factors is therefore important to increase awareness of stroke and its risk factors in women, including young women. We must move away from the stereotype that stroke primarily affects older men.

Clinicians should routinely ask about sex-specific risk factors for stroke, including complications of pregnancy, exposure to exogenous hormones, and reproductive span. Inquiring about these risk factors should be no different than asking about hypertension and glycemic control. Furthermore, clinical visits for screening and treatment of traditional vascular risk factors are an important opportunity for ensuring that both women and men are receiving optimal evidence-based therapy, as well as patient education on how certain vascular risk factors disproportionately affect women compared to men.

Research efforts to generate new and high-quality evidence on stroke in women should be supported and the proportion of women enrolled in clinical trials should match the proportion of women with disease in the population. Policymakers should strive to improve the population's brain health by addressing social determinants of health including living environments, nutrition, socioeconomic status, and gender inequality.

## References

1. GBD (2021) 2019 Stroke Collaborators. Global, regional, and national burden of stroke and its risk factors, 1990–2019: a systematic analysis for the Global Burden of Disease Study 2019. Lancet Neurol 20(10):795–820
2. Feigin VL, Nguyen G, Cercy K, Johnson CO, Alam T, Parmar PG et al (2018) Global, regional, and country-specific lifetime risks of stroke, 1990 and 2016. N Engl J Med 379(25):2429–2437
3. Vyas MV, Silver FL, Austin PC, Yu AYX, Pequeno P, Fang J et al (2021) Stroke incidence by sex across the lifespan. Stroke 52(2):447–451
4. Ekker MS, Verhoeven JI, Vaartjes I, van Nieuwenhuizen KM, Klijn CJM, de Leeuw FE (2019) Stroke incidence in young adults according to age, subtype, sex, and time trends. Neurology 92(21):e2444–e2454
5. Leppert MH, Ho PM, Burke J, Madsen TE, Kleindorfer D, Sillau S et al (2020) Young women had more strokes than young men in a large United States claims sample. Stroke 51(11):3352–3355
6. Leppert MH, Burke JF, Lisabeth LD, Madsen TE, Kleindorfer DO, Sillau S et al (2022) Systematic review of sex differences in ischemic strokes among young adults: are young women disproportionately at risk? Stroke 53(2):319–327
7. Ekker MS, Boot EM, Singhal AB, Tan KS, Debette S, Tuladhar AM et al (2018) Epidemiology, aetiology, and management of ischaemic stroke in young adults. Lancet Neurol 17(9):790–801
8. Bushnell CD, Reeves MJ, Zhao X, Pan W, Prvu-Bettger J, Zimmer L et al (2014) Sex differences in quality of life after ischemic stroke. Neurology 82(11):922–931

9. Yu AYX, Maclagan LC, Diong C, Austin PC, Kapral MK, Swartz RH et al (2020) Sex differences in care need and survival in patients admitted to nursing home poststroke. Can J Neurol Sci Le Journal Canadien Des Sciences Neurologiques 47(2):153–159
10. Carcel C, Wang X, Sandset EC, Delcourt C, Arima H, Lindley R et al (2019) Sex differences in treatment and outcome after stroke: pooled analysis including 19,000 participants. Neurology 93(24):e2170–e2180
11. Pandian JD, Gall SL, Kate MP, Silva GS, Akinyemi RO, Ovbiagele BI et al (2018) Prevention of stroke: a global perspective. Lancet 392(10154):1269–1278
12. Swartz RH, Cayley ML, Foley N, Ladhani NNN, Leffert L, Bushnell C et al (2017) The incidence of pregnancy-related stroke: a systematic review and meta-analysis. Int J Stroke 12(7):687–697
13. Liu S, Chan W-S, Ray JG, Kramer MS, Joseph KS (2019) Stroke and cerebrovascular disease in pregnancy. Stroke 50(1):13–20
14. Leffert LR, Clancy CR, Bateman BT, Bryant AS, Kuklina EV (2015) Hypertensive disorders and pregnancy-related stroke: frequency, trends, risk factors, and outcomes. Obstet Gynecol 125(1):124–131
15. Tate J, Bushnell C (2011) Pregnancy and stroke risk in women. Womens Health (Lond Engl) 7(3):363–374
16. Miller EC, Gatollari HJ, Too G, Boehme AK, Leffert L, Marshall RS et al (2017) Risk factors for pregnancy-associated stroke in women with preeclampsia. Stroke 48(7):1752–1759
17. Canoy D, Cairns BJ, Balkwill A, Wright FL, Khalil A, Beral V et al (2016) Hypertension in pregnancy and risk of coronary heart disease and stroke: a prospective study in a large UK cohort. Int J Cardiol 222:1012–1018
18. de Havenon A, Delic A, Stulberg E, Sheibani N, Stoddard G, Hanson H et al (2021) Association of preeclampsia with incident stroke in later life among women in the Framingham heart study. JAMA Netw Open 4(4):e215077
19. Wu P, Haththotuwa R, Kwok CS, Babu A, Kotronias RA, Rushton C, Zaman A, Fryer AA, Kadam U, Chew-Graham CA, Mamas MA (2017) Preeclampsia and future cardiovascular health: a systematic review and meta-analysis. Circ Cardiovasc Qual Outcomes 10(2):e003497
20. Hung SK, Lee MS, Lin HY, Chen LC, Chuang CJ, Chew CH et al (2022) Impact of hypertensive disorders of pregnancy on the risk of stroke stratified by subtypes and follow-up time. Stroke 53(2):338–344
21. Auger N, Fraser WD, Schnitzer M, Leduc L, Healy-Profitós J, Paradis G (2017) Recurrent preeclampsia and subsequent cardiovascular risk. Heart 103(3):235–243
22. Bushnell C, McCullough LD, Awad IA, Chireau MV, Fedder WN, Furie KL et al (2014) Guidelines for the prevention of stroke in women: a statement for healthcare professionals from the American Heart Association/American Stroke Association. Stroke 45(5):1545–1588
23. Force UPST (2021) Aspirin use to prevent preeclampsia and related morbidity and mortality: US preventive services task force recommendation statement. JAMA 326(12):1186–1191
24. Miller EC, Boehme AK, Chung NT, Wang SS, Lacey JV Jr, Lakshminarayan K et al (2019) Aspirin reduces long-term stroke risk in women with prior hypertensive disorders of pregnancy. Neurology 92(4):e305–e316
25. Liang C, Chung HF, Dobson AJ, Mishra GD (2022) Infertility, miscarriage, stillbirth, and the risk of stroke among women: a systematic review and meta-analysis. Stroke 53(2):328–337
26. Crump C, Sundquist J, Sundquist K (2021) Preterm delivery and long-term risk of stroke in women: a national cohort and cosibling study. Circulation 143(21):2032–2044
27. Crump C, Sundquist J, Sundquist K (2021) Stroke risks in adult survivors of preterm birth: national cohort and cosibling study. Stroke 52(8):2609–2617
28. Tobias DK, Stuart JJ, Li S, Chavarro J, Rimm EB, Rich-Edwards J et al (2017) Association of history of gestational diabetes with long-term cardiovascular disease risk in a large prospective cohort of US women. JAMA Intern Med 177(12):1735–1742
29. Tschiderer L, Seekircher L, Kunutsor SK, Peters SAE, O'Keeffe LM, Willeit P (2022) Breastfeeding is associated with a reduced maternal cardiovascular risk: systematic review and meta-analysis involving data from 8 studies and 1 192 700 Parous women. J Am Heart Assoc 11(2):e022746

30. Peters SAE, Yang L, Guo Y, Chen Y, Bian Z, Du J, Yang J, Li S, Li L, Woodward M, Chen Z (2017) Breastfeeding and the risk of maternal cardiovascular disease: a prospective study of 300 000 Chinese women. J Am Heart Assoc 6(6):e006081
31. Jacobson LT, Hade EM, Collins TC, Margolis KL, Waring ME, Van Horn LV et al (2018) Breastfeeding history and risk of stroke among parous postmenopausal women in the women's health initiative. J Am Heart Assoc 7(17):e008739
32. Roach RE, Helmerhorst FM, Lijfering WM, Stijnen T, Algra A, Dekkers OM (2015) Combined oral contraceptives: the risk of myocardial infarction and ischemic stroke. Cochrane Database Syst Rev (8):Cd011054
33. Viscoli CM, Brass LM, Kernan WN, Sarrel PM, Suissa S, Horwitz RI (2001) A clinical trial of estrogen-replacement therapy after ischemic stroke. N Engl J Med 345(17):1243–1249
34. Hendrix SL, Wassertheil-Smoller S, Johnson KC, Howard BV, Kooperberg C, Rossouw JE et al (2006) Effects of conjugated equine estrogen on stroke in the women's health initiative. Circulation 113(20):2425–2434
35. Canonico M, Carcaillon L, Plu-Bureau G, Oger E, Singh-Manoux A, Tubert-Bitter P et al (2016) Postmenopausal hormone therapy and risk of stroke: impact of the route of estrogen administration and type of progestogen. Stroke 47(7):1734–1741
36. Canoy D, Beral V, Balkwill A, Wright FL, Kroll ME, Reeves GK et al (2015) Age at menarche and risks of coronary heart and other vascular diseases in a large UK cohort. Circulation 131(3):237–244
37. El Khoudary SR, Aggarwal B, Beckie TM, Hodis HN, Johnson AE, Langer RD et al (2020) Menopause transition and cardiovascular disease risk: implications for timing of early prevention: a scientific statement from the American heart association. Circulation 142(25):e506–e532
38. Zhu D, Chung HF, Pandeya N, Dobson AJ, Hardy R, Kuh D et al (2019) Premenopausal cardiovascular disease and age at natural menopause: a pooled analysis of over 170,000 women. Eur J Epidemiol 34(3):235–246
39. Bushnell C, Howard VJ, Lisabeth L, Caso V, Gall S, Kleindorfer D et al (2018) Sex differences in the evaluation and treatment of acute ischaemic stroke. Lancet Neurol 17(7):641–650
40. Howard VJ, Madsen TE, Kleindorfer DO, Judd SE, Rhodes JD, Soliman EZ et al (2019) Sex and race differences in the association of incident ischemic stroke with risk factors. JAMA Neurol 76(2):179–186
41. Peters SAE, Carcel C, Millett ERC, Woodward M (2020) Sex differences in the association between major risk factors and the risk of stroke in the UK Biobank cohort study. Neurology 95(20):e2715–e2726
42. Madsen TE, Long DL, Carson AP, Howard G, Kleindorfer DO, Furie KL et al (2021) Sex and race differences in the risk of ischemic stroke associated with fasting blood glucose in regards. Neurology 97(7):e684–e694
43. Lip GY, Lane DA (2015) Stroke prevention in atrial fibrillation: a systematic review. JAMA 313(19):1950–1962
44. Avgil Tsadok M, Jackevicius CA, Rahme E, Humphries KH, Behlouli H, Pilote L (2012) Sex differences in stroke risk among older patients with recently diagnosed atrial fibrillation. JAMA 307(18):1952–1958
45. Jewett GA, Lindsay MP, Goia C, Zagorski B, Kamal N, Kapral MK et al (2019) National trends in hospital admission, case fatality, and sex differences in atrial fibrillation-related strokes. Int J Stroke:1747493019881349
46. Emdin CA, Wong CX, Hsiao AJ, Altman DG, Peters SA, Woodward M et al (2016) Atrial fibrillation as risk factor for cardiovascular disease and death in women compared with men: systematic review and meta-analysis of cohort studies. BMJ 532:h7013
47. Lip GY, Nieuwlaat R, Pisters R, Lane DA, Crijns HJ (2010) Refining clinical risk stratification for predicting stroke and thromboembolism in atrial fibrillation using a novel risk factor-based approach: the euro heart survey on atrial fibrillation. Chest 137(2):263–272
48. Avgil Tsadok M, Jackevicius CA, Rahme E, Humphries KH, Pilote L (2015) Sex differences in dabigatran use, safety, and effectiveness in a population-based cohort of patients with atrial fibrillation. Circ Cardiovasc Qual Outcomes 8(6):593–599

49. Thompson LE, Maddox TM, Lei L, Grunwald GK, Bradley SM, Peterson PN, Masoudi FA, Turchin A, Song Y, Doros G, Davis MB, Daugherty SL (2017) Sex differences in the use of oral anticoagulants for atrial fibrillation: a report from the National Cardiovascular Data Registry (NCDR®) PINNACLE Registry. J Am Heart Assoc 6(7):e005801
50. Connolly SJ, Eikelboom J, Joyner C, Diener HC, Hart R, Golitsyn S et al (2011) Apixaban in patients with atrial fibrillation. N Engl J Med 364(9):806–817
51. Connolly SJ, Ezekowitz MD, Yusuf S, Eikelboom J, Oldgren J, Parekh A et al (2009) Dabigatran versus warfarin in patients with atrial fibrillation. N Engl J Med 361(12):1139–1151
52. Giugliano RP, Ruff CT, Braunwald E, Murphy SA, Wiviott SD, Halperin JL et al (2013) Edoxaban versus warfarin in patients with atrial fibrillation. N Engl J Med 369(22):2093–2104
53. Granger CB, Alexander JH, McMurray JJ, Lopes RD, Hylek EM, Hanna M et al (2011) Apixaban versus warfarin in patients with atrial fibrillation. N Engl J Med 365(11):981–992
54. Gillis AM (2017) Atrial fibrillation and ventricular arrhythmias: sex differences in electrophysiology, epidemiology, clinical presentation, and clinical outcomes. Circulation 135(6):593–608
55. Kamel H, Okin PM, Elkind MS, Iadecola C (2016) Atrial fibrillation and mechanisms of stroke: time for a new model. Stroke 47(3):895–900
56. Saver JL, Carroll JD, Thaler DE, Smalling RW, MacDonald LA, Marks DS et al (2017) Long-term outcomes of patent foramen ovale closure or medical therapy after stroke. N Engl J Med 377(11):1022–1032
57. Mas JL, Derumeaux G, Guillon B, Massardier E, Hosseini H, Mechtouff L et al (2017) Patent foramen ovale closure or anticoagulation versus antiplatelets after stroke. N Engl J Med 377(11):1011–1021
58. Sondergaard L, Kasner SE, Rhodes JF, Andersen G, Iversen HK, Nielsen-Kudsk JE et al (2017) Patent foramen ovale closure or antiplatelet therapy for cryptogenic stroke. N Engl J Med 377(11):1033–1042
59. Kent DM, Saver JL, Kasner SE, Nelson J, Carroll JD, Chatellier G et al (2021) Heterogeneity of treatment effects in an analysis of pooled individual patient data from randomized trials of device closure of patent foramen ovale after stroke. JAMA 326(22):2277–2286
60. Nedeltchev K, Wiedmer S, Schwerzmann M, Windecker S, Haefeli T, Meier B et al (2008) Sex differences in cryptogenic stroke with patent foramen ovale. Am Heart J 156(3):461–465
61. Øie LR, Kurth T, Gulati S, Dodick DW (2020) Migraine and risk of stroke. J Neurol Neurosurg Psychiatry 91(6):593–604
62. Headache Classification Committee of the International Headache Society (IHS) (2013) The international classification of headache disorders, 3rd edn. (beta version). Cephalalgia 33(9):629–808
63. Yu AYX, Penn AM, Lesperance ML, Croteau NS, Balshaw RF, Votova K, Bibok MB, Penn M, Saly V, Hegedus J, Zerna C, Klourfeld E, Bilston L, Hong ZM, Coutts SB (2019) SpecTRA Study Group. Sex differences in presentation and outcome after an acute transient or minor neurologic event. JAMA Neurol 76(8):962–968
64. Yu AYX, Hill MD, Asdaghi N, Boulanger JM, Camden MC, Campbell BCV et al (2021) Sex differences in diagnosis and diagnostic revision of suspected minor cerebral ischemic events. Neurology 96(5):e732–e739
65. Schürks M, Rist PM, Bigal ME, Buring JE, Lipton RB, Kurth T (2009) Migraine and cardiovascular disease: systematic review and meta-analysis. BMJ 339:b3914
66. Kim YD, Jung YH, Caso V, Bushnell CD, Saposnik G (2017) Countries with women inequalities have higher stroke mortality. Int J Stroke 12(8):869–874
67. Ladhani NNN, Swartz RH, Foley N, Nerenberg K, Smith EE, Gubitz G et al (2018) Canadian stroke best practice consensus statement: acute stroke management during pregnancy. Int J Stroke 13(7):743–758
68. Swartz RH, Ladhani NNN, Foley N, Nerenberg K, Bal S, Barrett J et al (2018) Canadian stroke best practice consensus statement: secondary stroke prevention during pregnancy. Int J Stroke 13(4):406–419
69. Kremer C, Gdovinova Z, Bejot Y, Heldner MR, Zuurbier S, Walter S, Lal A, Epple C, Lorenzano S, Mono ML, Karapanayiotides T, Krishnan K, Jovanovic D, Dawson J, Caso V (2022)

European stroke organisation guidelines on stroke in women: management of menopause, pregnancy and postpartum. Eur Stroke J 7(2):I–XIX
70. Carcel C, Woodward M, Balicki G, Koroneos GL, Sousa DA, Cordonnier C et al (2019) Trends in recruitment of women and reporting of sex differences in large-scale published randomized controlled trials in stroke. Int J Stroke 1747493019851292
71. Strong B, Pudar J, Thrift AG, Howard VJ, Hussain M, Carcel C et al (2021) Sex disparities in enrolment in recent randomized clinical trials of acute stroke: a meta-analysis. JAMA Neurol 78(6):666–677

# Chapter 3
# The Effects of Sex Steroid Hormones on Cardiovascular Physiology in Females

**Nicole L. Tegg, Caitlynd Myburgh, and Colleen M. Norris**

**Abstract** There are fundamental sex-specific differences in cardiovascular (CV) physiology and cardiovascular disease (CVD) pathophysiology. Estrogen, a sex steroid hormone (SSH), has three receptors through which it enacts genomic and non-genomic actions. Estrogen has direct and indirect physiological effects on CV function. The roles of progesterone and testosterone on CV function are less understood, but both exert vasodilatory actions. This chapter will explore the relationship between CV physiology and the SSHs, estrogen, progesterone and testosterone in females.

**Keywords** Estrogen · GPER · Atherosclerosis · Oxidative stress · Hormone replacement therapy

## Estrogen Synthesis and Classification

It is well established that estrogens (17β estradiol, estriol and estrone) are synthesized in the ovaries, adrenal cortex and aromatized by testosterone in peripheral tissues. Since estrogen is a steroid hormone, it is able to easily cross the phospholipid bilayer of the plasma membrane, where it can bind to three types of estrogen receptors (ER), including ERα, ERβ and G-protein coupled receptors (GPER). ERs exist in

N. L. Tegg (✉) · C. M. Norris
Faculty of Nursing, University of Alberta, 3Rd Floor ECHA, Edmonton, AB, Canada
e-mail: nljohnso@ualberta.ca

C. Myburgh
The Kings University, Edmonton, AB, Canada

C. M. Norris
Heart Health and Stroke Strategic Clinical Network, Alberta Health Services, Edmonton, AB, Canada

Faculty of Medicine, School of Public Health Sciences, University of Alberta, Edmonton, AB, Canada

© The Author(s), under exclusive license to Springer Nature Switzerland AG 2023 
L. Kirshenbaum and I. Rabinovich-Nikitin (eds.), *Biology of Women's Heart Health*, Advances in Biochemistry in Health and Disease 26,
https://doi.org/10.1007/978-3-031-39928-2_3

many tissues, including the endometrium, breast, ovarian stroma cells, hypothalamus, kidney, brain, bone and more [1].

A small portion of ERα and ERβ receptors are found on the plasma membrane, but the majority are found in the cytosol where, upon estrogen binding, the receptors translocate into the nucleus. GPER is located on the plasma membrane and the endoplasmic reticulum. One of the pivotal functions of estrogen is its ability to modify gene transcription through interaction with sex steroid hormone receptors (SSHR) and a variety of coregulatory proteins [2]. Upon estrogen binding, ERs undergo conformational changes followed by subsequent ER dimerization and binding to various consensus estrogen response element sites on nuclear DNA [3]. Through this pathway, gene expression can thus be selectively activated or inhibited when co-regulators are recruited to the estrogen-ER complex [4].

## Genomic and Non-Genomic Effects of Estrogen

The binding of estrogen to nuclear ERα and ERβ enhances gene transcription and hence is known as the genomic effects of estrogen. Estrogen is also capable of initiating more rapid intracellular signaling pathways when acting on membrane-bound GPER, which has been proposed as a major route for the non-genomic effects [5], as shown in Fig. 3.1. The acute intracellular effects of binding to GPER include the opening of $K^+$ channels, closing of $CA^{2+}$ ion channels, and downstream activation of second messengers [5].

It is through these genomic and non-genomic mechanisms that all three receptors play a role in CV function [6, 7] through their regulation of processes such as cardiac hypertrophy, cardiac failure, ischemic heart diseases, vascular injury and atherosclerosis [8, 9]. More specifically, ERα can activate endothelial nitric oxide (NO) synthase (eNOS) [10, 11], inhibit vascular smooth muscle cell (VSMC) proliferation [12] and inhibit intimal-medial thickening [13]. ERβ supports normal vasodilation and blood pressure (BP) [2], reduces cardiac hypertrophy [14] and inhibits cardiac fibrosis [15]. An animal study identified that in females, ERβ mediated cardioprotection against ischemia and reperfusion injuries, [16], suggesting that ERβ may be predominantly responsible for the protective actions of estrogen following a vessel injury. The specific mechanisms through which GPER lowers blood pressure remain unclear. However, animal studies have demonstrated that GPER plays an important role in the regulation of blood pressure, progression of atherosclerosis and inflammation [17], plays a role in the NO-mediated vasodilatory effects of estrogen [18], reduces VSMC of $CA^{2+}$ sensitivity thereby inhibiting endothelin-initiated vasoconstriction [19], and has an antioxidant effect by suppressing oxidative stress (OS) through activation of the cyclic adenosine monophosphate pathway [20].

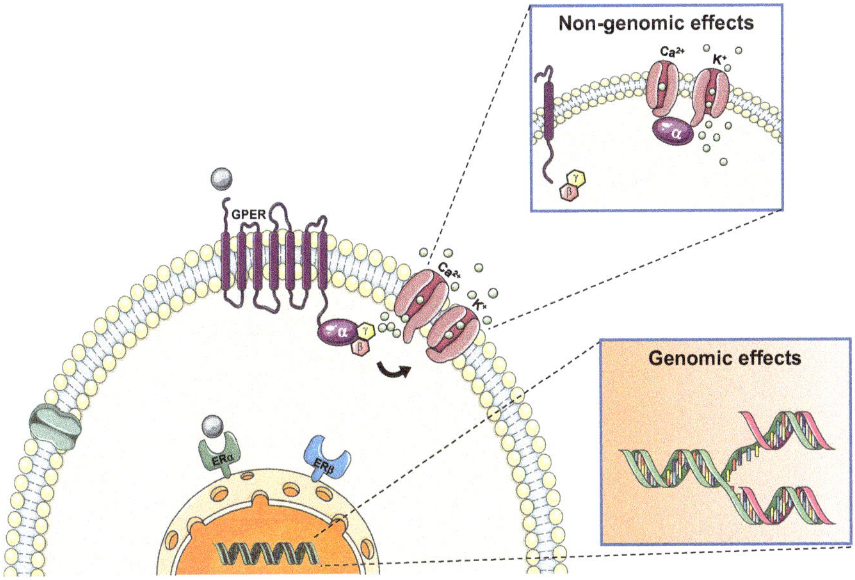


**Fig. 3.1** Genomic and Non-genomic Effects of Estrogen. The figure was generated using images from Servier Medical Art and modified using text and shapes. Servier Medical Art by Servier is licensed under a Creative Commons Attribution 3.0 Unported License. (https://creativecommons.org/licenses/by/3.0/)

## Estrogens and the Cardiovascular System

Estrogens can directly affect the CV system through several mechanisms, as shown in Fig. 3.2, including endothelial NO generation, cell proliferation, angiogenesis, and control of VSMC of $CA^{2+}$ and $K^+$ channels [21]. NO inhibits platelet aggregation, adhesion and proliferation of VSMCs and is a potent vasodilator [22]. NO further regulates vascular tone, myocardial contractility, endothelial integrity, permeability, and interactions between leukocytes and the endothelium [1]. Thereby, endothelium-dependent effects of estrogen, including rapid vasodilation, are related to increased bioavailability of NO via ER activation of eNOS [2, 23–27]. ERs in VSMCs affect several VSMC functions, such as contractility and growth [28]. Estradiol attenuates voltage-dependent T & L type of $CA^{2+}$ channel currents in VSMCs [29]. The activation of $K^+$ channels is one of the mechanisms through which estrogen produces an inhibitory effect on the VSMC [21]. Estrogens further affect the vascular system by modulating angiotensin II (ANG II) stimulated endothelin synthesis and inducing prostaglandin production [30, 31]. When investigating the direct effects in the large arteries, estrogens are associated with improved coronary blood flow, coronary vasodilatory responses and decreased coronary vascular resistance [32, 33].

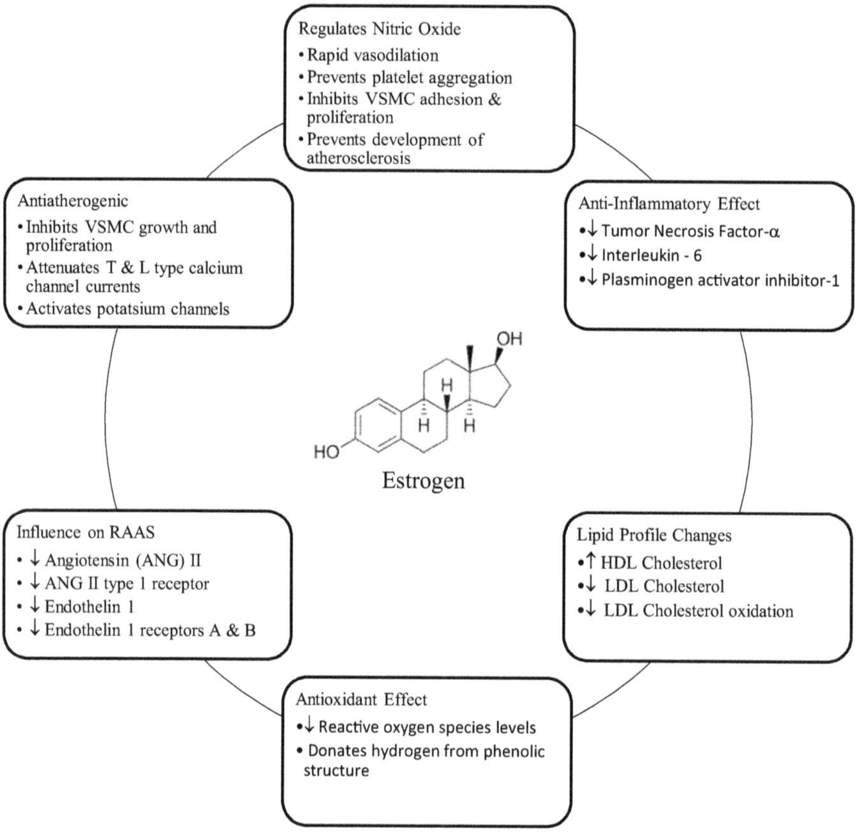


**Fig. 3.2** Physiological effects of estrogen

Indirect effects of estrogen on the CV system are through beneficial changes in the lipid profile, including increased high-density lipoprotein (HDL) cholesterol, decreased low-density lipoprotein (LDL) cholesterol, and a reduction in LDL oxidation [21].

During menstruation, pregnancy and estrogen supplementation, vasodilation and BP are affected by estrogen level fluctuations [34–36].

## The Role of Estrogen in Oxidative Stress, Inflammation and Endothelial Function

17β-estradiol influences physiological systems such as the sympathetic nervous system and renin–angiotensin–aldosterone system (RAAS), oxidative stress (OS), endothelial function and salt sensitivity [37].

Genetic factors can impact all these systems, which are related to a crucial inflammatory state and lead to the cardiac, vascular and renal damage found in hypertension [37].

Evidence indicates that estrogen is CV protective partly due to its ability to act on the RAAS and endothelin (ET) system. Estrogen's profound impact on these systems may underlie its long-term regulatory effects on BP [37, 38]. RAAS activity is reduced by estrogen as estrogen reduces the production of angiotensin II (ANG II) and levels of ANG II type 1 receptor (AT1R) [37]. Estrogen down-regulates AT1R in the kidney, adrenal cortex and VSMC [39, 40]. Premenopausal women appear mainly to have the depressor RAAS pathway activated [41]. In various animal studies, estrogen increased angiotensinogen synthesis and decreased the synthesis of RAAS enzymes renin and angiotensin-converting enzyme [42–44]. Estrogen also decreases the pro-hypertensive effects of endothelin 1(ET1) by modulating the production of ET1, and ET receptors type A and B [38]. Estrogen can also inhibit ET1 synthesis by increasing eNOS [45] or decreasing the production of angiotensin II [46, 47]. Additionally, estrogen alters cardiac, renal and vascular expression of ET receptors [38].

OS is associated with the development of hypertension [37]. OS occurs when reactive oxygen species (ROS) production rate exceeds that of the antioxidant defence system [37]. The higher levels of OS that accompany aging affect multiple stages of the NO biosynthetic pathway [48]. Estrogen has antioxidant properties due to its ability to act as a ROS scavenger by donating hydrogen from its phenolic structure [22]. A deficiency of SSHs combined with aging results in an over-production of ROS [49]. The increased ROS levels inactivate NO and increase peroxynitrite production, which then oxidizes tetrahydrobiopterin (BH4), causing uncoupling of eNOS. This results in reduced NO bioavailability and impaired endothelial function [49]. NO scavenging by ROS leads to reduced bioavailability of NO, which contributes to the age-associated decline in endothelial function [50].

Evidence has shown that compared to age-matched men, premenopausal women have lower levels of OS through the antioxidant actions of estrogen [51–53], but postmenopausal women show higher levels of OS [54]. Investigating this phenomenon and the effect of antioxidant therapy has shown that in estrogen-deficient postmenopausal women, vitamin C (antioxidant) infusion temporarily and reversibly reduced ROS, removed the suppression of endothelial function due to oxidative stress, and thereby increased brachial artery flow-mediated dilation (FMD) [55, 56].

Inflammation has a vital role in the CV system and underlies multiple CV pathologies. Alterations in the CV system lead to an inflammatory response, whereby cellular stress pathways become activated, and pro-inflammatory and anti-inflammatory factors are released [37]. During menopause, women have increased levels of inflammatory markers, including tumour necrosis factor-alpha, interleukin-6 and plasminogen activator inhibitor-1 [57, 58].

Endothelial dysfunction is associated with OS and vascular inflammation [37]. Endothelial dysfunction is characterized by decreased vasodilators such as NO and increased ET levels [37]. The reduced estrogen levels in postmenopausal women are associated with increased arterial stiffness, vascular and hemodynamic parameters and decreased endothelial function [59], highlighting an important relationship between estrogen and endothelial dysfunction.

Salt sensitivity is an exaggerated BP response to dietary salt consumption caused by impaired sodium excretion by the kidneys that results in extracellular fluid loading and increased cardiac output. Salt sensitivity is related to kidney malfunction, endothelial dysfunction and hyperactive sympathetic nervous activity [60], and is an important risk factor in the development of hypertension. Not only is this phenomenon more common amongst African ethnic groups, but postmenopausal women appear to be more salt sensitive than premenopausal women [61]. This is shown through the correlation between the removal of the ovaries with developing salt sensitivity [62], suggesting that changes in SSHs, namely estrogen, may be associated with salt sensitivity. Further aggravating this hormonal effect is the increased sympathetic nerve activity in the presence of metabolic syndrome and weight gain [63, 64], which are common in postmenopausal women.

## The Role of Estrogen in Lipid Profile Alterations, Obesity and Atherosclerosis

A sharp decline in estrogen is the most established hormonal shift seen in postmenopausal women [65], which is seen as a causal factor in the increased incidence of vascular disease [66]. A negative association exists between vascular calcification and serum estrogen levels [67]. Reduced estrogen levels following menopause are associated with lipid profile variations, increased abdominal fat, insulin resistance, and atherogenic lipoproteins [37, 48]. Compared to men, women generally have higher amounts of fat constituting overall body mass and deposit fat in their lower extremities and subcutaneously. Estrogen may be responsible for the fact that women have increased rates of non-esterified fatty acid reuptake into adipose tissue as well as increased fat oxidation during extended exercise [48].

Age is associated with increased plasma triglyceride levels (TGL) and significant variations in fasting TGL in both sexes; however, only women demonstrate an increased prevalence of hypercholesterolemia [48] and HDL levels are higher in women than men of all ages. Lipid abnormalities are important in the progression of atherogenesis and atherosclerosis [1, 2]. Following menopause, levels of LDL and TGL increase and HDL levels decrease [68]. Estrogen can inhibit lesion formation in atherosclerosis [69], impacting lipid profiles and stimulating NO production, as NO prevents the development of atherosclerosis [22]. However, once atherosclerotic plaques have formed, estrogens increase the expression of matrix metalloproteinases (MMPs) [70], which increase the risk of plaque rupture [71–73].

## Hormone Replacement Studies and the Timing Hypothesis

Studies have found improvements in endothelial function in postmenopausal women treated with estradiol, but function was not restored to a premenopausal state [56, 74]. Studies have shown that the FMD response produced by treatment with estradiol is similar to the response seen in perimenopausal women [56, 74, 75], suggesting that aging may be a contributing factor to endothelial dysfunction. One study found that 18 h after treatment with estradiol postmenopausal women aged 50–59 had apparent improvement in endothelial function, however women aged 60–79 showed no improvement [76].

Large clinical trials, such as the Heart and Estrogen/progestin Replacement Study (HERS) and HERS-II, did not find a protective effect of HRT but rather found adverse CV events. These findings may have been related to the type of HRT used, ERs, and other factors such as the patient's age or pre-existing CV conditions [48]. The HRT timing hypothesis in the prevention of atherosclerosis indicates that HRT needs to be initiated before advanced atherosclerosis. This hypothesis suggests that SSHs can alter the cell biology of the vessel walls and inflammatory cells that accumulate during atherosclerosis [2]. Several studies found improvements or a reversal of endothelial dysfunction when HRT was initiated early prior to the development of advanced atherosclerosis [24, 25, 34].

Estrogen replacement therapy (ERT) can reduce CVD in postmenopausal women [77]. However, the timing of the ERT within six years of menopause onset appears to be critical for effectiveness [78]. The Women's Health Initiative Study found long-term ERT reduced vascular calcification in postmenopausal women aged 50–59 years [79]. Another study found that brachial artery FMD increased 3 h following oral BH4 administration in estrogen-deficient postmenopausal women but had no effect in postmenopausal women on estradiol or premenopausal controls [74].

## The Influence of Estrogen on Endothelial Adaptations to Endurance Training

Regular exercise is encouraged as a strategy to reduce CVD risk. However, there may be a sex specificity for endothelial adaption in older adults in response to endurance training. One study found that endothelial function improved with endurance exercise training in estradiol-treated postmenopausal women but not in women given a placebo, indicating that estrogen may play an essential role in a woman's vascular adaptations to endurance exercise [56]. Exercise and estradiol both utilize intracellular signalling pathways to activate and phosphorylate eNOS in order to release NO [26, 80]. A synergistic relationship may exist between estrogen and exercise in modulating gene expression and intracellular signalling in endothelial cells; research is needed to explore if therapies which target ERs should be prescribed in conjunction with exercise for postmenopausal women [49].

## Progesterone

Progesterone is synthesized in the ovaries (corpus luteum) and adrenal cortex. Progesterone receptors (PR) exist in the cytosol and, upon binding, translocate into the nucleus. PRs include PR-A and PR-B. They are found in the uterus, ovaries, mammary glands, brain, pancreas, bone, and lower urinary tract tissues [22]. PRs are found in many CV system cells, with the most notable CV effect on vascular cells where estrogen induces PR expression [81] in an effort to facilitate the inhibitory effects of estrogen [82].

Progesterone is mainly vasodilatory, although studies have found conflicting results [22]. Anti-atherogenic properties include decreasing LDLs and increasing HDLs [83]. Progesterone inhibits the proliferation and migration of VSMC by reducing mitogen-activated protein kinase activity [84]. The role of progesterone on CV function modulation is not as clearly identified as the role of estrogen. This may be because most trials investigated progestin therapy in combination with estrogens [82].

## Testosterone

In women, androgens are synthesized in the ovaries, adrenal cortex and converted in peripheral tissues. Androgen receptors (AR) are found in the cytosol and, upon androgen binding, translocate into the nucleus, with a small portion of receptors found on the plasma membrane. The AR exists as ARα and ARβ. Testosterone can additionally bind to an intracellular receptor. The complex then binds to DNA in the nucleus. ARs are related to PRs, and high doses of progestins can block ARs [85].

Testosterone also has genomic and non-genomic effects similar to estrogen, and its vascular actions are mediated by endothelial-dependent and independent mechanisms [22]. Testosterone acts predominantly on VSMC to elicit vasorelaxant effects through the modulation of VSMC membrane ion channels. This is performed through the inactivation of voltage-operated L-type $CA^{2+}$ channels [86, 87] and activation of $K^+$ channels [88, 89]. Testosterone also induces vasorelaxation by stimulating NO production through neuronal NOS [90]. Other known functions of testosterone include inducing matrix remodelling, growth factor signalling and eccentric cardiac remodelling [91], collagen protein synthesis, and plays a role in the expression and activity of collagen-degrading enzymes such as MMPs. MMP-2 has a dominant role in cardiac remodelling [91].

In menopause, changes in androgens are correlated with insulin resistance, abdominal fat and atherogenic lipoproteins [48]. Abdominal obesity is a key determinant of the relationship between androgenic parameters and CVD risk; therefore, CVD risk evaluations in women should include androgens [92]. Women typically have better cardiac function and survival than men, but this difference is lost when comparing postmenopausal women and men [48]. These differences are thought to be partly due to estrogen being cardioprotective compared to testosterone, which is detrimental to cardiac function. In women with polycystic ovary syndrome (PCOS), a positive association has been shown between endothelial dysfunction and elevated androgen levels, which suggests a correlation with early-onset endothelial dysfunction seen in PCOS [93]. Overall the effects of testosterone on vessel wall pathology are still controversial.

## Conclusion

Estrogen has numerous effects on the CV system. Indirect effects include increased HDL, with decreased LDL and decreased LDL oxidation. Direct genomic effects include increased CV production of endothelial, smooth muscle and myocardial cells, the inhibition of VSMC and myocardial growth and remodelling and increased angiogenesis. Non-genomic effects include increases in endothelial cells and platelet NO production, inhibition of $CA^{2+}$ currents, activation of $K^+$ channels, and reduced vascular constriction. The roles of progesterone and testosterone are less well understood, but both appear to have vasodilatory actions.

## References

1. Genovefa DK (2015) "Women's Heart and Estrogens". Bentham Science Publishers
2. Mendelsohn ME (2000) Mechanisms of estrogen action in the cardiovascular system. J Steroid Biochem Mol Biol 74:337–343
3. Woodward HJ, Zhu D, Hadoke PWF, MacRae VE (2021) Regulatory role of sex hormones in cardiovascular calcification. Int J Mol Sci 22(9):4620
4. Murphy E (2011) Estrogen signaling and cardiovascular disease. Circ Res 109:687–696
5. Luo J, Liu D (2020) Does GPER really function as a G protein-coupled estrogen receptor in vivo? Front Endocrinol 11:148
6. Kuiper G, Enmark E, Pelto-Huikko M, Nilsson S, Gustafsson J (1996) Cloning of a novel receptor expressed in rat prostate and ovary. Proc Natl Acad Sci 93:5925–5930
7. Filardo EJ, Quinn JA, Frackelton AR Jr, Bland KI (2002) Estrogen action via the G protein-coupled receptor, GPR30: stimulation of adenylyl cyclase and cAMP-mediated attenuation of the epidermal growth factor receptor-to-MAPK signaling axis. Mol Endocrinol 16:70–84
8. Ueda K, Adachi Y, Liu P, Fukuma N, Takimoto E (2020) Regulatory actions of estrogen receptor signaling in the cardiovascular system. Front Endocrinol 10:909
9. Teoh J-p, Li X, Simoncini T, Zhu D, Fu X (2020) Estrogen-mediated gaseous signaling molecules in cardiovascular disease. Trends Endocrinol Metab 31:773–784
10. Dessy C, Feron O, Balligand J-L (2010) The regulation of endothelial nitric oxide synthase by caveolin: a paradigm validated in vivo and shared by the 'endothelium-derived hyperpolarizing factor.' Pflügers Archiv-Eur J Physiol 459:817–827
11. Fleming I (2010) Molecular mechanisms underlying the activation of eNOS. Pflügers Archiv-Eur J Physiol 459:793–806
12. Takahashi K, Ohmichi M, Yoshida M et al (2003) Both estrogen and raloxifene cause G1 arrest of vascular smooth muscle cells. J Endocrinol 178:319
13. Pare G, Krust A, Karas RH et al (2002) Estrogen receptor-α mediates the protective effects of estrogen against vascular injury. Circ Res 90:1087–1092

14. Fliegner D, Schubert C, Penkalla A et al (2010) Female sex and estrogen receptor-β attenuate cardiac remodeling and apoptosis in pressure overload. Am J Physiol-Regul, Integrative Comp Physiol 298:R1597–R1606
15. Pedram A, Razandi M, O'Mahony F, Lubahn D, Levin ER (2010) Estrogen receptor-β prevents cardiac fibrosis. Mol Endocrinol 24:2152–2165
16. Gabel SA, Walker VR, London RE, Steenbergen C, Korach KS, Murphy E (2005) Estrogen receptor beta mediates gender differences in ischemia/reperfusion injury. J Mol Cell Cardiol 38:289–297
17. Meyer MR, Fredette NC, Howard TA et al (2014) G protein-coupled estrogen receptor protects from atherosclerosis. Sci Rep 4:1–9
18. Meyer MR, Prossnitz ER, Barton M (2011) GPER/GPR30 and regulation of vascular tone and blood pressure. Immunol, Endocr Metab Agents Med Chem (Formerly Current Medicinal Chemistry-Immunology, Endocrine and Metabolic Agents) 11:255–261
19. Meyer MR, Field AS, Kanagy NL, Barton M, Prossnitz ER (2012) GPER regulates endothelin-dependent vascular tone and intracellular calcium. Life Sci 91:623–627
20. Ogola BO, Zimmerman MA, Sure VN et al (2019) G protein-coupled estrogen receptor protects from angiotensin II-induced increases in pulse pressure and oxidative stress. Front Endocrinol 10:586
21. Skafar DF, Xu R, Morales J, Ram J, Sowers JR (1997) Female sex hormones and cardiovascular disease in women. J Clin Endocrinol Metab 82:3913–3918
22. dos Santos RL, da Silva FB, Ribeiro RF Jr, Stefanon I (2014) Sex hormones in the cardiovascular system. Horm Mol Biol Clin Invest 18:89–103
23. Yu J, Akishita M, Eto M et al (2010) Androgen receptor-dependent activation of endothelial nitric oxide synthase in vascular endothelial cells: role of phosphatidylinositol 3-Kinase/Akt pathway. Endocrinol 151:1822–1828
24. Mendelsohn ME, Karas RH (1999) The protective effects of estrogen on the cardiovascular system. N Engl J Med 340:1801–1811
25. Edwards DP (2005) Regulation of signal transduction pathways by estrogen and progesterone. Annu Rev Physiol 67:335
26. Chambliss KL, Shaul PW (2002) Estrogen modulation of endothelial nitric oxide synthase. Endocrinol Rev 23:665–686
27. Nevzati E, Shafighi M, Bakhtian KD, Treiber H, Fandino J, Fathi AR (2015) Estrogen induces nitric oxide production via nitric oxide synthase activation in endothelial cells. Neurovascular events after subarachnoid hemorrhage. Springer, pp 141–145
28. Oparil S, Levine RL, Chen Y-F (1996) Sex hormones and the vasculature. Endocrinol Vascul 225–237
29. Zhang F, Ram JL, Standley PR, Sowers JR (1994) 17 beta-Estradiol attenuates voltage-dependent Ca2+ currents in A7r5 vascular smooth muscle cell line. Am J Physiol Cell Physiol 266:C975–C980
30. Ylikorkala O, Cacciatore B, Paakkari I, Tikkanen MJ, Viinikka L, Toivonen J (1998) The long-term effects of oral and transdermal postmenopausal hormone replacement therapy on nitric oxide, endothelin-1, prostacyclin, and thromboxane. Fertil Steril 69:883–888
31. Morey KA, Razand M, Pedram A, Hu R-M, Prins AB, Levin RE (1998) Oestrogen and progesterone inhibit the stimulated production of endothelin-1. Biochem J 330:1097–1105
32. Node K, Kitakaze M, Kosaka H et al (1997) Roles of NO and Ca2+-activated K+ channels in coronary vasodilation induced by 17β-estradiol in ischemic heart failure. FASEB J 11:793–799
33. Raddino R, Manca C, Poli E, Bolognesi R, Visioli O (1986) Effects of 17 beta-estradiol on the isolated rabbit heart. Arch Int Pharmacodyn Ther 281:57–65
34. Dubey RK, Oparil S, Imthurn B, Jackson EK (2002) Sex hormones and hypertension. Cardiovasc Res 53:688–708
35. Sader MA, Celermajer DS (2002) Endothelial function, vascular reactivity and gender differences in the cardiovascular system. Cardiovasc Res 53:597–604
36. Reckelhoff JF (2001) Gender differences in the regulation of blood pressure. Hypertens 37:1199–1208

37. Sabbatini AR, Kararigas G (2020) Estrogen-related mechanisms in sex differences of hypertension and target organ damage. Biol Sex Differ 11:31
38. Colafella KMM, Denton KM (2018) Sex-specific differences in hypertension and associated cardiovascular disease. Nat Rev Nephrol 14:185–201
39. Baiardi G, Macova M, Armando I, Ando H, Tyurmin D, Saavedra JM (2005) Estrogen upregulates renal angiotensin II AT1 and AT2 receptors in the rat. Regul Pept 124:7–17
40. Roesch DM, Tian Y, Zheng W, Shi M, Verbalis JG, Sandberg K (2000) Estradiol attenuates angiotensin-induced aldosterone secretion in ovariectomized rats. Endocrinology 141:4629–4636
41. Hilliard LM, Sampson AK, Brown RD, Denton KM (2013) The "his and hers" of the renin-angiotensin system. Curr Hypertens Rep 15:71–79
42. Schunkert H, Danser AJ, Hense H-W, Derkx FH, Ku¨ rzinger S, Riegger GnA (1997) Effects of estrogen replacement therapy on the renin-angiotensin system in postmenopausal women. Circ 95:39–45
43. Brosnihan KB, Li P, Ganten D, Ferrario CM (1997) Estrogen protects transgenic hypertensive rats by shifting the vasoconstrictor-vasodilator balance of RAS. Am J Physiol 273(6):R1908–1915
44. Brosnihan KB, Hodgin JB, Smithies O, Maeda N, Gallagher P (2008) Tissue-specific regulation of ACE/ACE2 and AT1/AT2 receptor gene expression by oestrogen in apolipoprotein E/oestrogen receptor-α knock-out mice. Exp Physiol 93:658–664
45. Akishita M, Kozaki K, Eto M et al (1998) Estrogen attenuates endothelin-1 production by bovine endothelial cells via estrogen receptor. Biochem Biophys Res Commun 251:17–21
46. Hong H-J, Liu J-C, Chan P et al (2004) 17β-estradiol downregulates angiotensin-II-induced endothelin-1 gene expression in rat aortic smooth muscle cells. J Biomed Sci 11:27–36
47. Chao H, Chen J, Chen C et al (2005) Inhibition of angiotensin II induced endothelin-1 gene expression by 17-β-oestradiol in rat cardiac fibroblasts. Heart 91:664–669
48. Guarner-Lans V, Rubio-Ruiz ME, Perez-Torres I, Banos de MacCarthy G (2011) Relation of aging and sex hormones to metabolic syndrome and cardiovascular disease. Exp Gerontol 46:517–523
49. Moreau KL (2019) Modulatory influence of sex hormones on vascular aging. Am J Physiol Heart Circ Physiol 316:H522–H526
50. Kojda G, Harrison D (1999) Interactions between NO and reactive oxygen species: pathophysiological importance in atherosclerosis, hypertension, diabetes and heart failure. Cardiovasc Res 43:652–671
51. Barp J, Araújo ASdR, Fernandes T et al (2002) Myocardial antioxidant and oxidative stress changes due to sex hormones. Braz J Med Biol Res 35:1075–1081
52. Ruiz-Larrea MB, Martín C, Martínez R, Navarro R, Lacort M, Miller NJ (2000) Antioxidant activities of estrogens against aqueous and lipophilic radicals; differences between phenol and catechol estrogens. Chem Phys Lipids 105:179–188
53. Martin C, Barturen K, Martinez R, Lacort M, Ruiz-Larrea M (1998) In vitro inhibition by estrogens of the oxidative modifications of human lipoproteins. J Physiol Biochem 54:195–202
54. Vassalle C, Sciarrino R, Bianchi S, Battaglia D, Mercuri A, Maffei S (2012) Sex-related differences in association of oxidative stress status with coronary artery disease. Fertil Steril 97(414–419):e412
55. Eskurza I, Monahan KD, Robinson JA, Seals DR (2004) Effect of acute and chronic ascorbic acid on flow-mediated dilatation with sedentary and physically active human ageing. J Physiol 556:315–324
56. Moreau KL, Stauffer BL, Kohrt WM, Seals DR (2013) Essential role of estrogen for improvements in vascular endothelial function with endurance exercise in postmenopausal women. J Clin Endocrinol Metab 98:4507–4515
57. Pfeilschifter J, Köditz R, Pfohl M, Schatz H (2002) Changes in proinflammatory cytokine activity after menopause. Endocr Rev 23:90–119
58. Sites CK, Toth MJ, Cushman M et al (2002) Menopause-related differences in inflammation markers and their relationship to body fat distribution and insulin-stimulated glucose disposal. Fertil Steril 77:128–135

59. Hildreth KL, Ozemek C, Kohrt WM, Blatchford PJ, Moreau KL (2018) Vascular dysfunction across the stages of the menopause transition is associated with menopausal symptoms and quality of life. Menopause 25:1011
60. Choi HY, Park HC, Ha SK (2015) Salt sensitivity and hypertension: a paradigm shift from kidney malfunction to vascular endothelial dysfunction. Electrolytes Blood Press 13:7–16
61. Tominaga T, Suzuki H, Ogata Y, Matsukawa S, Saruta T (1991) The role of sex hormones and sodium intake in postmenopausal hypertension. J Hum Hypertens 5:495–500
62. Schulman IH, Aranda P, Raij L, Veronesi M, Aranda FJ, Martin R (2006) Surgical menopause increases salt sensitivity of blood pressure. Hypertens 47:1168–1174
63. Seals DR, Esler MD (2000) Human ageing and the sympathoadrenal system. J Physiol 528:407–417
64. Esler M, Rumantir M, Wiesner G, Kaye D, Hastings J, Lambert G (2001) Sympathetic nervous system and insulin resistance: from obesity to diabetes. Am J Hypertens 14:304S–309S
65. Zhu D, Li X, Macrae VE, Simoncini T, Fu X (2018) Extragonadal effects of follicle-stimulating hormone on osteoporosis and cardiovascular disease in women during menopausal transition. Trends Endocrinol Metab 29:571–580
66. Dalal PK, Agarwal M (2015) Postmenopausal syndrome. Indian J Psychiatry 57:S222
67. Nakao J, Orimo H, Ooyama T, Shiraki M (1979) Low serum estradiol levels in subjects with arterial calcification. Atherosclerosis 34:469–474
68. Mendelsohn ME, Karas RH (2005) Molecular and cellular basis of cardiovascular gender differences. Sci 308:1583–1587
69. Bourassa P, Milos PM, Gaynor B, Breslow JL, Aiello RJ (1996) Estrogen reduces atherosclerotic lesion development in apolipoprotein E-deficient mice. Proc Natl Acad Sci 93:10022–10027
70. Ardans JA, Blum A, Mangan PR, Wientroub S, Cannon RO III, Wahl LM (2001) Raloxifene-mediated increase in matrix metalloproteinase-1 production by activated monocytes. Arterioscler Thromb Vasc Biol 21:1265–1268
71. Mudali S, Dobs AS, Ding J, Cauley JA, Szklo M, Golden SH (2005) Endogenous post-menopausal hormones and serum lipids: the atherosclerosis risk in communities study. J Clin Endocrinol Metab 90:1202–1209
72. Mueck AO, Seeger H, Wallwiener D (2002) Medroxyprogesterone acetate versus norethisterone: effect on estradiol-induced changes of markers for endothelial function and atherosclerotic plaque characteristics in human female coronary endothelial cell cultures. Menopause 9:273–281
73. Rosenfeld ME, Kauser K, Martin-McNulty B, Polinsky P, Schwartz SM, Rubanyi GM (2002) Estrogen inhibits the initiation of fatty streaks throughout the vasculature but does not inhibit intra-plaque hemorrhage and the progression of established lesions in apolipoprotein E deficient mice. Atherosclerosis 164:251–259
74. Moreau KL, Meditz A, Deane KD, Kohrt WM (2012) Tetrahydrobiopterin improves endothelial function and decreases arterial stiffness in estrogen-deficient postmenopausal women. Am J Physiol Heart Circ Physiol 302:H1211–H1218
75. Moreau KL, Hildreth KL, Meditz AL, Deane KD, Kohrt WM (2012) Endothelial function is impaired across the stages of the menopause transition in healthy women. J Clin Endocrinol Metab 97:4692–4700
76. Sherwood A, Bower JK, McFetridge-Durdle J, Blumenthal JA, Newby LK, Hinderliter AL (2007) Age moderates the short-term effects of transdermal 17β-estradiol on endothelium-dependent vascular function in postmenopausal women. Arterioscler Thromb Vasc Biol 27:1782–1787
77. Lopez-Pier MA, Lipovka Y, Koppinger MP, Harris PR, Konhilas JP (2018) The clinical impact of estrogen loss on cardiovascular disease in menopausal females. Med Res Arch 6(2):1663
78. Hodis HN, Mack WJ, Shoupe D et al (2015) Methods and baseline cardiovascular data from the early versus late intervention trial with estradiol testing the menopausal hormone timing hypothesis. Menopause 22:391

79. Manson JE, Allison MA, Rossouw JE et al (2007) Estrogen therapy and coronary-artery calcification. N Engl J Med 356:2591–2602
80. Zhang Q-J, McMillin SL, Tanner JM, Palionyte M, Abel ED, Symons JD (2009) Endothelial nitric oxide synthase phosphorylation in treadmill-running mice: role of vascular signalling kinases. J Physiol 587:3911–3920
81. Karas RH, van Eickels M, Lydon JP et al (2001) A complex role for the progesterone receptor in the response to vascular injury. J Clin Investig 108:611–618
82. Salerni S, Di Francescomarino S, Cadeddu C, Acquistapace F, Maffei S, Gallina S (2015) The different role of sex hormones on female cardiovascular physiology and function: not only oestrogens. Eur J Clin Invest 45:634–645
83. Smiley DA, Khalil RA (2009) Estrogenic compounds, estrogen receptors and vascular cell signaling in the aging blood vessels. Curr Med Chem 16:1863–1887
84. Orshal JM, Khalil RA (2004) Gender, sex hormones, and vascular tone. Am J Physiol-Regul Integr Comp Physiol 286:R233–R249
85. Wilson C, Maass R, Estrada M (2011) Cardiovascular effects of androgens. Basic Clin Endocrinol Up-to-Date. Chile: Intech pp 63–78
86. English K, Jones R, Jones T, Morice A, Channer K (2002) Testosterone acts as a coronary vasodilator by a calcium antagonistic action. J Endocrinol Invest 25:455–458
87. Hall J, Jones R, Jones T, Channer K, Peers C (2006) Selective inhibition of L-type Ca2+ channels in A7r5 cells by physiological levels of testosterone. Endocrinology 147:2675–2680
88. Cairrão E, Álvarez E, Santos-Silva AJ, Verde I (2008) Potassium channels are involved in testosterone-induced vasorelaxation of human umbilical artery. Naunyn Schmiedebergs Arch Pharmacol 376:375–383
89. Seyrek M, Yildiz O, Ulusoy HB, Yildirim V (2007) Testosterone relaxes isolated human radial artery by potassium channel opening action. J Pharmacol Sci 103:309–316
90. Deenadayalu V, Puttabyatappa Y, Liu AT, Stallone JN, White RE (2012) Testosterone-induced relaxation of coronary arteries: activation of BKCa channels via the cGMP-dependent protein kinase. Am J Physiol-Heart Circ Physiol 302:H115–H123
91. Hofmann U, Bonz A, Frantz S et al (2012) A collagen α2 (I) mutation impairs healing after experimental myocardial infarction. Am J Pathol 180:113–122
92. Mesch V, Siseles N, Maidana P et al (2008) Androgens in relationship to cardiovascular risk factors in the menopausal transition. Climacteric 11:509–517
93. Maturana MA, Breda V, Lhullier F, Spritzer PM (2008) Relationship between endogenous testosterone and cardiovascular risk in early postmenopausal women. Metabolism 57:961–965

# Chapter 4
# Spontaneous Coronary Artery Dissection (SCAD): An Overview of the Condition, Diagnostic Work Up and Management

**Jenny Y. Namkoong, Tracey J. F. Colella, Carolina Gonzaga Carvalho, Mina Madan, and Shuangbo Liu**

**Abstract** There has been increased awareness of Spontaneous Coronary Artery Dissection (SCAD) as a type of myocardial infarction over the last decade. However, as its underlying pathology and mechanisms are still being understood, and the best management principles for SCAD still being realized, there are little robust guidelines for those not subspecializing in SCAD patient management to help guide management. This book chapter targets general medical practitioners who are faced with taking care of the small but increasing SCAD population in the community, in partnership with SCAD specialists. It provides an updated understanding of SCAD including the "inside out" versus "outside in" hypothesis of pathophysiology, and the management principles to be mindful of in SCAD in comparison to traditional acute coronary syndrome (ACS) management. Uniquely, the chapter provides colloquial answers to frequently asked questions by patients about SCAD, and an in-depth review of the benefits of cardiac rehabilitation for SCAD patients.

J. Y. Namkoong · S. Liu (✉)
Section of Cardiology, Department of Internal Medicine, Max Rady College of Medicine, Rady Faculty of Health Sciences, University of Manitoba, Winnipeg, MB, Canada
e-mail: sliu@sbgh.mb.ca

St. Boniface Hospital, Winnipeg, MB, Canada

T. J. F. Colella · C. G. Carvalho
KITE—Toronto Rehabilitation Institute, University Health Network, Toronto, ON, Canada

T. J. F. Colella
Lawrence S. Bloomberg Faculty of Nursing, Rehabilitation Science Institute, University of Toronto, Toronto, Ontario, Canada, University of Toronto, Toronto, ON, Canada

C. G. Carvalho
Division of Physical Medicine and Rehabilitation, Temerty Faculty of Medicine, University of Toronto, Toronto, ON, Canada

M. Madan
Shulich Herat Centre, Sunnybrook Health Sciences Centre, Toronto, ON, Canada

© The Author(s), under exclusive license to Springer Nature Switzerland AG 2023

L. Kirshenbaum and I. Rabinovich-Nikitin (eds.), *Biology of Women's Heart Health*, Advances in Biochemistry in Health and Disease 26,
https://doi.org/10.1007/978-3-031-39928-2_4

**Keywords** SCAD · Pregnancy · Hypertension · Acute coronary syndrome (ACS) · MINOCA

## Introduction

Spontaneous Coronary Artery Dissection (SCAD) is a type of myocardial infarction (MI) that presents acutely with similar characteristics as the more common atherosclerotic MI. SCAD is the cause of up to 4% of acute coronary syndrome (ACS) presentations [1]. However, in females under the age of 50, it can represent up to 35% of ACS presentations [1–3].

There is an increasing awareness of SCAD as a unique type of heart attack, and an increasing need for medical practitioners to be equipped to manage SCAD patients in the community. Thus, the aim of this chapter is to review the current definition, classification and understanding of SCAD, and highlight key similarities and differences in diagnosis (Table 4.1), management (Table 4.2) and prognosis between SCAD ACS and atherosclerotic ACS.

In addition, a schematic (Fig. 4.1) and suggested outpatient partnered management (Table 4.3) of SCAD between SCAD specialists, rehabilitation specialists, general cardiologists, and primary care providers is provided. Commonly asked questions by patients are addressed (Table 4.5), including recommendations regarding pregnancy after SCAD. Furthermore, the role of cardiac rehabilitation in this population is highlighted [4].

## Definition and Classification of SCAD

The definition of SCAD is a 'spontaneous separation of the coronary artery wall, which is not iatrogenic and not related to atherosclerosis or trauma' [5]. It leads to myocardial infarction as the hematoma within the separated coronary artery wall compresses the lumen of the artery and prevents blood flow, causing ischemia.

Traditionally, SCAD was thought to result from a tear in the intima of the coronary artery, which allowed blood to flow into the intimal layer to create this false lumen that expanded and compressed the true lumen [6]. This is referred to as the "inside out" hypothesis but is now superseded largely by a modified hypothesis, comparatively named the "outside in" hypothesis which suggests that the primary incident is an intramural hematoma, rather than an intimal tear [2, 6]. The intramural hematoma is thought to be caused by spontaneous medial dissection or rupture of the vasa vasorum [2]. From there, it is believed that the underlying hemorrhage expands in either a focal, linear, or spiral fashion to compress into the lumen of the coronary artery [2, 6].

The way the coronary artery lumen is compressed is how SCAD is classified into its three types, an angiographic classification identified and labelled by the Canadian

**Table 4.1** Comparison of history, physical exam, and investigations for acute coronary syndrome from atherosclerotic plaque rupture and spontaneous coronary artery dissection

| Atherosclerotic ACS | SCAD ACS |
|---|---|
| *History* | |
| • Chest pain onset, timing, character<br>• Associated features<br>• Coronary artery disease risk factors including smoking, diabetes, hypertension, dyslipidemia, family history of premature coronary artery disease | • Chest pain onset, timing, character<br>• Associated features<br>• Coronary artery disease risk factors including smoking, diabetes, hypertension, dyslipidemia, family history of premature coronary artery disease<br>• Current or recent pregnancy<br>• Detailed obstetrics and gynecological history:<br>– Age of menarche<br>– Menopause onset<br>– Pregnancies<br>– Abortions<br>– Multiparity (twin) pregnancies<br>– Birth control<br>– Hormone replacement therapy<br>• Possible triggers (physical or emotional)<br>• History of SCAD, hypertension or migraines<br>• History of autoimmune disease or connective tissue disorders<br>• Previous coronary angiograms<br>• Family history of SCAD, vascular disorder (including fibromuscular dysplasia), autoimmune disease or connective tissue disorders |
| *Physical examination* | |
| • Cardiovascular exam<br>• Respiratory exam | • Cardiovascular exam<br>• Respiratory exam<br>• Bruits: carotid, renal and femoral arteries<br>• Signs of connective tissue disorders |
| *Investigations* | |
| • Troponin<br>• Creatinine<br>• Electrolytes<br>• Fasting cholesterol<br>• HbA1c | • Troponin<br>• Creatinine<br>• Electrolytes<br>• Fasting cholesterol<br>• HbA1c<br>• CT angiography of head/neck and abdomen/pelvis to assess for fibromuscular dysplasia<br>• Vascular or autoimmune work up |

**Table 4.2** Coronary angiogram findings, management, and course in hospital of acute coronary syndrome from atherosclerotic plaque rupture and spontaneous coronary artery dissection

| Atherosclerotic ACS | SCAD ACS |
|---|---|
| *Coronary angiogram* | |
| • Locate area of plaque rupture (culprit lesion) | • Rule out acute atherosclerotic plaque disruption<br>• Confirm SCAD diagnosis |
| *Revascularization* | |
| • Percutaneous coronary intervention (PCI) to stabilize culprit lesion | • Avoid revascularization if possible, unless high risk features<br>• Natural history of SCAD is that most (96%) heal by 30 days<br>• Higher complication rate than atherosclerotic ACS, with lower rate of PCI success |
| *Course in hospital* | |
| • First 24–48 h has highest arrhythmia risk | • 7% of SCAD patients treated conservatively will require intervention, generally within 5 days of initial presentation |
| *Medications* | |
| • Dual antiplatelet therapy<br>• Statin<br>• ACEI/ARB<br>• Beta-blockers<br>• Anti-anginal therapy | • Single antiplatelet (aspirin preferred)<br>• Beta blocker<br>• Blood pressure control<br>• Antianginal therapy may be helpful to control chronic chest pain syndromes arising after SCAD |

SCAD group (Fig. 4.2). Type 1 is an intimal tear with contrast staining of the false lumen, Type 2 represents a long diffuse and smooth narrowing angiographically, and Type 3 is a focal or tubular angiographic stenosis with both Type 2 and 3 resulting from an intra-mural hematoma [5].

## Pathophysiology of SCAD

Recent literature has suggested conceptualizing SCAD as “SCAB” (spontaneous coronary artery “bleed”) rather than “dissection”, as the presence of an intimal tear is not necessary for diagnosis [7]. Increasing understanding of SCAD suggests that as the “bleed” expands between the coronary artery layers, an intimal tear can occur to decompress the intramural pressure [6]. This newer pathophysiological theory is supported by the evidence that not all SCAD cases have an intimal tear, with recent cohort studies identifying an intimal tear in only 30% of cases [2]. Further, intra-coronary imaging studies in more severe SCAD cases have shown to be associated with an intimal tear [8], which supports the “outside in” hypothesis as a unifying mechanism with the tear occurring secondarily for decompression [8, 9].

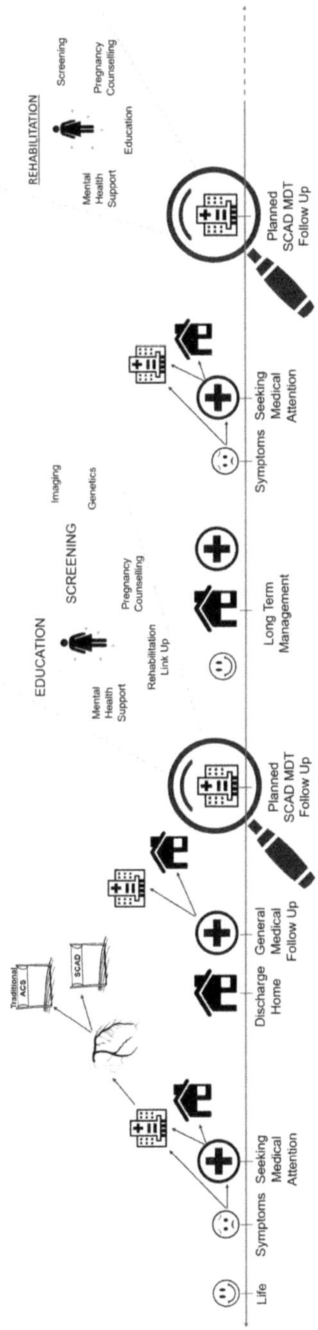


**Fig. 4.1** Partnered follow up with Family Practitioner and SCAD Hospital Multi-Disciplinary Team

**Table 4.3** Outpatient management following diagnosis of SCAD ACS (Correlate with Fig. 4.1)

| *1. Family physician follow up* | |
|---|---|
| Ideally within 4–6 weeks of hospital presentation | |
| Aim | • Validation of the significant health event<br>– High proportion of patients report anxiety and depression following SCAD, related to feeling extreme vulnerability from lack of understanding of SCAD and variability of information from health care providers<br>• Pregnancy is high risk and strongly not recommended (see below)<br>• Medical management<br>– Hypertension management is important<br>– Beta blockers—decreases SCAD recurrence<br>– No role for routine repeat coronary angiography<br>– If similar symptoms to initial presentation, need to seek medical attention, declare the prior SCAD-related MI to ED staff, have ECG and troponin measured<br>• Lifestyle<br>– Physical activity<br>Regular, moderate-intensity activity is recommended<br>Avoid heavy weightlifting or Valsalva maneuvers<br>– Sexual activity—expected back to usual within 4 to 6 weeks. However, strongly advised against pregnancy<br>– Food—healthy diet aimed at maintaining normal BMI, but SCAD is not a diet mediated disease<br>• Patients must make informed decisions about their future health<br>• Link in with SCAD support groups<br>– Online<br>Women@heart (www.womenheart.org) support group<br>SCAD Patient Guide 2022 [54] (www.sunnybrook.ca/SCAD/guide)<br>SCAD Alliance (www.scadalliance.org) |
| *2. Planned SCAD MDT follow up I* | |
| Ideally within 8–12 weeks of hospital presentation | |
| Aim | • Consultation reviewing SCAD and understanding of its mechanism<br>• One-time cross-sectional imaging from head to pelvis with CT angiography looking for:<br>– Fibromuscular dysplasia (tortuosity, aneurysms, stenoses, dissections, contour irregularities (beading))<br>• Pregnancy is high risk and is not recommended<br>– Due to the possibility of severe and potentially life-threatening SCAD during pregnancy, and recurrence rates of 17% for women with peripartum SCAD, pregnancy after SCAD-related MI is not recommended<br>– In the event of pregnancy—patients must seek involvement of SCAD expert and high-risk obstetrics team prior to pregnancy or as early as possible after pregnancy to consider their options<br>• Further cardiac investigations if indicated, which may include<br>– Echocardiogram;<br>– Cardiac MRI (rule out differentials)<br>– Coronary CT angiography or repeat coronary angiography—rarely required<br>• Review of medication<br>• Review of associated conditions<br>• Review of psychosocial state<br>• Confirm referral/attendance at cardiac rehab |

(continued)

**Table 4.3** (continued)

| *3. Long term management* | |
|---|---|
| Aim | • Risk stratifying your patient to likelihood of recurrence of SCAD, higher likelihood in those with<br>– Migraine headaches<br>– Poorly controlled hypertension<br>– Fibromuscular dysplasia diagnosis<br>– Known coronary artery tortuosity (from initial angiogram)<br>Current literature suggests that receiving coronary stents is not associated with prevention of recurrent SCAD in future. Further, if SCAD recurs, it tends to affect different coronary arteries from that which caused the initial event |
| *4. Seeking medical attention with unplanned chest pain presentation* | |
| Higher concern if within 5–6 days of initial presentation | |
| Aim | Determine need for hospital management<br>• Immediate hospital management if<br>– Hemodynamically unstable<br>– Acute ongoing chest pain<br>– Dynamic ECG changes, dynamic rise and fall of troponin<br>• Likely community management if:<br>– Hemodynamically stable<br>– Symptoms resolved<br>– Patient appears well<br>– No troponin elevation, or falling troponin compared with recent hospitalization |
| *5. Planned SCAD MDT follow up II* | |
| Aim | • SCAD education<br>– Increasing education as more research on SCAD conducted and medical understanding increases<br>• SCAD rehabilitation, involving:<br>– Graduated exercise regimen<br>• SCAD psychosocial support, involving:<br>– Cognitive behavioral therapy (CBT) as needed<br>– Peer support (Women@heart: www.womenheart.org)<br>• Education regarding SCAD and pregnancy |

## Epidemiology of SCAD

The mean age of SCAD patients is between 44 and 52 years, and it is uncommon in the extremes of age (under 25 years or over 80 years) [5, 10]. The female predominance is now well established in the literature, with a large meta-analysis of 2172 patients revealing 84% of SCAD patients as female [6].

There has been a temporal increase in the incidence of SCAD, with a rise in awareness paralleling a rise in diagnosis, particularly since 2012 [2]. Although it is likely still under-recognized at present, SCAD is believed to be the cause of up to 4% of ACS presentations [3]. Further, in administrative databases, SCAD represents 15–20% of myocardial infarctions during pregnancy or peripartum, a particularly high-risk period for known SCAD patients [1, 10, 11].

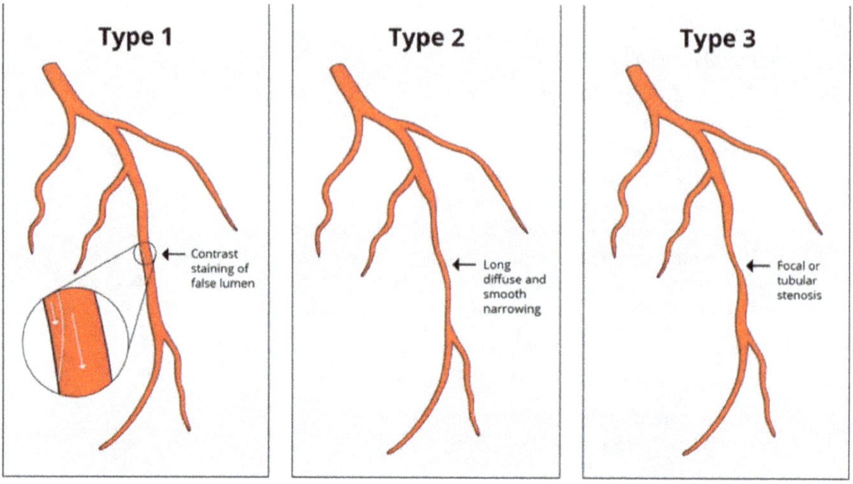


**Fig. 4.2** Classification of SCAD. *Figure from Saw, 2016, contemporary review of spontaneous coronary dissection* [5]

Recent data from the Canadian SCAD registry demonstrated that recurrence rates for SCAD are surprisingly low, 2.4% at 3 years [12]. Further, the 30-day mortality rate is 0.1% [13] and 3-year mortality rate is 1% [14] for SCAD patients.

## Risk Factors for SCAD

The most notable risk factor for poor outcome in SCAD is pregnancy and the post-partum state in a patient with prior SCAD. Hypertension also is a risk factor for SCAD-related myocardial infarction during pregnancy [15]. Amongst the SCAD patient population, 15–20% of patients are pregnant or post-partum at the time of diagnosis [1, 16]. However, the exact mechanism of association between hormones and SCAD is not yet understood, and the data is conflicting, as there does not appear to be a clear association between oral contraceptives or hormone replacement therapy and SCAD [16].

Secondly, hypertension has a high prevalence (45%) in SCAD patients [15]. There is a plausible explanation mechanistically, as hypertension can lead to chronic pathological changes in the arterial wall via increasing arterial wall stress, endothelial damage, triggering smooth muscle cell proliferation and breakdown of elastin fibers, that ultimately make the vessel more vulnerable to SCAD [15]. Hypertension is also independently associated with a higher risk of in-hospital mortality (OR 2.19, CI 1.86, 2.58) [17]. Furthermore, in a prospective Spanish registry, hypertensive SCAD patients had higher risk of increased severity of SCAD lesion (15 versus 7%, $p >$

0.05), higher risk of procedural complications (65 versus 41%, $p < 0.05$) and lower likelihood of procedural success (65 versus 88%, $p < 0.05$) [15, 18].

In addition, there should be clinical suspicion for SCAD in the presence of risk factors such as fibromuscular dysplasia (FMD). Between 15 and 70% of SCAD patients are found to have FMD with the wide range explained by variable screening protocol [1, 3, 6, 19]. Further risk factors may include connective tissue disorders, autoimmune and inflammatory disorders [1, 6], as well as microvascular dysfunction [15, 20], however there is not enough data to establish a clear association between these entities and SCAD at present. The role of genetics in the mechanism of SCAD is also not yet fully understood. A recent study noted a higher prevalence of fibrillin 1 (FBN1) in the plasma of SCAD patients compared with non-SCAD ACS patients [21]. This medium sized ($n = 70$ in each arm) study generated interest due to the mechanistic plausibility as fibrillin 1 is a component of the elastic tissue found in the media of coronary arteries [21]. Presently, there are no specific genetic screening tests for SCAD or fibromuscular dysplasia, but this is an active area of research [22].

## Clinical Presentation of SCAD

As a type of ACS, SCAD presents acutely with similar characteristics as atherosclerotic myocardial infarction. Although ACS presentations in young and middle-aged females with no atherosclerosis risk factors should lead to a high suspicion for SCAD, it must be emphasized that ACS by atherosclerotic plaque rupture is significantly more common than SCAD (50 versus 4%) [23]. Thus, atherosclerotic ACS must be ruled out by coronary angiography before the diagnosis of SCAD can be considered in any clinical presentation [10].

Similar to atherosclerotic ACS, most patients with SCAD (85–96%) experience chest pain or discomfort as their primary presenting complaint [10]. However, a small proportion of SCAD patients may have atypical symptoms including dyspnea (20%), back pain (14%), diaphoresis (21%), nausea and vomiting (24%) and presyncope (9%). Clinically, approximately 20–50% of SCAD patients present as STEMI, and up to 5% present with ventricular arrhythmia, and 2% with cardiogenic shock [10].

In addition, SCAD should be suspected in patients with associated extreme physical or emotional trigger, which is notable in 40% and 24% of SCAD presentations, respectively [9]. The postulated mechanism is a hyper-catecholaminergic state, which along with Valsalva-like maneuvers in physical or emotional states can damage the vasculature and arterial wall with shear force [6]. In the recently published study by the Canadian SCAD group, an isometric physical trigger was more commonly observed in men (40.2 versus 24.0%, $p = 0.007$), whilst emotional stress triggers have been found to be more likely in women (35 versus 60%, $p = 0.001$) [9, 24].

## What to Do if You Suspect Your Patient Has SCAD

A detailed history and physical examination are key to timely and accurate diagnosis of SCAD (Table 4.1). When there is a suspicion of SCAD, the assessment needs to include questions regarding previous SCAD, pregnancy, menstrual cycle, migraines, physical (particularly isometric) or emotional triggers, known FMD and vascular disorders. Similarly, the physical examination should extend beyond the cardiovascular exam to include auscultating for renal bruits and assessing for signs of connective tissue disorders.

Investigation and management are initially similar for suspected SCAD patients and atherosclerotic ACS patients (Table 4.2). This includes serial electrocardiograms (ECG) and troponin levels. Following that, coronary angiography is vital as it can visualize plaque rupture to identify traditional atherosclerotic ACS, or rule out plaque rupture to confirm the suspected diagnosis of SCAD [10].

We know that there are also ACS presentations that are neither plaque-rupture atherosclerotic ACS nor SCAD, and can be caused by myocardial infarction with no obstructive coronary arteries (MINOCA), coronary vasospasm or microvascular dysfunction. Presentations can also be ACS-mimics such as myocarditis, myopericarditis, takotsubo cardiomyopathy, or other non-cardiac diagnoses including pulmonary emboli or aortic dissection.

Having considered the wider differential for acute chest pain, the general principle is to treat the patient as atherosclerotic ACS until SCAD or an alternate differential is diagnosed on coronary angiography [15]. If SCAD is the true diagnosis, the course of management significantly changes, including avoiding invasive percutaneous coronary intervention (PCI) if possible [18]. However, PCI may still be considered in the presence of clinical high-risk features, including ongoing ischemia/chest pain, cardiogenic shock, sustained ventricular arrhythmia and left main dissection [25].

Further, coronary angiography assists in determining the type of SCAD and the amount of myocardium affected, which can also influence management and follow up frequency and strategy [26]. It is important to explain to the patient the risks of coronary angiography for informed consent, including vascular damage, stroke, myocardial infarction, arrhythmia, contrast risk and focusing specifically on radiation risk for young female patients. Also, the possibility of pregnancy must be assessed, and risks must be discussed.

Moreover, SCAD patients are expected to be admitted in hospital for a longer duration of time (5–7 days) than atherosclerotic ACS patients, as studies have shown that SCAD patients who deteriorate do so usually within 5 days of presentation, compared to atherosclerotic ACS where the highest risk period is the first 48 h [27].

## Conservative Versus Invasive Management

Management for revascularization differs between SCAD and atherosclerotic ACS patients. It is well established that if the clinical situation allows, conservative management without PCI is the preferred treatment strategy for SCAD patients, whereas for an atherosclerotic plaque rupture, PCI is recommended. This is explained by the natural history of SCAD, whereby 95% of SCAD are healed spontaneously by 30 days [25]. However, 5–7% of patients who were initially treated conservatively will require PCI in the same presentation, most often due to cardiogenic shock, ongoing ischemic chest pain, or proximal coronary artery involvement [8, 25].

Intervention is considered only for select clinical situations in SCAD because the presence of dissection and intramural hematoma adds to the complexity of the procedure, resulting in higher complication rates [2]. The success rate of PCI for SCAD is 50–70% in large cohort studies, which is much lower than that for the atherosclerotic ACS PCI population [1, 28, 29]. Further, although this may be impacted by a selection bias of the more severe SCAD patients receiving intervention, PCI for SCAD has not been independently associated with reduced infarct size, nor with reduced in-hospital MACE (Major adverse cardiovascular events) compared to conservatively managed SCAD groups [27, 30].

## Medical Management

The main debate in the acute medical management of SCAD ACS has been regarding single or dual antiplatelet therapy use. Currently, there are no randomized clinical data or established protocol [31], but there is expert consensus opinion to offer guidance. The most recent literature favors single antiplatelet therapy (SAPT) for SCAD ACS over dual antiplatelet therapy (DAPT) with the multicenter DIssezioni Spontanee COronariche (DISCO) registry from Italy and Spain ($n = 199$) showing patients on DAPT ($n = 132$, 66%) experienced significantly higher MACE (defined as all-cause death, non-fatal MI or unplanned PCI; 18.9 versus 6.0%. HR 2.62, 95% CI 1.22–5.61, $p = 0.013$) than those treated with SAPT ($n = 67$, 34%) [32]. This result was driven mainly by non-fatal MI (15.2 versus 3.0%, $p = 0.009$) and unplanned PCI (12.1 versus 1.5%, $p = 0.001$) [32]. The pathophysiological rationale behind this is increased bleeding and intramural hemorrhage due to DAPT, with no added benefit as there is no acute plaque rupture or disturbance. Similarly, the consensus is that anticoagulation is not necessary for SCAD ACS [1, 2, 33].

The key medication for the management of SCAD long term are beta blockers. A single center observational study has shown a significant decrease of SCAD recurrence with the use of beta blockers, by over 50%, over 3 years [33]. The second step of optimal medical management of SCAD is good blood pressure control, aiming for SBP < 130 mmHg. Angiotensin Converting Enzyme-Inhibitors (ACE-I) or Angiotensin Receptor Blockers (ARBs) are used to treat hypertension, which is a clear risk factor for both SCAD ACS and its recurrence, and are also indicated for the management of left ventricular dysfunction after SCAD-related myocardial infarction. If more blood pressure control is necessary, calcium channel blockers or diuretics may also be used as secondary agents, in line with treatment guidelines for essential hypertension [34].

Unlike atherosclerotic ACS, there is no evidence for the routine use of statins in SCAD [31]. Cholesterol-laden plaque is not implicated in the pathophysiology of SCAD ACS, while it is central to the pathophysiology of traditional atherosclerotic-myocardial infarction. This is a challenging concept for most cardiologists and family physicians to accept, as the use of statin therapy after ACS is firmly entrenched in the practice patterns of most physicians. Similarly, anti-anginal medications (nitrates, and calcium channel blockers) may be very useful for symptom control for chest pain syndromes after SCAD, but are not prescribed routinely [2].

## Outpatient Management of SCAD Patients

The optimal outpatient management of SCAD patients is continually being developed as SCAD becomes more understood. Currently, as a lesser-known form of ACS with minimal public awareness or education, many SCAD patients experience social isolation, anxiety and depression during this vulnerable period in their recovery [4].

Indeed, good quality clinical research on outpatient management of SCAD patients are lacking, with many current recommendations being derived from the larger vessel (aortic) dissection literature whilst direct experience with SCAD patients is still building. Thus, the partnered management of patients between SCAD specialists, general cardiologists, cardiac rehabilitation specialists and primary care providers is important for the communication of accurate information (Table 4.5), education of effective management strategies, and screening of associated pathologies [35]. Such management partnerships should be identified and linked in with the SCAD patient prior to hospital discharge.

This schematic (Fig. 4.1) is a timeline from the point of SCAD ACS presentation to diagnosis and management in the hospital, followed by discharge home and the subsequent outpatient avenues of follow up (Table 4.3) with both the primary care provider (Time points: 1, 3, 4), and the planned SCAD multi-disciplinary team (MDT).

The first follow up with the family physician (Point 1* in Fig. 4.1 and Table 4.3) after discharge from a SCAD admission to hospital is recommended to occur within 4–6 weeks. One of the key message that needs to be reiterated to young females of child-bearing age is that after SCAD ACS, pregnancy is not recommended due to the high risk of severe and potentially life-threatening recurrence of SCAD during pregnancy or post-partum [16, 36]. Thus, in the event of pregnancy despite such warning, patients must seek involvement of SCAD expert and the high-risk obstetrics team as early as possible to consider their options. Seeking expert opinion prior to conception is most ideal if pregnancy is desired despite awareness of its high risk.

Regarding exertion, SCAD patients are often advised to avoid high intensity physical exertion and Valsalva maneuvers, which involves not lifting anything heavier than 30lbs for women, or 50lbs for men. However, this should not preclude patients from regaining an active lifestyle with regular moderate activity post SCAD. Participation in cardiac rehabilitation as early as possible is vital in facilitating this.

Between 8 and 12 weeks after hospital discharge, a SCAD MDT follow up (Point 2* in Fig. 4.1) is ideal for further assessment and investigations for associated disorders. The most important and well documented association is FMD, and all patients who have experienced SCAD must have an initial screening test for FMD via a head-to-pelvis computed tomographic (CT) angiography looking for aneurysm, tortuosity, dissection, irregularity, or stenosis of arterial beds. This may have already been done during the inpatient stay if a CT angiography was available.

During the late follow up period, an unplanned chest pain presentation (Point 4* in Fig. 4.1) may occur. These need to be approached in a similar way to a new chest pain presentation, without presumption that it is related to SCAD as recurrence rates of SCAD are very low (2.4% over 3 years) [37, 38]. Thereby, hemodynamic stability, acute onset of symptoms and dynamic ECG changes and troponin levels are used to determine if the patient is best for in-hospital or community management. Notably, women are significantly more likely than men to be re-admitted to hospital for chest pain following prior SCAD (8 versus 1%, $p = 0.001$) [24].

Further, these outpatient scheduled reviews or unscheduled chest pain presentations are opportune to review medications and ensure patients are taking beta blockers long term following their SCAD ACS episode, assess their psychosocial state, and ensure referral to outpatient cardiac rehabilitation [26].

Longer term, after two to three years of consistent follow up by the SCAD MDT and a stable clinical state, shared decision making is advised to individualize the best method for ongoing patient access to SCAD resources and healthcare. Anecdotally, most patients and family physicians have felt comfortable at this stage to take ownership of the follow up long term, given the established relationship with the SCAD MDT and ability to re-refer if concerns arise.

## Cardiac Rehabilitation

Cardiovascular rehabilitation (CR) is a comprehensive secondary prevention program that includes patient education, nutritional counseling, exercise training and psychosocial interventions [39, 40]. Participation has been associated with up to a 50% reduction in morbidity and mortality, as well as improved quality of life following a cardiac event such as MI or cardiac procedure [41].

Unfortunately, less than 30% of eligible patients participate in CR [42] and women in particular, are less likely to be referred and participate in programming [43]. Younger age at diagnosis, lack of referral, caregiving/family responsibilities, transportation, scheduling and logistical difficulties with attending an onsite program have been identified as predominant barriers for enrollment in CR [26, 44, 45].

Despite the significant benefits associated with CR program completion, women diagnosed with SCAD are significantly under-represented in CR programs across Canada [12, 46]. Fear, anxiety and hesitancy regarding physical activity after SCAD are common and often lead patients to curtail or avoid physical activity altogether [11]. CR programs are recommended to safely support the resumption of moderate intensity physical activity after SCAD-related events [11, 47, 48]. Recommendations for physical activity should be tailored to the individual and may be best determined and guided during participation in CR [47, 48].

## *Cardiac Rehabilitation and SCAD Patients*

Referral to CR is considered the standard of care for all patients who have experienced a cardiac event or procedure, including those with a SCAD diagnosis [40].

Recent North American studies have shown improvements in aerobic capacity, body composition and mental health with no reported adverse events in SCAD patients completing CR [46, 48]. Structured rehabilitation (Table 4.4), particularly multi-dimensional interventions that offer comprehensive components (e.g. exercise, psycho-educational support, mindful living sessions, nutrition counselling, peer support networks) have produced both physical and emotional benefits [49].

Krittanawong et al. [49] examined the usefulness of CR in a large cohort of SCAD patients (48/50 US states, Canada, Europe, Australia and New Zealand) enrolled in the Mayo Clinic SCAD registry (n = 354). In this cohort, 66% of patients participated in > 10 CR sessions while 32% chose not to participate. A lack of recommendation to CR by a health care provider was cited as the primary reason for non-participation. An inherent reluctance for clinician referral may be attributed to the relative unfamiliarity with SCAD, a lack of perceived benefit in this population due to the younger age without traditional risk factors, and concerns that exercise training may prompt a recurrent SCAD event.

Wagers et al. [50] further reinforced the positive experiences and safety outcomes for SCAD survivors (n = 409) participating in a CR program. However, a focus on the 'typical' or traditional CR program was deemed by patients as "not a good fit" considering the patient's younger age, gender and/or prior activity levels. Some participants reported feelings of isolation during CR due to lack of younger counterparts to whom they could relate, while others felt their program was geared to the "standard heart attack" patient. Baechler et al. [26] interviewed 38 patients diagnosed with SCAD and observed that frustration with the lack of mental health resources was mentioned among the most common reported themes. Patients described that an ideal CR program should be specific for patients with SCAD and would include mental health support [26]. The importance of psychosocial support for SCAD survivors has been highlighted in several studies [51, 52] and reinforces that it should be accessible to every patient attending CR [39, 40], notably to women [53].

Neubeck et al. [52] performed an extensive systematic review of 28 studies and analyzed the physical and psychosocial recovery following discharge from hospital

**Table 4.4** Recommendations for cardiac rehabilitation (CR) programs treating SCAD survivors

| | |
|---|---|
| CR program duration | 6 months |
| Cardiovascular risk factors | • Traditional and non-traditional risk factors for cardiovascular disease should be assessed and risk profile shared with patients<br>• CR programs should follow the current guidelines on management of blood glucose, hypertension, body weight, physical activity, psychosocial wellbeing, sleep, healthy diet, and smoking cessation<br>• Note that SCAD ACS is not considered a risk factor to initiate statin therapy. SCAD is not a statin-indicated condition<br>• Statin therapy should be considered when LDL-cholesterol is very high, > 5.0 mmol/L or as recommended by the current guidelines for primary prevention of cardiovascular disease (if there is no established atherosclerosis) |
| Exercise structure | • One-hour weekly exercise class consisting of 15 min warm-up, 30 min cardiovascular exercise on aerobic machines, and 15 min cool-down |
| Frequency | • 30–40 min of moderate intensity physical activity 5–7 days/week (150 min/week) |
| Target exercise heart rate | • 50–70% of heart rate reserve based on the entrance exercise treadmill test |
| Blood pressure limit | • < 130/80 mm Hg |
| Target perceived exertion rate | • "Moderate" to "somewhat difficult" |
| Resistance training | • Focus on a proper breathing and lifting technique avoiding the Valsalva maneuver/straining |
| Weightlifting | • 2- to 12-pound free weights to increase muscle strength, starting with lighter weights and progressing with strength gain<br>• Avoid lifting weights > 30 pounds for women and > 50 pounds for men |
| Educational sessions | • 20 min educational session per week on risk factors, and treatment of heart disease and SCAD, and stress management, emphasizing women's heart disease when appropriate |
| Mental health | • Counselling, mindful living sessions, and peer-support from other SCAD survivors<br>• Peer Support from patients with lived experience has been identified as a key component in recovery following a cardiac event such as SCAD |
| Medications | • If medication changes (eg: beta-blocker dose), consider repeating cardiopulmonary test to support optimizations of exercise prescription |
| Health care team | • Physician, Kinesiologist, Physiotherapist, Dietitian, Psychologist, Social Worker, Registered Nurse |
| Recommendation to patients | • Stop or slow down if becoming extremely exhausted or feeling uncomfortable, or if symptoms arise, including but not limited to chest discomfort, chest pain, dyspnea, or dizziness |

**Table 4.5** What to tell my SCAD patients?

| *Have I had a heart attack?* |
|---|
| Yes, but a different kind of heart attack. A heart attack is when there is no flow through one of the heart arteries to the heart muscle, causing damage to the heart muscle. Most commonly, this occurs from cholesterol plaque buildup in the lining of a heart artery which ruptures and blocks the artery (and thus prevents the blood from flowing to the heart muscle)<br>However, with SCAD, the blood flow blocks from a bleed that occurs within the layers of the heart artery. This causes compression of the heart artery from the outside, leading to reduced blood flow or in some cases, a tear in the artery lining |
| *If I have had a heart attack, why did I not get a stent?* |
| Some people with SCAD require a stent because the compression of the heart artery from the outside is very severe. However, it is preferrable to avoid a stent as it can lead to other complications such as extension of the tear in the artery or making the amount of artery compression more extensive. You did not have a stent inserted because you were still able to get enough blood flow to your heart muscle<br>As the underlying problem is not a plaque within the heart artery, but a compression from the outside, all current research point to the best outcomes for SCAD if you can wait for the "bruise" in the wall of the heart artery to slowly dissolve itself over time |
| *Will I have SCAD again?* |
| A small proportion of people do get SCAD again. Understandably, this can be distressing<br>Having had SCAD however, you are in a better position to know what it feels like and seek medical attention early. Thankfully, most people do not get SCAD recurrence and most SCAD patients recover well and safely |
| *What do I need to avoid getting SCAD again?* |
| The main thing to do is ensure your blood pressure is under good control, through lifestyle changes and medication. Stay on the medications prescribed by your doctor, which may include beta blockers. They aim to reduce the pressure stress on your heart arteries and have been shown to lower the chance of SCAD recurrence<br>It is also important to avoid known physical triggers such as extreme heavy lifting that involve sudden Valsalva (bearing down) maneuvers |
| *Why am I getting so many mixed messages?* |
| Again, this is understandably confusing. Although the knowledge of SCAD is increasing over the years, particularly compared to regular heart attacks it is a relatively uncommon diagnosis. Thus, the understanding of it amongst the general community and the medical community is much less than for a traditional heart attack |
| *Can I get pregnant?* |
| Informed decision making around pregnancy is very important for SCAD patients. Pregnancy cannot be recommended because of the increased risk for another SCAD related heart attack which may be much more severe during pregnancy. Continuation of pregnancy, or the decision to terminate pregnancy must be a very carefully considered decision along with a specialized multidisciplinary team. This is because SCAD is the most common pregnancy-related heart attack and is thought to be related to the hormone changes that occur in pregnancy, particularly in the early post-partum period |

in 4167 patients (93.5% female) with SCAD. They observed lack of specific guidance about physical activity and high levels of psychosocial distress. These findings suggest ideal programs with tailored exercise and psycho-educational components including peer support networks require further examination to test feasibility and effectiveness specific to the unique needs of these patients. Moreover, it is imperative that care providers are aware of the safety and associated benefits of CR participation to improve the recovery trajectory for this understudied population.

## *Physical Activity After SCAD*

Fear and hesitancy regarding physical activity after SCAD are common among SCAD survivors and clinicians despite a lack of evidence for protection from recurrent SCAD [11]. Additionally, imposing physical restrictions to young individuals could lead to lifelong impact possibly promoting sedentarism, weight gain, and psychosocial issues. In this realm, CR programs can contribute substantially to overall physical and mental health [11].

As outlined previously, CR has demonstrated overall benefit and safety [46, 48–50]; therefore, pursuing regular, moderate exercise likely outweighs the theoretical risks of recurrent SCAD [11, 47, 48]. It is important to note that SCAD survivors, especially those who experienced recurrent SCAD or who have noncoronary aneurysms or dissections should avoid extreme endurance training, exercising to exhaustion, elite competitive sports, or vigorous exertion in extremes of ambient temperature [11, 47, 48].

The Vancouver General Hospital (VGH) SCAD-CR protocol has been adopted worldwide and provides safe parameters for patients initiating CR [48]. Each patient should have an individualized exercise prescription, considering clinical parameters, physical exercise prior to SCAD and personal goals. These considerations do not substantially differ from non-SCAD patients however, it is suggested to start physical activity at a lower level, as recommended by Chou et al. [48], then gradual progression should be targeted (Table 4.4). Further, SCAD survivors should avoid lifting or carrying heavy objects that require straining or prolonged Valsalva [11, 47, 48]. Nevertheless, there is no evidence that heavier loads with proper technique (no straining or Valsalva) are harmful [47].

The length of time between a SCAD event and CR initiation should be considered, and empirically exercise prescription could be more conservative in the first months post SCAD as suggested in Table 4.4, and less restrictive after 6–12 months. The presence of FMD can also represent an additional challenge. Patients with carotid or vertebral artery dissections should avoid resistance training including body weight exercises such as push-ups and sit-ups during the first 8–12 weeks after the acute dissection after which the recommendations would be similar as for SCAD [47].

## Future Directions and Opportunities

The evolving understanding of SCAD over the last ten years is promising to translate into partnered care and better outcomes for SCAD ACS patients. More robust data have clarified prior assumptions, and management strategies have become more unified globally. The concern for pregnancy following a SCAD ACS episode continues, particularly as recurrence in this scenario is more likely to be severe.

There are numerous future opportunities that need to be addressed in order to better understand the presentation, diagnosis, early recognition and treatment as well as long term management of patients diagnosed with SCAD.

Cardiac rehabilitation is safe and feasible, early studies have shown the beneficial effects on clinical cardiovascular parameters and mental health in SCAD survivors [46, 49, 50]. Future studies need to specifically examine the clinical experiences and preferences of SCAD patients in order to develop a better understanding of alternative, flexible program models, the integration of peer support networks and how best to tailor CR programming to meet the patient's psychosocial, physiological and educational needs.

It is crucial that care providers are educated and understand the differences between SCAD and atherosclerotic diseases, and how best to mitigate barriers to SCAD care. Consideration of family systems and roles as well as involvement of family members in SCAD education may help alleviate patient and family anxiety which may subsequently optimize recovery transitions.

## References

1. Kim ES (2020) Spontaneous coronary-artery dissection. N Engl J Med 383(24):2358–2370
2. Lewey J, El Hajj SC, Hayes SN (2022) Spontaneous coronary artery dissection: new insights into this not-so-rare condition. Annu Rev Med 73:339–354
3. Yang C, Alfadhel M, Saw J (2020) Spontaneous coronary artery dissection: latest developments and new frontiers. Curr Atheroscler Rep 22(9):1–8
4. Johnson AK et al (2020) Analysis of posttraumatic stress disorder, depression, anxiety, and resiliency within the unique population of spontaneous coronary artery dissection survivors. J Am Heart Assoc 9(9):e014372
5. Saw J, Mancini GJ, Humphries KH (2016) Contemporary review on spontaneous coronary artery dissection. J Am Coll Cardiol 68(3):297–312
6. Di Fusco SA, Rossini R, Zilio F, Pollarolo L, di Uccio FS, Iorio A, Lucà F, Gulizia MM, Gabrielli D, Colivicchi F (2022) Spontaneous coronary artery dissection: overview of pathophysiology. Trends Cardiovasc Med 32(2):92–100
7. Iismaa SE et al (2021) Spontaneous coronary artery dissection and fibromuscular dysplasia: vasculopathies with a predilection for women. Heart Lung Circ 30(1):27–35
8. Kotecha D et al (2021) Risks and benefits of percutaneous coronary intervention in spontaneous coronary artery dissection. Heart 107(17):1398–1406
9. Aslam A et al (2021) Spontaneous coronary artery dissection: an underdiagnosed clinical entity—a primer for cardiac imagers. Radiographics 41(7):1897–1915
10. Adlam D et al (2021) Spontaneous coronary artery dissection: pitfalls of angiographic diagnosis and an approach to ambiguous cases. JACC Cardiovasc Interv 14(16):1743–1756

11. Hayes SN et al (2020) Spontaneous coronary artery dissection: JACC state-of-the-art review. J Am Coll Cardiol 76(8):961–984
12. Saw J et al (2022) Canadian spontaneous coronary artery dissection cohort study: 3-year outcomes. J Am Coll Cardiol 80(17):1585–1597
13. Saw J et al (2019) Canadian spontaneous coronary artery dissection cohort study: in-hospital and 30-day outcomes. Eur Heart J 40(15):1188–1197
14. Adams C et al (2021) Mortality in spontaneous coronary artery dissection: a systematic review and meta-analysis. Catheter Cardiovasc Interv 98(7):1211–1220
15. Alfonso F, García-Guimaraes M, Alvarado T, Sanz-Ruiz R, Roura G, Amat-Santos IJ, Abdul-Jawad Altisent O, Tizón-Marcos H, Flores-Ríos X, Masotti M, Pérez-de Prado A, Ferre GF, Ruiz-Poveda FL, Valero E, Portero-Portaz JJ, Diez-Villanueva P, Salamanca J, Bastante T, Rivero F (2022) Clinical implications of arterial hypertension in patients with spontaneous coronary artery dissection. Coron Artery Dis 33(2):75–80
16. Tweet MS, Miller VM, Hayes SN (2019) The evidence on estrogen, progesterone, and spontaneous coronary artery dissection. JAMA Cardiology 4(5):403–404
17. Mahmoud AN et al (2018) Trends of incidence, clinical presentation, and in-hospital mortality among women with acute myocardial infarction with or without spontaneous coronary artery dissection: a population-based analysis. JACC Cardiovas Interv 11(1):80–90
18. Alfonso F, García-Guimaraes M (2021) Spontaneous coronary artery dissection: Where do we stand? Med Intensiva 45(6):371–374
19. Kim Y et al (2021) Clinical characteristics of spontaneous coronary artery dissection in young female patients with acute myocardial infarction in Korea. Korean J Intern Med 36(1):106
20. Sedlak T et al (2020) Coronary flow reserve in patients with prior spontaneous coronary artery dissection and recurrent angina. J Am Heart Assoc 9(16):e015834
21. Hui P et al (2020) The value of plasma fibrillin-1 level in patients with spontaneous coronary artery dissection. Int J Cardiol 302:150–156
22. Amrani-Midoun A, Adlam D, Bouatia-Naji N (2021) Recent advances on the genetics of spontaneous coronary artery dissection. Circ Genomic Precis Med 14(6):e003393
23. Nishiguchi T et al (2016) Prevalence of spontaneous coronary artery dissection in patients with acute coronary syndrome. Eur Heart J Acute Cardiovasc Care 5(3):263–270
24. McAlister C et al (2022) Differences in demographics and outcomes between men and women with spontaneous coronary artery dissection. Cardiovasc Interv 15(20):2052–2061
25. Hassan S et al (2019) Natural history of spontaneous coronary artery dissection with spontaneous angiographic healing. Cardiovasc Interv 12(6):518–527
26. Baechler CJ et al (2022) Spontaneous coronary artery dissection and evidence-based medicine. Am J Cardiol 171:65–68
27. Isogai T et al (2022) Factors associated with revascularization in women with spontaneous coronary artery dissection and acute myocardial infarction. Am J Cardiol 166:1–8
28. Gilhofer TS, Saw J (2019) Spontaneous coronary artery dissection: a review of complications and management strategies. Expert Rev Cardiovasc Ther 17(4):275–291
29. Hassan S et al (2021) Outcomes of percutaneous coronary intervention in patients with spontaneous coronary artery dissection. J Interv Cardiol 2021
30. García-Guimarães M et al (2022) Characteristics, acute results, and prognostic impact of percutaneous coronary interventions in spontaneous coronary artery dissection (from the Prospective Spanish Registry on SCAD [SR-SCAD]). Am J Cardiol 171:177–178
31. Seidl S, Rickli H, Rogowski S, Weilenmann D, Ammann P, Haager PK, Joerg L, Rohner F, Chronis J, Rigger J, Maeder MT (2021) Long-term follow-up of medically treated patients with spontaneous coronary artery dissection: a prospective, Swiss single-centre cohort study. Swiss Med Wkly 151:w30067
32. Cerrato E et al (2021) Antiplatelet therapy in patients with conservatively managed spontaneous coronary artery dissection from the multicentre DISCO registry. Eur Heart J 42(33):3161–3171
33. Kermali M et al (2021) Spontaneous coronary artery dissection: presentation and management options. Coron Artery Dis 32(2):152–163

34. Ferdinand KC, Nasser SA (2017) Management of essential hypertension. Cardiol Clin 35(2):231–246
35. Vaca KC, Tremmel JA, Edwards KS (2021) Preliminary support for group cognitive behavioral therapy (CBT) to reduce psychological distress in patients with spontaneous coronary artery dissection (SCAD). J Clin Psychol Med Settings 28(4):826–832
36. Tweet MS et al (2020) Association of pregnancy with recurrence of spontaneous coronary artery dissection among women with prior coronary artery dissection. JAMA Netw Open 3(9):e2018170–e2018170
37. Kok SN, Tweet MS (2021) Recurrent spontaneous coronary artery dissection. Expert Rev Cardiovasc Ther 19(3):201–210
38. Krittanawong C et al (2020) Recurrent spontaneous coronary artery dissection in the United States. Int J Cardiol 301:34–37
39. Hamm LF et al (2011) Core competencies for cardiac rehabilitation/secondary prevention professionals: 2010 update: position statement of the American Association of Cardiovascular and Pulmonary Rehabilitation. J Cardiopulm Rehabil Prev 31(1):2–10
40. Stone JA, Arthur HM (2005) Canadian guidelines for cardiac rehabilitation and cardiovascular disease prevention, 2004: executive summary. Can J Cardiol 21:3D–19D
41. Alter DA, Oh PI, Chong A (2009) Relationship between cardiac rehabilitation and survival after acute cardiac hospitalization within a universal health care system. Eur J Cardiovasc Prev Rehabil 16(1):102–113
42. Anderson L, Taylor RS (2014) Cardiac rehabilitation for people with heart disease: an overview of Cochrane systematic reviews. Cochrane Database Syst Rev (12):CD011273
43. Colella TJ et al (2015) Sex bias in referral of women to outpatient cardiac rehabilitation? A meta-analysis. Eur J Prev Cardiol 22(4):423–441
44. Galati A et al (2018) Cardiac rehabilitation in women: state of the art and strategies to overcome the current barriers. J Cardiovasc Med 19(12):689–697
45. Supervía M, Medina-Inojosa JR, Yeung C, Lopez-Jimenez F, Squires RW, Pérez-Terzic CM, Brewer LC, Leth SE, Thomas RJ (2017) Cardiac rehabilitation for women: a systematic review of barriers and solutions. Mayo Clin Proc 13:S0025-6196(17)30026-5
46. Silber TC et al (2015) Cardiac rehabilitation after spontaneous coronary artery dissection. J Cardiopulm Rehabil Prev 35(5):328–333
47. Tweet MS et al (2021) Physical activity and exercise in patients with spontaneous coronary artery dissection and fibromuscular dysplasia. Eur Heart J 42(37):3825–3828
48. Chou AY et al (2016) The first dedicated cardiac rehabilitation program for patients with spontaneous coronary artery dissection: description and initial results. Can J Cardiol 32(4):554–560
49. Krittanawong C et al (2016) Usefulness of cardiac rehabilitation after spontaneous coronary artery dissection. Am J Cardiol 117(10):1604–1609
50. Wagers TP et al (2018) Spontaneous coronary artery dissection (SCAD): female survivors' experiences of stress and support. J Cardiopulm Rehabil Prev 38(6):374
51. Bouchard K et al (2021) Recovering from spontaneous coronary artery dissection: patient-reported challenges and rehabilitative intervention needs. Health Psychol 40(7):472
52. Neubeck L, McHale S, Ross M, MacGillivray S, Galbraith M, Hanson C (2022) Spontaneous coronary artery dissection: a systematic review of physical and psychosocial recovery following discharge from hospital. Eur J Cardiovasc Nurs 21(7):665–676
53. de Melo Ghisi GL et al (2022) Women-focused cardiovascular rehabilitation: an international council of cardiovascular prevention and rehabilitation clinical practice guideline. Can J Cardiol
54. Petropoulos T, Madan M (2022) Spontaneous coronary artery dissection, patient guide 2022. Sunnybrook Health Sciences Centre

# Chapter 5
# Takotsubo Cardiomyopathy

**Samantha S. L. Liauw, Shuangbo Liu, and Alexandra Bastiany**

**Abstract** Takotsubo cardiomyopathy, also known as stress-induced cardiomyopathy, apical ballooning syndrome or broken heart syndrome, is characterized by acute, transient cardiac dysfunction following a significant catecholaminergic disturbance. It is a relatively newly described clinical phenomenon, with incompletely understood socio-environmental mediating factors and neurohormonal mechanisms implicated in the pathophysiology. It is more commonly seen in females. The diagnosis requires a thorough clinical history and the use of multimodality assessment commonly including angiography and echocardiography, with computed tomography and magnetic resonance imaging used in some cases. The prognosis of patients with Takotsubo cardiomyopathy and of patients with acute coronary syndrome are similar, with high burden of cardiogenic shock (6–20%) and mortality (1–8% in hospital, 3.5–5.6% annual rate). Treatment in the acute setting is mainly supportive with monitoring and treatment of life-threatening arrhythmias and cardiogenic shock. Longer term therapies addressing the complex underpinnings of this condition (i.e. left ventricular remodelling, neurohormonal pathways, and psychosocial and environmental factors) are still under investigation. This chapter reviews the epidemiology, pathophysiology, clinical manifestations, and diagnosis of takotsubo cardiomyopathy. The management and prognosis of takotsubo cardiomyopathy will also be reviewed.

**Keywords** Takotsubo cardiomyopathy · Acute coronary syndrome (ACS) · Arrythmia · Cardiogenic shock · Echocardiography

S. S. L. Liauw
Department of Medicine, University of Toronto, Toronto, ON, Canada

S. Liu
Department of Internal Medicine, Max Rady College of Medicine, University of Manitoba, Winnipeg, MB, Canada

A. Bastiany (✉)
Thunder Bay Regional Health Sciences Centre, Thunder Bay, ON, Canada

© The Author(s), under exclusive license to Springer Nature Switzerland AG 2023 
L. Kirshenbaum and I. Rabinovich-Nikitin (eds.), *Biology of Women's Heart Health*, Advances in Biochemistry in Health and Disease 26,
https://doi.org/10.1007/978-3-031-39928-2_5

## Introduction

Takotsubo cardiomyopathy is a clinical phenomenon, characterized by acute, transient cardiac dysfunction following a significant catecholaminergic disturbance. It is also known as stress-induced cardiomyopathy, apical ballooning syndrome, or broken heart syndrome. Takotsubo cardiomyopathy can mimic acute coronary syndrome (ACS). Although there are similarities in their clinical presentations, Takotsubo cardiomyopathy is differentiated from ACS in that the left ventricular myocardial dysfunction is not secondary to obstructive epicardial coronary artery disease, or acute plaque rupture [1]. The diagnosis hinges on left ventricular assessment with initial reports in the 1990s noting the resemblance to an octopus trap ("takotsubo", in Japanese). It is a relatively newly described phenomenon and much remains to be discovered regarding its pathophysiology, optimal management, and prognosis.

This chapter will begin with a review of the definition, epidemiology, and pathophysiology of takotsubo cardiomyopathy. This will be followed by a discussion on the clinical manifestations, diagnosis, management and prognosis.

## Definition and Diagnostic Criteria

Many different diagnostic criteria have been proposed, with no universal consensus [2]. Commonly cited criteria include the criteria of the Heart Failure Association Takotsubo Syndrome Taskforce of the European Society of Cardiology (ESC) [3], the InterTAK Diagnostic Criteria [4], and the Revised Mayo Clinic Criteria [5–7]. These three classifications are compared in Table 5.1. All three describe transient regional wall motion abnormalities that may be typical (i.e. apical ballooning-pattern), or atypical (i.e. midventricular, basal, or focal), the potential presence of triggers, and the presence of abnormal biomarkers. The criteria differ with regards to pheochromocytoma as a potential trigger (permitted in the InterTAK but not in the revised Mayo Clinic criteria). Moreover, documentation of the left ventricular dysfunction resolution is only mentioned in the Heart Failure Association-European Society of Cardiology criteria.

## Epidemiology and Predisposing Factors

Takotsubo cardiomyopathy occurs following an acute physical or emotional stressor in the majority of cases [8], although in large series such as the International Takotsubo Registry, 29% of patients had no evident trigger [9]. It is more commonly seen in females (9:1 female:male ratio) with an average age at presentation of 67–70 years [4]. Approximately 5–6% of women who present with suspected ST-segment elevation myocardial infarction, and approximately 2% of patients with suspected

**Table 5.1** Table of Heart Failure Association-European Society of Cardiology Criteria, INTERTAK and Revised Mayo criteria

| Component | Heart failure association-ESC criteria | InterTAK diagnostic criteria | Revised mayo clinic criteria |
|---|---|---|---|
| Characterization of ventricular myocardial dysfunction | Transient regional LV or RV WMA. The RWMA usually extend beyond a single epicardial vascular distribution, and often result in circumferential dysfunction of the ventricular segments involved | Transient LV dysfunction presenting as apical ballooning, midventricular, basal, or focal WMA. RV involvement can be present. The RWMA usually extends beyond a single epicardial vascular distribution; however, rare cases exist where the RWMA is present in the subtended myocardial territory of a single coronary artery | Transient dyskinesis of LV midsegments, with or without apical involvement. The RWMA extend beyond a single epicardial vascular distribution |
| Trigger | Frequently, but not always, preceded by a stressful trigger (emotional or physical) | An emotional, physical, or combined trigger can precede the takotsubo syndrome event, but this is not obligatory. May be triggered by SAH, stroke/TIA, seizure, pheochromocytoma, asthma/COPD exacerbation | A stressful trigger is often, but not always present |
| Predisposing clinical characteristics | | Post-menopausal women are predominantly affected | |
| Alternate diagnoses are excluded | Absence of other pathological conditions to explain the pattern of temporary LV dysfunction observed (e.g. hypertrophic cardiomyopathy, viral myocarditis) | No evidence of infectious myocarditis | Absence of pheochromocytoma and myocarditis |
| ECG | STE, STD, LBBB, TWI, and/or QTc prolongation) during the acute phase (3 months) | STE, STD, TWI, QTc prolongation or no changes | STE and/or TWI |

(continued)

**Table 5.1** (continued)

| Component | Heart failure association-ESC criteria | InterTAK diagnostic criteria | Revised mayo clinic criteria |
|---|---|---|---|
| Biomarkers | Positive but relatively small elevation in cardiac troponin (i.e., disparity between the troponin levels and the amount of dysfunctional myocardium present)<br>Significantly elevated BNP or NT-proBNP during the acute phase | Levels of cardiac biomarkers (troponin and creatine kinase) moderately elevated<br>Significant elevation of BNP | Modest elevation in the cardiac troponin levels |
| Coronary artery assessment | Absence of culprit atherosclerotic CAD, including acute plaque rupture, thrombus formation, and coronary dissection | Significant CAD is not a contradiction | Absence of obstructive CAD or absence of angiographic evidence of acute plaque rupture |
| Recovery | Recovery of ventricular systolic function on cardiac imaging at follow up (3–6 months) | | |

*BNP*: B-type natriuretic peptitide; *CAD*: coronary artery disease; *LBBB*: left bundle branch block; *NT-proBNP*: N-terminal-pro hormone BNP; *LV*: left ventricle; *RV*: right ventricle; *STD*: ST-segment depression; *STE*: ST-segment elevation; *TWI*: T-wave inversion; *RWMA*: regional wall motion abnormality

Adapted from Lyon AR, Bossone E, Schneider B, Sechtem U, Citro R, Underwood SR, et al. Current state of knowledge on Takotsubo syndrome: a Position Statement from the Taskforce on Takotsubo Syndrome of the Heart Failure Association of the European Society of Cardiology. Eur J Heart Fail. 2016;18(1):8–27; Ghadri JR, Wittstein IS, Prasad A, Sharkey S, Dote K, Akashi YJ, et al. International Expert Consensus Document on Takotsubo Syndrome (Part I): Clinical Characteristics, Diagnostic Criteria, and Pathophysiology. Eur Heart J. 2018;39(22):2032–46; Prasad A, Lerman A, Rihal CS. Apical ballooning syndrome (Tako-Tsubo or stress cardiomyopathy): a mimic of acute myocardial infarction. Am Heart J. 2008;155(3):408–17

acute coronary syndrome [10], are ultimately diagnosed with Takotsubo cardiomyopathy [4]. Younger patients are more likely to be male, who mostly present with atypical Takotsubo, and tend to have acute neurological or psychiatric disorders and in-hospital complications [1]. In a sample of 92 patients admitted to an intensive care unit, 26 (28%) were found to have left ventricular apical ballooning on screening echocardiograms [11]. The National Inpatient Sample, a U.S.-based database from approximately 1,000 hospitals, found a significant increase in Takotsubo rising from 315 cases in 2006 to 6230 cases in 2012 (incidence of 0.11 to 1.98 per year per 100,000 persons; $p < 0.001$). The authors suspected the main reason for the increased incidence to be the growing recognition of this condition [12].

Much remains to be understood about how social and environmental factors are related to Takotsubo cardiomyopathy. A large US-based retrospective study examined racial differences of Takotsubo cardiomyopathy outcome in over 97,000 patients (about 89,000 (91.8%) Caucasian, and about 8000 (8.2%) African American) [13]. In an unadjusted analysis, African American patients were more likely to have in-hospital complications such as cardiac arrest and invasive mechanical ventilation. These differences were eliminated after adjusting for age and gender. Based on their analysis, the authors proposed that the increased rate of complication seen in African Americans was explained by an increased proportion of African American men among patients with Takotsubo cardiomyopathy compared to Caucasian male patients. The study did not show mortality rate difference based on race.

A small study of fifty patients looked at Takotsubo presentation stratified by Hispanic and non-Hispanic groups. Patients who were Hispanic were more likely to present in the summer months compared to non-Hispanic, and were more likely to experience Taktosubo cardiomyopathy during nocturnal hours compared to the non-Hispanic group [14]. Other studies have also identified an increased incidence of Takotsubo cardiomyopathy during summer months [15, 16].

There are likely complex and interconnected social and environmental factors that mediate Takotsubo incidence and morbidity which are incompletely understood. As the pathophysiology of Takotsubo cardiomyopathy is thought to be related to increased level of catecholamines (see next section), it has been hypothesized that chronic medical and socioeconomic stressors, which disproportionately affect certain racial groups, contribute to worse outcomes [13].

## Pathophysiology

The pathophysiology of Takotsubo cardiomyopathy is not completely understood. Various mechanisms implicating central and autonomic nervous systems activation, catecholamine excess, and endothelial dysfunction have been hypothesized [17–19].

Patients with Takotsubo cardiomyopathy have been found to have elevated serum and myocardial catecholamine levels, as well as cardiac sympathetic hyperactivity [1, 18]. Sympathetic nervous system activity increases with age, while vagal tone and baroreflex sensitivity decreases. Cardiac alpha and beta-adrenergic receptors are exposed to three sources of catecholamines: (1) circulating norepinephrine and epinephrine, regulated by the hypothalamic-pituitary-axis, with input from the hypothalamus, cingulate gyrus and amygdala via the locus coeruleus; (2) locally released norepinephrine from sympathetic nerve terminals (primary source of circulating norepinephrine); and (3) local release of norepinephrine and epinephrine located in blood vessels and cardiac myocytes [17].

The mechanism by which increased catecholamines translates to left ventricular dysfunction is incompletely understood. However, myocardial biopsies of eight patients with Takotsubo cardiomyopathy during the phase of severe LV dysfunction showed focal mononuclear inflammatory cells, fibrosis, and characteristic contraction bands, all findings suggestive of catecholamine toxicity [1, 20]. Enhanced sympathetic stimulation and catecholamine excess leads to cAMP-mediated calcium overload, contractive dysfunction and myocardial stunning [10, 21]. Other postulated mechanisms include, oxygen supply–demand mismatch, disruption of ATP synthesis, altered cationic homeostasis, and increased free radicals [21].

A single-centre photon-emission/computed tomography (PET/CT) study demonstrated increased cerebral blood flow in the hippocampus, brainstem and basal ganglia, and decreased cerebral blood flow in the prefrontal cortex in three patients with Takotsubo cardiomyopathy [22]. This pattern of brain activation remained to some extent after left ventricular recovery, suggesting the role of an ongoing central mechanism in the pathogenesis of this entity, beyond the acute presentation. Furthermore, a retrospective case–control study of 41 patients with Takotsubo cardiomyopathy found higher baseline amygdala activity on 18F-FDG-PET/CT years prior to presentation, compared to individuals without Takotsubo cardiomyopathy. This suggests that chronically heightened stress-associated neurobiological activity may potentially impact the risk of subsequent Takotsubo cardiomyopathy [23].

Given the increased incidence in women, further hypothesized mechanisms include postmenopausal estrogen deprivation and its role in regulating sympathetic drive and microcirculatory function [1]. Estrogen is generally vascular protective, and it improves coronary blood flow. The reduction of estrogen during menopause is hypothesized to contribute to Takotsubo cardiomyopathy via endothelial dysfunction and possibly coronary spasm [21]. When compared to a similar group of women, postmenopausal women with a history of Takotsubo cardiomyopathy were found to have excessive vasoconstriction, impaired vasodilatation and increased sympathetic activation in response to stressful stimuli [24]. Few small studies have reported on the observation of epicardial coronary spasm in patients with Takotsubo syndrome and are hypothesis generating [25, 26]. Reports of coronary microvascular dysfunction in Takotsubo cardiomyopathy are mixed with case series reports of diminished Thrombolysis in Myocardial Infarction frame-count, myocardial blush grade, coronary flow reserve, and abnormal contrast echocardiography in some, but not in all patients presenting with Takotsubo cardiomyopathy [27]. Whether microvascular dysfunction represents a cause or epiphenomenon of Takotsubo cardiomyopathy, is currently the subject of debate [27].

Lastly, given rare reports of Familial Takotsubo cardiomyopathy, genetic predisposing factors have been proposed [4]. However, the existing literature is limited and consists of small sample size and retrospective study designs [4, 28–30]. The genetic basis for Takotsubo is the subject of ongoing research (ie. GENETIC study) [1].

## Clinical Presentation and Diagnostic Work Up

Patients with Takotsubo cardiomyopathy most commonly present with acute chest pain (55–68%) and dyspnea (18–30%) [10, 31, 32]. In a small, but not insignificant percentage of cases, patients present in cardiogenic shock or ventricular fibrillation (4% and 1.5%, respectively) [10]. Some patients are asymptomatic [33]. About one third of patients are reported to have an emotional stressor preceding the onset of Takotsubo cardiomyopathy, such as the death of a loved one or surprise birthday party [4]. One third of patients have a physical stressor, such as septic shock and vigorous housework. The remaining patients have no identifiable trigger preceding the event [4, 8]. The rate of preceding stressor may be underreported as some patients may not feel comfortable discussing sensitive topics during their acute presentation. Physical exam may be notable for tachycardia, hypoxia, increased work of breathing and other features of low output or congestive heart failure [2].

Systematic reviews of over 1000 patients with Takotsubo cardiomyopathy have reported similar frequencies of traditional cardiac risk factors between patients presenting with Takotsubo cardiomyopathy compared to an acute coronary syndrome presentation (hypertension: 42 versus 54%, diabetes: 11 versus 17%, dyslipidemia: 25 versus 33%, and smoking: 22 versus 42%) (8, 10). Takotsubo cardiomyopathy patients have a higher prevalence of psychiatric illness, with 42% percent of the 1750 patients in the International Takotsubo Registry having a history of psychiatric illness [9]. The presence of a psychiatric condition is represented in classification scores such as the Intertak Diagnostic Criteria (Table 5.1) [4]. There is also a relatively high prevalence of neurologic disease (27%), pulmonary disease (15%), malignancy (10%), chronic kidney disease (7%), and thyroid disease (6%) in patients presenting with Takotsubo cardiomyopathy [8, 9]. Rare cases of cocaine-induced cardiomyopathy [34] and pheochromocytoma have also been described [35].

Electrocardiographic (ECG) changes are non-specific and may show variable ST and T wave abnormalities, and long QT interval. Nearly half of patients had ST changes, and about half of patients had Q waves or T-wave changes [8]. In an international and collaborative systematic review of over 1000 subjects, 44% of patients had ST-segment elevation on presentation, with the most frequent localization being the precordial leads [4, 17]. Frangieh et al. reported ECG findings in 200 patients with Takotsubo cardiomyopathy and in 200 patients with myocardial infarction, within 12 h of symptom onset [36]. In patients who presented with inferior or anteroseptal ST-segment elevation, concurrent ST-segment depression in lead aVR was very specific for Takotsubo cardiomyopathy (98 and 100% specificity, respectively). In patients without ST-segment elevation, ST-segment depression in aVR had a specificity of 99% for Takotsubo cardiomyopathy [36]. Anterior ST-segment depression was specific for myocardial infarction. The evolution of ECG changes starts with ST-segment normalization, followed by progressive T-wave inversion, and QT interval prolongation over several days, with subsequent resolution of T-wave and QT abnormalities over days to weeks [37].

QTc prolongation was similar in frequency in Takotsubo cardiomyopathy and myocardial infarction in patients presenting with ST-elevation, but more prevalent in non-ST segment elevation (56% in Takotsubo cardiomyopathy vs. 34% in non-ST-segment myocardial infarction, $p = 0.005$) [36]. Case series report the mean QTc in patients with Takotsubo cardiomyopathy as ranging between 450 and 555 ms in patients without torsades de pointes (TdP) [38]. The mean presenting and maximal QTc in patients presenting with Takotsubo cardiomyopathy and TdP is 595 ms and 706 ms, respectively [38, 39].

Troponin levels usually peak within 24 hours of presentation and the peak levels are generally lower compared to patients with acute myocardial infarction [9]. The median initial troponin is 7.7 times the upper limit of normal (interquartile range 2.2–24) in the International Takotsubo Registry study [9]. Creatine kinase is typically only slightly increased with a medial CK of 0.85 the ULN (0.52–1.48 interquartile range) [9]. Plasma B-type natriuretic peptide and N-terminal prohormone of brain natriuretic peptide are substantially increased in patients with Takotsubo cardiomyopathy, usually peak within 48 h of symptom onset, and gradually normalise over months [33]. The presence of circulating microRNAs including miR-1, miR-16, miR-26am and miR-133a can differentiate Takotsubo cardiomyopathy from myocardial infarction [37].

In the work up of patients presenting with acute chest pain, non-specific ECG abnormalities, and abnormal cardiac biomarkers, one of the major differential diagnoses is acute myocardial infarction. It is critical to differentiate between the two entities as the treatment differs vastly. It can be challenging to differentiate between the two clinically, and coronary angiography is frequently performed to confirm the diagnosis. The InterTAK guidelines propose that in patients without ST-segment elevation, ongoing chest pain, clinical instability, and high pre-test probability of Takotsubo cardiomyopathy, initial testing with echocardiogram and subsequent non-invasive coronary CT-angiography is an option if a transthoracic echocardiography shows circumferential ballooning pattern [37].

Left ventricular angiography can identify wall motion abnormalities in either the typical apical-ballooning pattern (Fig. 5.1) or atypical pattern and aids in the diagnosis of Takotsubo cardiomyopathy. Coronary angiography usually reveals either normal coronary arteries or nonobstructive coronary artery disease in most patients. However, 10–29% of patients are found to have concomitant obstructive coronary atherosclerosis [9, 17, 40, 41]. Coronary artery disease, such as stable, non-obstructive fibroatheroma or flow-limiting lesions, is incidental, rather than the inciting culprit. If lesions are found on the angiogram, it should not correspond to the regional wall motion abnormality [9]. Left ventricular end diastolic pressure is frequently elevated (93% of patients) [9]. Specialized intracoronary imaging modalities such as optical coherence tomography (OCT) and intravascular ultrasound (IVUS) can be used to characterize indeterminate or intermediate lesions (See Case example 1). The role of physiologic testing for coronary microvascular dysfunction or spasm during acute presentation of Takotsubo is currently under investigation and predominantly performed in research settings, rather than as a routine clinical basis currently [42].

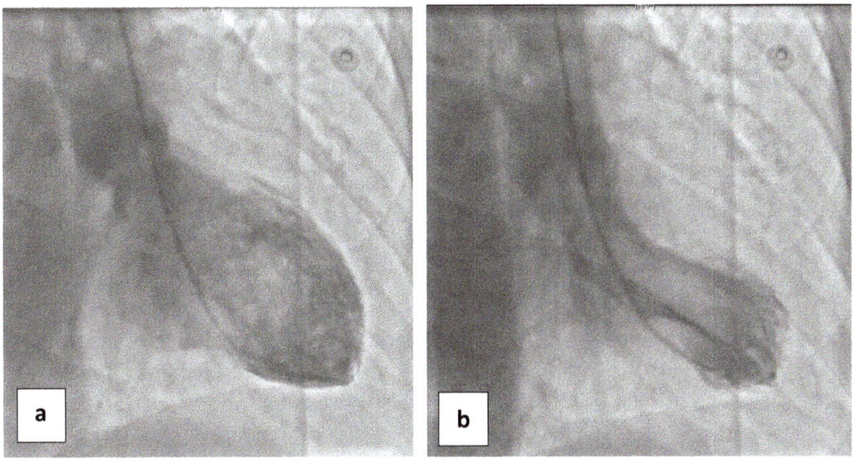


**Fig. 5.1** Left ventricular angiography in **a** diastole and **b** systole demonstrating apical akinesis in typical, apical-ballooning type Takotsubo cardiomyopathy

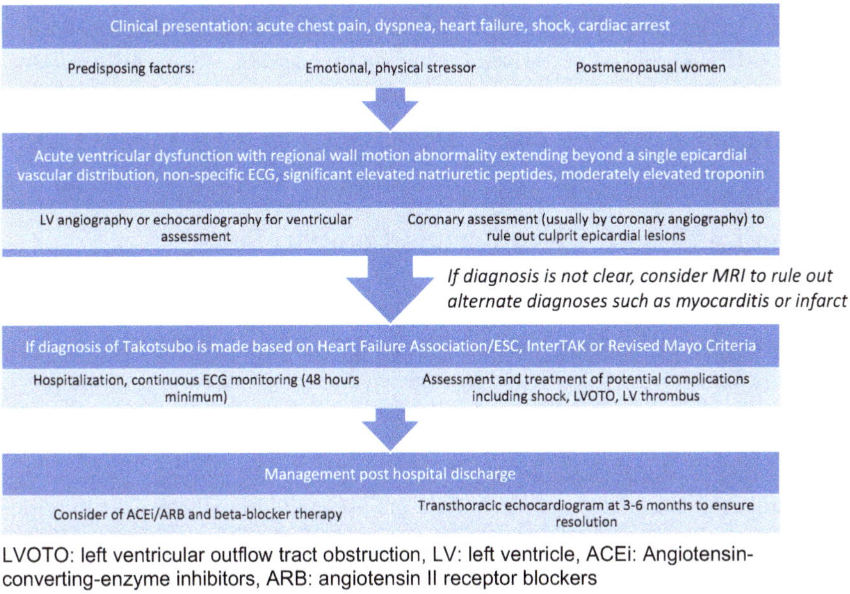


**Fig. 5.2** Clinical diagnostic and management pathway for a patient with Takotsubo cardiomyopathy. LVOTO: left ventricular outflow tract obstruction, LV: left ventricle, ACEi: Angiotensin-converting-enzyme inhibitors, ARB: angiotensin II receptor blockers

Transthoracic echocardiography is recommended in patients presenting with Takotsubo cardiomyopathy to establish the diagnosis and evaluate left ventricular function [37]. In Takotsubo cardiomyopathy, the extent of wall motion abnormality seen on echocardiogram often extends beyond a single coronary artery distribution, in contrast to an acute myocardial infarction wherein the wall motion abnormality corresponds to the myocardium supplied by the culprit coronary artery. The mean left ventricular ejection fraction is 28–54% [8, 9, 21, 31]. Four subtypes of the syndrome have been described, based on the pattern of wall motion abnormality. The apical ballooning subtype (81.7% of cases [43]) is characterized by hypo-, a-, or dyskinesia of mid-apical myocardial segments [4]. Approximately 20% of cases have associated left ventricular outflow tract obstruction (LVOTO) due to compensatory hyperdynamic contraction of the base [37]. Mitral regurgitation secondary to systolic anterior motion of the mitral leaflet or papillary muscle dysfunction can also be seen [37]. There may be an "apical nipple sign", representing a small zone of preserved contractility at the most distal apex [37]. The other subtypes are mid-ventricular (14.6%), basal (2.2%), and focal (1.5%) [33, 43]. Right ventricular involvement has also been described in one-third of the cases and may be a marker of worse prognosis [44, 45] (Fig. 5.2).

If there are any concerns for infectious myocarditis, cardiac magnetic resonance imaging (MRI) should be performed to rule out myocarditis. Takotsubo cardiomyopathy would be notable for myocardial edema and the absence of late gadolinium enhancement [37].

## In Hospital Risk Stratification and Complications

In-hospital and 30-day outcomes for Takotsubo cardiomyopathy patients are similar when compared with patients with acute coronary syndrome [9]. In the International Takotsubo Registry, the risk of a composite of catecholamine use, cardiogenic shock, the use of invasive or non-invasive ventilation, cardiopulmonary resuscitation, and death was 19% in patients with Takotsubo cardiomyopathy compared to 19.3% in a control group of patients with acute coronary syndrome.

Cardiogenic shock can occur in 6–20% of cases [37, 46], whereas death occurs in 1–8% of cases [6, 9, 21, 31]. Clinical characteristics associated with adverse in-hospital outcomes include male sex, age > 75 years, physical trigger, acute neurologic or psychiatric diseases, a high Charlson comorbidity index, tachycardia, systolic hypotension, initial troponin > 10 × the upper reference limit, elevated BNP and white blood cell count, admission LVEF < 45%, and right-ventricular involvement [37, 47–49]. A recent systematic review found a strong correlation between physical stressor and in-hospital mortality [31]. Other potential complications include left ventricular thrombi (1–2%) and left ventricular rupture (0.2%) [9, 33]. Arrhythmic complications include advanced atrioventricular block (2.9%), atrial fibrillation (4.7%), and ventricular arrhythmia/Torsade de Pointes (3.0–8.6%) [37]. Anterolateral T-wave inversion, QT interval prolongation and presence of J-waves are associated with sudden cardiac death and ventricular tachyarrhythmia [37].

Given the possibility of malignant arrhythmia, continuous ECG monitoring for at least 48 h is recommended for patients with suspected Takotsubo cardiomyopathy [37] (Fig. 5.2). A regular 12-lead ECG recording should also be performed to assess the QT interval duration [50]. In cases of additional risk factors for arrhythmias such as QT prolongation, monitoring should be extended as needed. The risk–benefit ratio of permanent implantable cardioverter-defibrillators for Takotsubo cardiomyopathy is currently debated [50].

The benefit of beta-blockers in the acute phase of Takotsubo cardiomyopathy remains unproven and while it may theoretically target elevated catecholamine levels, they carry the risk of bradycardia and pause-dependent torsade de pointes [37]. Therefore, beta-blockers should be used cautiously. Beta blockers and angiotensin-converting enzyme inhibitors (ACEi) can be considered if the ejection fraction is 45% or less [3]. Angiotensin-converting-enzyme inhibitors (ACEi) or angiotensin II receptor blockers (ARB) may potentially facilitate LV recovery [37] (Fig. 5.2).

Considering the overlap in clinical presentation with acute coronary syndrome, patients should receive guideline-based therapy for acute coronary syndrome with aspirin, P2Y12 inhibitor, anticoagulation, high-intensity statin, and pain management therapy. Once the diagnosis of Takotsubo cardiomyopathy is made, the antiplatelet therapy, as well as anticoagulation, and the statin can be discontinued. However, if coronary artery disease is seen on diagnostic work up, aspirin and statin should be continued. Anticoagulation should be continued in cases of documented left ventricular thrombus and can be considered in cases of severe left ventricular dysfunction with apical involvement. It is not recommended as a routine prophylactic strategy in all cases. Continuation for 3 months or until left ventricular function has resolved is recommended [33].

In cases of cardiogenic shock, evaluation for LVOTO should be performed either on invasive angiography or on echocardiography. If absence of LVOTO, inotropes may be used. However, their use is associated with 20% increased mortality rates [37]. The use of Calcium sensitizer levosimendan has been suggested as an alternative to catecholamine agents [37]. In the presence of LVOTO, patients should receive fluid resuscitation and cautious use of short-acting beta blockers. Inotropes and nitrates should be avoided [33]. An expert heart failure opinion should be obtained for assessment of possible mechanical support [3].

## Long Term Prognosis and Follow up

Left ventricular dysfunction appears to resolve within four to eight weeks. Follow up imaging at 3 months is recommended post discharge to ensure recovery of ventricular function [3]. Independent predictors of adverse outcomes in the subacute phase include decreased left-ventricular ejection fraction, increased LV filling pressure, atrial fibrillation, and moderate to severe mitral regurgitation at 4–6 weeks [37]. It was recently recognized that despite apparent early recovery, contractile dysfunction may develop with time due to cardiac inflammation, which could lead to global microscopic fibrosis, and could be detected as early as 4 months [51].

Regarding long-term clinical outcomes, a large systematic review of over 4000 patients found the annual mortality rate to be 3.5%(31). Most deaths (78%) are attributed to noncardiac causes [31]. The International Takotsubo Registry of 1750 patients found the rate of major adverse cardiac and cerebrovascular events to be 9.9% per patient-year, and the rate of death to be 5.6% per patient-year [9]. Physical trigger and neurologic disorder were associated with higher 5-year mortality when compared to emotional stressor as a reference group (HR 3.78; 95% CI: 2.21 to 6.44; $p < 0.001$ and 5.76; 95% CI 2.96 to 11.2; $p < 0.001$, respectively). Atypical ballooning subtypes were also associated with worse long-term outcome [31]. In a multivariable analysis, age, male sex, diabetes, pulmonary disease, and chronic kidney disease were associated with a higher risk of recurrence or death [52]. Patients may experience long term symptoms of dyspnea, lethargy, palpitations, fleeting chest pains that affect quality of life [1].

Studies reviewing the effect of commonly prescribed therapy for left ventricular dysfunction (primarily beta-blocker and ACEi/ARB) on long-term outcomes yield mixed results, and they are limited by their retrospective study design. The recurrence rate of Takotsubo cardiomyopathy was found to be 1.8% per patient-year in the International Takotsubo Registry [9]. A retrospective study of 519 patients followed over a median of 5.2 years found a recurrence rate of 7.5%. Treatment with beta-blockers was associated with lower risk of recurrence or death (HR 0.46, 95% CI 0.29 to 0.72; $p = 0.001$). There was no correlation between treatment with ACEi or ARB and the recurrence or death [52]. In a retrospective study of 825 patients, beta-blocker therapy upon discharge was associated with a significantly lower risk of all-cause death (adjusted HR 0.563; 95% CI 0.356 to 0.889; $p = 0.014$) at a median follow up of 24 months [53]. There was no significant difference in recurrence rate of Takotsubo between the two groups. However, another large retrospective study of 1750 patients found no survival benefit at 1-year from beta-blocker therapy upon discharge [9]. The use of ACEi or ARB was associated with improved survival at 1-year follow up [9]. One hypothesis to explain the discrepant findings of the above observational studies is that the use of ACEi/ARB may be more beneficial in earlier stages (within the first year) to facilitate ventricular recovery whereas beta-blocker therapy may offer benefits in recurrence and survival in the longer-term via other systemic pathways. In the absence of randomized controlled trials, current treatment recommendations based on expert opinion are for treatment with an ACEi or ARB, as well as beta-blocker for at least 3 months.

The benefit from estrogen supplementation, anti-depressants or other psychiatric drugs is controversial [37]. There are many ongoing randomized clinical trials investigating pharmacological treatment (NACRAM—*N*-Acetylcysteine and Ramipril

Takotsubo Syndrome Trial, BROKEN-SWEDE- HEART—Optimized Pharmacological Treatment for Broken Heart [Takotsubo] Syndrome, NCT04666454), exercise, and cognitive behavioral therapies (PLEASE study, NCT04425785, and E-SMINC study, NCT04178434) in improving the mortality and morbidity associated with Takotsubo cardiomyopathy.

## Conclusion

Takotsubo cardiomyopathy is an increasingly recognised condition consisting of a transient left ventricular dysfunction most likely due to a catecholamine surge. It disproportionately affects females. Attention to clinical nuances and coronary lesion characterization are necessary to differentiate between this entity and the main differential diagnosis of acute coronary syndrome. Although most patient have an excellent prognosis, Takotsubo cardiomyopathy is not a benign condition, as it can be associated with serious complications. Further research is needed and is currently underway to inform on the long-term management, and to improve Takotsubo-related morbidity and mortality.

Case example 1: Patient presenting with acute chest pain and suspected Takotsubo cardiomyopathy and bystander coronary artery disease. The differential includes acute coronary syndrome and further testing with cardiac MRI could be helpful to confirm the diagnosis.

1: Electrocardiogram: significant T-wave inversion anterolaterally and QTc prolongation.

2: Coronary angiography: Mild to moderate stenosis in the right-coronary artery.

3: Coronary angiography: Moderate stenosis in the left anterior descending (LAD).

4: Coronary angiography: Wire-induced pseudolesion in the LAD (black arrow).

5: Optical coherence tomography (OCT): showing pseudolesion mimicking red thrombus.

6: OCT: fibrocalcific plaque in mid LAD.

Transthoracic echocardiogram with contrast during diastole (7) and systole (8) showing distal 2/3 of the left ventricle is hypokinetic.

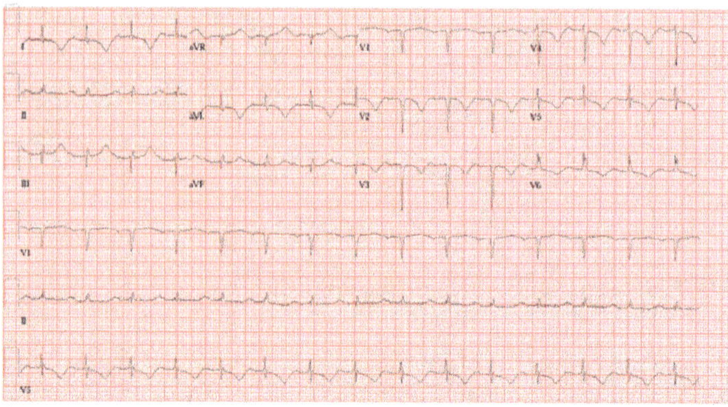


1

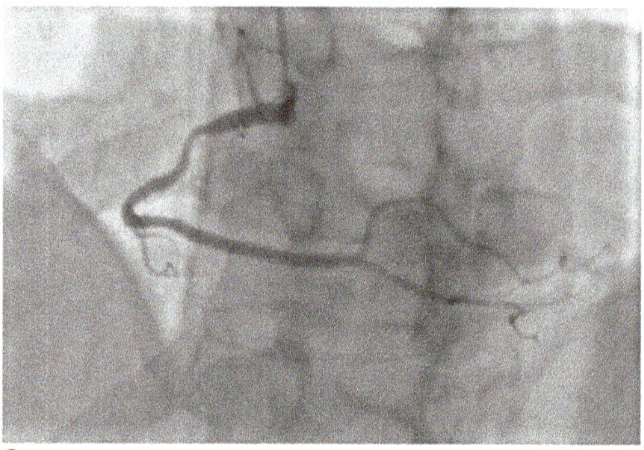

2

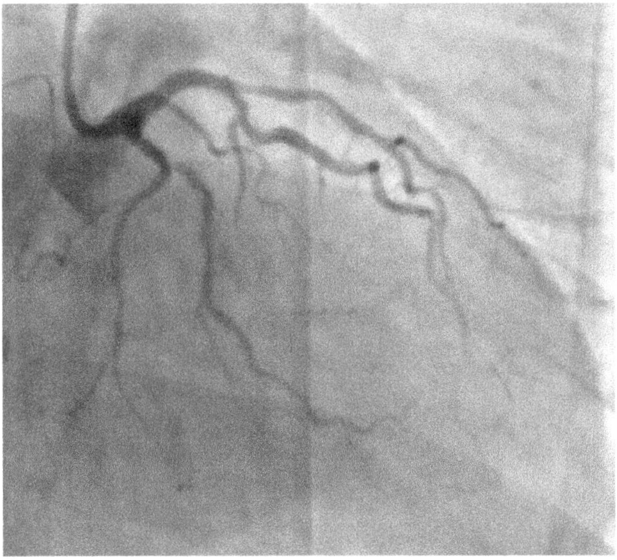

3

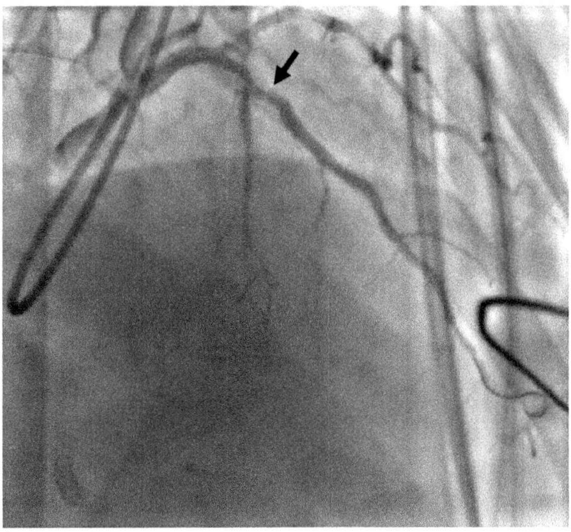

4

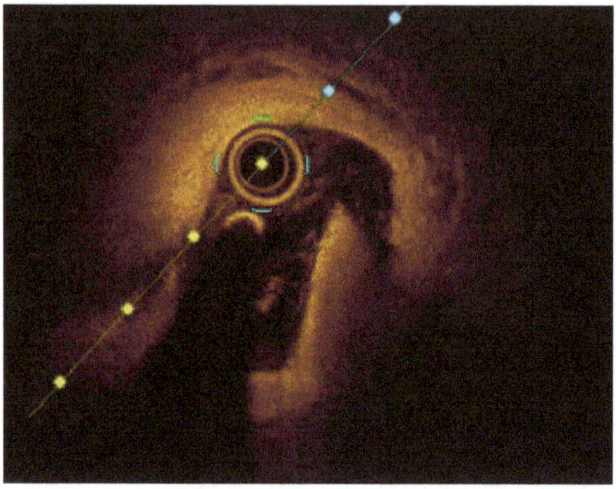

5

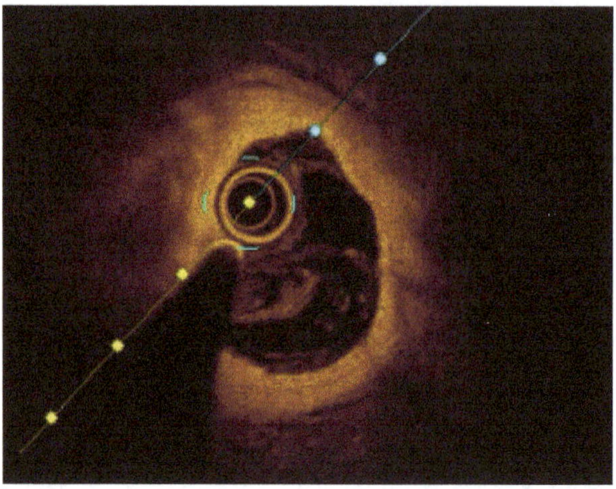

6

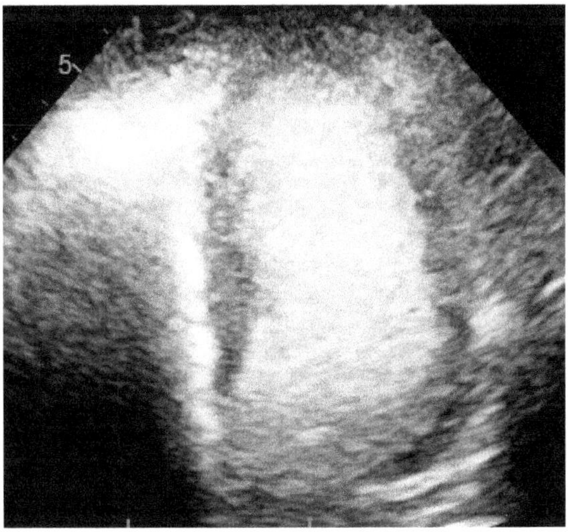


7

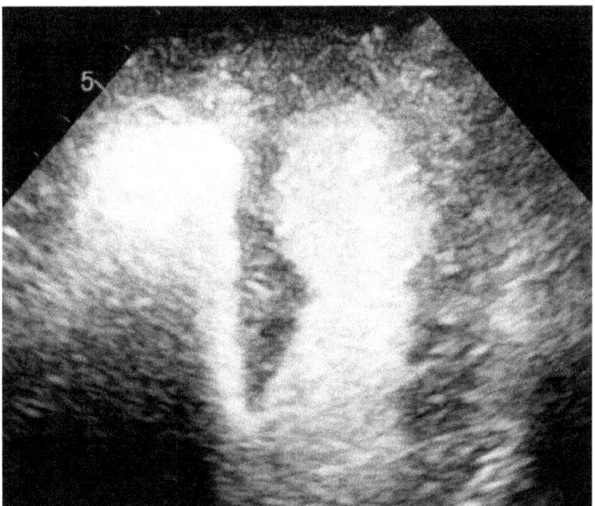


8

## References


1. Singh T, Khan H, Gamble DT, Scally C, Newby DE, Dawson D (2022) Takotsubo syndrome: pathophysiology, emerging concepts, and clinical implications. Circulation 145(13):1002–1019
2. Medina de Chazal H, Del Buono MG, Keyser-Marcus L, Ma L, Moeller FG, Berrocal D et al (2018) Stress cardiomyopathy diagnosis and treatment: JACC state-of-the-art review. J Am Coll Cardiol 72(16):1955–71
3. Lyon AR, Bossone E, Schneider B, Sechtem U, Citro R, Underwood SR et al (2016) Current state of knowledge on Takotsubo syndrome: a Position Statement from the Taskforce on Takotsubo Syndrome of the Heart Failure Association of the European Society of Cardiology. Eur J Heart Fail 18(1):8–27
4. Ghadri JR, Wittstein IS, Prasad A, Sharkey S, Dote K, Akashi YJ et al (2018) International expert consensus document on Takotsubo syndrome (part i): clinical characteristics, diagnostic criteria, and pathophysiology. Eur Heart J 39(22):2032–2046
5. Boyd B, Solh T (2020) Takotsubo cardiomyopathy: review of broken heart syndrome. JAAPA. 33(3):24–29
6. Bybee KA, Kara T, Prasad A, Lerman A, Barsness GW, Wright RS et al (2004) Systematic review: transient left ventricular apical ballooning: a syndrome that mimics ST-segment elevation myocardial infarction. Ann Intern Med 141(11):858–865
7. Prasad A, Lerman A, Rihal CS (2008) Apical ballooning syndrome (Tako-Tsubo or stress cardiomyopathy): a mimic of acute myocardial infarction. Am Heart J 155(3):408–417
8. Pelliccia F, Parodi G, Greco C, Antoniucci D, Brenner R, Bossone E et al (2015) Comorbidities frequency in Takotsubo syndrome: an international collaborative systematic review including 1109 patients. Am J Med 128(6):654.e11–9
9. Templin C, Ghadri JR, Diekmann J, Napp LC, Bataiosu DR, Jaguszewski M et al (2015) Clinical features and outcomes of Takotsubo (stress) cardiomyopathy. N Engl J Med 373(10):929–938
10. Gianni M, Dentali F, Grandi AM, Sumner G, Hiralal R, Lonn E (2006) Apical ballooning syndrome or Takotsubo cardiomyopathy: a systematic review. Eur Heart J 27(13):1523–1529
11. Park JH, Kang SJ, Song JK, Kim HK, Lim CM, Kang DH et al (2005) Left ventricular apical ballooning due to severe physical stress in patients admitted to the medical ICU. Chest 128(1):296–302
12. Minhas AS, Hughey AB, Kolias TJ (2015) Nationwide trends in reported incidence of Takotsubo cardiomyopathy from 2006 to 2012. Am J Cardiol 116(7):1128–1131
13. Zaghlol R, Dey AK, Desale S, Barac A (2020) Racial differences in takotsubo cardiomyopathy outcomes in a large nationwide sample. ESC Heart Fail. 7(3):1056–1063
14. Nascimento FO, Larrauri-Reyes MC, Santana O, Pérez-Caminero M, Lamas GA (2013) Comparison of stress cardiomyopathy in Hispanic and non-Hispanic patients. Rev Esp Cardiol (Engl Ed). 66(1):67–68
15. Regnante RA, Zuzek RW, Weinsier SB, Latif SR, Linsky RA, Ahmed HN et al (2009) Clinical characteristics and four-year outcomes of patients in the Rhode Island Takotsubo Cardiomyopathy Registry. Am J Cardiol 103(7):1015–1019
16. Deshmukh A, Kumar G, Pant S, Rihal C, Murugiah K, Mehta JL (2012) Prevalence of Takotsubo cardiomyopathy in the United States. Am Heart J 164(1):66–71.e1
17. Pelliccia F, Kaski JC, Crea F, Camici PG (2017) Pathophysiology of Takotsubo syndrome. Circ 135(24):2426–2441
18. Wittstein IS, Thiemann DR, Lima JA, Baughman KL, Schulman SP, Gerstenblith G et al (2005) Neurohumoral features of myocardial stunning due to sudden emotional stress. N Engl J Med 352(6):539–548
19. Paur H, Wright PT, Sikkel MB, Tranter MH, Mansfield C, O'Gara P et al (2012) High levels of circulating epinephrine trigger apical cardiodepression in a β2-adrenergic receptor/Gi-dependent manner: a new model of Takotsubo cardiomyopathy. Circ 126(6):697–706

20. Nef HM, Möllmann H, Kostin S, Troidl C, Voss S, Weber M et al (2007) Tako-Tsubo cardiomyopathy: intraindividual structural analysis in the acute phase and after functional recovery. Eur Heart J 28(20):2456–2464
21. Akashi YJ, Nef HM, Lyon AR (2015) Epidemiology and pathophysiology of Takotsubo syndrome. Nat Rev Cardiol 12(7):387–397
22. Suzuki H, Matsumoto Y, Kaneta T, Sugimura K, Takahashi J, Fukumoto Y et al (2014) Evidence for brain activation in patients with Takotsubo cardiomyopathy. Circ J 78(1):256–258
23. Radfar A, Abohashem S, Osborne MT, Wang Y, Dar T, Hassan MZO et al (2021) Stress-associated neurobiological activity associates with the risk for and timing of subsequent Takotsubo syndrome. Eur Heart J 42(19):1898–1908
24. Martin EA, Prasad A, Rihal CS, Lerman LO, Lerman A (2010) Endothelial function and vascular response to mental stress are impaired in patients with apical ballooning syndrome. J Am Coll Cardiol 56(22):1840–1846
25. Patel VI, Sobnosky S (2022) Multivessel coronary artery vasospasm-induced Takotsubo cardiomyopathy. Case Rep Cardiol 2022:2192863
26. Angelini P (2008) Transient left ventricular apical ballooning: a unifying pathophysiologic theory at the edge of Prinzmetal angina. Catheter Cardiovasc Interv 71(3):342–352
27. Y-Hassan S (2021) Coronary microvascular dysfunction in Takotsubo syndrome: cause or consequence. Am J Cardiovasc Dis 11(2):184–93
28. Pison L, De Vusser P, Mullens W (2004) Apical ballooning in relatives. Heart 90(12):e67
29. Kumar G, Holmes DR, Prasad A (2010) "Familial" apical ballooning syndrome (Takotsubo cardiomyopathy). Int J Cardiol 144(3):444–445
30. Ikutomi M, Yamasaki M, Matsusita M, Watari Y, Arashi H, Endo G et al (2014) Takotsubo cardiomyopathy in siblings. Heart Ves 29(1):119–122
31. Pelliccia F, Pasceri V, Patti G, Tanzilli G, Speciale G, Gaudio C et al (2019) Long-term prognosis and outcome predictors in Takotsubo syndrome: a systematic review and meta-regression study. JACC Heart Fail 7(2):143–154
32. Arcari L, Musumeci MB, Stiermaier T, El-Battrawy I, Möller C, Guerra F et al (2020) Incidence, determinants and prognostic relevance of dyspnea at admission in patients with Takotsubo syndrome: results from the international multicenter GEIST registry. Sci Rep 10(1):13603
33. Assad J, Femia G, Pender P, Badie T, Rajaratnam R (2022) Takotsubo syndrome: a review of presentation, diagnosis and management. Clin Med Insights Cardiol 16:11795468211065782
34. Gill D, Sheikh N, Ruiz VG, Liu K (2018) Case report: cocaine-induced Takotsubo cardiomyopathy. Hellenic J Cardiol 59(2):129–132
35. Chiang YL, Chen PC, Lee CC, Chua SK (2016) Adrenal pheochromocytoma presenting with Takotsubo-pattern cardiomyopathy and acute heart failure: a case report and literature review. Medicine (Baltimore) 95(36):e4846
36. Frangieh AH, Obeid S, Ghadri JR, Imori Y, D'Ascenzo F, Kovac M, Ruschitzka F, Lüscher TF, Duru F, Templin C (2016) InterTAK Collaborators. ECG criteria to differentiate between Takotsubo (Stress) cardiomyopathy and myocardial infarction. J Am Heart Assoc 5(6):e003418
37. Ghadri JR, Wittstein IS, Prasad A, Sharkey S, Dote K, Akashi YJ et al (2018) International expert consensus document on Takotsubo syndrome (part ii): diagnostic workup, outcome, and management. Eur Heart J 39(22):2047–2062
38. Behr ER, Mahida S (2009) Takotsubo cardiomyopathy and the long-QT syndrome: an insult to repolarization reserve. Europace 11(6):697–700
39. Madias C, Fitzgibbons TP, Alsheikh-Ali AA, Bouchard JL, Kalsmith B, Garlitski AC et al (2011) Acquired long QT syndrome from stress cardiomyopathy is associated with ventricular arrhythmias and torsades de pointes. Heart Rhythm 8(4):555–561
40. Winchester DE, Ragosta M, Taylor AM (2008) Concurrence of angiographic coronary artery disease in patients with apical ballooning syndrome (Takotsubo cardiomyopathy). Catheter Cardiovasc Interv 72(5):612–616
41. Kurisu S, Inoue I, Kawagoe T, Ishihara M, Shimatani Y, Nakama Y et al (2009) Prevalence of incidental coronary artery disease in Takotsubo cardiomyopathy. Coron Artery Dis 20(3):214–218

42. Morrow AJ, Nordin S, O'Boyle P, Berry C (2019) 'Acute micro-coronary syndrome': detailed coronary physiology in a patient with Takotsubo cardiomyopathy. BMJ Case Rep 12(8)
43. Ghadri JR, Cammann VL, Napp LC, Jurisic S, Diekmann J, Bataiosu DR et al (2016) Differences in the clinical profile and outcomes of typical and atypical Takotsubo syndrome: data from the international Takotsubo registry. JAMA Cardiol 1(3):335–340
44. Kagiyama N, Okura H, Tamada T, Imai K, Yamada R, Kume T et al (2016) Impact of right ventricular involvement on the prognosis of Takotsubo cardiomyopathy. Eur Heart J Cardiovasc Imaging 17(2):210–216
45. Eitel I, von Knobelsdorff-Brenkenhoff F, Bernhardt P, Carbone I, Muellerleile K, Aldrovandi A et al (2011) Clinical characteristics and cardiovascular magnetic resonance findings in stress (Takotsubo) cardiomyopathy. JAMA 306(3):277–286
46. Tsuchihashi K, Ueshima K, Uchida T, Oh-mura N, Kimura K, Owa M et al (2001) Transient left ventricular apical ballooning without coronary artery stenosis: a novel heart syndrome mimicking acute myocardial infarction. Angina Pectoris-Myocardial Infarction Investigations in Japan. J Am Coll Cardiol 38(1):11–18
47. Yerasi C, Koifman E, Weissman G, Wang Z, Torguson R, Gai J et al (2017) Impact of triggering event in outcomes of stress-induced (Takotsubo) cardiomyopathy. Eur Heart J Acute Cardiovasc Care 6(3):280–286
48. Murakami T, Yoshikawa T, Maekawa Y, Ueda T, Isogai T, Konishi Y et al (2014) Characterization of predictors of in-hospital cardiac complications of Takotsubo cardiomyopathy: multi-center registry from Tokyo CCU Network. J Cardiol 63(4):269–273
49. Böhm M, Cammann VL, Ghadri JR, Ukena C, Gili S, Di Vece D et al (2018) Interaction of systolic blood pressure and resting heart rate with clinical outcomes in Takotsubo syndrome: insights from the International Takotsubo Registry. Eur J Heart Fail 20(6):1021–1030
50. Möller C, Eitel C, Thiele H, Eitel I, Stiermaier T (2018) Ventricular arrhythmias in patients with Takotsubo syndrome. J Arrhythm 34(4):369–375
51. Scally C, Rudd A, Mezincescu A, Wilson H, Srivanasan J, Horgan G et al (2018) Persistent long-term structural, functional, and metabolic changes after stress-induced (Takotsubo) cardiomyopathy. Circ 137(10):1039–1048
52. Lau C, Chiu S, Nayak R, Lin B, Lee MS (2021) Survival and risk of recurrence of Takotsubo syndrome. Heart 107(14):1160–1166
53. Silverio A, Parodi G, Scudiero F, Bossone E, Di Maio M, Vriz O et al (2022) Beta-blockers are associated with better long-term survival in patients with Takotsubo syndrome. Heart 108(17):1369–1376

# Chapter 6
# Preeclampsia: Early and Long-Term Clinical Considerations

**Sarah Gibbs, Rachelle Govia, Jessica Cudmore, Laura Chisick, and Robin Ducas**

**Abstract** Preeclampsia is a hypertensive disorder of pregnancy characterized by new onset of hypertension after 20 weeks gestational age, in the setting of proteinuria and/or other end organ damage. It is a multisystem disorder and is caused by abnormal placentation and release of angiogenic factors with resultant maternal vascular dysfunction. Preeclampsia complicates 5% of pregnancies and the incidence has increased 25% in the last 20 years. Severe forms of preeclampsia can result in dysfunction of maternal neurologic, renal, cardiac, hepatic, pulmonary function, as well as haematologic disturbances and death. Fetal complications include severe growth restriction, preterm birth and stillbirth/neonatal death. Screening, timely diagnosis and management of preeclampsia are integral to optimizing outcomes, with definitive therapy being delivery of the fetus. However, preeclampsia may also be diagnosed in the postpartum period, highlighting the need for post-partum assessment. Though much work has been done in the antepartum diagnosis and management of preeclampsia, a growing body of evidence has shown an increased risk of long-term cardiovascular disease in patients who develop preeclampsia. Not only must healthcare providers be able to diagnose and manage preeclampsia, providers must also understand the role that preeclampsia plays in the lifelong cardiovascular risk of their patients.

**Keywords** Preeclampsia · Pregnancy · Hypertension · HELLP · Proteinuria

S. Gibbs
Department of Internal Medicine, University of Manitoba, Winnipeg, MB, Canada

R. Govia
Department of Obstetrics, Gynecology and Reproductive Science, Rady Faculty of Health Sciences, Max Rady College of Medicine, University of Manitoba, Winnipeg, MB, Canada

J. Cudmore · L. Chisick
Department of Internal Medicine, Section of General Internal Medicine, Rady Faculty of Health Sciences, Max Rady College of Medicine, University of Manitoba, Winnipeg, MB, Canada

R. Ducas (✉)
Department of Internal Medicine, Section of Cardiology, Rady Faculty of Health Sciences, Max Rady College of Medicine, University of Manitoba, Winnipeg, MB, Canada
e-mail: rducas@sbgh.mb.ca

© The Author(s), under exclusive license to Springer Nature Switzerland AG 2023

L. Kirshenbaum and I. Rabinovich-Nikitin (eds.), *Biology of Women's Heart Health*, Advances in Biochemistry in Health and Disease 26,
https://doi.org/10.1007/978-3-031-39928-2_6

## Introduction

Hypertensive disorders of pregnancy are one of the leading causes of both maternal and perinatal mortality globally. Preeclampsia is a hypertensive disorder of pregnancy characterized by new onset of hypertension after 20 weeks gestational age (GA), in the setting of proteinuria and/or other end organ damage. It complicates 5% of pregnancies and the incidence has increased by 25% in the last 20 years [1, 2]. Preeclampsia is a multisystem disorder and is caused by abnormal placentation and release of angiogenic factors with resultant maternal vascular dysfunction [3]. Severe forms of preeclampsia can result in dysfunction of maternal neurologic, renal, cardiac, hepatic, and pulmonary function, as well as haematologic disturbances and death. Fetal complications include severe growth restriction, preterm birth and stillbirth/neonatal death. Screening, timely diagnosis and management of preeclampsia are integral to optimizing both maternal and fetal outcomes, with definitive therapy being delivery of the fetus. However, preeclampsia may also be diagnosed in the postpartum period, highlighting the need for post-partum blood pressure assessment. Though much work has been done in the antepartum diagnosis and management of preeclampsia, a growing body of evidence has shown an increased risk of long-term cardiovascular disease in patients who develop preeclampsia. Not only must healthcare providers be able to diagnose and manage preeclampsia in the setting of pregnancy, providers must also understand the role that preeclampsia plays in the lifelong cardiovascular risk of their patients.

## Overview

Hypertension in a pregnant patient is defined as a systolic BP $\geq$ 140 mmHg and/or diastolic BP $\geq$ 90 mmHg, based on an average of at least two measurements [4].

There are four major categories of hypertensive disorders in pregnant women:

1. Isolated gestational hypertension—New hypertension after 20 weeks gestational age (GA) in the absence of end organ involvement.
2. Preeclampsia and spectrum conditions (preeclampsia, hemolysis elevated liver enzymes and low platelets (HELLP) syndrome and eclampsia).—New hypertension after 20 weeks GA with variable end-organ damage.
3. Chronic hypertension—hypertension antecedent to pregnancy or present prior to 20 weeks GA.
4. Preeclampsia superimposed on chronic hypertension—worsening/resistant hypertension with new onset proteinuria or end organ involvement after 20 weeks GA, in a woman with pre-existing hypertension.

## *Diagnosis of Preeclampsia*

Preeclampsia is diagnosed when a previously normotensive patient develops hypertension (> 140 mmHg systolic blood pressure or > 90 mmHg diastolic blood pressure) after 20 weeks of gestation along with evidence of end-organ damage.

For a diagnosis of preeclampsia, at least one of the following needs to be present in addition to new onset hypertension [5–8]:

- Proteinuria (protein:creatinine ratio $\geq$ 30 mg/mmol or albumin:creatinine ratio $\geq$ 8 mg/mmol, $\geq$ 0.3 g/day in a 24 h urine collection)
- Maternal end organ dysfunction
  - Neurological sequelae (headache, altered mental status, visual scotoma, clonus, eclampsia, blindness, stroke, posterior reversible encephalopathy syndrome (PRES)
  - Cardiovascular (myocardial dysfunction, pulmonary edema)
  - Hematological sequelae (thrombocytopenia, hemolysis, disseminated intravascular coagulation)
  - Renal insufficiency (creatinine > 90 μmol/L)
  - Liver involvement (elevated liver transaminases or right upper quadrant/epigastric pain)
- Uteroplacental dysfunction (fetal growth restriction, abnormal fetal dopplers, placental abruption, stillbirth)

HELLP syndrome is a severe form of preeclampsia with an increased risk of maternal and fetal complications [4]. HELLP syndrome is characterized by hemolysis (with microangiopathy blood smear/schistocytes/burr cells), elevated liver enzymes, and low platelet count.

Eclampsia is diagnosed in the setting of new onset tonic–clonic focal or multifocal seizures or coma in a patient with preeclampsia, and is the manifestation of severe neurologic involvement. Eclamptic seizures are defined as occurring in the absence of other causative conditions such as cerebral ischemia/infarction, epilepsy, intracranial hemorrhage or drug use [7].

## *Pathophysiology of Preeclampsia:*

Preeclampsia is a multisystem disorder of pregnancy with a complex pathophysiology that remains incompletely understood. Suboptimal trophoblast invasion and inadequate remodeling of the maternal spiral arteries in the early stages of placentation is thought to underlie this clinical syndrome [3, 9]. These abnormalities result in a reduction in prefusion to the utero-placental unit and subsequent placental ischemia/hypoxia which in turn causes an increase in angiogenic markers (including fms-like tyrosine kinase-1 and soluble endoglin) [9]. The increase in angiogenic markers, has been proposed to result in a decrease in vascular growth factor (VEGF) and placental growth factor, and subsequent maternal vascular endothelial dysfunction. Endothelial dysfunction may be accompanied by vasoconstriction, oxidative stress and micro-emboli which can affect multiple organ systems [3, 10]. The degree of maternal inflammatory response and placental ischemia as well as the evolving imbalance in angiogenic and antiangiogenic factors influence the clinical severity of the syndrome [3, 6–8, 11]. Additional mechanisms likely involved in the development of preeclampsia include immunologic aberrations in pregnancy and genetic factors [3]. Abnormal placentation early in pregnancy tends to be a key feature of early-onset preeclampsia (diagnosis < 34 weeks) which carries the highest risk of maternal fetal morbidity and mortality. In contrast, cases of late-onset preeclampsia (> 34 weeks) often demonstrate normal early placental development but develop due to abnormal placental perfusion, inflammation and oxidative stress later in pregnancy [12].

## *Outcomes and Burden of Preeclampsia:*

Globally, preeclampsia has been found to be present in 5% of pregnancies, noting some regional variation [13]. Hypertensive disorders of pregnancy are one of the leading causes of maternal death globally, with 13% being attributed to preeclampsia/eclampsia [14]. The most common cause of death from preeclampsia is cerebral hemorrhage from severe uncontrolled hypertension [2]. Significant morbidity for the mother also takes the form of increased rates of pulmonary edema (0.1–2.3%), renal failure requiring dialysis (2.3%) and HELLP syndrome (8.3%) [15–17]. Preeclampsia increases the rates of various conditions in the fetus, including prematurity, low birth weight, and death. It has been found that in women with preeclampsia the risk of preterm birth ranges from 20 to 55% and the risk of a low-birth-weight infant ranges between 23 and 34% [16, 18, 19]. Many studies globally have examined the cost to health care systems associated with preeclampsia; there is a significant increase in health care spending in patients with preeclampsia with the majority of extra cost (millions to billions of dollars annually, depending on country) attributed to caring for premature infants and increased use of health care services [20–22].

## *Risk Factors and Primary Prevention of Preeclampsia:*

There have been multiple risk factors associated with the development of preeclampsia (Table 6.1) [6–8, 23]. However, it is important to understand that most cases of preeclampsia occur in women with no overt risk factors. The identification of risk factors for preeclampsia is clinically important as it can guide care providers to initiate preventative therapy.

Given the burden and scope of disease, prevention of preeclampsia is a healthcare priority. In high-risk women (Table A), acetylsalicylic acid (ASA) has been shown

**Table 6.1** Risk factors for the development of preeclampsia [4, 8, 24]

| | "High risk" factor | "Moderate risk" factor |
|---|---|---|
| Maternal demographics | • Pre-pregnancy body mass index > 30 kg/m$^2$ | • Maternal age (≥ 35 years) |
| Maternal history | • Preexisting hypertension<br>• Type 1 or 2 diabetes<br>• Chronic renal disease<br>• Autoimmune disease (e.g. SLE)<br>• Antiphospholipid antibody syndrome | • Family history of preeclampsia<br>• Interpregnancy interval (> 10 years) |
| Reproductive history | • Previous hypertensive disorder of pregnancy/preeclampsia | • Prior placental abruption/fetal growth restriction or stillbirth |
| Current pregnancy | • Assisted reproductive technology | • Multifetal gestation<br>• Nulliparity |

*SLE* systemic lupus erythematosus
*Patients are typically considered "high risk" if they have at least one risk factor from the high-risk category or two or more from the moderate risk category

to reduce rates of preterm preeclampsia by 70%, in addition to reducing preterm birth and severe disease [6, 25, 26]. It is recommended to start ASA > 100 mg daily prior to 12 weeks GA and to continue to 36 weeks or delivery, depending on obstetrical plan [7, 26]. Exercise in pregnancy is recommended for all women without contraindications, to decrease the odds of developing preeclampsia. It has been demonstrated that 140 min of moderate-intensity exercise weekly can reduce the risk of developing preeclampsia by 25% [27]. In patients with low calcium intake (< 900 mg/day), calcium supplementation (> 500 mg/day) has been associated with reduced risk and severity of preeclampsia by up to 50% [28–30]. There is no clear role for vitamins C or E, metformin, statins, oral magnesium, low molecular weight heparin or folic acid in the prevention of preeclampsia currently, although studies assessing these are in progress [11].

## Antepartum Management of Preeclampsia

### *Screening for the Development of Preeclampsia*

Screening for preeclampsia begins at the first prenatal (or antenatal) visit by screening for maternal risk factors (Table 6.1). Integral to assessment is accurate blood pressure measurement. This is done with an appropriately sized cuff, while the patient is in the sitting position and the arm is placed at the level of the heart. Two measurements should be taken in the same arm, at least 15 min apart [8]. Blood pressure should be measured at each clinical encounter to identify pre-existing hypertension (< 20 weeks GA) or the development of a hypertensive disorder of pregnancy (> 20 weeks GA).

At 11–14 weeks, women should be screened again with blood pressure measurement, risk factor assessment, uterine artery pulsatility index and placental growth factor (PlGF) if available [6, 8, 11]. This integrated approach assessing maternal characteristic/risk factors, blood pressure, biomarkers and fetal flow characteristics demonstrates improved accuracy at predicting development of preeclampsia compared to maternal risk factors alone [6]. In addition to blood pressure monitoring screening for preeclampsia includes assessment for maternal proteinuria. Urine dipstick is a sufficient screening method for new onset proteinuria however, 24 h urine collection or spot urine PCR should be used to confirm proteinuria if the urine dipstick is > + 1 or if preeclampsia is suspected [8].

Though the diagnosis of preeclampsia is typically made in the prenatal period, a diagnosis of hypertensive disorders of pregnancy and/or preeclampsia can be made up to 6 weeks post-partum, highlighting the need for appropriate post-partum follow-up [8]. For women who did not develop hypertension or preeclampsia antenatally, routine post-partum care is typically done with assessment at 6 weeks post-partum. Women with preeclampsia are typically seen frequently in the immediate post-partum period, as blood pressures tend to rise in the first week post-partum. Patient education is critical as blood pressures still need to be checked and medications should not be stopped until it is safe to do so. Thereafter, the frequency of appointments tends to wane as blood pressures settle. Patients without pre-existing hypertension are slowly weaned off medication as the cardiovascular changes of pregnancy and delivery subside.

## *Clinical Presentation of Preeclampsia and Assessment of Complications*

Preeclampsia affects multiple organ systems; as such, clinical and laboratory assessment is necessary for both diagnosis, identification of severe disease and management. Affected organ systems include the central nervous system, cardiovascular, renal, hepatic, haematologic and the fetal-placental unit (Fig. 6.1). During pregnancy, management of preeclampsia is based on the severity of blood pressure elevation in addition to the degree of end-organ damage that develops, including complications in the fetus. Adverse conditions are features of preeclampsia that increase the risk of negative maternal or fetal outcomes and require urgent management (typically expedient delivery) in order to mitigate severe complications. (Fig. 6.1) [8].

## *Antepartum Management of Preeclampsia:*

Once the diagnosis of preeclampsia has been established, frequent clinical assessment is recommended [5, 11]. Further monitoring is dependent on maternal and fetal status

| Organ System | Adverse Clinical Signs and Symptoms | Severe Complications *Indications for Delivery* |
|---|---|---|
|  | Headache<br>Visual changes | Eclampsia<br>PRES<br>Cortical blindness/retinal detachment<br>GCS <13<br>TIA/Stroke |
|  | Low platelets<br>Elevated INR/PTT<br>Elevated WBC | Platelets <50<br>Cytopenia requiring transfusion<br>DIC |
|  | Dyspnea/chest pain<br>Oxygen saturation <97% | Hypertensive emergency<br>Heart failure<br>Need for intubation/inotropes<br>Myocardial ischemia/infarction |
| 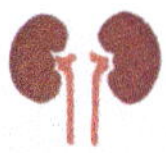 | Elevated creatinine<br>Elevated uric acid | Acute kidney injury<br>New start dialysis |
|  | Nausea/vomiting<br>RUQ/epigastric pain<br>Abnormal liver enzymes<br>Low albumin | Hepatic dysfunction (INR >2)<br>Hepatic rupture/hematoma |
|  | Abnormal fetal heart rate<br>Growth restriction | Placental abruption<br>Intrauterine fetal death |

**Fig. 6.1** Clinical presentation and indications for delivery in preeclampsia. DIC = disseminated intravascular coagulopathy; GCS = Glasgow coma scale; INR = international normalised ratio; PRES = posterior reversible encephalopathy syndrome; PTT = prothrombin time; RUQ = right upper quadrant; TIA = transient ischemic attack; WBC = white blood cell

and close clinical follow up is used to help guide indications for medical therapy and planning for delivery.

With regards to blood pressure management, blood pressure > 140/90 mmHg typically can be managed with oral medications to a target blood pressure of systolic < 140 and diastolic < 85 mmHg. Oral antihypertensive medications of choice typically include labetalol, methyldopa and/or nifedipine [23]. When blood pressure is severely elevated (> 160/ > 110 mmHg), intravenous medications (labetalol or hydralazine) or shorter acting oral medications (nifedipine) may be used in order to bring the blood pressure down more rapidly [7, 8, 11]. Though antihypertensive therapy in preeclampsia is typically given to mitigate maternal adverse outcomes, planning for delivery is one of the most important aspects of preeclampsia management as this is the definitive treatment. Management of preeclampsia needs to be individualized. Important considerations include the gestational age, disease severity, end-organ involvement and patient preference. Women with a diagnosis of preeclampsia < 24 weeks GA should receive counselling regarding the pros and cons of continuing the pregnancy. In general, expectant management would be recommended between 24 and 33 + 6 weeks GA [8, 11]. Once a patient reaches 34 weeks GA, there is insufficient evidence to recommend continued expectant management [5, 8, 11]. For women > 37 weeks GA, delivery is typically recommended [5]. Regardless of GA, women with preeclampsia who have developed severe complications require immediate delivery (Fig. 6.1). If preterm delivery is clinically indicated, consideration should be given to administering antenatal corticosteroids to women presenting < 34 + 6 weeks GA in order to accelerate fetal lung maturity [8, 11, 31]. It is important to note that exercise (beyond typical activities of daily living) is considered contraindicated in women with established preeclampsia and relatively contraindicated in women with gestational hypertension [32]. Women with gestational hypertension should speak with their health care provider regarding participation in moderate- to vigorous activities.

## *Management During Delivery*

In the absence of obstetrical indications, most women with preeclampsia may have a trial of vaginal delivery with diligent intrapartum blood pressure monitoring and pharmacologic management. In cases of abnormal fetal testing, labour may not be tolerated and delivery by caesarean section is indicated. It is important to avoid excess IV fluid to reduce the risk of pulmonary edema. Antihypertensives are to be continued during labour and it is suggested to continue active management of the 3rd stage of labour with oxytocin, especially if the patient has thrombocytopenia or a coagulopathy. However, it is important to avoid ergometrine for women with gestational hypertension and preeclampsia as it increases the risk of uncontrolled hypertension [8].

## *Management and Prevention of Eclampsia:*

For women having preeclampsia with severe features or those with a formal diagnosis of eclampsia magnesium sulfate (MgSO4) should be given to prevent initial or recurrent seizures [11]. Antihypertensive therapy is also used to prevent stroke. Severe features warranting MgSO4 administration include: severe hypertension, headaches/visual disturbances, RUQ/epigastric pain, platelets < 100, progressive renal insufficiency, and /or elevated liver enzymes. A loading dose of MgSO4 is typically given, followed by an infusion. Dose adjustments should be made in renal insufficiency or oliguria [2]. All patients should be monitored for signs of magnesium toxicity every 1–2 h (including absent deep tendon reflexes, respiratory distress or altered level of consciousness) [8, 11].

## Post-partum Diagnosis and Management of Preeclampsia

### *Diagnosis of Post-partum Preeclampsia*

Though preeclampsia is typically diagnosed antepartum, about 5% of cases can present in the post-partum period, (typically within the first week post-partum, but can be seen up to 6 weeks) and can be responsible for significant maternal morbidity; as some studies have shown that up to 50% of patients who develop eclampsia will do so in the postpartum period, with approximately 26% occurring > 48 h post-partum [33–36]. It is important to ensure that blood pressure is measured 3–7 days post-partum as this is the anticipated peak secondary to the extravascular fluid redistribution [11]. Target blood pressure remains < 140/90 mmHg (target in patients with diabetes < 130/80 mmHg) [37]. Options for medical therapy include those used in antepartum management in addition to captopril and enalapril, which may be used with breastfeeding [8]. If breastfeeding is not pursued standard/guideline directed antihypertensive therapy may be used [37]. It is important to continue antihypertensive treatment for women with antenatal preeclampsia and those who delivered preterm, in the post-partum period. Women with ongoing hypertension at > 6 weeks post-partum, should be screened for pre-existing hypertension or a secondary cause [5, 8].

### *Recurrence and Secondary Prevention*

The recurrence rate of preeclampsia in a subsequent pregnancy is approximately 15% [38]. Importantly, early onset preeclampsia has a recurrence rate of roughly 50%. Women with a history of preeclampsia who are seen in preconception counselling or in early subsequent pregnancy should be counselled on the risk of recurrent preeclampsia and offered the established primary prevention interventions [7, 8].

### *Lifelong Cardiovascular Risk and Outcomes*

In 1964, Epstein showed for the first time ever that women who had developed preeclampsia were at increased risk of developing cardiovascular disease later in life [39]. Since that time, a robust body of literature has evolved to show that there is a strong connection between preeclampsia and long-term maternal cardiovascular risk. One of the landmark studies exploring the link between hypertensive disorders of pregnancy and long-term cardiovascular risk was the CHAMPS study. This was a retrospective Canadian population-based cohort study involving over 1 million women without pre-existing cardiovascular disease before their first delivery. In these women, 7% developed a maternal placental syndrome (hypertensive disorders of pregnancy, abruption or infarction of the placenta). They found a doubling of the risk of premature cardiovascular disease in women who had developed a maternal placental syndrome compared with those who had not. The mean and maximum age at the time of first cardiovascular event in this group was 38 years and 60 years, respectively, which was significantly earlier than in women who had not developed maternal placental syndromes [40]. Numerous other works have shown similar findings in women with a history of preeclampsia. A robust metanalysis of over 50 studies and 10 million women demonstrated at a twofold higher incidence of cardiovascular events (including: death, myocardial infarction, stroke, hypertension, diabetes and dyslipidemia) in women with previous preeclampsia compared to those with previous normotensive pregnancy. In addition, this meta-analysis highlighted a fourfold higher burden of cardiovascular disease/outcomes in women who had early onset preeclampsia (preeclampsia requiring delivery before or at 34 weeks gestational age) [1].

Preeclampsia is associated with fourfold increase risk of hypertension [41], with 30% of women having hypertension at 2-years post-delivery and 25% with metabolic syndrome at that time [42, 43]. Women who have developed preeclampsia have at least twice the risk of type 2 diabetes and dyslipidemia [44]. Early onset heart failure and dysrhythmias are also more common in women with maternal placental syndromes (including preeclampsia), and the mean age at composite outcome in another retrospective cohort study of > 1 million women was 37.8 years [45]. Similar associations are also seen between preeclampsia and dementia, chronic kidney disease, seizures, and even overall death from any cause [41].

## *Factors Associated with Lifelong Risk*

Our current understanding of the role preeclampsia plays in long term adverse health risk is incomplete. It is unknown if the antecedents of preeclampsia were present in women long before pregnancy, emphasizing the importance of a comprehensive history in these women. It is possible that these women had subclinical risk factors, such as increased peripheral vascular resistance, childhood obesity or a strong family history of vascular disease, which predisposed them to developing preeclampsia and then subsequently overt cardiovascular disease later in life [3]. Alternatively, preeclampsia could be the first "hit" on a phenotype which then becomes susceptible to cardiovascular and metabolic disease later on. In this hypothesis, the direct effects of endothelial dysfunction in pregnancy may have initiated and then accelerated atherosclerosis in these individuals [46]. Further research is required to develop a better understanding of the role preeclampsia plays in life-long disease.

## *Screening for the Development of Cardiovascular Disease*

Post-partum management of women with preeclampsia has evolved over a number of years, and now hypertension during pregnancy is recognized as a major cardiovascular risk factor amongst many national and international associations. As of 2016, the Canadian Cardiovascular Society's Dyslipidemia Guidelines were changed to include screening for all women regardless of age if they had a hypertensive disorder in pregnancy [47]. As of 2019, the UK NICE guidelines made similar recommendations, advising women who have had a hypertensive disorder of pregnancy that this is associated with an increased risk of cardiovascular disease later in life, and advising these women to avoid smoking, and to maintain a healthy lifestyle and weight [5]. Finally, the American Heart Association recommends that these women undergo cardiovascular screening within 3 months after delivery [48].

Currently, there is no clear consensus on when is the optimal time to screen patients for cardiovascular disease after the diagnosis of preeclampsia. The Mother's Clinic in Kingston, Ontario is one of the longest running Canadian post-partum risk factor reduction clinics, whereby women are seen 6 months post-partum and both modifiable (e.g. smoking) and non-modifiable (e.g. total cholesterol) risk factors are assessed. Women are then given a sense of their lifetime cardiovascular disease risk estimate based on 5 different risks factors (total cholesterol, systolic blood pressure, diastolic blood pressure, elevated fasting glucose and smoking) [49]. However, no consensus has yet been reached on the best way to assess and characterize risk in all individuals, nor the optimal timing of assessment/intervention post-partum. Furthermore, many post-partum cardiovascular risk reduction clinics are challenged by high attrition, lack of proven effectiveness, and low patient engagement [50].

## *Long Term Pharmacologic Management:*

The literature exploring long term pharmacologic management of women who have developed preeclampsia is underway but far from robust. Women have long been underrepresented in the cardiovascular literature and only recently has there been concerted efforts to address the specific needs of this population [51]. Currently, the pharmacologic management of women who have experienced hypertensive disorders of pregnancy is focused mainly on early and aggressive treatment of their other cardiovascular risk factors including: diabetes, hypertension, chronic kidney disease and dyslipidemia. Without literature specific to this population of women, these conditions are treated in the usual fashion with the same targets as the general population.

## Statins

The role of statin therapy in the post-partum period is controversial. Current recommendations support screening at-risk women with a lipid panel in late post-partum period and focus primarily on lifestyle modifications to optimize lipid profile with the decision to start a statin guided by cardiovascular age estimates [52]. Though there has been some early work using pravastatin in an animal model, demonstrating improved cardiac remodelling and cardiac output post-partum [53], much more research in this area is required to understand the role that statins might play in the long term management of women who develop preeclampsia.

## Acetylsalicylic Acid and Angiotensin Converting Enzyme Inhibitors

The use of ASA and angiotensin converting enzyme inhibitors in the post-partum period, after pregnancy complicated by hypertensive disorder have been evaluated. In one small placebo controlled trial, women who took ASA for 2 months after pregnancy complicated by preeclampsia, had increased arterial flow mediated dilation indicating an improvement in endothelial function [54]. Another small feasibility randomized controlled trial showed improved diastolic dysfunction and left ventricular remodelling in women with a history of preeclampsia who took enalapril for 6 months post-partum [55]. Both studies had small sample sizes and were testing feasibility only. Larger studies are needed to determine if these interventions impact long term cardiovascular risk.

With an increasing interest and awareness around the cardiovascular health of women, we can expect more robust research in this area of medicine in the coming years. This will help to formalize and guide the management of women who have experienced preeclampsia and hypertensive disorders of pregnancy, in order to optimize their health and reduce their long-term cardiovascular risk.

## Summary

Preeclampsia remains one of the leading causes of adverse maternal and fetal pregnancy outcomes. It complicates millions of pregnancies globally, however, with a growing understanding of risk factors, improved screening programs and interventions for prevention, health care providers have tools to help optimize pregnancy outcomes for both mothers and their offspring. Women who have developed preeclampsia are at a significantly increased risk of adverse multisystem outcomes and premature death after pregnancy. Though the mechanisms for the increased burden of disease are not entirely well understood, it is imperative for healthcare providers and patients to recognize development and history of preeclampsia as a maker of increased lifelong risk. Continued research is required to evaluate strategies to reduce the increased lifelong risk of adverse events in these women.

## References

1. Dall'Asta A, D'Antonio F, Saccone G, Buca D, Mastantuoni E, Liberati M et al (2021) Cardiovascular events following pregnancy complicated by pre-eclampsia with emphasis on comparison between early-and late-onset forms: systematic review and meta-analysis. Ultrasound Obstet Gynecol 57(5):698–709
2. Witcher PM (2018) Preeclampsia: acute complications and management priorities. AACN Adv Crit Care 29(3):316–326
3. Ives CW, Sinkey R, Rajapreyar I, Tita ATN, Oparil S (2020) Preeclampsia-pathophysiology and clinical presentations: JACC state-of-the-art review. J Am Coll Cardiol 76(14):1690–1702
4. Magee LA, Brown MA, Hall DR, Gupte S, Hennessy A, Karumanchi SA et al (2022) The 2021 international society for the study of hypertension in pregnancy classification, diagnosis & management recommendations for international practice. Pregnancy Hypertens 27:148–169
5. Hypertension in pregnancy: diagnosis and management (2019) London: National Institute for Health and Care Excellence (NICE)
6. Poon LC, Shennan A, Hyett JA, Kapur A, Hadar E, Divakar H et al (2019) The international federation of Gynecology and obstetrics (FIGO) initiative on pre-eclampsia: a pragmatic guide for first-trimester screening and prevention. Int J Gynaecol Obstet 145(Suppl 1):1–33
7. Gestational hypertension and preeclampsia: ACOG practice bulletin, number 222 (2020). Obstet Gynecol 135(6):e237–e60
8. Magee LA, Smith GN, Bloch C, Cote AM, Jain V, Nerenberg K et al (2022) Guideline No. 426: hypertensive disorders of pregnancy: diagnosis, prediction, prevention, and management. J Obstet Gynaecol Can 44(5):547–71 e1
9. Sircar M, Thadhani R, Karumanchi SA (2015) Pathogenesis of preeclampsia. Curr Opin Nephrol Hypertens 24(2):131–138
10. El-Sayed AAF (2017) Preeclampsia: A review of the pathogenesis and possible management strategies based on its pathophysiological derangements. Taiwan J Obstet Gynecol 56(5):593–598
11. Magee LA, Nicolaides KH, von Dadelszen P (2022) Preeclampsia. N Engl J Med 386(19):1817–1832
12. Steegers EA, von Dadelszen P, Duvekot JJ, Pijnenborg R (2010) Pre-eclampsia. Lancet 376(9741):631–644
13. Abalos E, Cuesta C, Grosso AL, Chou D, Say L (2013) Global and regional estimates of preeclampsia and eclampsia: a systematic review. Eur J Obstet Gynecol Reprod Biol 170(1):1–7
14. Say L, Chou D, Gemmill A, Tuncalp O, Moller AB, Daniels J et al (2014) Global causes of maternal death: a WHO systematic analysis. Lancet Glob Health 2(6):e323–e333
15. Gasem T, Al Jama FE, Burshaid S, Rahman J, Al Suleiman SA, Rahman MS (2009) Maternal and fetal outcome of pregnancy complicated by HELLP syndrome. J Matern Fetal Neonatal Med 22(12):1140–1143
16. Li X, Zhang W, Lin J, Liu H, Yang Z, Teng Y et al (2018) Risk factors for adverse maternal and perinatal outcomes in women with preeclampsia: analysis of 1396 cases. J Clin Hypertens (Greenwich) 20(6):1049–1057
17. Tuffnell DJ, Jankowicz D, Lindow SW, Lyons G, Mason GC, Russell IF et al (2005) Outcomes of severe pre-eclampsia/eclampsia in Yorkshire 1999/2003. BJOG 112(7):875–880
18. Bilano VL, Ota E, Ganchimeg T, Mori R, Souza JP (2014) Risk factors of pre-eclampsia/eclampsia and its adverse outcomes in low- and middle-income countries: a WHO secondary analysis. PLoS ONE 9(3):e91198

19. Leon LJ, McCarthy FP, Direk K, Gonzalez-Izquierdo A, Prieto-Merino D, Casas JP et al (2019) Preeclampsia and cardiovascular disease in a large UK pregnancy cohort of linked electronic health records: a CALIBER study. Circulation 140(13):1050–1060
20. Liu A, Wen SW, Bottomley J, Walker MC, Smith G (2009) Utilization of health care services of pregnant women complicated by preeclampsia in Ontario. Hypertens Pregnancy 28(1):76–84
21. Fox A, McHugh S, Browne J, Kenny LC, Fitzgerald A, Khashan AS et al (2017) Estimating the cost of preeclampsia in the healthcare system: cross-sectional study using data from SCOPE study (screening for pregnancy end points). Hypertension 70(6):1243–1249
22. Stevens W, Shih T, Incerti D, Ton TGN, Lee HC, Peneva D et al (2017) Short-term costs of preeclampsia to the United States health care system. Am J Obstet Gynecol 217(3):237–248 e16
23. Brown MA, Magee LA, Kenny LC, Karumanchi SA, McCarthy FP, Saito S et al (2018) Hypertensive disorders of pregnancy: ISSHP classification, diagnosis, and management recommendations for international practice. Hypertens 72(1):24–43
24. Force USPST, Davidson KW, Barry MJ, Mangione CM, Cabana M, Caughey AB et al (2021) Aspirin use to prevent preeclampsia and related morbidity and mortality: US preventive services task force recommendation statement. JAMA 326(12):1186–1191
25. Rolnik DL, Wright D, Poon LC, O'Gorman N, Syngelaki A, de Paco MC et al (2017) Aspirin versus placebo in pregnancies at high risk for preterm preeclampsia. N Engl J Med 377(7):613–622
26. Roberge S, Bujold E, Nicolaides KH (2018) Aspirin for the prevention of preterm and term preeclampsia: systematic review and metaanalysis. Am J Obstet Gynecol 218(3):287–293 e1
27. Davenport MH, Ruchat SM, Poitras VJ, Jaramillo Garcia A, Gray CE, Barrowman N et al (2018) Prenatal exercise for the prevention of gestational diabetes mellitus and hypertensive disorders of pregnancy: a systematic review and meta-analysis. Br J Sports Med 52(21):1367–1375
28. Organization WH (2013) Guideline: calcium supplementation in pregnant women: World Health Organization
29. Hofmeyr GJ, Lawrie TA, Atallah AN, Torloni MR (2018) Calcium supplementation during pregnancy for preventing hypertensive disorders and related problems. Cochrane Database Syst Rev 10:CD001059
30. Woo Kinshella ML, Sarr C, Sandhu A, Bone JN, Vidler M, Moore SE, Elango R, Cormick G, Belizan JM, Hofmeyr GJ, Magee LA, von Dadelszen P (2022) PRECISE Network. Calcium for pre-eclampsia prevention: a systematic review and network meta-analysis to guide personalised antenatal care. BJOG. 129(11):1833–1843
31. Skoll A, Boutin A, Bujold E, Burrows J, Crane J, Geary M et al (2018) No. 364-antenatal corticosteroid therapy for improving neonatal outcomes. J Obstet Gynaecol Can 40(9):1219–1239
32. Mottola MF, Davenport MH, Ruchat SM, Davies GA, Poitras VJ, Gray CE et al (2018) 2019 Canadian guideline for physical activity throughout pregnancy. Br J Sports Med 52(21):1339–1346
33. Al-Safi Z, Imudia AN, Filetti LC, Hobson DT, Bahado-Singh RO, Awonuga AO (2011) Delayed postpartum preeclampsia and eclampsia: demographics, clinical course, and complications. Obstet Gynecol 118(5):1102–1107
34. James AH, Bushnell CD, Jamison MG, Myers ER (2005) Incidence and risk factors for stroke in pregnancy and the puerperium. Obstet Gynecol 106(3):509–516
35. Chames MC, Livingston JC, Ivester TS, Barton JR, Sibai BM (2002) Late postpartum eclampsia: a preventable disease? Am J Obstet Gynecol 186(6):1174–1177
36. Matthys LA, Coppage KH, Lambers DS, Barton JR, Sibai BM (2004) Delayed postpartum preeclampsia: an experience of 151 cases. Am J Obstet Gynecol 190(5):1464–1466
37. Rabi DM, McBrien KA, Sapir-Pichhadze R, Nakhla M, Ahmed SB, Dumanski SM et al (2020) Hypertension Canada's 2020 comprehensive guidelines for the prevention, diagnosis, risk assessment, and treatment of hypertension in adults and children. Can J Cardiol 36(5):596–624
38. van Oostwaard MF, Langenveld J, Schuit E, Papatsonis DN, Brown MA, Byaruhanga RN et al (2015) Recurrence of hypertensive disorders of pregnancy: an individual patient data metaanalysis. Am J Obstet Gynecol 212(5):624 e1–17

39. Epstein FH (1964) Late vascular effects of toxemia of pregnancy. N Engl J Med 271:391–395
40. Ray JG, Vermeulen MJ, Schull MJ, Redelmeier DA (2005) Cardiovascular health after maternal placental syndromes (CHAMPS): population-based retrospective cohort study. Lancet 366(9499):1797–1803
41. Coutinho T, Lamai O, Nerenberg K (2018) Hypertensive disorders of pregnancy and cardiovascular diseases: current knowledge and future directions. Curr Treat Options Cardiovasc Med 20(7):56
42. Veerbeek JH, Hermes W, Breimer AY, van Rijn BB, Koenen SV, Mol BW et al (2015) Cardiovascular disease risk factors after early-onset preeclampsia, late-onset preeclampsia, and pregnancy-induced hypertension. Hypertens 65(3):600–606
43. Hermes W, Franx A, van Pampus MG, Bloemenkamp KW, Bots ML, van der Post JA et al (2013) Cardiovascular risk factors in women who had hypertensive disorders late in pregnancy: a cohort study. Am J Obstet Gynecol 208(6):474 e1–8
44. Brouwers L, van der Meiden-van Roest AJ, Savelkoul C, Vogelvang TE, Lely AT, Franx A et al (2018) Recurrence of pre-eclampsia and the risk of future hypertension and cardiovascular disease: a systematic review and meta-analysis. BJOG 125(13):1642–1654
45. Ray JG, Schull MJ, Kingdom JC, Vermeulen MJ (2012) Heart failure and dysrhythmias after maternal placental syndromes: HAD MPS Study. Heart 98(15):1136–1141
46. Rana S, Lemoine E, Granger JP, Karumanchi SA (2019) Preeclampsia: pathophysiology, challenges, and perspectives. Circ Res 124(7):1094–1112
47. Anderson TJ, Gregoire J, Pearson GJ, Barry AR, Couture P, Dawes M et al (2016) 2016 Canadian cardiovascular society guidelines for the management of Dyslipidemia for the prevention of cardiovascular disease in the adult. Can J Cardiol 32(11):1263–1282
48. Cho L, Davis M, Elgendy I, Epps K, Lindley KJ, Mehta PK et al (2020) Summary of updated recommendations for primary prevention of cardiovascular disease in women: JACC state-of-the-art review. J Am Coll Cardiol 75(20):2602–2618
49. Lloyd-Jones DM, Leip EP, Larson MG, D'Agostino RB, Beiser A, Wilson PW et al (2006) Prediction of lifetime risk for cardiovascular disease by risk factor burden at 50 years of age. Circ 113(6):791–798
50. Sia WW, Montgomery-Fajic E, Germaine D, Wilkie J, Khurana R, Marnoch C et al (2012) OS106. The postpartum preeclampsia clinic (PPPEC)—an interdisciplinary clinic for cardiovascular risk reduction for women with preeclampsia. Pregnancy Hypertens 2(3):237
51. Melloni C, Berger JS, Wang TY, Gunes F, Stebbins A, Pieper KS et al (2010) Representation of women in randomized clinical trials of cardiovascular disease prevention. Circ Cardiovasc Qual Outcomes 3(2):135–142
52. Pearson GJ, Thanassoulis G, Anderson TJ, Barry AR, Couture P, Dayan N et al (2021) 2021 Canadian cardiovascular society guidelines for the management of Dyslipidemia for the prevention of cardiovascular disease in adults. Can J Cardiol 37(8):1129–1150
53. Kraker K, O'Driscoll JM, Schutte T, Herse F, Patey O, Golic M et al (2020) Statins reverse postpartum cardiovascular dysfunction in a rat model of preeclampsia. Hypertens 75(1):202–210
54. Hashemi M, Baktash F, Heshmat-Ghahdarijani K, Zarean E, Bahrani S (2016) Evaluation the effect of low-dose aspirin on endothelial dysfunction in preeclamptic patients. J Res Med Sci 21:131
55. Ormesher L, Higson S, Luckie M, Roberts SA, Glossop H, Trafford A et al (2020) Postnatal Enalapril to improve cardiovascular function following preterm preeclampsia (PICk-UP): a randomized double-blind placebo-controlled feasibility trial. Hypertens 76(6):1828–1837

# Chapter 7
# Myocardial Infarction with Non-obstructive Coronary Artery Disease: An Overview of Definition, Classification, Diagnostic Workup and Management

**Laura Albak, Arjun K. Gupta, Mahwash Saeed, and Shuangbo Liu**

**Abstract** Myocardial infarction in the absence of obstructive coronary artery disease (MINOCA) represents 5–10% of patients presenting with acute myocardial infarction. MINOCA is a working diagnosis and coronary angiography, intravascular imaging and cardiac magnetic resonance imaging are important diagnostic tests. MINOCA is not a benign condition and can be associated with major adverse cardiac events. Therefore, diagnosis of the underlying etiology of MINOCA is imperative and impacts treatment and prognosis. The objective of this chapter is to provide a definition of MINOCA, examine the epidemiology, provide a diagnostic schema, review the pathophysiologic mechanisms of various causes of MINOCA and evaluate the prognosis of these patients. Several case examples are utilized to illustrate different causes and management of MINOCA.

**Keywords** MINOCA · Acute myocardial infarction · Non-ST-segment-elevation · Coronary artery disease (CAD) · Electrocardiogram

## Introduction

Myocardial infarction in the absence of obstructive coronary artery disease (MINOCA) was first described in the 1930s by Gross et al. after identifying extensive myocardial infarction (MI) in patients with patent coronary arteries [1]. Since that time, several landmark angiographic studies have found that 5–10% of patients presenting with acute myocardial infarction (AMI) had nonobstructive coronary artery disease [2]. This finding is supported by a contemporary meta-analysis

L. Albak · A. K. Gupta · M. Saeed · S. Liu (✉)
Section of Cardiology, Department of Internal Medicine, Rady Faculty of Health Sciences, Max Rady College of Medicine, University of Manitoba, Winnipeg, MB, Canada
e-mail: sliu@sbgh.mb.ca

© The Author(s), under exclusive license to Springer Nature Switzerland AG 2023 
L. Kirshenbaum and I. Rabinovich-Nikitin (eds.), *Biology of Women's Heart Health*, Advances in Biochemistry in Health and Disease 26,
https://doi.org/10.1007/978-3-031-39928-2_7

reporting an overall prevalence of MINOCA to be 6% in patients presenting with AMI [3]. MINOCA is a 'working diagnosis' comprised of a heterogenous group of underlying conditions, often divided into coronary and noncoronary mechanisms [4]. As such, a diagnosis of MINOCA should prompt further evaluation to identify its underlying pathophysiology [5].

MINOCA is often underdiagnosed, however, recognition and accurate diagnosis of MINOCA and the underlying cause is important for treatment and prognosis. The diagnostic work up can include multiple imaging modalities, beginning with coronary angiogram and extending to cardiac magnetic resonance [4]. Based on the ACTION (Acute Coronary Treatment and Intervention Outcomes Network) registry, MINOCA is not benign as the frequency of in-hospital major adverse cardiovascular events (MACE) in MINOCA patients is 4.9% [6]. As highlighted in the ACUITY trial, while patients presenting with non-ST-segment-elevation MI without obstructive coronary artery disease (CAD) have low rates of subsequent MI (3.6%, hazard ratio (HR) 0.35, 95% confidence interval (CI) 0.12–1.98) with no unplanned revascularization, overall 1-year mortality (5.2%, HR 3.44, 95%CI 1.05–11.28) is still significant [7]. In contrast to patients who present with AMI from obstructive CAD, treatment of MINOCA often does not require revascularization, and management should target the underlying cause [8].

The aim of this chapter is to define MINOCA and describe the patient population. We will also provide a diagnostic schematic, identify causes, and appropriate management strategies of this important diagnosis.

## Definition

The diagnostic criterion for MINOCA was first outlined in 2017 by the European Society of Cardiology (ESC). A diagnosis of MINOCA requires: (1) AMI defined by the Third Universal Definition of MI [9], (2) non-obstructive coronary arteries on angiography (<50% stenosis), and (3) no obvious alternative diagnosis for the acute presentation [5]. AMI defined by the Third Universal Definition is a detection of rise and/or fall of cardiac biomarker values (preferably cardiac troponin) with at least one value above the 99th percentile upper reference limit and at least one of the following: (1) symptoms of ischemia, (2) new significant ST-segment or T wave changes or new left bundle branch block, (3) pathological Q waves on electrocardiogram (ECG), (4) imaging of new loss of viable myocardium or new regional wall motion abnormality, or (5) intracoronary thrombus identified by angiography or autopsy [9].

As the ESC criteria is based on the Third Universal Definition, the diagnosis of AMI is dependent on the elevation of a cardiac biomarker such as troponin [9]. However, an elevated troponin can result from myocardial injury from ischemic and nonischemic mechanisms, and therefore is not disease specific. This lack of specificity prompted the Fourth Universal Definition of Myocardial Infarction, requiring that MI should be due to ischemic mechanisms [10]. Similarly, in 2019, the American Heart Association (AHA) updated the definition of MINOCA, reserving its use for patients with an ischemic basis for their clinical presentation [11]. It is critical to exclude (1) clinically overt causes for elevated troponin (e.g., sepsis, pulmonary embolism), (2) clinically missed obstructive disease (e.g., complete occlusion of a small coronary artery subsegment from plaque disruption or embolism or an unnoticed >50% distal stenosis of a coronary artery), and (3) myocyte injury secondary to subtle non-ischemic mechanisms (e.g., myocarditis) [11]. This statement was supported by ESC in 2020, with consensus to exclude myocarditis and Takotsubo syndrome from a final diagnosis of MINOCA [12].

## Epidemiology

A meta-analysis of current literature comprising of 176,502 patients estimated that the prevalence of MINOCA ranges from 1–14% of patients presenting with AMI [3]. MINOCA occurs more commonly in females than males (10.5% versus 3.4%; $p < 0.0001$) as shown in large AMI registeries [6]. The mean age of MINOCA patients is 55 years (95%CI 51–59 years) [3]. Studies have also shown that patients with MINOCA tend to be younger than MI-CAD patients (males 63 versus 66 years and females 54 versus 59 years, $p < 0.001$) [6]. In addition, females with MINOCA tend to be younger than males with MINOCA (median 54 versus 63 years, $p < 0.001$).

Traditional CAD risk factors occur less frequently in patients with MINOCA when compared to patients with MI-CAD. Patients were less likely to be diabetic (16.8% versus 29.9%, $p < 0.001$), hypertensive (55.2% versus 61.4%, $p < 0.001$) and dyslipidemic (36.1% versus 61.4%, $p < 0.001$) [13]. However, 75% of MINOCA patients had a common cardiac risk factor [6]. Conversely, patients with MINOCA had higher rates of prior heart failure (8.6% versus 4.9%), atrial fibrillation or flutter (8.2% versus 4.1%), or chronic lung disease (14.5% versus 9.2%) [6] compared to MI-CAD.

The clinical presentation of MINOCA is indistinguishable from AMI secondary to acute atherosclerotic plaque rupture, with similar symptoms and signs [5]. Chest pain is the most common presenting symptom for both MI-CAD and MINOCA patients (87.3% versus 86.3%; $p = 0.63$) [14]. However, MINOCA is more common in patients presenting with non-ST-elevation myocardial infarction (NSTEMI) than ST-elevation myocardial infarction (STEMI, 8.9% versus 2.2%; $p < 0.0001$) [6].

## Diagnostics

Figure 7.1 provides a clinical flowchart for the diagnosis of MINOCA. The initial diagnostic evaluation in patients with suspected AMI is coronary angiogram. If nonobstructive CAD (<50% stenosis) is identified, consideration of the clinical context and exclusion of other causes of myocardial injury is necessary [11]. If AMI remains the diagnosis of exclusion (step 1), review of the coronary angiogram

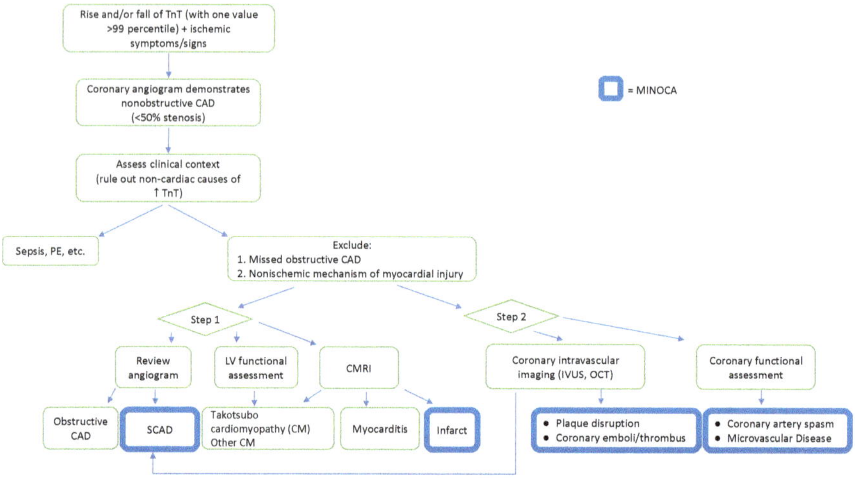


**Fig. 7.1** Diagnostic flow chart of workup of myocardial infarction in the absence of obstructive coronary artery disease (MINOCA). (TnT = Troponin; CAD = coronary artery disease; PE = pulmonary embolism; CMRI = cardiac magnetic resonance imaging; IVUS = intravascular ultrasound; OCT = optical coherence tomography; SCAD = spontaneous coronary artery dissection)

is recommended to identify any overlooked coronary lesions (including obstructive CAD and spontaneous coronary artery dissection). Left ventricular (LV) functional assessment (echocardiography or left ventricular angiogram) is helpful to delineate non-coronary and non-ischemic mechanisms of myocardial injury such as Takotsubo syndrome or other cardiomyopathies [11]. Cardiac magnetic resonance imaging (CMR) is a key investigation in MINOCA. Although not always readily available, it can assess for myocarditis, Takotsubo syndrome, and other cardiomyopathies, in addition to confirming AMI [11].

Further investigations (step 2), including coronary intravascular imaging and functional assessment, may be required to identify the underlying cause of MINOCA. Coronary intravascular imaging consists of optical coherence tomography (OCT) and intravascular ultrasound (IVUS), both of which are helpful for identifying plaque disruption, coronary dissection, and thromboembolism. OCT is a light wave based imaging technique which creates cross-sectional images of tissue with high resolution [15]. It provides more detailed coronary plaque morphology, such as plaque rupture, erosion, and calcified nodules compared to coronary angiography [16]. IVUS is an important adjunct during coronary angiography as it directly images the vessel wall and aids in the characterization of plaque composition, distribution, morphology and extent [17]. Lastly, coronary functional assessments may be needed if the above tests do not reveal a cause. Functional assessment includes provocative testing with acetylcholine (Ach) to help diagnose coronary vasospasm and coronary microvascular dysfunction [18, 19].

## Pathophysiologic Mechanisms

MINOCA should be considered a working diagnosis that requires evaluation of the underlying cause so that treatment can be directed as appropriate. Both atherosclerotic and nonatherosclerotic mechanisms can lead to MINOCA. The pathophysiology, diagnosis and management of these entities are described below.

### *Atherosclerotic*

*Plaque disruption* can occur at a site of non-obstructive atherosclerosis. Plaque rupture, erosion and calcific nodules can trigger thrombus formation leading to AMI via distal embolization, superimposed coronary spasm or transient complete thrombosis with spontaneous thrombolysis [11]. Using intravascular ultrasound (IVUS), plaque rupture and plaque erosion has been identified in over 40% of patients with MINOCA [20, 21]. CMR can demonstrate myocardial edema with or without necrosis, which would suggest transiently compromised flow in a larger vessel. A small, well-defined area of late gadolinium enhancement (LGE) on CMR may suggest distal embolization as the most likely mechanism [20].

ESC and AHA guidelines recommend medical management over percutaneous coronary intervention (PCI) with dual antiplatelet therapy for one year, followed by lifetime single antiplatelet therapy for suspected or confirmed plaque disruption [5, 11]. This is supported by acceptable 1-year revascularization rate of 5.7% demonstrated in a small study evaluating patients with MINOCA secondary to plaque erosion who received medical management only [22]. As this mechanism of injury occurs in the setting of atherosclerotic disease, statin therapy is also recommended [23].

### *Nonatherosclerotic*

Nonatherosclerotic mechanisms of MINOCA include coronary artery dissection, vasospasm, microvascular dysfunction, and thromboembolism.

#### Spontaneous Coronary Artery Dissection (SCAD)

*Spontaneous coronary artery dissection* causes myocardial injury by intimal disruption leading to a false lumen or intramural hematoma, causing compression of the true lumen resulting in coronary insufficiency [24]. SCAD is not associated with atherosclerosis or trauma and is not iatrogenic in nature [24]. It is more common in females and is responsible for up to 25% of all AMI cases in females under 50 years of age [25]. Predisposing conditions include fibromuscular dysplasia, hormonal therapy and peripartum state [26].

Coronary angiography is considered first-line for imaging modality. IVUS and OCT can aid in the diagnosis and both are considered safe with very low complication rates [17, 27].

Conservative management is recommended as PCI may cause propagation of the dissection [28]. Medical management includes single antiplatelet agent and beta-blockers [24]. When comparing single versus dual antiplatelet (DAPT) regiments, risk of MACE is higher in DAPT group at 12-month follow up (18.9% versus 6.0%; HR 2.62, 95%CI 1.22–5.61; $p = 0.013$) [29]. However, the conservative approach may not be appropriate in high-risk patients with ongoing ischemia, left main artery dissection or hemodynamic instability. Urgent PCI (with balloon angioplasty or cutting balloon) or coronary artery bypass grafting (CABG) can be considered based on the expertise of the operators and stability of the patient [24].

It is recommended to monitor patients in-hospital for a minimum of 48 h for medication adjustments and evidence of complications [24]. Risk of dissection extension and new recurrent SCAD is low (5–10%), however it is important to detect early as it can lead to further ischemia [28, 30]. Repeat angiography is warranted if there is evidence of ischemia clinically, electrocardiographically, or presence of significant arrhythmia. Emergency revascularization may be undertaken to relieve ischemia if feasible [24].

## Epicardial Coronary Vasospasm

*Epicardial coronary vasospasm* is defined as transient total or subtotal coronary artery occlusion (>90% constriction) with angina and ischemic ECG changes that occur spontaneously or in response to provocative stimuli [18]. This transient occlusion can result in MI [31], sudden cardiac death [32], and/or arrhythmia [33]. Most patients also have a component of endothelial dysfunction [34]. Additional mechanisms of vasospasm include smooth muscle hyperreactivity [35] in response to drugs or toxins (e.g., cocaine, fluorouracil) [11] and autonomic dysfunction [36]. Cigarette smoking has also been identified as a preventable risk factor of coronary spasm [37].

Coronary angiography may help confirm the diagnosis if there is >75% coronary artery diameter reduction that is transient and reversible [4]. Provocative testing is often required to confirm the diagnosis. This involves administering a stimulus (ACh or ergonovine) during coronary angiography and monitoring for chest pain, ischemic ECG changes and >90% vasoconstriction on angiography [18].

Management involves avoiding known triggers, controlling cardiovascular risk factors to promote vascular health, and pharmacological therapy with calcium channel blockers (CCB) and nitrates [4]. Vasodilation with CCB has been associated with improved long-term survival compared to no treatment with these drugs [38], whereas nitrates are used for symptom control. Statins have also been shown to improve clinical outcomes due to inhibition of vascular smooth muscle contraction [39].

## Coronary Microvascular Dysfunction

*Coronary microvascular dysfunction (CMD)* is defined as abnormal microvascular resistance that is clinically evident as inappropriate coronary blood flow response, impaired myocardial perfusion and/or myocardial ischemia that cannot be accounted for by abnormalities in the epicardial coronary arteries [40]. The coronary microvasculature (vessels <0.5 mm diameter) is responsible for >70% of coronary resistance in the absence of obstructive coronary disease [40]. Patients with CMD usually present with stable angina [40], and this clinical entity is responsible for up to 30–50% of patients presenting with chest pain and nonobstructive CAD [19].

CMD is more commonly diagnosed in females [11], and its' risk factors are similar to typical coronary atherosclerotic disease, including hypertension, hypercholesterolemia, diabetes and cigarette smoking [40]. Pathogenesis of CMD is divided into endothelium-dependent or -independent dysfunction [11]. Invasive physiologic testing during coronary angiography can demonstrate impaired coronary flow and the type of dysfunction. Alternatively, CMD can be diagnosed with a combination of (1) non-invasive testing such as positron emission tomography or CMR demonstrating ischemia and (2) the absence of obstructive epicardial CAD [40].

Coronary microvascular dysfunction is characterized by any of (1) impaired coronary flow reserve (CFR) <2.0 in response to vasodilator stimuli such as adenosine; (2) microvascular spasm evidenced by chest pain and/or ECG signs of ischemia but no epicardial spasm after acetylcholine provocation; (3) impaired microvascular blood flow measured by corrected thrombolysis in myocardial infarction (TIMI) frame count >25; and (4) abnormal coronary microvascular resistance indices (index of microvascular resistance >25U) [19].

CFR defines the vasodilator capacity of the coronary vascular tree [41] and is measured during coronary angiography using a Doppler flow wire [42]. Intracoronary acetylcholine provocation testing produces vasodilation in healthy endothelium and vasoconstriction in presence of endothelial dysfunction [43]. Delayed flow of angiographic contrast reflects increased distal coronary resistance, and is measured by the TIMI frame count [19]. Index of microvascular resistance (IMR) quantitatively reflects the coronary microvascular resistance. It is derived from thermodilution technique using a coronary pressure wire with temperature sensor at the tip [41]. Baseline mean transit time is measured after injection of room-temperature normal saline down a coronary artery. Intravenous adenosine is then administered to induce steady state maximal hyperemia, followed by repeat normal saline injections, and the hyperemic mean transit time is measured. IMR is calculated as the distal coronary pressure at maximal hyperemia divided by the inverse of hyperemic mean transit time [44].

Management options are limited in CMD, as conventional antianginal vasodilator drugs are less effective on the microvasculature compared to large epicardial vessels [40]. Antianginal therapies, including CCB and beta-blockers can help alleviate symptoms, in comparison, nitrates may be less effective [40]. Therapies to improve endothelial function (e.g. L-arginine, statin, enalapril), promote microvascular vasodilation (e.g. dipyridamole, ranolazine) or stimulate visceral analgesic effect (imipramine, aminophylline) can also be considered [11].

### Coronary Thromboembolism

*Coronary thromboembolism* causing MINOCA usually involves the microcirculation [45] or partial lysis of an epicardial coronary thrombus resulting in nonobstructive angiographic disease [11]. Thrombosis may result from hereditary or acquired thrombotic disorders, and emboli may occur from coronary or systemic arterial thrombi [45]. Inherited disorders include factor V Leiden, elevated factor VIII/von Willebrand factor [11], and protein S and C deficiencies [45]. In a systematic review of 8 studies that evaluated inherited thrombotic disorders in MINOCA patients, 14% of patients that underwent screening had evidence of an inherited thrombotic disorder [3]. Acquired thrombotic disorders include antiphospholipid syndrome, myeloproliferative disorders, thrombotic thrombocytopenic purpura (TTP) and heparin-induced thrombocytopenia (HIT) [11].

Coronary emboli can also occur in predisposing hypercoagulable states such as atrial fibrillation and valvular heart disease [5]. Non-thrombotic sources may also lead to coronary embolization, including valvular vegetations, cardiac tumors, calcified valves and iatrogenic air emboli [45]. Paradoxical embolism secondary to right-to-left shunts (e.g. patent foramen ovale, atrial septal defect or coronary arteriovenous fistula) is a rare cause of coronary emboli [45].

The prevalence of coronary thromboembolism in MINOCA is low [5], with prevalence of de novo coronary embolism being 2.9% [46, 47]. Often, angiographically small pruned coronary vessels obstructed with thrombi or emboli may not be recognized, aortic valve disease may not be appreciated and thrombophilia disorders may not be screened for [5]. Overall, diagnostic work-up includes laboratory analysis for thrombophilia disorders, echocardiography to assess for intracardiac thrombus, mass, valvular disease and right-left shunts, and coronary angiography with intravascular imaging (IVUS, OCT) [5]. Long-term management is based on underlying etiology, usually involving systemic anticoagulation [4].

## Prognosis

MINOCA is not a benign condition, and prognosis is heterogeneous due to the diverse mechanisms of the underlying causes. In comparison to MI-CAD, most studies demonstrate MINOCA patients have better outcomes [3, 48]. However, this conclusion is not consistent across all studies. For example, in the VIRGO study (Variation in Recovery: Role of Gender on Outcomes of Young AMI Patients), in comparison to MI-CAD patients, MINOCA patients had similar 1-month and 1-year mortality rate (1-month: 1.7% versus 1.1% [$p = 0.43$]; 12-month: 2.3% versus

0.6% [$p = 0.68$], respectively) and similar quality-of-life measures [14]. The Korean Infarct Registry also demonstrated that MINOCA patients and MI-CAD patients with single- or double-vessel angiographic disease had similar risk of MACE at 12 months (7.8% versus 12.2%, $p = 0.359$) [49]. Additionally, risk of recurrent events (incidence of mortality or non-fatal MI) in MINOCA patients is significantly higher than the general population without cardiovascular disease (4.6% versus 2.2%, $p < 0.001$) [50].

A large meta-analysis demonstrated a pooled in-hospital mortality rate of 0.9% (95%CI, 0.5–1.3%) and a 12 month mortality rate of 4.7% (95%CI, 2.6–6.9%) [3] in MINOCA patients. In comparison to MI-CAD patients (in-hospital mortality rate: 3.2%, 95%CI, 1.8–4.6% and 12-month mortality rate: 6.7%, 95%CI, 4.3–9.0%), mortality rates were lower in MINOCA patients, however the findings are still concerning and demonstrate the gravity of this condition. In the SWEDEHEART study, which examined long-term outcomes in MINOCA patients at a mean of 4.1 years follow up, rate of mortality was 13.4% with 7.1% experiencing another AMI, 4.3% had an ischemic stroke, 6.4% were hospitalized for heart failure and 3.6% were hospitalized for bleeding [51].

## Conclusion

MINOCA is an important clinical entity that has diverse pathophysiological causes. Prompt recognition by healthcare professionals is paramount to provide optimal evaluation and diagnosis. Appropriate management and prognosis are based on the underlying mechanism of the syndrome.

## Case Examples

Case 1:

A 45-year-old female presented to the emergency department with an acute onset of chest pain. She had no relevant past medical history or cardiac risk factors. She described her pain as a heavy retrosternal pressure, which was 7/10 severity. The pain radiated to her back and was associated with diaphoresis, nausea, and emesis. The onset of pain had occurred while showering. She noted recent increase in personal stress.

She was hemodynamically stable on presentation, and a 12-lead ECG demonstrated lateral ST-segment depression and T-wave inversions. Her serum troponin value was elevated, and she was thus diagnosed with a NSTEMI. Treatment was initiated with dual antiplatelet therapy, anticoagulation with low-molecular weight heparin and high-intensity statin.

She was referred to the cardiac catheterization lab (CCL). The LV ejection fraction was mildly reduced at 59%, with visualized hypokinesis of an area of distal anterolateral wall and apex. Coronary angiogram demonstrated mild narrowing of the entire length of the left anterior descending artery (LAD), save for a prominent segment of severe stenosis in the mid portion, along with moderate narrowing of the origin of the second diagonal branch. Otherwise, coronary arteries were angiographically normal. Given the finding of severe single vessel coronary artery disease, intracoronary OCT imaging of the LAD artery was performed, which confirmed a double lumen with intramural hematoma surrounding the true lumen, consistent with SCAD, with no evidence of atherosclerosis (Fig. 7.2). An abdominal aortogram was performed for non-selective renal angiography, which revealed no evidence of renal fibromuscular dysplasia (FMD).

She was admitted to the hospital for medical management. She did not have further episodes of chest pain or arrhythmias. Her subsequent management included aspirin and treatment with beta-blockers. The patient was referred for follow-up with the SCAD clinic and further imaging including computerized tomography (CT) scan of Circle of Willis and CT angiography to assess for iliofemoral FMD.

Case 2:

A 47-year-old male with past medical history of ischemic heart disease with a previous inferoposterior STEMI treated with right coronary artery (RCA) percutaneous coronary intervention (PCI), hypertension, dyslipidemia, and active tobacco smoking, presented to a rural emergency department with acute onset retrosternal chest pain while playing baseball. He described his pain as similar to his previous acute myocardial infarction, 7/10 severity, and associated with diaphoresis.

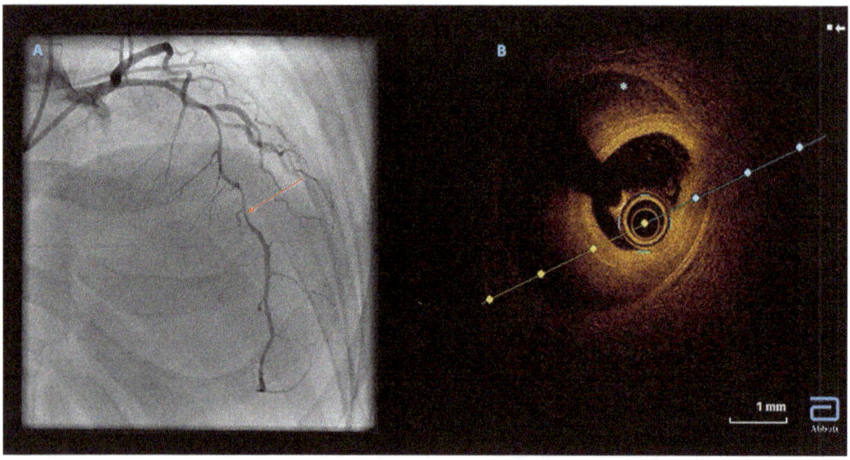


**Fig. 7.2** Panel A demonstrates the left anterior descending artery with diffuse luminal narrowing (marked by red arrow) consistent with type 2 SCAD (Panel A). Panel B is the OCT of the left anterior descending artery with an intramural hematoma (denoted by *) and compression of the true lumen

He was hemodynamically stable on presentation. A 12 lead ECG demonstrated ST-segment elevation in the inferior leads. The patient was diagnosed with an inferior STEMI. Given the geographic distance from a PCI-capable centre, treatment was initiated with thrombolysis with tenecteplase, dual antiplatelet therapy, low-molecular weight heparin and high-intensity statin.

The patient was then transferred to nearest coronary interventional centre for facilitated PCI. Coronary angiogram demonstrated widely patent RCA drug eluting stent (DES), a moderate to severe mid left circumflex (LCx) lesion that did not appear acute in nature, and mild atherosclerotic disease elsewhere. Intracoronary OCT imaging was used to further evaluate the LCx artery. This demonstrated lipidic plaque with plaque rupture and thrombus, therefore identifying the mid-LCx as the culprit vessel (Fig. 7.3). PCI with intracoronary OCT guidance was performed with deployment of one DES. Subsequent medical management included dual antiplatelet therapy with aspirin and ticagrelor, high-intensity statin, beta-blocker and angiotensin II receptor blocker treatment.

Case 3:

A 60-year-old female with past medical history of diabetes mellitus, hypertension, pancreatitis, remote follicular lymphoma and active tobacco smoking, presented to emergency department with acute epigastric pain and generalized weakness. She had recent endoscopic ultrasound-guided cystogastrostomy for pancreatic pseudocyst. Since the procedure, she had ongoing epigastric pain, nausea, poor oral intake and diarrhea.

She was hemodynamically stable on presentation. Her serum troponin was elevated and 12 Lead ECG demonstrated significant ST-segment elevation in the anteroseptal leads. She was diagnosed with a STEMI. Treatment was initiated with

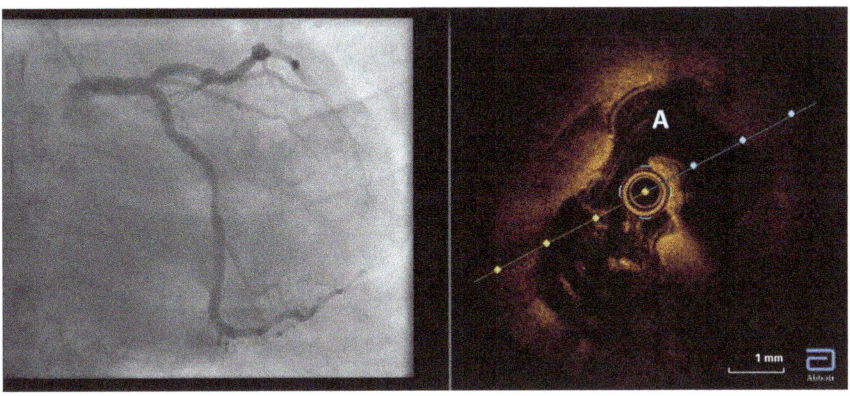


**Fig. 7.3** The image on the left demonstrates the left circumflex artery upon re-presentation with inferior STEMI with moderate mid-vessel disease. The image on the right is the OCT demonstrating plaque rupture and associated hemorrhage (denoted by 'A')

dual antiplatelet therapy, anticoagulation with low-molecular weight heparin, and high-intensity statin.

She was referred urgently to the CCL. Coronary angiogram demonstrated minor luminal irregularities with no flow-limiting lesions. Left ventriculogram was performed demonstrating an ejection fraction of 30%. Regional wall motion abnormalities were identified with hyperkinetic anterobasal and posterobasal wall segments, akinetic anterolateral and diaphragmatic wall segments and dyskinetic apex (Fig. 7.4). These findings were consistent with Takotsubo cardiomyopathy.

Post-procedure, the patient was diagnosed with starvation ketosis and managed medically with intravenous fluids and subcutaneous insulin. The patient was referred for follow-up with general cardiology LV reassessment with a transthoracic echocardiogram.

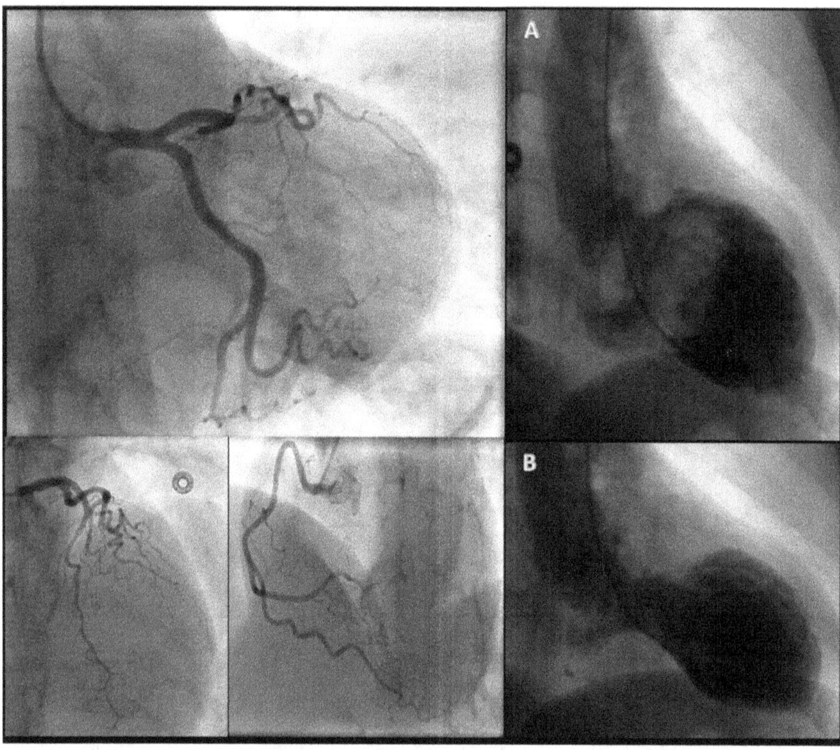


**Fig. 7.4** The panels on the left demonstrate the angiographically normal epicardial coronary arteries. The images on the left demonstrate left ventriculography in diastole (**a**) and systole (**b**). There is basal hyperkinesis with anterior, apical and lateral akinesis consistent with Takotsubo cardiomyopathy

## References

1. Gross H (1939) Myocardial infarction without significant lesions of coronary arteries. Arch Intern Med 64(2):249. https://doi.org/10.1001/archinte.1939.00190020035003
2. DeWood MA, Spores J, Notske R et al (1980) Prevalence of total coronary occlusion during the early hours of transmural myocardial infarction. N Engl J Med 303(16):897–902. https://doi.org/10.1056/NEJM198010163031601
3. Pasupathy S, Air T, Dreyer RP, Tavella R, Beltrame JF (2015) Systematic review of patients presenting with suspected myocardial infarction and nonobstructive coronary arteries. Circ 131(10):861–870. https://doi.org/10.1161/CIRCULATIONAHA.114.011201
4. Cheema AN, Yanagawa B, Verma S, Bagai A, Liu S (2021) Myocardial infarction with nonobstructive coronary artery disease (MINOCA): a review of pathophysiology and management. Curr Opin Cardiol 36(5):589–596. https://doi.org/10.1097/HCO.0000000000000886
5. Agewall S, Beltrame JF, Reynolds HR et al (2017) ESC working group position paper on myocardial infarction with non-obstructive coronary arteries. Eur Heart J 38(3):143–153. https://doi.org/10.1093/eurheartj/ehw149
6. Smilowitz NR, Mahajan AM, Roe MT et al (2017) Mortality of myocardial infarction by sex, age, and obstructive coronary artery disease status in the ACTION registry-GWTG (acute coronary treatment and intervention outcomes network registry-get with the guidelines). Circ Cardiovasc Qual Outcomes 10(12):1–8. https://doi.org/10.1161/CIRCOUTCOMES.116.003443
7. Planer D, Mehran R, Ohman EM et al (2014) Prognosis of patients with non-ST-segment-elevation myocardial infarction and nonobstructive coronary artery disease: propensity-matched analysis from the acute catheterization and urgent intervention triage strategy trial. Circ Cardiovasc Interv 7(3):285–293. https://doi.org/10.1161/CIRCINTERVENTIONS.113.000606
8. Singh T, Chapman AR, Dweck MR, Mills NL, Newby DE (2021) MINOCA: a heterogenous group of conditions associated with myocardial damage. Heart 107(18):1458–1464. https://doi.org/10.1136/heartjnl-2020-318269
9. Vojáček J, Janský P, Janota T (2013) Third universal definition of myocardial infarction. Cor Vasa 55(3):228–235. https://doi.org/10.1016/j.crvasa.2012.12.004
10. Thygesen K, Alpert JS, Jaffe AS et al (2018) Fourth universal definition of myocardial infarction (2018). J Am Coll Cardiol 72(18):2231–2264. https://doi.org/10.1016/j.jacc.2018.08.1038
11. Tamis-Holland JE, Jneid H, Reynolds HR et al (2019) Contemporary diagnosis and management of patients with myocardial infarction in the absence of obstructive coronary artery disease: a scientific statement from the American Heart Association. Circ 139(18):E891–E908. https://doi.org/10.1161/CIR.0000000000000670
12. Collet JP, Thiele H, Barbato E et al (2021) 2020 ESC Guidelines for the management of acute coronary syndromes in patients presenting without persistent ST-segment elevation. Eur Heart J 42(14):1289–1367. https://doi.org/10.1093/eurheartj/ehaa575
13. Pizzi C, Xhyheri B, Costa GM et al (2016) Nonobstructive versus obstructive coronary artery disease in acute coronary syndrome: a meta-analysis. J Am Heart Assoc 5(12):1–14. https://doi.org/10.1161/JAHA.116.004185
14. Safdar B, Spatz ES, Dreyer RP et al (2018) Presentation, clinical profile, and prognosis of young patients with myocardial infarction with nonobstructive coronary arteries (MINOCA): results from the VIRGO study. J Am Heart Assoc 7(13). https://doi.org/10.1161/JAHA.118.009174
15. Aumann S, Donner S, Fischer J (2019) High resolution imaging in microscopy and ophthalmology, pp 59–85. https://doi.org/10.1007/978-3-030-16638-0
16. Cruz Ferreira R, Pereira-Da-Silva T, Patrício L, Bezerra H, Costa M (2016) Coronary optical coherence tomography: a practical overview of current clinical applications. Rev Port Cardiol 35(2):105–112. https://doi.org/10.1016/j.repc.2015.09.016
17. Bangalore S, Bhatt DL (2013) Coronary intravascular ultrasound. Circulation 127(25):868–874. https://doi.org/10.1161/CIRCULATIONAHA.113.003534

18. Beltrame JF, Crea F, Kaski JC et al (2017) International standardization of diagnostic criteria for vasospastic angina. Eur Heart J 38(33):2565–2568. https://doi.org/10.1093/eurheartj/ehv351
19. Ong P, Camici PG, Beltrame JF et al (2018) International standardization of diagnostic criteria for microvascular angina. Int J Cardiol 250:16–20. https://doi.org/10.1016/j.ijcard.2017.08.068
20. Reynolds HR, Srichai MB, Iqbal SN et al (2011) Mechanisms of myocardial infarction in women without angiographically obstructive coronary artery disease. Circ 124(13):1414–1425. https://doi.org/10.1161/CIRCULATIONAHA.111.026542
21. Ouldzein H, Elbaz M, Roncalli J et al (2012) Plaque rupture and morphological characteristics of the culprit lesion in acute coronary syndromes without significant angiographic lesion: analysis by intravascular ultrasound. Ann Cardiol Angeiol (Paris) 61(1):20–26. https://doi.org/10.1016/j.ancard.2011.07.011
22. Xing L, Yamamoto E, Sugiyama T et al (2017) EROSION study (effective anti-thrombotic therapy without stenting: intravascular optical coherence tomography-based management in plaque erosion): a 1-year follow-up report. Circ Cardiovasc Interv 10(12):1–8. https://doi.org/10.1161/CIRCINTERVENTIONS.117.005860
23. Roffi M, Patrono C, Collet JP et al (2016) 2015 ESC Guidelines for the management of acute coronary syndromes in patients presenting without persistent ST-segment elevation: task force for the management of acute coronary syndromes in patients presenting without persistent ST-segment elevation of. Eur Heart J 37(3):267–315. https://doi.org/10.1093/eurheartj/ehv320
24. Hayes SN, Kim ESH, Saw J, Adlam D, Arslanian-Engoren C, Economy KE, Ganesh SK, Gulati R, Lindsay ME, Mieres JH, Naderi S, Shah S, Thaler DE, Tweet MS, Wood MJ (2018) American Heart Association Council on Peripheral Vascular Disease; Council on Clinical Cardiology; Council on Cardiovascular and Stroke Nursing; Council on Genomic and Precision Medicine; and Stroke Council. Spontaneous Coronary Artery Dissection: Current State of the Science: A Scientific Statement From the American Heart Association. Circ 137(19):e523–e557
25. Saw J, Aymong E, Mancini GBJ, Sedlak T, Starovoytov A, Ricci D (2014) Nonatherosclerotic coronary artery disease in young women. Can J Cardiol 30(7):814–819. https://doi.org/10.1016/j.cjca.2014.01.011
26. Saw J, Starovoytov A, Humphries K et al (2019) Canadian spontaneous coronary artery dissection cohort study: in-hospital and 30-day outcomes. Eur Heart J 40(15):1188–1197. https://doi.org/10.1093/eurheartj/ehz007
27. Imola F, Mallus M, Ramazzotti V et al (2010) Safety and feasibility of frequency domain optical coherence tomography to guide decision making in percutaneous coronary intervention. EuroIntervention 6(5):575–581. https://doi.org/10.4244/EIJV6I5A97
28. Tweet MS, Eleid MF, Best PJM et al (2014) Spontaneous coronary artery dissection: revascularization versus conservative therapy. Circ Cardiovasc Interv 7(6):777–786. https://doi.org/10.1161/CIRCINTERVENTIONS.114.001659
29. Cerrato E, Giacobbe F, Quadri G et al (2021) Antiplatelet therapy in patients with conservatively managed spontaneous coronary artery dissection from the multicentre DISCO registry. Eur Heart J 42(33):3161–3171. https://doi.org/10.1093/eurheartj/ehab372
30. Saw J, Aymong E, Sedlak T et al (2014) Spontaneous coronary artery dissection association with predisposing arteriopathies and precipitating stressors and cardiovascular outcomes. Circ Cardiovasc Interv 7(5):645–655. https://doi.org/10.1161/CIRCINTERVENTIONS.114.001760
31. Da Costa A, Isaaz K, Faure E, Mourot S, Cerisier A, Lamaud M (2001) Clinical characteristics, aetiological factors and long-term prognosis of myocardial infarction with an absolutely normal coronary angiogram: a 3-year follow-up study of 91 patients. Eur Heart J 22(16):1459–1465. https://doi.org/10.1053/euhj.2000.2553
32. Lanza GA, Sestito A, Sgueglia GA et al (2007) Current clinical features, diagnostic assessment and prognostic determinants of patients with variant angina. Int J Cardiol 118(1):41–47. https://doi.org/10.1016/j.ijcard.2006.06.016
33. Prinzmetal M, Kennamer R, Merliss R, Wada T, Bor N (1959) Angina pectoris I. A variant form of angina pectoris. Preliminary report. Am J Med 27(3):375–388. https://doi.org/10.1016/0002-9343(59)90003-8

34. Cox ID, Bøtker HE, Bagger JP, Sonne HS, Kristensen BO, Kaski JC (1999) Elevated endothelin concentrations are associated with reduced coronary vasomotor responses in patients with chest pain and normal coronary arteriograms. J Am Coll Cardiol 34(2):455–460. https://doi.org/10.1016/S0735-1097(99)00224-7
35. Okumura K, Yasue H, Matsuyama K et al (1996) Diffuse disorder of coronary artery vasomotility in patients with coronary spastic angina: hyperreactivity to the constrictor effects of acetylcholine and the dilator effects of nitroglycerin. J Am Coll Cardiol 27(1):45–52. https://doi.org/10.1016/0735-1097(95)00432-7
36. Lanza GA, Pedrotti P, Pasceri V, Lucente M, Crea F, Maseri A (1996) Autonomic changes associated with spontaneous coronary spasm in patients with variant angina. J Am Coll Cardiol 28(5):1249–1256. https://doi.org/10.1016/S0735-1097(96)00309-9
37. Takaoka K, Yoshimura M, Ogawa H et al (2000) Comparison of the risk factors for coronary artery spasm with those for organic stenosis in a Japanese population: role of cigarette smoking. Int J Cardiol 72(2):121–126. https://doi.org/10.1016/S0167-5273(99)00172-2
38. Yasue H, Takizawa A, Nagao M et al (1988) Long-term prognosis for patients with variant angina and influential factors. Circ 78(1):1–9. https://doi.org/10.1161/01.CIR.78.1.1
39. Piao ZH, Jeong MH, Li Y et al (2016) Benefit of statin therapy in patients with coronary spasm-induced acute myocardial infarction. J Cardiol 68(1):7–12. https://doi.org/10.1016/j.jjcc.2015.09.013
40. Beltrame JF, Crea F, Camici P (2009) Advances in coronary microvascular dysfunction. Hear Lung Circ 18(1):19–27. https://doi.org/10.1016/j.hlc.2008.11.002
41. Albadri A, Mavromatis K, Noel Bairey Merz C (2019) The role of coronary reactivity testing in women with no obstructive coronary artery disease. Curr Opin Cardiol 34(6):656–662. https://doi.org/10.1097/HCO.0000000000000682
42. Barbato E, Aarnoudse W, Aengevaeren WR et al (2004) Validation of coronary flow reserve measurements by thermodilution in clinical practice. Eur Heart J 25(3):219–223. https://doi.org/10.1016/j.ehj.2003.11.009
43. Mohri M, Koyanagi M, Egashira K et al (1998) Angina pectoris caused by coronary microvascular spasm. Lancet 351(9110):1165–1169. https://doi.org/10.1016/S0140-6736(97)07329-7
44. Luo C, Long M, Hu X et al (2014) Thermodilution-derived coronary microvascular resistance and flow reserve in patients with cardiac syndrome X. Circ Cardiovasc Interv 7(1):43–48. https://doi.org/10.1161/CIRCINTERVENTIONS.113.000953
45. Scalone G, Niccoli G, Crea F (2019) Editor's choice-pathophysiology, diagnosis and management of MINOCA: an update. Eur Hear J Acute Cardiovasc Care 8(1):54–62. https://doi.org/10.1177/2048872618782414
46. Vidal-Perez R, Casas CAJ, Agra-Bermejo RM et al (2019) Myocardial infarction with non-obstructive coronary arteries: a comprehensive review and future research directions. World J Cardiol 11(12):305–315. https://doi.org/10.4330/wjc.v11.i12.305
47. Shibata T, Kawakami S, Noguchi T et al (2015) Prevalence, clinical features, and prognosis of acute myocardial infarction attributable to coronary artery embolism. Circulation 132(4):241–250. https://doi.org/10.1161/CIRCULATIONAHA.114.015134
48. Pasupathy S, Tavella R, Beltrame JF (2015) The what, when, who, why, how and where of myocardial infarction with non-obstructive coronary arteries (MINOCA). Circ J 80(1):11–16. https://doi.org/10.1253/circj.CJ-15-1096
49. Kang WY, Jeong MH, Ahn YK et al (2011) Are patients with angiographically near-normal coronary arteries who present as acute myocardial infarction actually safe? Int J Cardiol 146(2):207–212. https://doi.org/10.1016/j.ijcard.2009.07.001
50. Barr PR, Harrison W, Smyth D, Flynn C, Lee M, Kerr AJ (2018) Myocardial infarction without obstructive coronary artery disease is not a benign condition (ANZACS-QI 10). Hear Lung Circ 27(2):165–174. https://doi.org/10.1016/j.hlc.2017.02.023
51. Lindahl B, Baron T, Erlinge D et al (2017) Medical therapy for secondary prevention and long-term outcome in patients with myocardial infarction with nonobstructive coronary artery disease. Circ 135(16):1481–1489. https://doi.org/10.1161/CIRCULATIONAHA.116.026336

# Chapter 8
# Blood Pressure: Changes Over a Woman's Life, the Effect of Estrogen, and Special Considerations in Women

**Daniel Esau and Beth L. Abramson**

**Abstract** Hypertension is an important risk factor for cardiovascular disease in women and is one of the leading contributors to morbidity and mortality in women. Although the majority of women who become hypertensive are post-menopausal, hypertension in the pre-menopausal years remains an important public health issue. Hypertensive disorders in pregnancy are of particular importance and lead to significant mortality in women world-wide. In this chapter, we will discuss the role of estrogen and progesterone on blood pressure control and examine how a woman's blood pressure may change during pregnancy and in the post-menopausal years. We will also discuss factors contributing to hypertension that are unique to women, including the use of oral contraceptives, pregnancy related conditions, polycystic ovary syndrome (PCOS), and the loss of estrogen during menopause. Finally, we will discuss the underrepresentation of women in clinical trials on hypertension and sex-specific variations in hypertension treatment (both pharmacologic and non-pharmacologic).

**Keywords** Blood-pressure · Menopause · Estrogen · Polycystic ovary syndrome · Pregnancy · Hormone replacement therapy

## Introduction

Hypertension implies a blood pressure that is higher than normal. Defining what blood pressure is "normal" is difficult, since there is variation among major guidelines regarding the level at which elevated blood pressure becomes pathologic [1–4]. The blood pressure level at which hypertension is diagnosed is also different depending

D. Esau
Division of Community Internal Medicine, University of British Columbia, Victoria, Canada

B. L. Abramson (✉)
St. Michael's Hospital, University of Toronto, Toronto, Canada
e-mail: Beth.Abramson@unityhealth.to

© The Author(s), under exclusive license to Springer Nature Switzerland AG 2023

L. Kirshenbaum and I. Rabinovich-Nikitin (eds.), *Biology of Women's Heart Health*, Advances in Biochemistry in Health and Disease 26,
https://doi.org/10.1007/978-3-031-39928-2_8

on other comorbidities (for example, the presence of diabetes), and thus we can infer that a blood pressure reading that may be considered "normal" (or at least not pathological) in one individual may be considered abnormal and pathologic in another [1]. Hypertension is often divided into primary and secondary hypertension. Secondary hypertension is due to an underlying cause such as renovascular disease, renal failure, pheochromocytoma, aldosteronism, hypercortisolism, or monogenic forms, while primary (or essential) hypertension is a diagnosis of exclusion when secondary causes are not present [5]. Not all patients with a diagnosis of hypertension require pharmacologic treatment, and separate, higher, blood pressure cut-offs are often used to determine when pharmacotherapy is recommended [1–3]. Regardless, there is a clear and strong correlation with increasing blood pressure and the risk of cardiovascular disease, renal disease, and overall mortality. Data gathered through the Global Burden of Disease study has shown that hypertension is the leading cause of attributable deaths among women world-wide, and is second only to child and maternal malnutrition in disability-adjusted life years amongst level 2 risk factors [6]. Similarly, in 2018 in the United States, there were 95,876 deaths primarily attributed to high blood pressure, 52% of which were women [7].

Data from 2007–2017 has found that approximately 23% of Canadians (nearly 5.8 million people) carry a diagnosis of hypertension, with no significant change in prevalence over the same period [8, 9]. From 2015 to 2018 in the United States, 42.8% of women and 51.7% of men had a diagnosis of hypertension [7]. In early adulthood mean systolic blood pressure is higher in men than women, and men have higher rates of hypertension [7, 10, 11]. Although the prevalence of hypertension increases with age in both sexes, there is a precipitous increase in hypertension rates in women in the fourth and fifth decade of life, such that by age 65 women are slightly more likely to have hypertension then men [7, 11].

Historically women have had lower rates of hypertension control relative to men [12–14]. However, more contemporary data has indicated that women now have higher rates of hypertension awareness, treatment, and control than men across all ethnic groups in the United States as well as in low-, middle, and high-income countries [7, 15–17]. Although this may be a promising sign of increasing familiarity from both patients and care providers for the diagnosis and treatment of hypertension in women, there remains a significant gap in the treatment and control of hypertension in both sexes, with only 50.4% of hypertensive adults being treated and only 21.6% having their hypertension controlled in the United States [7]. In Canada, awareness, treatment, and control of hypertension seems better, at 83.5%, 78.9%, and 65.4%, respectively [9]. However, there are concerns about widening gender disparities in Canada, as data from 2017 showed a significant decrease in hypertension awareness, treatment, and control in women compared to previous years [9].

Furthermore, several studies have reported that women are at a higher relative risk of developing complications of hypertension than men. In adults with stage 1 hypertension, women were more likely to develop microalbuminuria and left ventricular hypertrophy than men [18]. Although men remain at higher absolute risk of cardiovascular events, the association of cardiovascular events with blood pressure may be steeper in women than in men [19, 20]. This would imply that for an equivalent decrease in blood pressure, a larger proportion of cardiovascular events are prevented in women than in men [19]. This remains an area of debate, and several clinical trials and meta-analyses have not found any sex-specific differences in clinical outcomes or treatment effect with the use of antihypertensive agents [21–24].

## Blood Pressure Effects of Endogenous Estrogens and Progesterone

Although cardiovascular disease remains a leading cause of death among women, women have lower rates of cardiovascular disease than men overall [7], and it is likely that the effects of estrogen contribute to the lower rates of cardiovascular disease in the pre-menopausal years [25]. Blood pressure is also clearly dependent on the effect of hormones, and as estrogens and progesterone levels change throughout a woman's life, so too does the risk of hypertension. During menopause, circulating levels of estrogen and progesterone fall, and the risk of hypertension in women is greatly increased in the post-menopausal years [7, 26]. Clinicians must also consider the effects of exogenous hormones on blood pressure, such as with the use of combined estrogen-progesterone contraceptives [27].

Estrogens and progesterone are sex hormones that play key roles in the sexual and reproductive development in women. They also play important roles in regulating blood pressure and promoting vascular health. Estrogen acts on vascular and cardiomyocyte estrogen receptors (ERs), including ERα, ERβ, and G protein-coupled ER and functions through both genomic and non-genomic pathways [25, 28]. The overall effects of endogenous estrogen are favourable in preventing cardiovascular disease and lowering blood pressure, but the effects of exogenous estrogens (such as those used in oral contraceptive pills (OCP) or menopausal hormone therapy (MHT)) are complex and more controversial [25, 29–31].

Endogenous estrogens reduce blood pressure by facilitating nitric oxide-mediated vasodilatory pathways and inhibiting vasoconstrictive pathways mediated by the sympathetic nervous system and angiotensin, while progesterone promotes natriuresis through its anti-mineralocorticoid effects [25, 26, 28, 32, 33]. Estrogen inhibits pro-fibrotic genes and stimulates neoangiogenesis, both of which function to reduce blood vessel inflammation and maintain arterial compliance [25, 34]. Estrogen is also involved in regulation of several pathways in the Renin–Angiotensin–Aldosterone system (RAAS) [35, 36]. In physiologic levels, these effects function to prevent high blood pressure, but as estrogen is lost during menopause this protective regulation is lost. For example, surgical menopause (due to bilateral oophorectomy) is associated with increased likelihood of salt-sensitivity of blood pressure [37], and animal models suggest that a loss of estrogen leads to increased expression of angiotensin subtype 1 (AT1) receptor in the kidneys and salt-induced hypertension [38].

The effects of exogenous hormones are distinct from endogenous estrogen and progesterone, and both OCP and MHT have in general been shown to increase blood pressure [26, 27, 39–43], although the specific drug used, the dose, and the route of administration are all important considerations in predicting this effect [26, 44, 45]. Exogenous estrogens, when taken orally and allowed to pass through the liver during first-pass metabolism, stimulate the synthesis of angiotensinogen which, in susceptible women, may lead to salt and water retention and significant blood pressure elevation through activation of RAAS [32, 33, 46]. When exogenous estrogens are delivered through an oral route, they undergo first pass metabolism in the liver, which leads to supra-physiologic levels in the liver and increased hepatic synthesis of proinflammatory and procoagulant molecules [26, 32, 47]. Non-oral routes of hormone administration, such as the estrogen/progestin transdermal patch, bypass the first-pass metabolism in the liver which may improve tolerability, minimize side effects, and potentially minimize negative cardiovascular effects [32, 47]. Although endogenous progesterone has affinity for the mineralocorticoid receptor and antagonizes the effects of aldosterone, most synthetic, exogenous progestogens do not share this effect [32, 33]. One exception is drospirenone, which shares similar properties to endogenous progesterone including affinity for the mineralocorticoid receptor and has been shown to decrease blood pressure in post-menopausal women treated with 1 mg 17$\beta$-estradiol/2 mg drospirenone for hormone replacement therapy [33].

## Hypertension in Girls and Adolescent Women

The development of elevated blood pressure in childhood and teenage years contributes to the early development of atherosclerosis and left ventricular hypertrophy and is also associated with the development of hypertension in adulthood [48, 49]. There is thus increasing attention on detection of both elevated blood pressure and hypertension in children and adolescence. The prevalence of hypertension in children and adolescents is around 3.5%, while an estimated 7.1% have elevated blood pressure [50]. Although there is no clear sex difference in the prevalence of hypertension in children and teenagers [50], boys are more likely than girls to have severe obesity [51]. Around 20% of children and adolescence in the United States are obese, and the largest driver of primary hypertension in children and teenagers is obesity [50]. However, sociodemographic status, ethnicity, and family history of hypertension are also important factors [50, 52]. Traditionally, most cases of hypertension in children and adolescence were thought to be secondary to another condition and up to 80–85% of cases were reported to have an identifiable—and potentially treatable—cause, with hypertension most frequently being renal or renovascular in etiology [53, 54]. In hypertensive children it is therefore prudent to perform at minimum a urinalysis and blood work to examine renal function and electrolytes [48]. A renal ultrasound is frequently ordered, and further testing such as a sleep study and drug screen are ordered if clinically indicated [48]. However, more recent data has argued that the prevalence of primary (often obesity related) hypertension is increasing and is now an equally important cause of hypertension in children and adolescence [50, 54, 55]. It should be noted that although an increased prevalence of obesity has contributed to an increase in primary hypertension, pediatric patients with secondary hypertension also have high rates of obesity, and the presence of obesity cannot be used to predict whether a child has primary or secondary hypertension [52, 55].

In general, the younger the age of onset of hypertension, the more likely that secondary causes (renal or endocrine disorders) are contributing. The use of illicit substances, diet and herbal supplements, as well as prescription drugs such as oral contraceptives and stimulants, should be considered [48]. Turner syndrome is the most common chromosomal abnormality in females, occurring in 1:2500 women, and is associated with an up to 40% risk of developing hypertension as a child [56]. Other conditions that need to be considered in hypertensive pediatric patients include congenital heart disease, such as coarctation of the aorta (especially in girls with Turner syndrome) [57], Takayasu's arteritis [58], and fibromuscular dysplasia [59].

## Hypertension in Women of Reproductive Age

The prevalence of hypertension in women of reproductive age is reported to be 7.7–9% [60, 61], with hypertension being more prevalent in women with diabetes, chronic kidney disease, and a higher body mass index. There is also a higher prevalence of hypertension as women age and among non-Hispanic black women (compared to non-Hispanic white women) [7, 60]. There are several important considerations that arise in women of this age group, including the use of oral contraceptives, polycystic ovary syndrome, and hypertensive disorders in pregnancy.

## Oral Contraceptives and Hypertension

Although oral contraceptives (OCP) are intended to reproduce the properties of endogenous estrogens and progesterone, oral contraceptives have generally been found to increase blood pressure [27, 39, 62]. In one meta-analysis, a linear dose–response relationship was found with an increased risk of hypertension of 13% for every 5-year increment of OCP use [39].

As previously discussed, non-oral routes of hormone administration, such as transdermal patches or intravaginal devices, avoid first-pass metabolism in the liver [32, 47], and this theoretically should mitigate the risk of hypertension seen with oral estrogens. The use of an estrogen/progestin transdermal patch was found to cause lower activation of the RAAS system compared to OCP in healthy premenopausal women [46]. However, even if systemic absorption is low and first-pass metabolism is avoided, there may still be a negative effect on blood pressure. The use of the combined hormonal contraceptive vaginal ring was associated with a small, but significant, increase in 24-h blood pressure compared to baseline (mean 24-h BP increased by $2.69 \pm 5.35$ mmHg) [63]. Similarly, levonorgestrel-releasing intrauterine devices have been reported to have either neutral effect [64] or to mildly increased blood pressure [65].

Unlike with estrogen containing contraceptives, progesterone only oral contraceptive are not associated with an increased blood pressure [66]. Furthermore, the use of oral combined contraceptives containing drospirenone, a progesterone like molecule with affinity for the mineralocorticoid receptor [33], may counteract or minimize the hypertensive effect of estrogen. After 6 months of use, there was no difference in blood pressure measurements or serum aldosterone and renin levels between hypertensive women treated with 20 mcg of ethinyl estradiol and 3 mg of drospirenone compared to those treated with nonhormonal contraceptive [67]. Although more research is needed, the use of non-oral contraceptive, progesterone only contraceptive, and contraceptives involving drospirenone have shown promising results in mitigating the increased blood pressure seen with oral contraceptive use [62, 68].

While the absolute risk remains low, women using OCPs are at an approximately 1.6-fold increased risk of myocardial infarction or ischemic stroke, although this risk is minimized when the dose of estrogen is minimized [69]. Though certainly an important contributor, the risk of cardiovascular events with the use of OCP is unlikely to be due to blood pressure effects alone, as exogenous estrogen has procoagulant and prothrombotic effects [70]. For example, a relative risk increase (with low absolute risk) of venous thromboembolism with OCP use is well described [27, 70]. To minimize risk, it is therefore important to ensure that women do not have risk factors for cardiovascular disease before starting OCP. Ensuring a woman does not have hypertension is paramount as women who did not have their blood pressure monitored prior to initiation of oral contraceptive pills were at higher risk for ischemic stroke and myocardial infarction than women who did have their blood pressure measured [27]. Because of these risks, the use of combined or estrogen-based contraception continues to be contraindicated in women with pre-existing hypertension regardless of the route of administration, and, depending on the severity of hypertension, is considered category 3 (the known risks likely exceed the benefit) or category 4 (unacceptable risk) according to the World Health Organization medical eligibility criteria [31]. Progesterone-only contraceptives and the copper-bearing or levonorgestrel-releasing IUD remain safe options in the majority of hypertensive women [31]. In normotensive women, blood pressure should be measured prior to the first dose of OCP and then annually thereafter to ensure that hypertension does not develop [31].

## Polycystic Ovary Syndrome and Hypertension

Polycystic ovary syndrome (PCOS) is a common heterogeneous endocrine syndrome that develops during a woman's reproductive years [71]. PCOS is associated with an increased risk of hypertension as well as several other cardiovascular risk factors including obesity, insulin resistance, diabetes, metabolic syndrome, and dyslipidemia

[61, 71–73]. Women with PCOS are predisposed to developing early atherosclerosis and cardiovascular disease [72, 74].

Women of reproductive age with PCOS have a higher risk of hypertension than those without PCOS (Relative risk 1.7, 95% CI: 1.43–2.07) [61, 75]. There has long been debate regarding whether this increased risk is due to an underlying effect of PCOS or simply secondary to the effect of obesity, which is highly prevalent in women with PCOS [61, 76]. Arguing for an independent effect of PCOS on the development of hypertension, the presence of hyperadrogenism is one of the major criteria used to diagnose PCOS and may play a pathogenic role in the development of hypertension and other cardiovascular risk factors in women with PCOS [61, 77]. Androgens may increase arterial blood pressure in women with PCOS by altering the components of the RAAS [36, 75]. PCOS is also characterized by insulin resistance, which contributes to salt retention and interferes with endothelium-dependent vasodilation mechanisms and thereby contributes to the development of hypertension [75, 76]. Furthermore, although there is an increased risk for hypertension in pre-menopausal women with PCOS, this risk is normalized somewhat with aging, and a recent review found no significant difference in risk of hypertension in women with or without PCOS once they reached menopausal age, although the prevalence of hypertension in these groups was high (40% and 49% in menopausal/post-menopausal women without and with PCOS, respectively) [61]. Similarly, although younger women with PCOS are at increased risk of cardiovascular events compared to their peers, as women age the risk in women with PCOS becomes similar to that of women without PCOS [61, 78, 79]. This trend may be due to a progressive decrease in androgen production with age, which occurs both in the general population and in women with PCOS [61, 80, 81].

Treatment of hypertension in women with PCOS is multifactorial. Lifestyle modifications including weight loss, diet, and exercise can improve hypertension, reduce cardiovascular risk, reduce androgenicity, and improve insulin resistance in women with PCOS [82, 83]. Metformin is widely used in PCOS as an insulin sensitizer and may improve weight loss compared to lifestyle modifications alone [84]. There is also evidence that metformin can reduce blood pressure in women with PCOS [85], likely mediated through its insulin sensitizing effects [76]. Treatment of hyperandrogenism is undertaken with the use of combination of oral contraceptives and/or spironolactone. Although estrogen-based contraceptives may be useful in treating hyperandrogenism in PCOS [86], they are generally contraindicated in hypertensive women (with or without PCOS), as discussed previously [31]. Spironolactone is a mineralocorticoid receptor antagonist used as an antihypertensive agent in a variety of clinical situations but may be especially suited to treating hypertension associated with PCOS due to its antiandrogenic effects, which were initially found serendipitously in women with PCOS who were being treated for hypertension with this agent [87, 88]. The bulk of the evidence for spironolactone is from treatment of hypertension in the general population [89], but one small study did show a significant reduction in blood pressure in women with PCOS with the use of spironolactone [90]. In addition, spironolactone has other favorable effects in PCOS and has been reported to normalize endothelial function and improve cholesterol levels [91], hirsutism, insulin sensitivity and menstrual cycle frequency [92].

Other than the theoretical benefit of spironolactone in PCOS, there is little evidence that any specific agents are preferable for the treatment of hypertension. A small case series reported that telmisartan, an angiotensin II receptor antagonist (ARB), decreased androgen levels and improved menstrual patterns in women with PCOS and hypertension [93]. However, in general, women with PCOS and hypertension are treated with pharmacotherapy with the same considerations regarding medication choice as any other woman with hypertension.

## Hypertensive Disorders in Pregnancy

Hypertensive disorders are the most common medical complication of pregnancy. In Canada, hypertension in pregnancy affects approximately 7.6% of all pregnancies [94]. Hypertension during pregnancy may be chronic (in women with pre-existing hypertension who become pregnant) or may develop during pregnancy (either gestational hypertension or preeclampsia) [95], and is defined as a systolic blood pressure ≥ 140 mmHg and/or a diastolic blood pressure ≥ 90 mmHg based on the average of at least two measurements taken at least 15 min apart [96]. The rates of hypertensive disorders in pregnancy appear to be increasing: data from the United States shows an increase from approximately 5.3% of deliveries in 1993 to 9.1% of deliveries in 2014 [7], while worldwide the incidence of hypertensive disorders in pregnancy have increased 10.9% from 1990 to 2019 [97]. Hypertensive disorders in pregnancy are associated with significant maternal, fetal, and neonatal morbidity and mortality, with increasing rates of placental abruption, intrauterine growth restriction, prematurity, intrauterine deaths, and intracerebral hemorrhage in pregnant women with hypertension [95, 97, 98]. In 2019, approximately 28 thousand women died from hypertensive disorders in pregnancy, the majority in low- and middle-income countries [97]. The detection and treatment of hypertensive disorders in pregnancy is therefore an important global health concern.

Chronic hypertension is defined as hypertension that precedes pregnancy or that develops within the first 20 weeks of pregnancy. Gestational hypertension is defined as a new diagnosis of hypertension occurring after 20 weeks' gestation without evidence of proteinuria or hematologic abnormalities [95]. In general, pregnancy outcomes in women with both chronic hypertension and gestational hypertension are good if blood pressure is controlled, although there remains an increased risk of preterm birth and small-for-gestational-age neonates in women with chronic hypertension [98]. However, around one quarter of these women will progress to preeclampsia, a more serious condition that is associated with an increased risk of significant mortality and morbidity [95, 98]. Severe hypertension (with or without preeclampsia) is also a concern, and is associated with fetal and maternal complications, including hemorrhagic stroke [99]. In women with hypertension in pregnancy, blood pressure control is important in preventing severe hypertension, preeclampsia, and preterm birth [99, 100].

Preeclampsia is a clinical diagnosis that requires elevated blood pressure in conjunction with proteinuria or other severe features [96, 101]. Chronic hypertension is one of the strongest predictors of a woman's risk of progression to preeclampsia (RR 5.1, 95% CI 4.0–6.5) [102], but several other factors such as obesity, diabetes, advanced maternal age, and genetic susceptibility play a role [101, 102]. It is important that preeclampsia is detected, diagnosed, and treated, since it can progress rapidly to serious complications including death of both the fetus and the mother [101]. Treatment of preeclampsia is beyond the scope of this chapter, but, in addition to blood pressure control, often involves delivery.

Medications are used to treat all types of hypertensive disorders in pregnancy. Because many antihypertensive medications are not suitable for use in pregnancy (either because of demonstrated teratogenicity/fetopathy or a lack of evidence for use in pregnancy), women with chronic hypertension who are planning to become pregnant or who become pregnant often have their antihypertensive medications changed. Most notably, angiotensin-converting enzyme inhibitors (ACE-I) and angiotensin receptor blockers (ARBs) are contraindicated at all stages of pregnancy due to demonstrated fetopathy [103–105]. Acceptable antihypertensive agents include labetalol, methyldopa, nifedipine, diltiazem, prazosin, clonidine, and hydralazine (although the last three are generally considered second or third line agents) [95, 96, 98, 105]. Although hypertension is of critical importance in pregnancy, it is also important to avoid excessive lowering of blood pressure. Low blood pressure in pregnancy has been associated with low fetal birth weight via placental hypoperfusion [99, 106]. Because some women with chronic hypertension experience a fall in their blood pressure in the first and second trimesters, antihypertensive medications are often stopped early in pregnancy with plans to restart if the blood pressure rises again in the second or third trimester [98].

Data relating to optimal blood pressure in pregnancy is scarce. The CHIP trial was a randomized clinical trial that compared "tight" (target dBP of 85 mmHg) versus "less tight" (target dBP of 100 mmHg) blood pressure control in women with mild or moderate nonproteinuric chronic or gestational hypertension [100]. No difference was reported between "tight" and "less tight" blood pressure control in the risk of pregnancy loss or overall maternal complications [100], but "less-tight" control was associated with a significantly higher frequency of severe maternal hypertension, which, in a later analysis, was shown to be associated with an increased risk of perinatal and maternal outcomes independently of the co-occurrence of preeclampsia [107]. A recent meta-analysis of 40 randomized controlled trials with 6355 patients found that treatment to a sBP of $< 130$ mmHg reduced the risk of severe hypertension by approximately one-third compared to a sBP of $\geq 140$ mmHg, and also reported a decreased risk for preeclampsia, placental abruption, and preterm birth but an increase in small-for-gestational-age infants with the use of blood pressure lowering medications [99]. Because of a paucity of data and because both high and low blood pressure can lead to complications, it is not surprising that the optimal target blood pressure in pregnancy remains somewhat controversial: the International

Society for the Study of Hypertension in Pregnancy recommending targeting 110–140/80–85 mmHg [95], the American College of Obstetricians and Gynecologists recommending targeting a systolic blood pressure ≥ 120 and < 160 mmHg and a diastolic pressure of ≥ 80 mmHg and < 110 mmHg [108], while Hypertension Canada recommends targeting a diastolic pressure of 85 mmHg [105]. However, following the publication of these guidelines, the recently published results from the Chronic Hypertension and Pregnancy (CHAP) project, which investigated treatment of mild chronic hypertension at a gestational age of less than 23 weeks with a target of < 140/90 mmHg versus only initiating pharmacotherapy in cases of severe hypertension (≥ 160/105 mmHg), have provided support for the lower blood pressure target with less adverse pregnancy outcomes (risk ratio 0.82; $p < 0.001$) reported without impairing fetal growth [109].

It is now recognized that women with a history of hypertensive disorders during pregnancy are at increased risk of cardiovascular events in later life [110–112]. Women with a history of preeclampsia or gestational hypertension have higher BMIs, higher blood pressure, and unfavorable lipid profiles compared to parous women who did not have hypertensive disorders or pregnancy [113, 114]. In addition to the elevated risk of arterial stiffness and chronic hypertension compared to women without a history of hypertensive disorders in pregnancy, an increased risk of coronary artery disease, stroke, heart failure, aortic stenosis, and mitral regurgitation in women with previous hypertensive disorders in pregnancy has been reported [110, 112]. This increased risk was partially (50–65%), but not completely explained by the development of chronic hypertension in these women [110]. Given these findings, a history of hypertensive disorder of pregnancy is an important sex-specific risk factors for women that should be assessed when clinicians are performing a cardiovascular risk assessment in women. Furthermore, the occurrence of a hypertensive disorder of pregnancy in a woman should be a trigger for clinicians to thoroughly perform a cardiovascular risk assessment and consider risk mitigating strategies.

## Hypertension and Menopausal Hormone Therapy

Menopause is associated with a significant rise in the prevalence of hypertension in women [7, 26]. Although the prevalence of hypertension increases with age, there is a large increase in the prevalence of hypertension in women from the pre-menopausal years (30.3% of women aged 35–44) to the menopausal years (50.9% of women aged 45–54) [7]. From the age of 20–64 the prevalence of hypertension is greater in men than women, but after the age of 65, the trend reverses and an equal or higher proportion of women have hypertension than men [7]. As estrogen and progesterone levels fall during menopause, women lose the protective, anti-hypertensive effects of these hormones [26, 28, 36]. Women undergoing menopause have higher blood pressure values and higher cardiovascular risk than premenopausal women of the same age [115, 116]. Similarly, women who go through early menopause (at < 45 years of age) are at higher risk of coronary heart disease, while women experiencing menopause from 50 to 54 years of age had a lower risk [117].

Another marker of vascular disease is arterial compliance, which is defined as the change in arterial blood volume attributed to a given change in pulsatile blood pressure [118]. Lower arterial compliance plays a role in the development of isolated systolic hypertension as well as refractory hypertension [34]. As previously described, estrogen has anti-fibrotic and anti-inflammatory affects that preserve vascular heath [25, 34]. Women have more estrogen receptors in their aorta than men [119]. Therefore, women may be more prone to developing arterial stiffness as they lose the effect of estrogen in the post-menopausal years. In one study, lower arterial compliance was associated with an increased burden of coronary artery plaque and calcification in post-menopausal women, but not in men [118]. The loss of ovarian hormones during menopause also leads to an increased risk of salt sensitivity, which could contribute to the development of hypertension [37, 38]. In addition to the increased incidence of hypertension post-menopause, both normotensive and hypertensive postmenopausal women may develop a blunted nocturnal blood pressure dip [120, 121]. This "non-dipping" pattern, defined as a < 10% drop in nocturnal systolic blood pressure [121], has been associated with negative cardiovascular outcomes and target organ damage, especially in older women [121, 122].

The cardiovascular benefits and risks of menopausal hormone therapy remains a complex and controversial subject [25, 29]. The study of menopausal hormone therapy has had a difficult course, and although an in-depth review of this subject is beyond the scope of this chapter, it is notable that studies have varied significantly in their results, with some studies showing positive effects to MHT in postmenopausal women while others have shown neutral or negative effects [30]. In 2002, the Women's Health Initiative (WHI) randomized clinical trial of oral estrogen and progestin that was stopped three years earlier due to increased health risks that exceeded the benefits of the therapy, including an increased risk of cardiovascular events, venous thromboembolism (VTE), breast cancer, and stroke [43]. A 2017 Cochrane review involving 43,637 women came to a similar conclusion regarding the continuous use of combined (estrogen and progesterone based) or estrogen based MHT in postmenopausal women, and although a subgroup analysis of women aged 50–59 found only an increased risk of VTE (and not cardiovascular disease or malignancy) the authors concluded that a small difference in risk could not be excluded due to low power [123]. A 2004 meta-analysis found a reduction in all-cause mortality (OR 0.61, CI: 0.39–0.95) with the use of MHT in women under the age of 60 years, but did not find any difference in mortality in women older than 60 years or overall [124]. When long-term follow-up from the WHI trial was published, short-term MHT was not associated with a risk of all-cause, cardiovascular, or cancer mortality when used from a median of 5.6 years (conjugated equine estrogens (CEE) and medroxyprogesterone (MPA)) or 7.2 years (CEE alone) [125]. These studies would suggest that the cardiovascular risk of MHT is lowest when started early in the postmenopausal period, and this has led to the development of the "critical timing hypothesis", which states that the risks of MHT vary depending on the age (or time since menopause) that therapy is initiated [29, 30]. A recent scientific statement from the Society of

Obstetricians and Gynaecologist of Canada highlights that in appropriately selected, symptomatic women, the early and short-term use of menopausal hormone therapy does not confer increased cardiovascular risk, but women who initiate menopausal hormone therapy 10 or more years after menopause are at increased risk of adverse cardiac events [126]. This statement is in agreement with the recommendations of several other international societies [30, 127, 128].

In terms of the effect on blood pressure, both oral CEE and CEE plus MPA mildly increase blood pressure in postmenopausal women [42, 43]. A subsample analysis of the WHI published in 2018 reported that the use of CEE was associated with an 18% increased incidence of hypertension in post-menopausal women, an effect which was noted to dissipate once MHT was stopped [40]. In contrast, transdermal estrogen has been reported to decrease sympathetic activity, decrease salt sensitivity, and reduce blood pressure in postmenopausal women [41, 44, 45]. The WHI observational study involved 93 676 postmenopausal women followed for an average of 10.4 years, and found no significant difference in cardiovascular events or all-cause mortality between oral estradiol, CEE, or transdermal estradiol, although there was a nonsignificant trend towards favoring transdermal estrogen [129]. Although more study is needed, a transdermal administration may be preferable to oral hormone replacement in postmenopausal women, especially in patients with hypertension.

## Sex-Specific Differences in Clinical Trials and Differences in Treating Hypertension in Women

Given the sex-specific differences in physiology and clinical presentation of hypertension, it is important to consider whether the treatment approach used for hypertension is affective in women as well as men. Many studies do not perform subgroup analysis to examine the outcome of interest in women [11], and even if this is done, studies may not be powered sufficiently to look for sex-specific outcome differences.

Women are underrepresented in several of the clinical trials examining the effectiveness of antihypertensive agents [11]. Some of the first clinical trials investigating treatment of hypertension were performed in the United States Veterans Affairs (VA) system and did not enroll any women [130, 131]. More recently, the SPRINT trial, which found a benefit to intensive versus conservative blood pressure management in non-diabetic patients at high cardiovascular risk, enrolled only 35.6% women [132]. The ACCOMPLISH trial, which found superior reduction in cardiovascular events in high-risk patients with a combination of benzapril-amlodipine compared to benzapril-hydrochlorothiazide, enrolled 39.5% women [133].

The underrepresentation of women in cardiovascular clinical trials is a well-known issue [134], and although examples do exist of underrepresentation in hypertension trials, there are also examples where women are equally [21, 135, 136] and even overrepresented [22]. However, even when women are equally represented only a few large trials have performed subgroup analyses to examine whether sex-specific differences in outcomes exist. In the ALLHAT trial, which enrolled 47% women, no difference in CVD outcome was detected when the data were examined separately by sex [21]. Similarly, no sex-specific difference in outcomes were noted in the HYVAT trial, which enrolled 61% women [22]. In 2009, one review aimed to examine sex specific outcomes in 67 separate hypertension trials, and concluded that "evidence from trials of antihypertensive treatment benefits specific to women is weak, but in studies where the analysis was adjusted for gender, the results appear similar for women and men", although no study at that time had the main objective of comparing treatment effect between men and women [23]. Similarly, a large meta-analysis of over 190,000 patients did not find any sex-specific difference in the response of women or men to any particular hypertensive agent, nor was there any difference in achieved blood pressure or on the reduction in cardiovascular events seen with an equivalent drop in blood pressure [24].

Another concern is that blood pressure targets recommended by major guidelines do not differentiate based on sex, despite the known differences in the pathophysiology of hypertension and normal blood pressure differences between men and women. For example, normotensive women have lower blood pressure than age-matched normotensive men [137], and yet blood pressure targets do not reflect this. Furthermore, despite the aforementioned lack of sex-specific differences in treatment outcomes of hypertension in major clinical trials [23, 24, 138], some authors have reported that the relationship between elevated blood pressure and cardiovascular events is more pronounced in women than in men [19, 20]. The potentially contradictory nature of these results may be partly explained by the use of 24-h blood pressure monitoring in trials that found a sex-specific risk difference, rather than the more typical in-office blood pressure used in many clinical trials. Specifically, the relationship between night-time blood pressure and cardiovascular events was steeper in women [19], and as discussed previously, both seemingly normotensive and hypertensive post-menopausal women may develop a "non-dipping" nocturnal blood pressure [120, 121]. Thus, to approach an acceptably low relative risk of cardiovascular events compared to normotensive patients for each sex, more aggressive target blood pressures may be needed in women than in men. Hermida et al. estimate that, for a 24-h blood pressure target in men of < 135/85 mmHg, the equivalent threshold for women is < 125/80 mmHg [20].

Sex-specific variations in the prescribed antihypertension therapy have been reported. Several [138–142], but not all [143], studies report that hypertensive women were more likely to be prescribed diuretics or β-blocker therapy, while men were more likely to be prescribed ACE-I. It should be noted that one possible reason for the lower use of ACE-I and ARBs in women is the avoidance of these medications in women of child-bearing age, although it is unlikely that this would completely explain this discrepancy. Thiazide diuretics have also been reported to reduce the risk of bone fractures and osteoporosis [144, 145], which may partially explain a preferential selection of these agents in women.

Non-pharmacologic therapy for hypertension have been well studied in women. Dietary interventions (such as the DASH diet) have been shown to lower blood pressure and reduce calculated cardiovascular risk [135, 146, 147]. The combination of a DASH diet with exercise and cognitive-behavioral therapy for weight loss led to an average of 16.1/9.9 mmHg blood pressure reduction in overweight and obese participants, 67% of whom were women [148]. Similarly, both reducing salt intake to less than 100 mmol/day and adopting the DASH diet reduce blood pressure, but there is a greater effect from combining both salt restriction and the DASH diet, with a systolic blood pressure reduction of 11.5 mmHg in hypertensive patients [147]. Salt restriction has long been studied as a means to prevent hypertension, and a recent study of nearly 21,000 people found that replacing salt with a salt substitute (75% sodium chloride and 25% potassium chloride by mass) lead to a significant decrease in strokes, major cardiovascular events, and death over a mean follow-up of 4.74 years [149]. Because of the upregulation to RAAS after menopause [35], salt restriction may be especially beneficial in treating and preventing hypertension in postmenopausal women, and a reduction in blood pressure with salt restriction has been reported in this population [149, 150]. Interestingly, a recent study from Japan found that the effects of a lower-salt and higher-potassium diet lead to greater blood pressure reduction in women of all ages (pre- or post-menopausal) compared to men [151]. In people who drink more than two alcoholic beverages per day, reducing alcohol consumption has been shown to lower blood pressure [152]. Regular exercise is also beneficial in reducing blood pressure in women [153–155]. In addition to lowering blood pressure, combined aerobic and resistance exercises reduced arterial stiffness in postmenopausal women [156].

## Future Directions and Unanswered Questions

There remain several unanswered questions regarding hypertension in women. The optimal blood pressure target for blood pressure in pregnancy remains controversial, and there are efforts ongoing to try to clarify this target. Another question is whether the relation between cardiovascular events and hypertension in women is steeper than in men. There have been conflicting results regarding this question [19, 20, 23, 24, 138], but, if present, this sex-specific difference would imply that differential blood pressure targets for men and women should be employed [20]. No current guidelines recommend using sex-specific blood pressure targets. Because the pathophysiology of hypertension may be different in women than in men, with, for example, a greater importance on arterial stiffness [34, 118] and the effect on salt intake [37, 38, 149–151] in women, it is important to consider whether certain medications or interventions may be more or less effective in treating hypertension in women. Although the majority of studies up to this point have shown no clinical difference in the effect of any particular antihypertensive on blood pressure in women or men [23, 24, 138], future studies should be mindful to incorporate sex-specific analyses so that any difference in treatment response does not go unnoticed (Fig. 8.1).

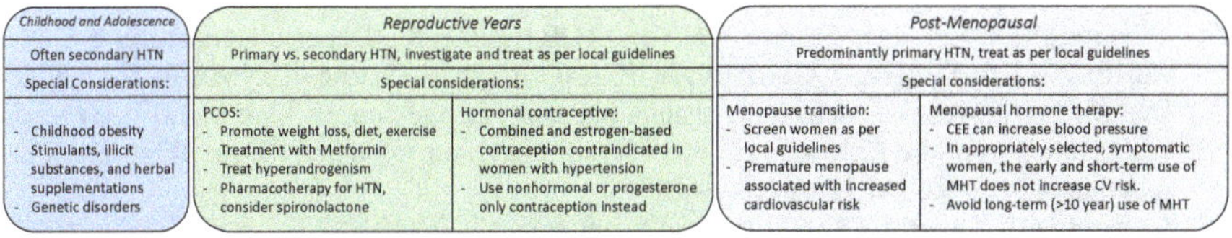


**Fig. 8.1** Blood pressure across a woman's lifespan. Certain special considerations discussed in this text are highlighted. HTN = hypertension, CV = cardiovascular, MHT = menopausal hormone therapy

## Conclusion

Hypertension remains a leading cause of morbidity and mortality in women worldwide [6]. A woman's blood pressure is determined by numerous factors, including regulation by neurohormonal, renal, and endocrine systems. The effects of estrogens on a woman's blood pressure are important, and, although hardly the only regulator of blood pressure, can explain some of the changes in blood pressure that occur during pregnancy and menopause. Clinicians should be aware of the potential effect of hormonal contraception on a woman's blood pressure, which is regulated through the effects of exogenous estrogens.

Although significant strides have been made in improving the care of women with hypertension, there are several unanswered questions. Underrepresentation of women was prevalent in early cardiovascular trials, and it is important that researchers do not repeat the mistakes of the past. As further studies are done, there should be a focus on ensuring adequate recruitment of women and in powering studies so that sex-specific outcome differences can be analyzed.

Special considerations in women include PCOS, hypertensive disorders in pregnancy, and the use of exogenous estrogens for contraception or menopausal hormone therapy. Knowledge of the blood pressure changes that occur during these conditions is paramount to the care of women, and clinicians should be on the lookout for hypertension at all stages of a woman's life.

## References

1. Rabi DM, Mcbrien KA, Sapir-Pichhadze R, Nakhla M, Ahmed SB, Dumanski SM et al (2020) Hypertension Canada's 2020 comprehensive guidelines for the prevention, diagnosis, risk assessment, and treatment of hypertension in adults and children. Can J Cardiol 36(5):596–624
2. Williams B, Mancia G, Spiering W, Agabiti Rosei E, Azizi M, Burnier M et al (2018) ESC/ESH Guidelines for the management of arterial hypertension. Eur Heart J 39(33):3021–3104
3. Unger T, Borghi C, Charchar F, Khan NA, Poulter NR, Prabhakaran D et al (2020) 2020 international society of hypertension global hypertension practice guidelines. Hypertens 75(6):1334–1357

4. Whelton PK, Carey RM, Aronow WS, Casey DE, Collins KJ, Dennison Himmelfarb C et al (2018) 2017 ACC/AHA/AAPA/ABC/ACPM/AGS/APhA/ASH/ASPC/NMA/PCNA Guideline for the prevention, detection, evaluation, and management of high blood pressure in adults: a report of the American college of cardiology/American heart association task force on clinical Pr. Hyperten 71(6):e13–e115
5. Carretero OA, Oparil S (2000) Essential hypertension. Circ 101(3):329–335
6. Murray CJL, Aravkin AY, Zheng P, Abbafati C, Abbas KM, Abbasi-Kangevari M et al (2020) Global burden of 87 risk factors in 204 countries and territories, 1990–2019: a systematic analysis for the global burden of disease study 2019. The Lancet 396(10258):1223–1249
7. Virani SS, Alonso A, Aparicio HJ, Benjamin EJ, Bittencourt MS, Callaway CW et al (2021) Heart disease and stroke statistics—2021 update. Circ 143:e254–e743
8. Robitaille C, Dai S, Waters C, Loukine L, Bancej C, Quach S et al (2012) Diagnosed hypertension in Canada: incidence, prevalence and associated mortality. Can Med Assoc J 184(1):E49–E56
9. Leung AA, Williams JVA, Mcalister FA, Campbell NRC, Padwal RS, Tran K et al (2020) Worsening hypertension awareness, treatment, and control rates in Canadian women between 2007 and 2017. Can J Cardiol 36(5):732–739
10. Pimenta E (2012) Hypertension in women. Hypertens Res 35(2):148–152
11. Ramirez LA, Sullivan JC (2018) Sex differences in hypertension: where we have been and where we are going. Am J Hypertens 31(12):1247–1254
12. Abramson BL, Melvin RG (2014) Cardiovascular risk in women: focus on hypertension. Can J Cardiol 30(5):553–559
13. Samad Z, Wang TY, Frazier CG, Shah SH, Dolor RJ, Newby LK (2008) Closing the gap: treating hypertension in women. Cardiol Rev 16(6):305–313
14. Ong KL, Cheung BMY, Man YB, Lau CP, Lam KSL (2007) Prevalence, awareness, treatment, and control of hypertension among United States adults 1999–2004. Hypertens 49(1):69–75
15. Ahmad A, Oparil S (2017) Hypertension in women. Hypertens 70(1):19–26
16. Mills KT, Bundy JD, Kelly TN, Reed JE, Kearney PM, Reynolds K et al (2016) Global disparities of hypertension prevalence and control. Circ 134(6):441–450
17. Peters SAE, Muntner P, Woodward M (2019) Sex differences in the prevalence of, and trends in, cardiovascular risk factors, treatment, and control in the United States, 2001 to 2016. Circ 139(8):1025–1035
18. Palatini P, Mos L, Santonastaso M, Saladini F, Benetti E, Mormino P et al (2011) Premenopausal women have increased risk of hypertensive target organ damage compared with men of similar age. J Womens Health (Larchmt) 20(8):1175–1181
19. Boggia J, Thijs L, Hansen TW, Li Y, Kikuya M, Björklund-Bodegård K et al (2011) Ambulatory blood pressure monitoring in 9357 subjects from 11 populations highlights missed opportunities for cardiovascular prevention in women. Hypertens 57(3):397–405
20. Hermida RC, Ayala DE, Mojón A, Fontao MJ, Chayán L, Fernández JR (2013) Differences between men and women in ambulatory blood pressure thresholds for diagnosis of hypertension based on cardiovascular outcomes. Chronobiol Int 30(1–2):221–232
21. Group TAOACFTACR (2002) Major outcomes in high-risk hypertensive patients randomized to angiotensin-converting enzyme inhibitor or calcium channel blocker vs diuretic: the antihypertensive and lipid-lowering treatment to prevent heart attack trial (ALLHAT). JAMA: J Am Med Ass 288(23):2981–2997
22. Beckett NS, Peters R, Fletcher AE, Staessen JA, Liu L, Dumitrascu D et al (2008) Treatment of hypertension in patients 80 years of age or older. N Engl J Med 358(18):1887–1898
23. Ljungman C, Mortensen L, Kahan T, Manhem K (2009) Treatment of mild to moderate hypertension by gender perspective: a systematic review. J Womens Health (Larchmt) 18(7):1049–1062
24. Turnbull F, Woodward M, Neal B, Barzi F, Ninomiya T, Chalmers J et al (2008) Do men and women respond differently to blood pressure-lowering treatment? Results of prospectively designed overviews of randomized trials. Eur Heart J 29(21):2669–2680

25. Iorga A, Cunningham CM, Moazeni S, Ruffenach G, Umar S, Eghbali M (2017) The protective role of estrogen and estrogen receptors in cardiovascular disease and the controversial use of estrogen therapy. Biol Sex Differ 8(1):33
26. Ashraf MS, Vongpatanasin W (2006) Estrogen and hypertension. Curr Hypertens Rep 8(5):368–376
27. Curtis KM, Mohllajee AP, Martins SL, Peterson HB (2006) Combined oral contraceptive use among women with hypertension: a systematic review. Contracept 73(2):179–188
28. Ouyang P, Michos ED, Karas RH (2006) Hormone replacement therapy and the cardiovascular system. J Am Coll Cardiol 47(9):1741–1753
29. Srivaratharajah K, Abramson BL (2019) Hypertension in menopausal women: the effect and role of estrogen. Menopause 26(4):428–430
30. El Khoudary SR, Aggarwal B, Beckie TM, Hodis HN, Johnson AE, Langer RD et al (2020) Menopause transition and cardiovascular disease risk: implications for timing of early prevention: a scientific statement from the American heart association. Circ 142(25)
31. Medical Eligibility Criteria for Contraceptive Use. 5th ed. World Health Organization, Geneva (2015)
32. Oelkers WKH (1996) Effects of estrogens and progestogens on the renin-aldosterone system and blood pressure. Steroids 61(4):166–171
33. Oelkers WH (2005) Drospirenone in combination with estrogens: for contraception and hormone replacement therapy. Climacteric 8 Suppl 3:19–27
34. Coutinho T (2014) Arterial stiffness and its clinical implications in women (1916–7075 (Electronic))
35. O'Donnell E, Floras JS, Harvey PJ (2014) Estrogen status and the renin angiotensin aldosterone system. Am J Phys-Regulatory, Integrat Comp Phys 307(5):R498–R500
36. Komukai K, Mochizuki S, Yoshimura M (2010) Gender and the renin-angiotensin-aldosterone system. Fundam Clin Pharmacol 24(6):687–698
37. Schulman IH, Aranda P, Raij L, Veronesi M, Aranda FJ, Martin R (2006) Surgical menopause increases salt sensitivity of blood pressure. Hypertens 47(6):1168–1174
38. Harrison-Bernard LM, Schulman IH, Raij L (2003) Postovariectomy hypertension is linked to increased renal AT1Receptor and salt sensitivity. Hypertens 42(6):1157–1163
39. Liu H, Yao J, Wang W, Zhang D (2017) Association between duration of oral contraceptive use and risk of hypertension: a meta-analysis. J Clin Hypert 19(10):1032–1041
40. Swica Y, Warren MP, Manson JE, Aragaki AK, Bassuk SS, Shimbo D et al (2018) Effects of oral conjugated equine estrogens with or without medroxyprogesterone acetate on incident hypertension in the Women's Health Initiative hormone therapy trials. Menopause 25(7):753–761
41. Yoon B-K, Sung J, Song Y-M, Kim S-M, Son K-A, Yoo JH et al (2021) Effects of menopausal hormone therapy on ambulatory blood pressure and arterial stiffness in postmenopausal Korean women with grade 1 hypertension: a randomized, placebo-controlled trial. Clin Hypert 27(1)
42. Shimbo D, Wang L, Lamonte MJ, Allison M, Wellenius GA, Bavry AA et al (2014) The effect of hormone therapy on mean blood pressure and visit-to-visit blood pressure variability in postmenopausal women. J Hypertens 32(10):2071–2081
43. Writing Group For The Women SHII (2002) Risks and benefits of estrogen plus progestin in healthy postmenopausal women: principal results from the women's health initiative randomized controlled trial. JAMA: J Am Med Assoc 288(3):321–333
44. Vongpatanasin W, Tuncel M, Mansour Y, Arbique D, Victor RG (2001) Transdermal estrogen replacement therapy decreases sympathetic activity in postmenopausal women. Circ 103(24):2903–2908
45. Scuteri A, Lakatta EG, Anderson DE, Fleg JL (2003) Transdermal 17β-oestradiol reduces salt sensitivity of blood pressure in postmenopausal women. J Hypertens 21(12):2419–2420
46. Odutayo A, Cherney D, Miller J, Ahmed SB, Lai V, Dunn S et al (2015) Transdermal contraception and the renin-angiotensin-aldosterone system in premenopausal women. Am J Phys Renal Phys 308(6)(1522–1466 (Electronic)):535–540

47. Goodman M (2002) Are all estrogens created equal? a review of oral versus transdermal therapy. J Womens Health 2012(21):161–169
48. Flynn JT, Kaelber DC, Baker-Smith CM, Blowey D, Carroll AE, Daniels SR et al (2017) Clinical practice guideline for screening and management of high blood pressure in children and adolescents. Pediatr 140(3):e20171904
49. Wenger NK, Arnold A, Bairey Merz CN, Cooper-DeHoff RM, Ferdinand KC, Fleg JL et al (2018) Hypertension across a woman's life cycle. J Am Coll Cardiol 71(16):1797–1813
50. Hardy ST, Sakhuja S, Jaeger BC, Urbina EM, Suglia SF, Feig DI et al (2021) Trends in blood pressure and hypertension among US children and adolescents, 1999–2018. JAMA Netw Open 4(4):e213917
51. Pinhas-Hamiel O, Hamiel U, Bendor CD, Bardugo A, Twig G, Cukierman-Yaffe T (2022) The global spread of severe obesity in toddlers, children, and adolescents: a systematic review and meta-analysis. Obes Facts 15(2):118–134
52. Baracco R, Kapur G, Mattoo T, Jain A, Valentini R, Ahmed M et al (2012) Prediction of primary vs secondary hypertension in children. J Clin Hypert 14(5):316–321
53. Arar MY, Hogg RJ, Arant BS, Seikaly MG (1994) Etiology of sustained hypertension in children in the Southwestern United States. Pediatr Nephrol 8(2):186–189
54. Flynn J (2013) The changing face of pediatric hypertension in the era of the childhood obesity epidemic. Pediatr Nephrol 28(7):1059–1066
55. Kapur G, Ahmed M, Pan C, Mitsnefes M, Chiang M, Mattoo TK (2010) Secondary hypertension in overweight and stage 1 hypertensive children: a midwest pediatric nephrology consortium report. J Clin Hypert 12(1):34–39
56. Mavinkurve M, O'Gorman CS (2015) Cardiometabolic and vascular risks in young and adolescent girls with Turner syndrome. BBA Clinical 3:304–309
57. Doshi AR, Chikkabyrappa S (2018) Coarctation of aorta in children. Cureus 10(12):e3690
58. Kothari S (2001) Takayasu's arteritis in children—a review. Images Paediatr Cardiol 3(4):4–23
59. Green R, Gu X, Kline-Rogers E, Froehlich J, Mace P, Gray B et al (2016) Differences between the pediatric and adult presentation of fibromuscular dysplasia: results from the US Registry. Pediatr Nephrol 31(4):641–650
60. Bateman BT, Shaw KM, Kuklina EV, Callaghan WM, Seely EW, Hernández-Díaz S (2012) Hypertension in women of reproductive age in the United States: NHANES 1999–2008. PLoS ONE 7(4):e36171
61. Amiri M, Ramezani Tehrani F, Behboudi-Gandevani S, Bidhendi-Yarandi R, Carmina E (2020) Risk of hypertension in women with polycystic ovary syndrome: a systematic review, meta-analysis and meta-regression. Reproduct Biol Endocrinol 18(1)
62. Ribeiro CCM, Shimo AKK, Lopes MHBDM, Lamas JLT (2018) Effects of different hormonal contraceptives in women's blood pressure values. Revista Brasileira de Enfermagem 71(suppl 3):1453–1459
63. Cagnacci A, Zanin R, Napolitano A, Arangino S, Volpe A (2013) Modification of 24-h ambulatory blood pressure and heart rate during contraception with the vaginal ring: a prospective study. Contracept 88(4):539–543
64. Kayikcioglu F, Gunes M, Ozdegirmenci O, Haberal A (2006) Effects of levonorgestrel-releasing intrauterine system on glucose and lipid metabolism: a 1-year follow-up study. Contracept 73(5):528–531
65. Rönnerdag M, Odlind V (1999) Health effects of long-termuse of the intrauterine levonorgestrel-releasing system, A follow-up study over 12 years of continuous use. Acta Obstet Gynecol Scand 78(8):716–721
66. Hussain SF (2004) Progestogen-only pills and high blood pressure: is there an association? Contracept 69(2):89–97
67. Morais TLD, Giribela C, Nisenbaum MG, Guerra G, Mello N, Baracat E et al (2014) Effects of a contraceptive containing drospirenone and ethinylestradiol on blood pressure, metabolic profile and neurohumoral axis in hypertensive women at reproductive age. Eur J Obstet Gynecol Reproduct Biol 182:113–117

68. Palacios S, Regidor P-A, Colli E, Skouby SO, Apter D, Roemer T et al (2020) Oestrogen-free oral contraception with a 4 mg drospirenone-only pill: new data and a review of the literature. Eur J Contracept Reprod Health Care 25(3):221–227
69. Roach RE, Helmerhorst FM, Lijfering WM, Stijnen T, Algra A, Dekkers OM (2015) Combined oral contraceptives: the risk of myocardial infarction and ischemic stroke. Cochrane Database Syst Rev (8):CD011054
70. Gialeraki A, Valsami S, Pittaras T, Panayiotakopoulos G, Politou M (2018) Oral contraceptives and HRT risk of thrombosis. Clin Appl Thromb Hemost 24(2):217–225
71. Carmina E, Lobo RA (1999) Polycystic ovary syndrome (PCOS): arguably the most common endocrinopathy is associated with significant morbidity in women. J Clin Endocrinol Metab 84(6):1897–1899
72. Ding D-C, Tsai I-J, Wang J-H, Lin S-Z, Sung F-C (2018) Coronary artery disease risk in young women with polycystic ovary syndrome. Oncotarget 9(9):8756–8764
73. Joham AE, Boyle JA, Zoungas S, Teede HJ (2015) Hypertension in reproductive-aged women with polycystic ovary syndrome and association with obesity. Am J Hypertens 28(7):847–851
74. Dahlgren E, Janson PO, Johansson S, Lapidus L, Odén A (1992) Polycystic ovary syndrome and risk for myocardial infarction: evaluated from a risk factor model based on a prospective population study of women. Acta Obstet Gynecol Scand 71(8):599–604
75. Macut D, Mladenović V, Bjekić-Macut J, Livadas S, Stanojlović O, Hrnčić D et al (2020) Hypertension in polycystic ovary syndrome: novel insights. Curr Hypertens Rev 16(1):55–60
76. Bentley-Lewis R, Seely E, Dunaif A (2011) Ovarian hypertension: polycystic ovary syndrome. Endocrinol Metab Clin North Am 40(2):433–449
77. Rosenfield RL, Ehrmann DA (2016) The pathogenesis of polycystic ovary syndrome (PCOS): the hypothesis of PCOS as functional ovarian hyperandrogenism revisited. Endocr Rev 37(5):467–520
78. Carmina E (2009) Cardiovascular risk and events in polycystic ovary syndrome. Climacteric 12(sup1):22–25
79. Azziz R (2017) Does the risk of diabetes and heart disease in women with polycystic ovary syndrome lessen with age? Fertil Steril 108(6):959–960
80. Labrie F, BéLanger A, Cusan L, Gomez J-L, Candas B (1997) Marked decline in serum concentrations of adrenal c19 sex steroid precursors and conjugated androgen metabolites during aging. J Clin Endocrinol Metab 82(8):2396–2402
81. Carmina E, Campagna AM, Lobo RA (2013) Emergence of ovulatory cycles with aging in women with polycystic ovary syndrome (PCOS) alters the trajectory of cardiovascular and metabolic risk factors. Hum Reprod 28(8):2245–2252
82. Wild RA, Carmina E, Diamanti-Kandarakis E, Dokras A, Escobar-Morreale HF, Futterweit W et al (2010) Assessment of cardiovascular risk and prevention of cardiovascular disease in women with the polycystic ovary syndrome: a consensus statement by the androgen excess and polycystic ovary syndrome (AE-PCOS) society. J Clin Endocrinol Metab 95(5):2038–2049
83. Harrison CL, Lombard CB, Moran LJ, Teede HJ (2011) Exercise therapy in polycystic ovary syndrome: a systematic review. Hum Reprod Update 17(2):171–183
84. Naderpoor N, Shorakae S, De Courten B, Misso ML, Moran LJ, Teede HJ (2015) Metformin and lifestyle modification in polycystic ovary syndrome: systematic review and meta-analysis. Hum Reprod Update 21(5):560–574
85. Morley LC, Tang T, Yasmin E, Norman RJ, Balen AH (2017) Insulin-sensitising drugs (metformin, rosiglitazone, pioglitazone, D-chiro-inositol) for women with polycystic ovary syndrome, oligo amenorrhoea and subfertility. Cochrane Database of Syst Rev 2018(2)
86. Soares GM, Vieira CS, De Paula MW, Dos Reis RM, De Sá MFS, Ferriani RA (2009) Metabolic and cardiovascular impact of oral contraceptives in polycystic ovary syndrome. Int J Clin Pract 63(1):160–169
87. Ober KP, Hennessy JF (1978) Spironolactone therapy for hirsutism in a hyperandrogenic woman. annals of internal medicine 89(5_Part_1):643–644
88. Christy NA, Franks AS, Cross LB (2005) Spironolactone for hirsutism in polycystic ovary syndrome. Ann Pharmacother 39(9):1517–1521

89. Batterink J, Stabler S, Tejani A, Fowkes C (2010) Spironolactone for hypertension. Cochrane Database Syst Rev 2010(8)
90. Armanini D, Castello R, Scaroni C, Bonanni G, Faccini G, Pellati D et al (2007) Treatment of polycystic ovary syndrome with spironolactone plus licorice. Eur J Obst Gynecol Reproduct Biol 131(1):61–67
91. Bajuk Studen K, Šebeštjen M, Pfeifer M, Preželj J (2011) Influence of spironolactone treatment on endothelial function in non-obese women with polycystic ovary syndrome. Eur J Endocrinol 164(3):389–395
92. Ganie MA, Khurana ML, Eunice M, Gulati M, Dwivedi SN, Ammini AC (2004) Comparison of efficacy of spironolactone with metformin in the management of polycystic ovary syndrome: an open-labeled study. J Clin Endocrinol Metab 89(6):2756–2762
93. Jensterle M, Janez A, Vrtovec B, Meden-Vrtovec H, Pfeifer M, Prezelj J et al (2007) Decreased androgen levels and improved menstrual pattern after angiotensin II receptor antagonist telmisartan treatment in four hypertensive patients with polycystic ovary syndrome: case series. Croat Med J 48(6):864–870
94. Centre for Surveillance and Applied Research PHAoC. Canadian Chronic Disease Indicators Data Tool, 2021 Edition Ottawa (ON): Public Health Agency of Canada; 2021 [Available from: https://health-infobase.canada.ca/ccdi/
95. Brown MA, Magee LA, Kenny LC, Karumanchi SA, McCarthy FP, Saito S et al (2018) Hypertensive disorders of pregnancy: ISSHP classification, diagnosis, and management recommendations for international practice. Hypertens 72(1):24–43
96. Magee LA, Pels A, Helewa M, Rey E, Von Dadelszen P (2014) Diagnosis, evaluation, and management of the hypertensive disorders of pregnancy. Pregnancy Hyper: An Int J Women's Cardiovasc Health 4(2):105–145
97. Wang W, Xie X, Yuan T, Wang Y, Zhao F, Zhou Z et al (2021) Epidemiological trends of maternal hypertensive disorders of pregnancy at the global, regional, and national levels: a population-based study. BMC Pregnancy Childbirth 21(1)
98. Vest AR, Cho LS (2012) Hypertension in pregnancy. Cardiol Clin 30(3):407–423
99. Abe M, Arima H, Yoshida Y, Fukami A, Sakima A, Metoki H, Tada K, Mito A, Morimoto S, Shibata H, Mukoyama M (2022) Optimal blood pressure target to prevent severe hypertension in pregnancy: a systematic review and meta-analysis. Hypertens Res 45(5):887–899
100. Magee LA, Singer J, von Dadelszen P, Group CS (2015) Less-tight versus tight control of hypertension in pregnancy. N Engl J Med 372(24):2367–2368
101. Rana S, Lemoine E, Granger JP, Karumanchi SA (2019) Preeclampsia. Circ Res 124(7):1094–1112
102. Bartsch E, Medcalf KE, Park AL, Ray JG (2016) High risk of pre-eclampsia identification group. Clinical risk factors for pre-eclampsia determined in early pregnancy: systematic review and meta-analysis of large cohort studies. BMJ 353:i1753
103. Cooper WO, Hernandez-Diaz S, Arbogast PG, Dudley JA, Dyer S, Gideon PS et al (2006) Major congenital malformations after first-trimester exposure to ACE inhibitors. N Engl J Med 354(23):2443–2451
104. Bullo M, Tschumi S, Bucher BS, Bianchetti MG, Simonetti GD (2012) Pregnancy outcome following exposure to angiotensin-converting enzyme inhibitors or angiotensin receptor antagonists: a systematic review. Hypertens 60(2):444–450
105. Butalia S, Audibert F, Côté AM, Firoz T, Logan AG, Magee LA et al (2018) Hypertension Canada's 2018 guidelines for the management of hypertension in pregnancy. Can J Cardiol 34(5):526–531
106. Von Dadelszen P, Magee LA (2002) Fall in mean arterial pressure and fetal growth restriction in pregnancy hypertension: an updated metaregression analysis. J Obstet Gynaecol Can 24(12):941–945
107. Magee LA, Von Dadelszen P, Singer J, Lee T, Rey E, Ross S et al (2016) The CHIPS randomized controlled trial (control of hypertension in pregnancy study). Hypertens 68(5):1153–1159
108. Bulletins—Obstetrics ACoOaGCoP (2019) ACOG practice bulletin no. 203: Chronic hypertension in pregnancy. Obstet Gynecol 133(1):e26–e50

109. Tita AT, Szychowski JM, Boggess K, Dugoff L, Sibai B, Lawrence K, Hughes BL, Bell J, Aagaard K, Edwards RK, Gibson K, Haas DM, Plante L, Metz T, Casey B, Esplin S, Longo S, Hoffman M, Saade GR, Hoppe KK, Foroutan J, Tuuli M, Owens MY, Simhan HN, Frey H, Rosen T, Palatnik A, Baker S, August P, Reddy UM, Kinzler W, Su E, Krishna I, Nguyen N, Norton ME, Skupski D, El-Sayed YY, Ogunyemi D, Galis ZS, Harper L, Ambalavanan N, Geller NL, Oparil S, Cutter GR, Andrews WW (2022) Chronic Hypertension and Pregnancy (CHAP) Trial Consortium. Treatment for mild chronic hypertension during pregnancy. N Engl J Med 386(19):1781–1792
110. Honigberg MC, Zekavat SM, Aragam K, Klarin D, Bhatt DL, Scott NS et al (2019) Long-term cardiovascular risk in women with hypertension during pregnancy. J Am Coll Cardiol 74(22):2743–2754
111. Tooher J, Thornton C, Makris A, Ogle R, Korda A, Horvath J, Hennessy A (2016) Hypertension in pregnancy and long-term cardiovascular mortality: a retrospective cohort study. Am J Obstet Gynecol 214(6):722.e1–e6
112. Garovic VD, Bailey KR, Boerwinkle E, Hunt SC, Weder AB, Curb D et al (2010) Hypertension in pregnancy as a risk factor for cardiovascular disease later in life. J Hypertens 28(4):826–833
113. Magnussen EB, Vatten LJ, Smith GD, Romundstad PR (2009) Hypertensive disorders in pregnancy and subsequently measured cardiovascular risk factors. Obstet Gynecol 114(5):961–970
114. Hermes W, Franx A, Van Pampus MG, Bloemenkamp KWM, Bots ML, Van Der Post JA et al (2013) Cardiovascular risk factors in women who had hypertensive disorders late in pregnancy: a cohort study. Am J Obstet Gynecol 208(6):474.e1–e8
115. Mercuro G, Zoncu S, Saiu F, Mascia M, Melis GB, Rosano GMC (2004) Menopause induced by oophorectomy reveals a role of ovarian estrogen on the maintenance of pressure homeostasis. Maturitas 47(2):131–138
116. Gordon T, Kannel WB, Hjortland MC, McNamara PM (1978) Menopause and coronary heart disease. Ann Intern Med 89(2):157–161
117. Muka T, Oliver-Williams C, Kunutsor S, Laven JSE, Fauser BCJM, Chowdhury R et al (2016) Association of age at onset of menopause and time since onset of menopause with cardiovascular outcomes, intermediate vascular traits, and all-cause mortality. JAMA Cardiol 1(7):767
118. Coutinho T, Yam Y, Chow BJW, Dwivedi G, Inácio J (2017) Sex differences in associations of arterial compliance with coronary artery plaque and calcification burden. J Am Heart Ass 6(8)
119. Nakamura Y, Suzuki T, Sasano H (2005) Estrogen actions and in situ synthesis in human vascular smooth muscle cells and their correlation with atherosclerosis. J Steroid Biochem Mol Biol 93(2–5):263–268
120. Schillaci G, Verdecchia P, Borgioni C, Ciucci A, Porcellati C (1998) Early cardiac changes after menopause. Hypertens 32(4):764–769
121. Routledge FS, Mcfetridge-Durdle JA, Dean CR (2009) Stress, menopausal status and nocturnal blood pressure dipping patterns among hypertensive women. Can J Cardiol 25(6):e157–e163
122. Mcsweeney JC, Rosenfeld AG, Abel WM, Braun LT, Burke LE, Daugherty SL et al (2016) Preventing and experiencing ischemic heart disease as a woman: state of the science. Circ 133(13):1302–1331
123. Marjoribanks J, Farquhar C, Roberts H, Lethaby A, Lee J (2017) Long-term hormone therapy for perimenopausal and postmenopausal women. Cochrane Database of Syst Rev (1)
124. Salpeter SR, Walsh JME, Greyber E, Ormiston TM, Salpeter EE (2004) Mortality associated with hormone replacement therapy in younger and older women. J Gen Intern Med 19(7):791–804
125. Manson JE, Aragaki AK, Rossouw JE, Anderson GL, Prentice RL, Lacroix AZ et al (2017) Menopausal hormone therapy and long-term all-cause and cause-specific mortality. JAMA 318(10):927

126. Abramson BL, Black DR, Christakis MK, Fortier M, Wolfman W (2021) Guideline No. 422e: Menopause and cardiovascular disease. J Obstet Gynaecol Can 43(12):1438–1443.e1
127. The 2017 hormone therapy position statement of The North American Menopause Society (2018). Menopause 25(11):1362–1387
128. De Villiers TJ, Gass MLS, Haines CJ, Hall JE, Lobo RA, Pierroz DD et al (2013) Global consensus statement on menopausal hormone therapy. Climacteric 16(2):203–204
129. Shufelt CL, Merz CNB, Prentice RL, Pettinger MB, Rossouw JE, Aroda VR et al (2014) Hormone therapy dose, formulation, route of delivery, and risk of cardiovascular events in women. Menopause 21(3):260–266
130. Effects of treatment on morbidity in hypertension (1967). Results in patients with diastolic blood pressures averaging 115 through 129 mm Hg. JAMA 202(11):1028–1034
131. Effects of treatment on morbidity in hypertension II (1970). Results in patients with diastolic blood pressure averaging 90 through 114 mm Hg. JAMA 213(7):1143–1152
132. SPRINT Research Group; Wright JT Jr, Williamson JD, Whelton PK, Snyder JK, Sink KM, Rocco MV, Reboussin DM, Rahman M, Oparil S, Lewis CE, Kimmel PL, Johnson KC, Goff DC Jr, Fine LJ, Cutler JA, Cushman WC, Cheung AK, Ambrosius WT (2015) A randomized trial of intensive versus standard blood-pressure control. N Engl J Med 373(22):2103–2116
133. Jamerson K, Weber MA, Bakris GL, Dahlöf B, Pitt B, Shi V et al (2008) Benazepril plus amlodipine or hydrochlorothiazide for hypertension in high-risk patients. N Engl J Med 359(23):2417–2428
134. Melloni C, Berger JS, Wang TY, Gunes F, Stebbins A, Pieper KS et al (2010) Representation of women in randomized clinical trials of cardiovascular disease prevention. Circ Cardiovascul Qual Outcomes 3(2):135–142
135. Appel LJ, Moore TJ, Obarzanek E, Vollmer WM, Svetkey LP, Sacks FM et al (1997) A clinical trial of the effects of dietary patterns on blood pressure. N Engl J Med 336(16):1117–1124
136. Zhang W, Zhang S, Deng Y, Wu S, Ren J, Sun G et al (2021) Trial of intensive blood-pressure control in older patients with hypertension. N Engl J Med 385(14):1268–1279
137. Reckelhoff JF (2001) Gender differences in the regulation of blood pressure. Hypertens 37(5):1199–1208
138. Turnbull F, Woodward M, Anna V (2010) Effectiveness of blood pressure lowering: evidence-based comparisons between men and women. Expert Rev Cardiovascular Therapy 8:199
139. Li C, Engström G, Hedblad B, Janzon L (2006) Sex-specific cardiovascular morbidity and mortality in a cohort treated for hypertension. J Hypertens 24(8)
140. Klungel OH, de Boer A, Paes AH, Seidell JC, Bakker A (1998) Sex differences in antihypertensive drug use: determinants of the choice of medication for hypertension. J Hypertens 16(10):1545–1553
141. Lloyd-Jones DM, Evans JC, Levy D (2005) Hypertension in adults across the age spectrum. JAMA 294(4):466
142. Ljungman C, Kahan T, Schiöler L, Hjerpe P, Hasselström J, Wettermark B et al (2014) Gender differences in antihypertensive drug treatment: results from the Swedish Primary Care Cardiovascular Database (SPCCD). J Am Soc Hypertens 8(12):882–890
143. Gu Q, Burt VL, Paulose-Ram R, Dillon CF (2008) Gender differences in hypertension treatment, drug utilization patterns, and blood pressure control among us adults with hypertension: data from the national health and nutrition examination survey 1999–2004. Am J Hypertens 21(7):789–798
144. Van Der Burgh AC, Oliai Araghi S, Zillikens MC, Koromani F, Rivadeneira F, Van Der Velde N et al (2020) The impact of thiazide diuretics on bone mineral density and the trabecular bone score: the Rotterdam study. Bone 138:115475
145. Aung K, Htay T (2011) Thiazide diuretics and the risk of hip fracture. Cochrane Database Syst Rev (10):CD005185
146. Siervo M, Lara J, Chowdhury S, Ashor A, Oggioni C, Mathers JC (2015) Effects of the dietary approach to stop hypertension (DASH) diet on cardiovascular risk factors: a systematic review and meta-analysis. Br J Nutr 113(1):1–15

147. Sacks FM, Svetkey LP, Vollmer WM, Appel LJ, Bray GA, Harsha D et al (2001) Effects on blood pressure of reduced dietary sodium and the dietary approaches to stop hypertension (DASH) diet. N Engl J Med 344(1):3–10
148. Blumenthal JA (2010) Effects of the DASH diet alone and in combination with exercise and weight loss on blood pressure and cardiovascular biomarkers in men and women with high blood pressure. Arch Intern Med 170(2):126
149. Neal B, Wu Y, Feng X, Zhang R, Zhang Y, Shi J et al (2021) Effect of salt substitution on cardiovascular events and death. N Engl J Med 385(12):1067–1077
150. Seals DR, Tanaka H, Clevenger CM, Monahan KD, Reiling MJ, Hiatt WR et al (2001) Blood pressure reductions with exercise and sodium restriction in postmenopausal women with elevated systolic pressure: role of arterial stiffness. J Am Coll Cardiol 38(2):506–513
151. Murao S, Takata Y, Yasuda M, Osawa H, Kohi F (2018) The influence of sodium and potassium intake and insulin resistance on blood pressure in normotensive individuals is more evident in women. Am J Hypertens 31(8):876–885
152. Roerecke M, Kaczorowski J, Tobe SW, Gmel G, Hasan OSM, Rehm J (2017) The effect of a reduction in alcohol consumption on blood pressure: a systematic review and meta-analysis. The Lancet Public Health 2(2):e108–e120
153. Kelley GA (1999) Aerobic exercise and resting blood pressure among women: a meta-analysis. Prev Med 28(3):264–275
154. Huang G, Shi X, Gibson CA, Huang SC, Coudret NA, Ehlman MC (2013) Controlled aerobic exercise training reduces resting blood pressure in sedentary older adults. Blood Press 22(6):386–394
155. Cornelissen VA, Verheyden B, Aubert AE, Fagard RH (2010) Effects of aerobic training intensity on resting, exercise and post-exercise blood pressure, heart rate and heart-rate variability. J Hum Hypertens 24(3):175–182
156. Son W-M, Sung K-D, Cho J-M, Park S-Y (2017) Combined exercise reduces arterial stiffness, blood pressure, and blood markers for cardiovascular risk in postmenopausal women with hypertension. Menopause 24(3)
157. Chronic Hypertension and Pregnancy (CHAP) Project (CHAP) ClinicalTrials.gov: U.S. National Library of Medicine; 2022 [Available from: https://clinicaltrials.gov/ct2/show/NCT02299414

# Chapter 9
# Gestational Diabetes as a Risk Factor for Cardiovascular Disease

**Jamie L. Benham and Jennifer M. Yamamoto**

**Abstract** Gestational diabetes (GDM), defined as glucose intolerance with first recognition in pregnancy, is one of the most common medical conditions in pregnancy. It is well known that gestational diabetes is associated with a number of obstetric and neonatal complications; however, GDM is also an important risk factor for conditions that are diagnosed far past the postpartum period. One of these, which has become increasingly recognized, is the higher risk of cardiovascular disease among people who have been diagnosed with GDM. Evidence suggests that people with GDM have a twofold higher risk of cardiovascular disease and that this risk is evident within the first decade following pregnancy. As cardiovascular disease is the leading cause of death worldwide, it is imperative that the factors that contribute to cardiovascular disease in women are delineated to inform prevention and treatment strategies. A diagnosis of GDM offers women, clinicians, and policy makers a unique opportunity to implement effective screening and treatment strategies to reduce cardiovascular disease in this population. Much more research is needed to identify the best evidence-based practices for cardiovascular disease protection in this population. Nonetheless, it is essential that people with GDM and their healthcare providers recognize this risk and the importance of continued screening, treatment, and follow-up. This chapter will discuss the relationship between GDM and cardiovascular disease. Specifically, we will review: key highlights regarding GDM epidemiology, diagnosis, and outcomes; potential pathophysiologic mechanisms for the development of cardiovascular disease following GDM; the epidemiology of

J. L. Benham · J. M. Yamamoto (✉)
Department of Medicine, Cumming School of Medicine, University of Calgary, Calgary, Canada
e-mail: Jennifer.Yamamoto@umanitoba.ca

J. L. Benham
Department of Community Health Sciences, Cumming School of Medicine, University of Calgary, Calgary, Canada

J. M. Yamamoto
Department of Internal Medicine, Faculty of Health Sciences, University of Manitoba, Winnipeg, Canada

Children's Hospital Research Institute of Manitoba, Winnipeg, Canada

© The Author(s), under exclusive license to Springer Nature Switzerland AG 2023

L. Kirshenbaum and I. Rabinovich-Nikitin (eds.), *Biology of Women's Heart Health*, Advances in Biochemistry in Health and Disease 26,
https://doi.org/10.1007/978-3-031-39928-2_9

cardiovascular disease following GDM; and the screening for and prevention of cardiovascular disease following GDM.

**Keywords** Gestational diabetes · Pregnancy · Glucose · Type 2 diabetes · Postpartum

## Gestational Diabetes

### *Definitions and Epidemiology*

It is estimated that globally up to 17% of pregnancies are complicated by hyperglycemia in pregnancy [1]. Gestational diabetes (GDM), typically defined as glucose intolerance with first recognition in pregnancy, represents ~85% of hyperglycemic disorders in pregnancy. Its prevalence can vary significantly across populations based on the diagnostic criteria used and GDM risk factors present. However, it remains one of the commonest metabolic conditions affecting pregnant people. The prevalence of GDM continues to rise due to both person-level factors such as increasing obesity and advanced maternal age as well as diagnostic considerations such as screening practices and the glucose thresholds used in the diagnostic criteria.

### *Pathophysiology of Gestational Diabetes*

Pregnancy is a time of substantial physiologic adaptation. The considerable hormonal changes that occur across pregnancy are essential to facilitate maternal–fetal nutrient transport and support fetal growth. The increase in a number of hormones such as human placental lactogen, progesterone, estrogen, prolactin, growth hormone, and corticotropin releasing hormone during pregnancy lead to a steep rise in insulin resistance typically beginning mid-gestation [2]. Additionally, excessive endogenous glucose production occurs as a result of hepatic gluconeogenesis and maternal glucose uptake inhibition. Gestational diabetes develops from the inability of pancreatic β-cells to meet the higher demands due to the combination of increased insulin resistance and excessive endogenous glucose production in the latter half of pregnancy.

## *Gestational Diabetes Diagnosis and Risk Factors*

The diagnostic criteria debate is contentious and there remains no global consensus on the optimal diagnostic criteria for diagnosing GDM. Thus, international guidelines differ in terms of which criteria should be used. Likely the most commonly used diagnostic criteria are from the International Association of the Diabetes and Pregnancy Study Group (IADPSG), which diagnoses GDM based on a one-step 75 g 2-h oral glucose tolerance test (OGTT) [3]. While the screening and diagnostic criteria differ, almost all major international diagnostic criteria use dynamic glucose testing (e.g., 75 g OGTT, 50 g glucose challenge, 100 g OGTT) [4]. Most guidelines recommend screening for GDM occur at ~24–28 weeks' gestation [5–7]. There remains controversy about the role for screening for GDM in early pregnancy as we currently lack published large randomized controlled trials to demonstrate the risks and/or benefits of early screening. Additionally, the recommendations for who should be screened for GDM also differ with some guidelines recommending risk factor-based screening and others recommending universal screening for GDM. Risk factors for GDM are summarized in Box 9.1.

**Box 9.1: Risk Factors for Gestational Diabetes**

$\geq$ 35 years of age

From a high-risk group ethnic group (African, Arab, Asian, Hispanic, Indigenous, or South Asian)

Having a body mass index $\geq$ 30 kg/m$^2$

Previous GDM

Previous birthweight > 4 kg

Family history of type 2 diabetes (first-degree relative)

Polycystic Ovary Syndrome

Acanthosis nigricans

Using corticosteroids

Having prediabetes

*Adapted from the 2018 Diabetes Canada Clinical Practice Guidelines [5]

## *Complications of Gestational Diabetes*

The Hyperglycemia and Adverse Pregnancy Outcomes Study published in 2008 was a landmark study that established that lesser levels of hyperglycemia than overt diabetes were associated with an increased risk of complications [8]. Specifically, it demonstrated that the risk of large for gestational age (birthweight >90th centile), cord-blood C-peptide levels >90th centile, primary caesarean delivery, and neonatal hypoglycemia increased as glucose levels increased with no obvious threshold.

While GDM is associated with a higher risk of pregnancy complications, large randomized controlled trials and high-quality meta-analyses have demonstrated treatment decreases the risk of a number of pregnancy complications such as primary caesarean delivery, shoulder dystocia, and large infant size (macrosomia and large for gestational age) [4, 9, 10].

While the pregnancy and neonatal complications associated with GDM are well-recognized, there are a number of metabolic long-term risks for people with GDM and their offspring associated with GDM. Type 2 diabetes is perhaps the most well recognized, with GDM being associated with a sevenfold increased risk of developing type 2 diabetes [11]. This is substantial for not only the large number of individuals being diagnosed with overt diabetes, but also in terms of the young age at diagnosis; within 10 years of delivery, about half of people with previous GDM will develop diabetes or pre-diabetes [12]. Increasingly recognized is the higher risk of future cardiovascular disease in people with GDM. A meta-analysis including over 5 million people demonstrated a twofold increased risk of cardiovascular events in people with previous GDM compared to those without GDM [13].

## Pathogenesis of Gestational Diabetes as a Risk Factor for Cardiovascular Disease

The pathogenesis of GDM as a risk factor for cardiovascular disease is still being elucidated however, it is likely complex and multifactorial. This includes mechanisms such as the higher risk of type 2 diabetes, shared metabolic risk factors for GDM and cardiovascular disease, and other mechanisms such as subclinical inflammation, placental factors, and adipocyte dysfunction.

### *Risk of Type 2 Diabetes*

Type 2 diabetes is a strong risk factor for cardiovascular disease. This may be particularly important in people with previous GDM given their young age at diagnosis. In fact, a younger age at diabetes diagnosis of type 2 diabetes confers an even stronger risk of all-cause mortality and macrovascular disease [14]. There are multiple pathogenic mechanisms that lead to diabetes causing cardiovascular disease. One important mechanism is that diabetes, a pro-inflammatory state, causes vascular inflammation and subsequent cardiovascular disease [15].

### *Shared Risk Factors*

Gestational diabetes has many shared risk factors for the development of cardiovascular disease such as metabolic syndrome, obesity, dyslipidemia, hypertension, hypertensive disorders in pregnancy, polycystic ovary syndrome, and higher levels of deprivation. These shared risk factors likely explain part of risk of cardiovascular disease incurred with a diagnosis of GDM.

### *Other Mechanisms*

There are a number of other postulated mechanisms that may further explain how GDM can lead to cardiovascular disease. These include subclinical inflammation, adipocyte dysfunction, placental factors, and subclinical alternations in cardiac structure or function. Interestingly, there is emerging evidence to suggest that a number of these mechanisms occur prior to pregnancy and/or persist after pregnancy [16–19]. Additional research is needed to further elucidate the pathophysiology of GDM leading to cardiovascular disease and determine the potential role for screening for biomarkers or subclinical disease may act as a risk factor for cardiovascular disease in people with previous GDM.

## Epidemiology

The risk of cardiovascular disease is increased twofold in women affected by GDM compared with women without a history of GDM [13]. For women who subsequently develop type 2 diabetes, this risk is fourfold [20]. Some studies have reported that the risk of cardiovascular disease in women with GDM may be related to the development of type 2 diabetes [21], a known risk factor for cardiovascular disease, however, even without progression to type 2 diabetes, people with GDM have a higher risk of cardiovascular disease [13]. This increased risk is evident as early as the first decade after delivery [13, 22], and increases over time [22] (Fig. 9.1).

## Cardiovascular Disease Risk Factors

Gestational diabetes is one of many factors that may contribute to the development of cardiovascular disease among affected women. Not only do women with GDM have an almost eightfold increased risk of developing type 2 diabetes [11, 23], they are also more likely to be diagnosed with hypertension [24, 25] and dyslipidemia [26] following a pregnancy affected by GDM. Other cardiovascular risk factors that

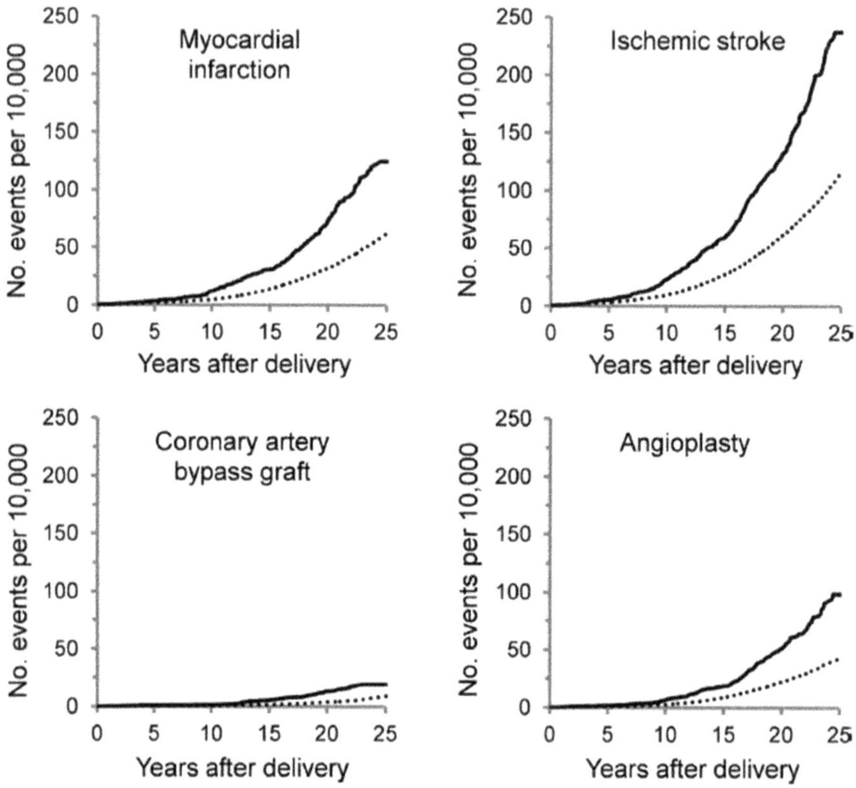


**Fig. 9.1** Cumulative incidence of cardiovascular hospitalization per 10,000 women over time*. (*Solid line, women with gestational diabetes, dotted, women without gestational diabetes. Results are shown for four outcomes that were representative of other cardiovascular disorders.) Reproduced from McKenzie-Sampson et al. [22] with permission from Springer Nature

may contribute to their risk of developing cardiovascular disease include age, family history, smoking and having overweight or obesity.

A cross-sectional study compared the cardiovascular disease risk profiles of women with GDM who did and did not develop type 2 diabetes using data from the Third National Health and Nutrition Examination Survey [27]. The authors found that while there were no differences in cardiovascular risk profiles between individuals with GDM without subsequent diabetes and those without a history of GDM, individuals who developed diabetes following GDM had a higher risk of metabolic syndrome and had more cardiovascular disease risk factors [27].

Participating in healthy lifestyle behaviours may play a mitigating role in the development of cardiovascular disease in this population. A prospective cohort study of women from the Nurse's Health Study II reported that the risk of cardiovascular disease among women with GDM was not elevated among individuals who participated in healthy lifestyle behaviours including regular physical activity, not smoking, maintaining a healthy weight, and following a good quality diet [20].

## Postpartum Screening

Cardiovascular risk factors including the development of type 2 diabetes are often identifiable in the first year postpartum [28]. Clinical practice guidelines recommend postpartum screening for diabetes in all women affected by GDM in pregnancy [6]. The reported proportion of women who complete postpartum screening for type 2 diabetes is highly variable with a systematic review reporting published estimates ranged from 5.7 to 57.9% in the first three years postpartum [29]. Although some studies have reported that the proportion of women completing postpartum screening tests for diabetes following a pregnancy affected by GDM has increased over time, less than half of women are being screened for this important cardiovascular disease risk factor [29, 30]. Postpartum screening rates for other cardiovascular disease risk factors is also poor.

Reasons for not screening for diabetes following a pregnancy affected by GDM are likely multifactorial with various health systems factors, healthcare provider factors, and patient factors contributing [31]. A survey of primary care providers and women affected by GDM in Ottawa, Ontario, Canada found that >95% of respondents believed that postpartum screening for diabetes was important, yet only 37% of primary care providers reported having their patients complete screening [32].

Women affected by GDM are asked to make many adaptations to treat the condition during pregnancy including monitoring blood glucose levels frequently, making lifestyle changes, and taking medication if required, and often have regular clinical follow-up with a diabetes team to navigate these changes. In the postpartum period, mothers experience competing demands and are less focused on their own health [33]. This may contribute to less adherence to postpartum recommendations regarding screening and lifestyle changes.

## Preventing Cardiovascular Disease Following Gestational Diabetes

Despite the growing body of evidence demonstrating an association between GDM and risk of cardiovascular disease, there are a number of challenges when considering risk reduction in this population [13]. It is becoming increasingly noted that women have been historically excluded from or underrepresented in cardiovascular studies and trials. This may be particularly true for women of reproductive age who are at risk of pregnancy or who may be breastfeeding as these are common exclusion criteria for clinical trials. Thus, there remains a paucity of clinical trial evidence guiding the evaluation of cardiovascular risk and cardiovascular risk reduction strategies in this population [34]. Nonetheless, pregnancy and the postpartum period are a critical window of opportunity to reduce cardiovascular risk as this is often a time of engagement with the healthcare team.

As we await more clinical trials and studies to guide us, there are a number of strategies we can use to reduce the risk of cardiovascular disease following GDM. These include the prevention or delay of type 2 diabetes onset, the management of other cardiovascular risk factors as well as a variety of other considerations.

## *Preventing or Delaying Type 2 Diabetes Following Gestational Diabetes*

While the risk of cardiovascular disease following GDM is not exclusively mediated through type 2 diabetes, preventing or delaying type 2 diabetes onset remains of paramount importance in cardiovascular risk reduction. It has been increasingly recognized that early onset type 2 diabetes (< 40 years) is an aggressive phenotype for the development of macro and microvascular complications [35, 36]. Given that half of people with GDM develop diabetes or prediabetes within 10 years postpartum, most people who develop type 2 diabetes following GDM will do so at a young age.

There is evidence for both lifestyle and pharmacotherapy for preventing or delaying type 2 diabetes onset. The landmark Diabetes Prevention Program (DPP) demonstrated that intensive lifestyle intervention and metformin reduced the risk of type 2 diabetes by 58% and 31% respectively in people at high risk for diabetes [37]. An analysis of the 10-year follow-up data examined the 350 women with a history of GDM who participated in the DPP [38]. This study demonstrated that both the intensive lifestyle intervention and metformin were effective in reducing progression to overt diabetes. Unlike the larger study population, metformin was found to be as efficacious as lifestyle therapy in the prevention of diabetes in this population.

A number of studies, trials, and systematic reviews have been published since the DPP examining strategies to reduce diabetes risk specifically following GDM. Overall, they suggest that both lifestyle and pharmacologic interventions are effective in delaying the onset of type 2 diabetes [39]. However, there are a number of things to highlight when translating this into clinical practice. The first is the presence of publication bias, with the presence of small positive trials and the absence small negative trials being noted in a recent high-quality review on lifestyle intervention trials [40]. The second is if clinical trial results can be effectively translated into real-world settings. The postpartum period is a busy and often stressful time in which many people find prioritizing their own health challenging. Intensive lifestyle therapy may not be a realistic or affordable option for many people in the postpartum period or the years to follow. Pharmacologic therapy may be a more feasible option for many people postpartum but evidence remains limited to metformin at this time. Ongoing studies in other classes of diabetes medication such as dipeptidyl peptidase 4 (DPP-4) inhibitors, sodium-glucose cotransporter 2 (SGLT-2) inhibitors, and glucagon-like peptide-1 (GLP-1) agonists may demonstrate other effective therapies.

There is a large body of evidence that has demonstrated that breastfeeding is associated with a lower risk of developing type 2 diabetes in both people with previous GDM and their offspring [41–43]. While we lack randomized controlled trial evidence to support causation, given the many benefits of breastfeeding, it is recommended in people who are able and willing to do so [5].

## *Treating Cardiovascular Disease Risk Factors—Postpartum Considerations*

As GDM has been recognized as a pregnancy associated risk factor for cardiovascular disease, ongoing cardiovascular disease risk surveillance and treatment post-GDM is warranted. This includes but is not limited to screening for hypertension, dyslipidemia, microalbuminuria, obesity, and smoking [34]. Effective screening strategies and treatments for cardiovascular risk factors are detailed elsewhere however, there are a number special considerations to consider in the postpartum population.

Since the higher cardiovascular disease risk after GDM is evident within the first decade following pregnancy, screening for and treatment of cardiovascular risks should occur early and often [13, 34]. Studies examining postpartum screening for overt diabetes indicate that few people are being screened for diabetes alone postpartum. As few as 1 in 2 people with GDM are being screened for diabetes within the year after pregnancy and only 1 in 3 are receiving the guideline recommended screening test [44]. There remains much work to do to improve postpartum screening for diabetes in addition to screening for other cardiovascular risk factors. Ongoing studies of postpartum cardiovascular risk prevention strategies may offer more insight into the best strategy to manage cardiovascular risks in the postpartum period.

Risk prediction is an important tenet in the prevention of cardiovascular disease however, this may be especially challenging in women. Unfortunately, current risk prediction tools have low validity in young women. Additional research is need to develop and validate risk prediction tools in young females and consider pregnancy associated cardiovascular risk factors.

## *Safe Effective Contraception*

Any discussion of risk management in the postpartum period must include safe effective contraception for those at risk of pregnancy. Many key cardiovascular risk reduction medications such as angiotensin-converting enzyme (ACE) inhibitors, angiotensin II receptor blockers, SGLT-2 inhibitors, and statins may be unsafe or lack

**Table 9.1** Lower risk contraceptive options for people at risk of pregnancy with risk factors for or established cardiovascular disease by 1-year failure rate with typical use

| Low failure rate | Moderate failure rate | High failure rate |
|---|---|---|
| (<1%) | (6–12%) | (18–28%) |
| *Long-acting reversible contraception* | Progestin-only* | Condoms* |
| Copper IUD* | | |
| Levonorgestrel IUD* | | |
| Subdermal implant* | | |
| Vasectomy** | | |

* Defined by the World Health Organisation Medical Eligibility Criteria for Contraceptive Use (MEC) categories 1 (no restriction) and 2 (advantages generally outweigh the theoretical or proven risks) [50]; ** where applicable

adequate safety data for their use in the setting of pregnancy or lactation. A number of studies highlight the lack of safe effective contraception in people at risk of pregnancy with diabetes and misconceptions about contraception in people with cardiovascular disease or risk factors [45, 46]. While cardiovascular risk should always be considered and discussed with people when choosing a contraception method, there are a number of options still available. Lower risk yet effective options include the intrauterine device (copper or levonorgestrel), subdermal implant, progestin-only pills, and vasectomy where applicable [47–50]. While barrier methods are low risk, they have a high failure rate when used alone and therefore should be considered after or in addition to other options [47] (Table 9.1).

## Future Directions

Gestational diabetes is a common endocrine disorder affecting pregnancy and is associated with an increased risk of developing cardiovascular risk factors and cardiovascular disease. Robust studies evaluating this association over time are needed along with evidence informed strategies to improve education about and prevention of the long-term cardiovascular sequelae of GDM. Until then, it is important to discuss screening and risk reduction strategies of type 2 diabetes and cardiovascular disease in all people with GDM.

## References


1. Guariguata L, Linnenkamp U, Beagley J, Whiting DR, Cho NH (2014) Global estimates of the prevalence of hyperglycaemia in pregnancy. Diabetes Res Clin Pract 103(2):176–185
2. Garcia-Patterson A, Gich I, Amini SB, Catalano PM, de Leiva A, Corcoy R (2010) Insulin requirements throughout pregnancy in women with type 1 diabetes mellitus: three changes of direction. Diabetologia 53(3):446–451
3. Metzger BE, Gabbe SG, Persson B, Buchanan TA et al (2010) International association of diabetes and pregnancy study groups recommendations on the diagnosis and classification of hyperglycemia in pregnancy. Diabetes Care 33(3):676–682
4. Pillay J, Donovan L, Guitard S, Zakher B, Gates M, Gates A et al (2021) Screening for gestational diabetes: updated evidence report and systematic review for the us preventive services task force. JAMA 326(6):539–562
5. Diabetes Canada Clinical Practice Guidelines Expert Committee, Feig DS, Berger H, Donovan L, Godbout A, Kader T et al (2018) Diabetes and pregnancy. Can J Diabetes 42(Suppl 1):S255–S282
6. National Institute for Health and Care Excellence. Diabetes in pregnancy: management from preconception to the postnatal period UK2020. Available from: https://www.nice.org.uk/guidance/ng3
7. American Diabetes A. 2. Classification and diagnosis of diabetes: standards of medical care in diabetes-2021. Diabetes Care 44(Suppl 1):S15–S33
8. Hapo Study Cooperative Research Group, Metzger BE, Lowe LP, Dyer AR, Trimble ER, Chaovarindr U et al (2008) Hyperglycemia and adverse pregnancy outcomes. N Engl J Med 358(19):1991–2002
9. Landon MB, Spong CY, Thom E, Carpenter MW, Ramin SM, Casey B et al (2009) A multicenter, randomized trial of treatment for mild gestational diabetes. N Engl J Med 361(14):1339–1348
10. Crowther CA, Hiller JE, Moss JR, McPhee AJ, Jeffries WS, Robinson JS et al (2005) Effect of treatment of gestational diabetes mellitus on pregnancy outcomes. N Engl J Med 352(24):2477–2486
11. Bellamy L, Casas JP, Hingorani AD, Williams D (2009) Type 2 diabetes mellitus after gestational diabetes: a systematic review and meta-analysis. Lancet 373(9677):1773–1779
12. Lowe WL Jr, Scholtens DM, Lowe LP, Kuang A, Nodzenski M, Talbot O et al (2018) Association of gestational diabetes with maternal disorders of glucose metabolism and childhood adiposity. JAMA 320(10):1005–1016
13. Kramer CK, Campbell S, Retnakaran R (2019) Gestational diabetes and the risk of cardiovascular disease in women: a systematic review and meta-analysis. Diabetologia 62(6):905–914
14. Nanayakkara N, Curtis AJ, Heritier S, Gadowski AM, Pavkov ME, Kenealy T et al (2021) Impact of age at type 2 diabetes mellitus diagnosis on mortality and vascular complications: systematic review and meta-analyses. Diabetologia 64(2):275–287
15. Pickup JC (2004) Inflammation and activated innate immunity in the pathogenesis of type 2 diabetes. Diabetes Care 27(3):813–823
16. Gunderson EP, Quesenberry CP Jr, Jacobs DR Jr, Feng J, Lewis CE, Sidney S (2010) Longitudinal study of prepregnancy cardiometabolic risk factors and subsequent risk of gestational diabetes mellitus: the CARDIA study. Am J Epidemiol 172(10):1131–1143
17. Quotah OF, Poston L, Flynn AC, White SL (2022) Metabolic profiling of pregnant women with obesity: an exploratory study in women at greater risk of gestational diabetes. Metabolites 12(10)
18. Retnakaran R, Qi Y, Connelly PW, Sermer M, Hanley AJ, Zinman B (2010) Low adiponectin concentration during pregnancy predicts postpartum insulin resistance, beta cell dysfunction and fasting glycaemia. Diabetologia 53(2):268–276
19. Oliveira AP, Calderon IM, Costa RA, Roscani MG, Magalhaes CG, Borges VT (2015) Assessment of structural cardiac abnormalities and diastolic function in women with gestational diabetes mellitus. Diab Vasc Dis Res 12(3):175–180

20. Tobias DK, Stuart JJ, Li S, Chavarro J, Rimm EB, Rich-Edwards J et al (2017) Association of history of gestational diabetes with long-term cardiovascular disease risk in a large prospective cohort of US women. JAMA Intern Med 177(12):1735–1742
21. Shah BR, Retnakaran R, Booth GL (2008) Increased risk of cardiovascular disease in young women following gestational diabetes mellitus. Diabetes Care 31(8):1668–1669
22. McKenzie-Sampson S, Paradis G, Healy-Profitos J, St-Pierre F, Auger N (2018) Gestational diabetes and risk of cardiovascular disease up to 25 years after pregnancy: a retrospective cohort study. Acta Diabetol 55(4):315–322
23. You H, Hu J, Liu Y, Luo B, Lei A (2021) Risk of type 2 diabetes mellitus after gestational diabetes mellitus: a systematic review and meta-analysis. Indian J Med Res 154(1):62–77
24. Tobias DK, Hu FB, Forman JP, Chavarro J, Zhang C (2011) Increased risk of hypertension after gestational diabetes mellitus: findings from a large prospective cohort study. Diabetes Care 34(7):1582–1584
25. Daly B, Toulis KA, Thomas N, Gokhale K, Martin J, Webber J et al (2018) Increased risk of ischemic heart disease, hypertension, and type 2 diabetes in women with previous gestational diabetes mellitus, a target group in general practice for preventive interventions: a population-based cohort study. PLoS Med 15(1):e1002488
26. Retnakaran R, Shah BR (2022) Mediating effect of vascular risk factors underlying the link between gestational diabetes and cardiovascular disease. BMC Med 20(1):389
27. Kim C, Cheng YJ, Beckles GL (2008) Cardiovascular disease risk profiles in women with histories of gestational diabetes but without current diabetes. Obstet Gynecol 112(4):875–883
28. Smith G, Louis J, Saade G (2019) Pregnancy and the postpartum period as an opportunity for cardiovascular risk identification and management. Obstet Gynecol 134(4)
29. Jones EJ, Hernandez TF, Edmonds JK, Edmonds JF, Ferranti EP, Ferranti EP (2019) Continued disparities in postpartum follow-up and screening among women with gestational diabetes and hypertensive disorders of pregnancy: a systematic review (1550-5073 (Electronic))
30. Carson MP, Frank MI, Keely E (2013) Original research: postpartum testing rates among women with a history of gestational diabetes–systematic review. Prim Care Diabetes 7(3):177–186
31. Nielsen KK, Kapur A, Damm P, de Courten M, Bygbjerg IC (2014) From screening to postpartum follow-up—the determinants and barriers for gestational diabetes mellitus (GDM) services, a systematic review. BMC Pregnancy Childbirth 14(1):41
32. Keely E, Clark H, Karovitch A, Graham I (2010) Screening for type 2 diabetes following gestational diabetes: family physician and patient perspectives. Can Fam Physician 56(6):558–563
33. Lie ML, Hayes L, Lewis-Barned NJ, May C, White M, Bell R (2013) Preventing type 2 diabetes after gestational diabetes: women's experiences and implications for diabetes prevention interventions. Diabet Med 30(8):986–993
34. Nerenberg KA, Cooke CL, Smith GN, Davidge ST (2021) Optimising women's cardiovascular health after hypertensive disorders of pregnancy: a translational approach to cardiovascular disease prevention. Can J Cardiol 37(12):2056–2066
35. Today Study Group, Bjornstad P, Drews KL, Caprio S, Gubitosi-Klug R, Nathan DM et al (2021) Long-term complications in youth-onset type 2 diabetes. N Engl J Med 385(5):416–426
36. Yamamoto JM, Murphy HR (2023) Treating to target glycaemia in type 2 diabetes pregnancy. Curr Diabetes Rev 19(2):e010222200742
37. Knowler WC, Barrett-Connor E, Fowler SE, Hamman RF, Lachin JM, Walker EA et al (2002) Reduction in the incidence of type 2 diabetes with lifestyle intervention or metformin. N Engl J Med 346(6):393–403
38. Aroda VR, Christophi CA, Edelstein SL, Zhang P, Herman WH, Barrett-Connor E et al (2015) The effect of lifestyle intervention and metformin on preventing or delaying diabetes among women with and without gestational diabetes: the Diabetes Prevention Program outcomes study 10-year follow-up. J Clin Endocrinol Metab 100(4):1646–1653
39. Hedeager Momsen AM, Hotoft D, Ortenblad L, Friis Lauszus F, Krogh RHA, Lynggaard V et al (2021) Diabetes prevention interventions for women after gestational diabetes mellitus: an overview of reviews. Endocrinol Diabetes Metab 4(3):e00230

40. Retnakaran M, Viana LV, Kramer CK (2023) Lifestyle intervention for the prevention of type 2 diabetes in women with prior gestational diabetes: a systematic review and meta-analysis. Diabetes Obes Metab 25(5):1196–1202
41. Martens PJ, Shafer LA, Dean HJ, Sellers EA, Yamamoto J, Ludwig S et al (2016) Breast-feeding initiation associated with reduced incidence of diabetes in mothers and offspring. Obstet Gynecol 128(5):1095–1104
42. Gunderson EP, Hurston SR, Ning X, Lo JC, Crites Y, Walton D et al (2015) Lactation and progression to type 2 diabetes mellitus after gestational diabetes mellitus: a prospective cohort study. Ann Intern Med 163(12):889–898
43. Jager S, Jacobs S, Kroger J, Fritsche A, Schienkiewitz A, Rubin D et al (2014) Breast-feeding and maternal risk of type 2 diabetes: a prospective study and meta-analysis. Diabetologia 57(7):1355–1365
44. Butalia S, Donovan L, Savu A, Johnson J, Edwards A, Kaul P (2017) Postpartum diabetes testing rates after gestational diabetes mellitus in Canadian women: a population-based study. Can J Diabetes 41(6):613–620
45. Forde R, Patelarou EE, Forbes A (2016) The experiences of prepregnancy care for women with type 2 diabetes mellitus: a meta-synthesis. Int J Womens Health 8:691–703
46. Murphy HR, Temple RC, Ball VE, Roland JM, Steel S, Zill EHR et al (2010) Personal experiences of women with diabetes who do not attend pre-pregnancy care. Diabet Med 27(1):92–100
47. O'Kelly AC, Michos ED, Shufelt CL, Vermunt JV, Minissian MB, Quesada O et al (2022) Pregnancy and reproductive risk factors for cardiovascular disease in women. Circ Res 130(4):652–672
48. Trussell J (2011) Contraceptive failure in the United States. Contracept 83(5):397–404
49. Lindley KJ, Bairey Merz CN, Davis MB, Madden T, Park K, Bello NA et al (2021) Contraception and reproductive planning for women with cardiovascular disease: JACC focus seminar 5/5. J Am Coll Cardiol 77(14):1823–1834
50. World Health Organization (2015) Medical eligibility criteria for contraceptive use 5th ed. Available from: https://www.who.int/publications/i/item/9789241549158

# Chapter 10
# Menopause and the Bridge to Cardiovascular Disease

**Sarah Rouhana and W. Glen Pyle**

**Abstract** Before menopause women are protected against cardiovascular disease morbidity and mortality, relative to age-matched men. After menopause cardiovascular disease mortality rises sharply in women to match or exceed levels in men. A higher rate of cardiovascular mortality is seen in women who experience menopause at an early age which supports the idea that cardiovascular risk is primarily driven by ovarian failure and not age. While the dramatic shift in cardiovascular disease risk is associated with the loss of estrogens in menopause, there are multiple biological and sociological phenomenon driving the threat. Shifts in adipose patterning, vasomotor symptoms and the damage to endothelial function, along with increased levels of inflammation, are all postmenopausal changes that pose threats to cardiovascular health. The inequitable treatment of women with cardiovascular disease compounds biological hazards, as does a lack of education and training for women's cardiovascular health. Estrogen replacement therapy offers some potential benefits for postmenopausal women but the knowledge gaps in our understanding of how estrogens impact cardiovascular health specifically, and postmenopausal well-being generally, limits its use. Research into the cardiovascular impact of menopause, the mechanisms of estrogen replacement therapy, and improved training for healthcare professionals could improve the health of post-menopausal women.

**Keywords** Menopause · Cardiovascular disease · Estrogens · Hormone replacement therapy · Ischemic heart disease

S. Rouhana
Laboratory of Molecular Cardiology, Department of Biomedical Sciences, University of Guelph, Guelph, ON, Canada

W. Glen Pyle (✉)
IMPART Investigator Team, Dalhousie Medicine, Saint John, NB, Canada
e-mail: gpyle@uoguelph.ca

© The Author(s), under exclusive license to Springer Nature Switzerland AG 2023

L. Kirshenbaum and I. Rabinovich-Nikitin (eds.), *Biology of Women's Heart Health*, Advances in Biochemistry in Health and Disease 26,
https://doi.org/10.1007/978-3-031-39928-2_10

# Introduction

Prior to menopause, women are relatively protected against cardiovascular disease, as evidenced by a lower incidence of cardiovascular disease morbidity and mortality when compared to men of the same age [1]. However, protection against cardiovascular disease is starkly diminished in women after menopause, as seen by the sharp rise of cardiovascular mortality later in life that matches or exceeds levels reported in the male population [2]. Despite the relative protection enjoyed by premenopausal women against cardiovascular disease mortality, cardiovascular disease in general—and ischemic heart disease specifically—is the leading cause of death of women globally [3–5].

The reasons for increased cardiovascular disease mortality in women after menopause are poorly understood. There are, however, a number of biological and sociological risk factors that change or emerge after menopause that contribute to this phenomenon. The rise in cardiovascular mortality reported in postmenopausal women is not explained by any single risk factor and is likely the result of interactions between several of these elements.

This chapter will outline the physiological changes that occur during menopause, provide mechanisms to explain the biological basis for the postmenopausal rise in cardiovascular mortality, and discuss sociological factors that compound biological risk.

# Setting the Stage: Defining Key Elements of Menopause

## (a) Menopause

Menopause is a physiological process that typically occurs over several years and results in a decline in ovarian function and the end of menstrual cycles. Some women experience iatrogenic menopause following the surgical removal of the ovaries or ovarian failure in association with chemotherapy, radiation therapy, or other medical treatments. The removal of ovaries (oophorectomy) in association with hysterectomy was once a relatively common occurrence, but current guidelines recommend retaining the ovaries unless there are medical concerns, such as a genetic risk of ovarian cancer [6, 7]. The onset of menopause is defined as the time 12 months after the final menstrual period as a result of ovarian follicle depletion, or at the time of bilateral oophorectomy [8].

Natural menopause can take up to 10 years to occur [9], with a mean age of onset of 51 years of age. However, age of menopause is impacted by a wide range of factors including genetics [10, 11]; diet and exercise; socio-economic status; and ethnicity, giving rise to variations across the world [12]. In much of the world the expanded life expectancy now means that women spend more than one-third of their lives in a postmenopausal state.

Menopause is more than a transition into a post-reproductive phase: it impacts health and quality of life. Estrogens—a key group of hormones produced by the ovaries—affect systems in the body beyond the reproductive organs, including bone, brain, adipose, muscle, blood vessels, and heart. The ovarian estrogens also regulate estrogen production by extragonadal tissues like adipose. The decline in circulating estrogens creates an imbalance with other hormones and regulatory factors in the body including testosterone and glucocorticoids, whose actions are normally countered by estrogens. Overall, the loss of estrogens and hormonal homeostasis creates a physiological environment that is associated with an increased risk for a number of health conditions including dementia, obesity, osteoarthritis, cardiovascular disease, and sarcopenia [13].

(b) **Perimenopause**

A standardized set of criteria developed by 'The Stages of Reproductive Aging Workshop + 10' (STRAW + 10) has divided the menopausal transition into 4 periods: late reproductive phase, early menopausal transition, late menopausal transition, and early postmenopause [8]. The postmenopausal stage has historically garnered the most attention in terms of research and medical interventions because of the rise in health conditions associated with this phase. Although the postmenopausal period is when most health conditions associated with menopause overtly manifest, the groundwork for risks is likely set much earlier, during the early and late menopausal transitions, more commonly referred to as "perimenopause".

Perimenopause is broadly defined as a period of irregular menstrual cycles coupled with amenorrhea that describes the time between the reproductive stage and menopause [8]. Early menopausal transition starts when women experience persistent differences in their menstrual cycle duration of 7 or more days. Cycle variability increases and there are periods of amenorrhea which last 60 days or more in the late menopausal transition. Perimenopause is completed at the end of a 12-month period of amenorrhea. In total, perimenopause can have a normal duration of 2–8 years [14].

Perimenopause is not simply a smooth and gradual decline in ovarian hormones. It is marked by highly variable and increasing levels of follicle stimulating hormone (FSH), as well as low levels of antimüllerian hormone (AMH) [9]. Perhaps the most well-known hormonal change in perimenopause are decreasing but punctuated levels of 17β-estradiol [8]. FSH levels rise up to 6 years before menopause and the decline in estrogens does not start until 2 years before the final menstrual period [15, 16]. Circulating levels of testosterone do not change significantly during the perimenopausal phase, but the declining levels of estrogens produces an imbalance between estrogens and androgens that creates a state of relative androgenic excess [17].

(c) **Early/Premature Menopause**

Menopause typically does not occur until middle age, an age that also corresponds with an increased cardiovascular disease risk in men. Because ageing is a risk factor for cardiovascular disease, it is difficult to quantify how much of the increased risk seen in postmenopausal women is a natural process of ageing and how much is the consequence of ovarian failure.

The case for ovarian involvement is evident in women who experience early menopause or premature ovarian insufficiency. Early menopause is defined as an onset that occurs before age 45, and premature ovarian insufficiency results in menopause before 40 years of age. These conditions affect 12 and 4% of women, respectively [18]. Meta-analyses of observational studies found premature ovarian insufficiency is associated with an increased rate of cardiovascular disease generally, and ischaemic heart disease specifically [19–21]. A review of 32 observational cohort, case–control, and cross-sectional studies including 310,329 women shows early menopause creates a higher risk of coronary heart disease and cardiovascular disease mortality, but not stroke [22]. Zhu and colleagues reported a similarly tight relationship between early and premature menopause with coronary heart disease, but they also report an elevated risk of stroke in women who enter menopause earlier than expected [20]. Overall, the data clearly show an elevated risk for cardiovascular disease development and mortality in women who experience either early menopause or premature ovarian insufficiency, compared to women of a similar age.

Hormone replacement therapy studies show cardiovascular disease development is reduced when women with premature ovarian insufficiency are treated until the typical age of menopause [23, 24], supporting the argument that the loss of ovarian function is a critical risk factor for cardiovascular disease. A review of 15 observational studies across 5 countries reported women with premature or early menopause who used hormone therapy for more than 10 years had a reduced risk of cardiovascular disease compared to women who did not [20]. Recommendations published in the Journal of Obstetrics and Gynaecology Canada recommend "women with premature or early-onset menopause appear to be at an increased risk of adverse cardiovascular outcomes, and this risk may be prevented by the use of menopausal hormone therapy until the average age of menopause" [25].

Together, these studies uncouple the risk of cardiovascular disease development in postmenopausal women from ageing, and establish a link between the decline in ovarian estrogen production and risk factors for cardiovascular disease. These data, coupled with studies demonstrating an ability of estrogen replacement to mitigate some cardiovascular disease hazard, make a strong case for a key role of ovarian failure in the increased rate of cardiovascular disease seen in postmenopausal women.

## Emergence of Risk

The physiological changes that accompany menopause create a favourable environment for the development of cardiovascular disease and co-morbidities that worsen or promote cardiovascular illness. Diabetes [26, 27], hypertension [28, 29], and dyslipidemia [30] are among the co-morbidities observed at higher rates in postmenopausal women. Metabolic syndrome, which is characterized by increased abdominal obesity, insulin resistance, elevated blood pressure, and increased inflammation, is more prevalent in women after menopause [31, 32]. These comorbidities exacerbate injury associated with cardiovascular events like myocardial infarction and complicate the clinical management of patients, giving rise to higher adverse events and mortality rates. How they arise is a reflection of the diverse impact estrogens have on the body and the systematic effects of menopause.

### (a) **Adipose**

Regional adiposity is a risk factor in cardiovascular disease. The loss of estrogens coupled with a relative increase in androgens after menopause creates a hormonal milieu in which body fat distribution is altered in women [33, 34]. Premenopausal adipose distribution is primarily subcutaneous in the gluteofemoral region and is associated with a lower risk of metabolic diseases and other pre-cursors to cardiovascular disease [35–37]. After menopause, there is a shift towards an increase in visceral storage of adipose [38, 39]. A key role for the location of adipose deposits in determining risk of disease was demonstrated in a study of normal weight postmenopausal women that showed visceral obesity was associated with increased mortality, but increased subcutaneous levels were not [40]. The adjustment in adipose distribution towards higher levels in abdominal and intraperitoneal regions more closely resembles the male pattern of adipose distribution and creates an elevated risk for cardiovascular risk factors like diabetes [41], and cardiovascular conditions such as hypertension [41] and coronary artery disease [42].

Sex differences in adipose patterning has long been known, but the mechanisms underlying shifts in adipose deposition with menopause remains poorly understood. There is emerging evidence for interactions between sex steroids and the adipose microenvironment to drive the alterations in adipose distribution that follow menopause. Abildgaard and colleagues reported menopause is associated with metabolic dysfunction in visceral and subcutaneous adipose [43]. Both visceral and subcutaneous adipose showed adipocyte hypertrophy, increased inflammation, hypoxia, and fibrosis, while visceral adipocytes exhibited a decrease in insulin sensitivity. Subcutaneous adipose tissue dysfunction causes an inability to store fats in this location and creates a spill-over that is taken up by visceral adipose. Price et al. previously speculated that a decline in circulating progesterone coupled with increased production of estrone in the gluteofemoral region decreases lipoprotein lipase activity regionally and explains the decreased ability to store fats in this adipose site [44]. After menopause, the visceral adipose tissue is the primary source of estrogens and increased local levels may create a cellular milieu that promotes visceral expansion [45]. Despite studies offering data to support a local mechanism of adipose regulation after menopause, the details of local adipose storage both before and after menopause remain largely unknown.

Increased postmenopausal obesity may also be centrally mediated. In mouse models, central nervous system estrogen receptor-α (ERα) influences abdominal obesity and physical activity, two areas that are commonly affected in postmenopausal women. Deletion of ERα in hypothalamic steroidogenic factor-1 neurons decreased energy expenditure in the form of physical activity and increased deposition of fat in the abdomen [46]. Deletion of the same estrogen signaling target in pro-opiomelanocortin neurons of the hypothalamus stimulated appetite. Interestingly, ERα reductions in ventromedial hypothalamic nucleus steroidogenic factor-1 neurons produces a similar hypoactive state, possibly through decreased SNS activity, which would also increase lipid storage [47]. The postmenopausal reduction in circulating estrogens produces less stimulant for central nervous system ERα, providing a feasible mechanism to explain some body composition alterations.

(b) **Vasomotor Symptoms**

Estrogens help maintain a healthy, functioning endothelium, which is diminished with menopause. The onset of endothelial dysfunction begins early in menopause [48–50] and may set the stage for later coronary artery lesions. The process of atherosclerosis and vascular dysfunction is enhanced with other menopausal changes, including increased activation of the renin–angiotensin–aldosterone [51, 52] and sympathetic nervous systems [53–55].

Vasomotor symptoms (VMS) are the cardinal symptom of the menopausal transition. Over 70% of women report VMS [56] which include hot flushes and night sweats. The intensity and frequency of VMS vary widely, and are impacted by ethnic background, genetics, diet, and level of physical activity [57]. In addition to impacting quality of life by disrupting sleep and mood, VMS appear to be a harbinger of future cardiovascular risk.

Women who experience more severe symptoms, or for a prolonged period of time, are at increased risk of developing cardiovascular disease [58, 59]. A study of 11,725 women who had frequent VMS before age 50 found an increased risk of coronary artery disease [60]. The SWAN [61, 62], WISE [63], and WHI-OS [59] studies all found VMS were associated with preclinical signs of cardiovascular disease, although KEEPS [64] found no association in women at low risk for cardiovascular disease.

Early onset [65] and prolonged VMS duration [66] are individually linked to subclinical changes in vascular structure and function that are risk factors for cardiovascular disease. How menopausal VMS increase the risk of cardiovascular disease is not definitively known, although there are several plausible mechanisms. VMS are a consequence of a more narrow thermoeneutral zone that arises during perimenopause. The constricted tolerance for temperature variations triggers a physiological response to dissipate the perceived increase in heat, largely through the autonomic nervous system. Increased SNS and/or decreased PNS activity work to mediate this effect, but the impact is systematic, including the cardiovascular system. Chronic SNS activation in association with VMS could lead to increased blood pressure and the risk for cardiovascular disease. Increased SNS activity is also linked to dyslipidemia [67] which may explain unhealthy lipid profiles that can develop in women with VMS, independent of BMI and other known CVD risk factors [68, 69]. Some [70], but not all [71] studies have demonstrated a positive correlation between VMS and increased levels of pro-inflammatory cytokines such as IL-6, IL-8, and TNF-α [72, 73], which promote cardiovascular disease including endothelial dysfunction.

Sleep disturbances and mood imbalances that are linked to VSM may also provide connections between the menopausal effects of VMS and the increased risk for CVD.

Studies have shown that reduced or disrupted sleep is associated with subclinical CVD [74, 75]. VSM also negatively impacts mood, which creates CVD risk [76, 77].

Whether VSM increase cardiovascular risk directly or are a coincidental occurrence in women who develop chronic conditions that have been linked to this menopausal phenomenon has been difficult to determine, given the number of other postmenopausal changes that occur simultaneously. For example, women with obesity [78, 79]—a known cardiovascular risk factor—are more likely to experience severe VMS.

(c) **Inflammation**

Both preclinical animal models of menopause and human studies have reported increased levels of circulating proinflammatory cytokines supporting a heightened level of inflammation [80–82]. Fernandes and colleagues were the first to report changes in myocardial levels of cytokines in a mouse model of menopause [83]. The links between inflammation and cardiovascular disease are well established [84–86, 90–93].

Estrogens—in particular 17β-estradiol—have direct impact on inflammatory cytokine levels by supressing expression of pro-inflammatory IL-1 [88], IL-6 [88–90], and TNF-α [89, 90], and stimulating release of anti-inflammatory IL-10 [91, 92]. These effects are mediated through ER which are expressed in a number of cell types including macrophages, B cells, T-cells, neutrophils, and monocytes [93]. The anti-inflammatory actions of estrogens occur when hormonal levels are high, but estrogens can be pro-inflammatory at low concentrations [94]. Progesterone, which is also lost during menopause, promotes a predominantly anti-inflammatory cytokine profile [95].

Estrogens can also impact immune function and inflammation through indirect pathways. Metabolically stressed cells secrete inflammatory cytokines and chemotactic molecules to recruit immune cells such as macrophages, lymphocytes, and eosinophils to the dysfunctional tissue. The loss of estrogens and an ability to protect against metabolic stress with menopause increases the likelihood of energetic disturbances which create a pro-inflammatory environment. The disruption in metabolism that occurs with menopause and promotes a pro-inflammatory state is referred to as "metainflammation" [96].

(d) **The Inequity of Medicine**

Menopause is associated with numerous biological changes that individually and in combination increase the risk of cardiovascular disease. But beyond the biological changes, social and medical management factors that influence care also contribute to the elevated risk of morbidity and mortality that accompanies menopause.

Myocardial infarctions with obstructive coronary artery disease (type 1) are 3 times more common in men than women, while myocardial infarctions with no obstructive coronary arteries (MINOCAs, type 2) are more prevalent in women [2]. The lack of coronary obstructions associated with MINOCA leads to a more optimistic prognosis and less aggressive treatment. This benign prediction stands in contrast to clinical trials showing that clinical outcomes with MINOCA patients are similar to those with obstructive coronary artery disease [97]. The disproportionate representation of postmenopausal women who experience MINOCA as compared to similarly aged men may contribute to the higher rates of acute myocardial infarction mortality, but the relatively small number of MINOCA cases as a proportion of all acute myocardial infarctions limits the ability to use this issue to explain a significant amount of the sex-dependent discrepancies in mortality.

Beyond the risks specifically associated with MINOCA, studies have consistently shown that women are less likely to be given timely treatment following acute myocardial infarctions. A study of 192,204 adults over the age of 60 diagnosed with STEMI found that women are less likely to receive Percutaneous Coronary Intervention (PCI) or coronary artery bypass grafts, and that the gap between the sexes for revascularization increased between 2005 and 2014 [98]. While the presence of co-morbidities that increase the risk of bleeding may explain some of the differences in PCI between the sexes, women who did receive revascularization therapy underwent procedures after a more significant delay than men. These data suggest that even when revascularization therapy is not contraindicated because of co-morbidities, there is still a difference in interventional therapies between the sexes. The combination of fewer interventions and delayed treatment was associated with a higher in-hospital mortality in older women as compared to men.

Medical management of patients at risk of cardiovascular disease or who have previously had a cardiovascular event is critical to limit the risk of recurrence or a first-time event. Guideline therapies for individuals at risk or recovering from cardiovascular disease generally do not differ between the sexes. The lack of sex-specific guidelines are the result of a paucity of clinical and preclinical research involving female patients or animals, as opposed to a need to treat similar pathophysiological mechanisms between the sexes [99, 100]. Despite similarities in treatment plans, women are generally less likely than men to receive pharmaceutical treatments according to guidelines to manage cardiovascular risk in both primary and secondary care setting [1, 101, 102].

## Hormone Replacement Therapy (HRT): Restoring the "Protection"?

Throughout its history HRT has undergone extensive investigation with paradoxical findings: some studies report beneficial effects on women's health, while others find no benefits or even increased risk for a variety of health conditions.

### (a) A Brief History of HRT

In the 1960s, American gynecologist Robert Wilson published the book "Feminine Forever" in which he advocated for the use of estrogen supplements to avoid menopausal "symptoms" that led to a "horror of…living decay" [103]. Wilson's work generated significant philosophical debate about the view of menopause as a medical condition which persisted decades after publication [104].

In 1975 two reports linking HRT to an increased risk of endometrial cancer presented the first significant obstacle to the clinical use of exogenous estrogens to mitigate the symptoms and risks of menopause [105, 106]. Concerns about a link to cancer were addressed by lowering the dose of estrogens and adding progesterone [107].

Over the next several decades preclinical research and observational studies produced a remarkably consistent body of evidence demonstrating clear cardiovascular benefits of HRT for postmenopausal women. However, in 1998 the Heart and Estrogen/progestin Replacement Study (HERS) yielded data questioning the benefits of HRT and offered evidence of some increased risk [108]. These concerns were hotly debated until 2002 when the Women's Health Initiative (WHI) released the results of a trial where data showed a 29% increase in heart attacks alongside a 26% increased risk of breast cancer [109]. The clear risks of HRT coupled with no obvious cardiovascular benefits shown in the WHI study seemed to mark the end of HRT.

(b) **Solving the Paradox**

Despite the clearly negative results of WHI and other clinical trials investigating the health benefits of HRT, a closer examination of the findings raised a number of issues that may see the re-emergence of estrogen replacement therapy as a tool to reduce risk and improve the health of postmenopausal women. Among the variables that differ across studies and may explain some of the variations in outcomes are the timing and duration of estrogen replacement therapy; dose and route of estrogen administration; as well as the formulation of estrogens used.

In general, studies have found that initiating hormone therapy within 10 years of the onset of menopause (< 60 years of age) reduces rates of myocardial infarctions, cardiac deaths, and all-cause mortality [110, 111]. Initiation of therapy within 6 years is beneficial in slowing the progression of atherosclerosis, but not when started after 10 years postmenopause [112]. A Cochrane review of placebo-controlled RCTs found HRT reduced coronary artery deaths and all-cause mortality when HRT was started within 10 years of menopause, with a neutral impact beyond 10 years [113]. Even in studies that found negative results, women who initiated treatment at earlier ages demonstrated benefits with respect to cardiovascular disease risk and mortality [113]. Together, these studies suggest that early intervention with HRT appears to have cardiovascular benefits, but initiating or continuing treatment beyond a decade after menopause is not widely supported.

Vaginally administered estrogens have beneficial cardiovascular outcomes, while transdermal estrogens may increase the risk of cardiovascular death [113]. Interestingly, vaginal estrogens are weaker than those administered using dermal patches, suggesting that low dose estrogens may have more benefits than higher doses, and that route of delivery impacts cardiovascular outcomes.

Estrogen is not a singular compound, but rather a group of hormones. Hormone replacement therapies are often comprised primarily of 17β-estradiol along with other estrogens [114]. The variable formulations of HRT are likely to contain agonists that differentially activate the various estrogen receptors in the body, creating diverse and even paradoxical effects.

Although the data are not clear in terms of fully understanding the risks and benefits of estrogen replacement therapy, on the whole, most medical societies support the use of hormone replacement therapy to relive menopausal symptoms for women who experience early menopause or premature ovarian insufficiency; postmenopausal women under 60 years of age; and those who are at low-to-moderate CVD risk [115–117]. While there are suggestions that HRT may be protective, there are contraindications which preclude its universal application. Women with a history of venous thromboembolism, or histories of cardiovascular disease or breast cancer should be excluded.

The effectiveness of hormonal replacement therapy has been a hotly debated topic with evidence supporting a cardiovascular advantage for use, as well as no benefit against cardiovascular disease or even an increased risk of negative outcomes. These discrepant findings and the inability to adequately explain their existence underscores the knowledge gap that exists concerning our understanding of how estrogens work in the body; how ageing impacts their effects; and how these sex hormones respond to stressors.

(c) **Mechanisms of HRT**

The three known estrogen receptors—α (ERα), β (ERβ), and G protein-coupled estrogen receptor 1 (GPER)—are expressed in numerous cell types. Depending on their distribution and level of expression they can induce and initiate different cascades and actions. The beneficial effects of estrogens seen on the vascular system are mainly due to an ERα mediated response [118, 119]. The activation of vascular ERα releases arterial vasodilatation actors and modulates the inflammatory response through genomic mechanisms [120]. In fact, with the hormonal changes occurring with menopause, it is believed that there is a shift toward a higher expression of ERβ which is associated with higher inflammation and endothelial dysfunction, possibly explaining the negative results of clinical trials when HRT was applied well after menopause.

Several studies have shown the beneficial effects of estrogens in downregulating vascular inflammation through the decrease of cytokines, intracellular adhesion molecules (VCAM-1 and ICAM-1), and the accumulation of leucocytes in the endothelium [94, 121–123]. In addition, the estrogen-dependent decrease in TNF-α and IL-1β in different cell types has been proposed as a driving force behind the anti-inflammatory and vasculo-protective actions [94, 124]. Some studies have, on the contrary, shown an increase in inflammatory cytokines as a result of estrogen excess [125]. However, it seems that this connection is only consolidated in the context of auto-immune diseases such rheumatoid arthritis and atherosclerosis [126].

Other effects of estrogen deficiency have highlighted their role as antioxidants. Estrogen loss leads to increased production of reactive oxygen species which promotes LDL oxidation [127]. LDL oxidation is a key step in the process of atherosclerosis by inducing the formation of foam cells on blood vessel walls. This disease is driven by an immune-mediated inflammation and increased production of proinflammatory cytokines, both of which damage vascular endothelium [128]. The endothelial damage arising from inflammation and oxidative stress leads to a decrease in endothelial-derived nitric oxide production, which impairs vascular function. The introduction of exogenous estrogens can prevent the oxidation of LDL and reduce endothelial damage [127, 129].

The balance between ERα and ERβ signaling is a crucial factor in maintaining the regulation of inflammation, oxidative stress, and other potentially damaging processes which contribute to cardiovascular disease. Most of the beneficial effects of estrogens on the cardiovascular system are mediated through ERα which promotes vasodilation and reduces platelet aggregation. By contrast, ERβ activation is highly expressed in conditions such as oxidative stress, hypoxia, and inflammation, and restrains the beneficial effects of estrogen through ERα activation [130]. Of some note is the recent finding that in a mouse model of menopause the responsiveness of cardiac estrogen receptors is significantly altered during the perimenopausal transition [83]. These findings support the concept that estrogen replacement therapy has time-dependent benefits and risks, and that the remodelling of organ systems with menopause creates a fundamentally different physiology whose regulation by estrogens may be altered. The idea that a simple replacement of a complex group of hormones with a relatively static delivery system in an altered physiological environment ignores the complexity and diversity of estrogen regulation and the profound changes associated with menopause.

## Women's Health Education

Following prescribed treatment plans and implementing lifestyle modifications to reduce risk is highly correlated with understanding cardiovascular disease [131]. A number of cardiovascular organizations and societies have undertaken public education campaigns to increase awareness of cardiovascular disease in women. These initiatives have been tremendously successful in making both women and men aware of the cardiovascular disease risks faced by women; the unique symptoms that may occur in women with cardiovascular disease; and empowering women to advocate for appropriate care [132].

Despite the success of public education campaigns, there remain significant knowledge gaps in both the general public and amongst healthcare practitioners concerning cardiovascular disease and women. In 2013, the Canadian Women's Heart Health Centre (CWHHC) surveyed Canadian women to determine their knowledge and understanding of heart health [133]. The CWHHC found 75% of women had low to medium level knowledge of cardiovascular disease and risk factors. A greater understanding of cardiovascular disease was associated with a university education and higher household income. Of some concern were the findings that those whose knowledge of cardiovascular disease was lowest, tended to have the greatest overestimation of their understanding, and that 60% of women deemed to be at high risk for cardiovascular disease described their risk as low or moderate. On the positive side, women who were over 55 and most likely to be postmenopausal were the most likely to discuss heart disease risks and prevention with their healthcare providers.

More recent studies and surveys conducted in other countries have reported similar results concerning public understanding of cardiovascular disease in women [132, 134]. Thus, while the efforts of research societies and medical organizations have dramatically improved awareness of cardiovascular disease in women, there remains a significant deficiency in public education.

A similarly concerning knowledge deficiency has been identified among healthcare professionals. A 2004 survey found 70% of medical school trainees received no formal education or training on sex or gender-based medicine [135]. A more recent review of medical school curriculum in Canada shows that the problem persists, as evidenced by a general lack of material on women's health or sex differences [136]. The lack of education and training on women's health has resulted in a lack of confidence among healthcare professionals when it comes to treating women [137]. Only 39% of primary healthcare providers recognized cardiovascular disease as a top health concern for women, and less than half of primary care physicians and cardiologists felt well prepared to manage cardiovascular disease in women [132].

## Path Forward

Menopause presents a complex and intricate series of biological stressors on a women's cardiovascular system (Fig. 10.1). The addition of sociological issues which complicate treatment creates an environment ripe for cardiovascular risk. Despite the bleak picture, there are several opportunities to advance and improve women's cardiovascular health, specifically by targeting the challenges posed by menopause.

Discovery and preclinical research have advanced our understanding of the role estrogens play in the body and how their loss impacts health and disease. The emergence of a new animal model of menopause to create and study the transitional perimenopausal phase offers an important opportunity to recapitulate the hormonal changes that characterizes menopause [138]. The more physiologically accurate model will allow for an investigation of the Timing Theory—the idea that menopause creates distinct windows of risk and that HRT may be most effective when applied within distinct phases. Studies using this model have already shown that cardiovascular changes occur during the perimenopausal period, and have identified potential targets to mitigate risk [83, 139].

The initial public education campaigns by professional societies and advocacy groups should be lauded for their success in improving knowledge about the risk of cardiovascular disease faced by women. However, there remain knowledge gaps in both the public and healthcare professional groups that require a new approach. In an effort to improve healthcare training and correct the inequity of treatment that negatively impacts women, the Canadian Women's Heart Health Alliance developed an accredited physician education program that specifically deals with women and cardiovascular disease [140]. The women's heart health curriculum targets physician specialists in internal medicine and cardiology; nurses; medical trainees; cardiac rehabilitation specialists; and women at risk for cardiovascular disease. A survey

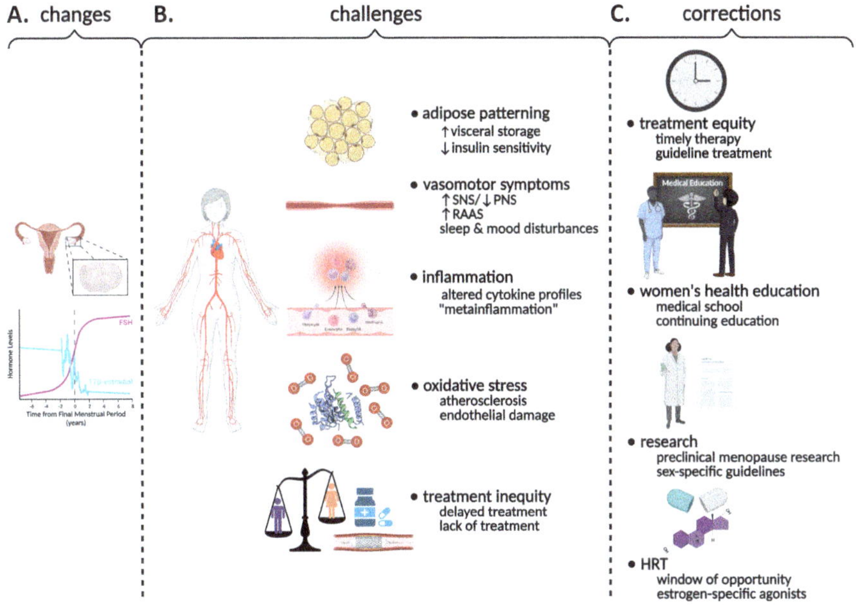


**Fig. 10.1** ***Cardiovascular risk in postmenopausal women*****. a** The decline in ovarian function throughout the perimenopausal phase presents a stress to the cardiovascular system that is driven largely by the punctuated decline in estrogens. **b** Systemic challenges including changes in adipose patterning, vascular dysfunction, inflammation, and oxidative stress cause damage and dysfunction across the cardiovascular system. Social inequities including unequal access to timely and guideline-driven medical care exacerbates biological risk. **c** The application of treatments according to recommended guidelines improves outcomes for postmenopausal women with cardiovascular disease. Public and professional education programs whose content is based on preclinical and clinical research dedicated to understanding women's cardiovascular physiology including the changes produced by menopause are critical for improved care and prevention strategies. Research into the mechanisms of HRT may reshape its delivery or advance alterative therapies that mitigate cardiovascular disease risk after menopause. Figure created with BioRender.com

found most healthcare professionals are willing to pursue additional training to address their knowledge gaps concerning sex and cardiovascular disease, which bodes well for these educational initiatives.

The concerning results of the WHI and other studies which found no cardiovascular benefits of HRT were a stark reminder of how little is known about estrogens, their role in cardiovascular physiology, and their potential as therapeutic tools. Preclinical research investigating the physiological changes that occur during menopause coupled with studies designed to understand the fundamental role of estrogens in regulating cardiovascular function could be used to re-design HRT. The use of specific estrogen receptor agonists/antagonists for more precise prophylactic treatment would limit unwanted side effects and increase effectiveness. Identifying the lowest effective dose and the best route of treatment would also limit side effects and provide the most impactful interventions.

Women make up slightly more than half the global population and on average spend nearly one-third of their lives in menopause. The cardiovascular risk posed by this phase of life is significant and profoundly impacts quality of life as well as life expectancy. Addressing the unique challenges of women's cardiovascular health demands a focused and significant investment of time and resources. Without a dedicated undertaking women will continue to receive substandard care. These costs are both economic and social, and reach beyond the women directly affected by cardiovascular disease to profoundly impact the societies in which they live.

**Acknowledgements** Glen Pyle is a Heart and Stroke Foundation of Canada Senior Career Investigator for Improving the Heart and Brain Health for Women in Canada. His research is supported with funding from the Heart and Stroke Foundation of Canada, Health Canada, Canadian Institutes of Health Research, and the Natural Sciences and Engineering Research Council of Canada.

## References

1. Irani SS, Alonso A, Benjamin EJ, Bittencourt MS, Callaway CW, Carson AP, Chamberlain AM, Chang AR, Cheng S, Delling FN, Djousse L, Elkind MSV, Ferguson JF, Fornage M, Khan SS, Kissela BM, Knutson KL, Kwan TW, Lackland DT, Lewis TT, Lichtman JH, Longenecker CT, Loop MS, Lutsey PL, Martin SS, Matsushita K, Moran AE, Mussolino ME, Perak AM, Rosamond WD, Roth GA, Sampson UKA, Satou GM, Schroeder EB, Shah SH, Shay CM, Spartano NL, Stokes A, Tirschwell DL, VanWagner LB, Tsao CW (2020) American Heart Association Council on Epidemiology and Prevention Statistics Committee and Stroke Statistics Subcommittee. Heart disease and stroke statistics-2020 update: a report from the American Heart Association. Circ 141(9):e139–e596
2. Maas AHEM et al (2021) Cardiovascular health after menopause transition, pregnancy disorders, and other gynaecologic conditions: a consensus document from European cardiologists, gynaecologists, and endocrinologists. Eur Heart J 42:967–984
3. Wilkins E, Wilson L, Wickramasinghe K, Bhatnagar P, Leal J, Luengo-Fernandez R, Burns R, Rayner M, Townsend N (2017) European cardiovascular disease statistics, 2017 edn. European Heart Network, pp 1–192
4. Kaptoge S et al (2019) World Health Organization cardiovascular disease risk charts: revised models to estimate risk in 21 global regions. Lancet Glob Heal 7:e1332–e1345
5. Abbafati C et al (2020) Global burden of 369 diseases and injuries in 204 countries and territories, 1990–2019: a systematic analysis for the Global Burden of Disease Study 2019. Lancet 396:1204–1222
6. Novetsky AP, Boyd LR, Curtin JP (2011) Trends in bilateral oophorectomy at the time of hysterectomy for benign disease. Obstet Gynecol 118:1280–1286
7. Mikhail E et al (2015) National trends of adnexal surgeries at the time of hysterectomy for benign indication, United States, 1998–2011. Am J Obstet Gynecol 213(713):e1–713.e13
8. Harlow SD et al (2012) Executive summary of the stages of reproductive aging workshop 10: addressing the unfinished agenda of staging reproductive aging. J Clin Endocrinol Metab 97:1159–1168
9. Randolph JF et al (2003) Reproductive hormones in the early menopausal transition: relationship to ethnicity, body size, and menopausal status. J Clin Endocrinol Metab 88:1516–1522
10. Ruth KS et al (2021) Genetic insights into biological mechanisms governing human ovarian ageing. Nat 596:393–397

11. Bae H et al (2019) Genetic associations with age of menopause in familial longevity. Menopause 26:1204–1212
12. Gold EB (2011) The timing of the age at which natural menopause occurs. Obstet Gynecol Clin North Am 38:425–440
13. Van Dijk GM, Kavousi M, Troup J, Franco OH (2015) Health issues for menopausal women: the top 11 conditions have common solutions. Maturitas 80:24–30
14. Brambilla DJ, Mckinlay SM, Johannes CB (1994) Defining the perimenopause for application in epidemiologic investigations. Am J Epidemiol 140:1091–1095
15. Randolph JF et al (2011) Change in follicle-stimulating hormone and estradiol across the menopausal transition: effect of age at the final menstrual period. J Clin Endocrinol Metab 96:746–754
16. Burger HG et al (1999) Prospectively measured levels of serum follicle-stimulating hormone, estradiol, and the dimeric inhibins during the menopausal transition in a population-based cohort of women. J Clin Endocrinol Metab 84:4025–4030
17. Hall JE (2015) Endocrinology of the menopause. Endocrinol Metab Clin North Am 44:485–496
18. Golezar S, Ramezani Tehrani F, Khazaei S, Ebadi A, Keshavarz Z (2019) The global prevalence of primary ovarian insufficiency and early menopause: a meta-analysis. Climacteric 22(4):403–411
19. Roeters van Lennep JE, Heida KY, Bots ML, Hoek A (2016) Collaborators of the Dutch Multidisciplinary Guideline Development Group on Cardiovascular Risk Management after Reproductive Disorders. Cardiovascular disease risk in women with premature ovarian insufficiency: a systematic review and meta-analysis. Eur J Prev Cardiol 23(2):178–186
20. Zhu D et al (2019) Age at natural menopause and risk of incident cardiovascular disease: a pooled analysis of individual patient data. Lancet Public Health 4:553–564
21. Tao XY, Zuo AZ, Wang JQ, Tao FB (2016) Effect of primary ovarian insufficiency and early natural menopause on mortality: a meta-analysis. Climacteric 19(1):27–36
22. Muka T, Oliver-Williams C, Kunutsor S, Laven JSE, Fauser BCJM, Chowdhury R, Kavousi M, Franco O (2016) Association of age at onset of menopause and time since onset of menopause with cardiovascular outcomes, intermediate vascular traits, and all-cause mortality. JAMA Cardiol 1:767–776
23. Webber L et al (2016) ESHRE Guideline: management of women with premature ovarian insufficiency. Hum Reprod 31:926–937
24. Webber L, Anderson RA, Davies M, Janse F, Vermeulen N (2017) HRT for women with premature ovarian insufficiency: a comprehensive review. Hum Reprod Open (2):hox007
25. Abramson BL, Black DR, Christakis MK, Fortier M, Wolfman W (2021) Guideline No. 422e: menopause and cardiovascular disease. J Obstet Gynaecol Canada 43:1438–1443
26. Ren Y et al (2019) Association of menopause and type 2 diabetes mellitus. Menopause 26:325–330
27. Slopien R et al (2018) Menopause and diabetes: EMAS clinical guide. Maturitas 117:6–10
28. Pollow DP, Uhlorn J, Husband N, Brooks HL (2019) Regulation of postmenopausal hypertension. Sex Differ Cardiovasc Physiol Pathophysiol 105–118
29. Lima R, Wofford M, Reckelhoff JF (2012) Hypertension in postmenopausal women. Curr Hypertens Rep 14:254
30. Wietlisbach V, Marques-Vidal P, Kuulasmaa K, Karvanen J, Paccaud F (2013) The relation of body mass index and abdominal adiposity with dyslipidemia in 27 general populations of the WHO MONICA Project. Nutr Metab Cardiovasc Dis 23:432–442
31. Carr MC (2003) The emergence of the metabolic syndrome with menopause. J Clin Endocrinol Metab 88:2404–2411
32. Janssen I, Powell LH, Crawford S, Lasley B, Sutton-Tyrrell K (2008) Menopause and the metabolic syndrome: the study of women's health across the nation. Arch Intern Med 168:1568–1575
33. Rebuffe-Scrive M, Cullberg G, Lundberg PA, Lindstedt G, Bjorntorp P (1989) Anthropometric variables and metabolism in polycystic ovarian disease. Horm Metab Res 21:391–397

34. Killinger DW, Perel E, Daniilescu D, Kharlip L, Lindsay WRN (1990) Influence of adipose tissue distribution on the biological activity of androgens. Ann NY Acad Sci 595:199–211
35. Hill JH, Solt C, Foster MT (2018) Obesity associated disease risk: the role of inherent differences and location of adipose depots. Horm Mol Biol Clin Investig 33(2):/j/hmbci.2018.33.issue-2/hmbci-2018-0012/hmbci-2018-0012.xml.
36. Piché ME, Vasan SK, Hodson L, Karpe F (2018) Relevance of human fat distribution on lipid and lipoprotein metabolism and cardiovascular disease risk. Curr Opin Lipidol 29:285–292
37. Diaz-Canestro C, Xu A (2021) Impact of different adipose depots on cardiovascular disease. J Cardiovasc Pharmacol 78:S30–S39
38. Svendsen OL, Hassager C, Christiansen C (1995) Age- and menopause-associated variations in body composition and fat distribution in healthy women as measured by dual-energy X-ray absorptiometry. Metabolism 44:369–373
39. Lovejoy JC, Champagne CM, De Jonge L, Xie H, Smith SR (2008) Increased visceral fat and decreased energy expenditure during the menopausal transition. Int J Obes (Lond) 32:949–958
40. Sun Y, Liu B, Snetselaar LG, Wallace RB, Caan BJ, Rohan TE, Neuhouser ML, Shadyab AH, Chlebowski RT, Manson JE, Bao W (2019) Association of normal-weight central obesity with all-cause and cause-specific mortality among postmenopausal women. JAMA Netw Open 2(7):e197337
41. Kissebah AH, Peiris AN (1989) Biology of regional body fat distribution: relationship to non-insulin-dependent diabetes mellitus. Diabetes Metab Rev 5:83–109
42. Lapidus L et al (1984) Distribution of adipose tissue and risk of cardiovascular disease and death: a 12 year follow up of participants in the population study of women in Gothenburg, Sweden. Br Med J (Clin Res Ed) 289:1257–1261
43. Abildgaard J, Ploug T, Al-Saoudi E, Wagner T, Thomsen C, Ewertsen C, Bzorek M, Pedersen BK, Pedersen AT, Lindegaard B (2021) Changes in abdominal subcutaneous adipose tissue phenotype following menopause is associated with increased visceral fat mass. Sci Rep 11:14750
44. Price TM et al (1998) Estrogen regulation of adipose tissue lipoprotein lipase—possible mechanism of body fat distribution. Am J Obstet Gynecol 178:101–107
45. Bracht JR et al (2020) The role of estrogens in the adipose tissue milieu. Ann NY Acad Sci 1461:127–143
46. Xu Y et al (2011) Distinct hypothalamic neurons mediate estrogenic effects on energy homeostasis and reproduction. Cell Metab 14:453–465
47. Musatov S et al (2007) Silencing of estrogen receptor alpha in the ventromedial nucleus of hypothalamus leads to metabolic syndrome. Proc Natl Acad Sci USA 104:2501–2506
48. Celermajer DS et al (1994) Aging is associated with endothelial dysfunction in healthy men years before the age-related decline in women. J Am Coll Cardiol 24:471–476
49. Moreau KL, Hildreth KL, Meditz AL, Deane KD, Kohrt WM (2012) Endothelial function is impaired across the stages of the menopause transition in healthy women. J Clin Endocrinol Metab 97:4692–4700
50. Moreau KL, Hildreth KL, Klawitter J, Blatchford P, Kohrt WM (2020) Decline in endothelial function across the menopause transition in healthy women is related to decreased estradiol and increased oxidative stress. GeroScience 42:1699–1714
51. Gersh FL, O'Keefe JH, Lavie CJ, Henry BM (2021) The renin-angiotensin-aldosterone system in postmenopausal women: the promise of hormone therapy. Mayo Clin Proc 96:3130–3141
52. Yanes LL et al (2010) Postmenopausal hypertension: role of the Renin-Angiotensin system. Hypertens. (Dallas, Tex. 1979) 56:359–363
53. Mercuro G et al (2000) Evidence of a role of endogenous estrogen in the modulation of autonomic nervous system. Am J Cardiol 85:787–789
54. Barnes JN et al (2014) Aging enhances autonomic support of blood pressure in women. Hypertens (Dallas, Tex. 1979) 63:303–308
55. Vongpatanasin W (2009) Autonomic regulation of blood pressure in menopause. Semin Reprod Med 27:338–345

56. Thurston RC (2018) Vasomotor symptoms: natural history, physiology, and links with cardiovascular health. Climacteric 21:96–100
57. Talaulikar V (2022) Menopause transition: physiology and symptoms. Best Pract Res Clin Obstet Gynaecol 81:3–7
58. El Khoudary SR, Thurston RC (2018) Cardiovascular implications of the menopause transition: endogenous sex hormones and vasomotor symptoms. Obstet Gynecol Clin North Am 45:641–661
59. Szmuilowicz ED et al (2011) Vasomotor symptoms and cardiovascular events in postmenopausal women. Menopause 18:603–610
60. Herber-Gast GCM, Brown WJ, Mishra GD (2015) Hot flushes and night sweats are associated with coronary heart disease risk in midlife: a longitudinal study. BJOG 122:1560–1567
61. Thurston RC et al (2021) Menopausal vasomotor symptoms and risk of incident cardiovascular disease events in SWAN. J Am Heart Assoc 10:1–17
62. Thurston RC et al (2011) Hot flashes and carotid intima media thickness among midlife women. Menopause 18:352–358
63. Thurston RC et al (2017) Menopausal symptoms and cardiovascular disease mortality in the Women's Ischemia Syndrome Evaluation (WISE). Menopause 24:126–132
64. Wolff EF et al (2013) Self-reported menopausal symptoms, coronary artery calcification, and carotid intima-media thickness in recently menopausal women screened for the Kronos early estrogen prevention study (KEEPS). Fertil Steril 99:1385–1391
65. Thurston RC et al (2016) Trajectories of vasomotor symptoms and carotid intima media thickness in the study of women's health across the nation. Stroke 47:12–17
66. Thurston RC, Kuller LH, Edmundowicz D, Matthews KA (2010) History of hot flashes and aortic calcification among postmenopausal women. Menopause 17:256–261
67. Valensi P (2021) Autonomic nervous system activity changes in patients with hypertension and overweight: role and therapeutic implications. Cardiovasc Diabetol 20
68. Sassarini J, Fox H, Ferrell W, Sattar N, Lumsden MA (2011) Vascular function and cardiovascular risk factors in women with severe flushing. Clin Endocrinol (Oxf) 74:97–103
69. Thurston RC et al (2012) Vasomotor symptoms and lipid profiles in women transitioning through menopause. Obstet Gynecol 119:753–761
70. Bechlioulis A et al (2012) Increased vascular inflammation in early menopausal women is associated with hot flush severity. J Clin Endocrinol Metab 97:E760–E764
71. Chedraui P et al (2011) Pro-inflammatory cytokine levels in postmenopausal women with the metabolic syndrome. Gynecol Endocrinol 27:685–691
72. Huang WY et al (2017) Circulating interleukin-8 and tumor necrosis factor-α are associated with hot flashes in healthy postmenopausal women. PLoS One 12
73. Gordon JL et al (2016) Cardiovascular, hemodynamic, neuroendocrine, and inflammatory markers in women with and without vasomotor symptoms. Menopause 23:1189–1198
74. Liu X, Yan G, Bullock L, Barksdale DJ, Logan JG (2021) Sleep moderates the association between arterial stiffness and 24-hour blood pressure variability. Sleep Med 83:222–229
75. Kundel V et al (2021) Sleep duration and vascular inflammation using hybrid positron emission tomography/magnetic resonance imaging: results from the Multi-Ethnic Study of Atherosclerosis. J Clin Sleep Med 17:2009–2018
76. Mulvagh SL et al (2021) The Canadian women's heart health alliance atlas on the epidemiology, diagnosis, and management of cardiovascular disease in women—Chapter 4: sex- and gender-unique disparities: CVD across the lifespan of a woman. CJC Open 4:115–132
77. Kahl KG, Stapel B, Correll CU (2022) Psychological and psychopharmacological interventions in psychocardiology. Front Psychiatry 13:831359
78. Saccomani S et al (2017) Does obesity increase the risk of hot flashes among midlife women?: a population-based study. Menopause 24:1065–1070
79. Anderson DJ et al (2020) Obesity, smoking, and risk of vasomotor menopausal symptoms: a pooled analysis of eight cohort studies. Am J Obstet Gynecol 222(478):e1–478.e17
80. Pfeilschifter J, Koditz R, Pfohl M, Schatz H (2002) Changes in proinflammatory cytokine activity after menopause. Endocr Rev 23:90–119

81. Cioffi M et al (2002) Cytokine pattern in postmenopause. Mat 41:187–192
82. Deswal A et al (2001) Cytokines and cytokine receptors in advanced heart failure: an analysis of the cytokine database from the Vesnarinone Trial (VEST). Circ 103:2055–2059
83. Fernandes RD et al (2019) Cardiac changes during the peri-menopausal period in a VCD-induced murine model of ovarian failure. Acta Physiol 227
84. Lu X, Crowley SD (2022) The immune system in hypertension: a lost shaker of salt 2021 Lewis K. Dahl Memorial Lecture. Hypertens (Dallas, Tex. 1979) 79:1339–1347
85. Anzai A, Ko S, Fukuda K (2022) Immune and inflammatory networks in myocardial infarction: current research and its potential implications for the clinic. Int J Mol Sci 23:5214
86. Evans BR, Yerly A, van der Vorst EPC, Baumgartner I, Bernhard SM, Schindewolf M, Döring Y (2022) Inflammatory mediators in atherosclerotic vascular remodeling. Front Cardiovasc Med 9:868934
87. Mikkelsen RR et al (2022) Immunomodulatory and immunosuppressive therapies in cardiovascular disease and type 2 diabetes mellitus: a bedside-to-bench approach. Eur J Pharmacol 925:174998
88. Miller AP et al (2004) Estrogen modulates inflammatory mediator expression and neutrophil chemotaxis in injured arteries. Circulat 110:1664–1669
89. Frink M, Thobe BM, Hsieh YC, Choudhry MA, Schwacha MG, Bland KI, Chaudry IH (2007) 17beta-Estradiol inhibits keratinocyte-derived chemokine production following trauma-hemorrhage. Am J Physiol Lung Cell Mol Physiol 292(2):L585–591
90. Kan WH, Hsu JT, Ba ZF, Schwacha MG, Chen J, Choudhry MA, Bland KI, Chaudry IH (2008) MAPK-dependent eNOS upregulation is critical for 17beta-estradiol-mediated cardioprotection following trauma-hemorrhage. Am J Physiol Heart Circ Physiol 294(6):H2627-2636
91. Klein PW, Easterbrook JD, Lalime EN, Klein SL (2008) Estrogen and progesterone affect responses to malaria infection in female C57BL/6 mice. Gend Med 5:423–433
92. Brunsing RL, Prossnitz ER (2011) Induction of interleukin-10 in the T helper type 17 effector population by the G protein coupled estrogen receptor (GPER) agonist G-1. Immunol 134:93–106
93. Rosenzweig R, Gupta S, Kumar V, Gumina RJ, Bansal SS (2021) Estrogenic bias in T-Lymphocyte biology: Implications for cardiovascular disease. Pharmacol Res 170:105606
94. Straub RH (2007) The complex role of estrogens in inflammation. Endocr Rev 28:521–574
95. Klein SL, Flanagan KL (2016) Sex differences in immune responses. Nat Rev Immunol 16:626–638
96. Hotamisligil GS (2006) Inflammation and metabolic disorders. Nat 444:860–867
97. Safdar B, Spatz ES, Dreyer RP, Beltrame JF, Lichtman JH, Spertus JA, Reynolds HR, Geda M, Bueno H, Dziura JD, Krumholz HM, D'Onofrio G (2018) Presentation, clinical profile, and prognosis of young patients with Myocardial Infarction with Nonobstructive Coronary Arteries (MINOCA): results from the VIRGO study. J Am Heart Assoc 7(13):e009174
98. Ye G et al (2022) Sex differences and temporal trends in revascularization and outcomes of ST-elevation myocardial infarction in older adults in the United States. Arch Med Res 53:441–450
99. Pacheco C et al (2022) The Canadian women's heart health alliance atlas on the epidemiology, diagnosis, and management of cardiovascular disease in women—Chapter 5: sex- and gender-unique manifestations of cardiovascular disease. CJC Open 4:243–262
100. Vallabhajosyula S, Verghese D, Desai VK, Sundaragiri PR, Miller V (2021) Sex differences in acute cardiovascular care: a review and needs assessment. Cardiovasc Res 118:667–685
101. Koopman C, Vaartjes I, Heintjes EM, Spiering W, van Dis I, Herings RMC, Bots M (2013) Persisting gender differences and attenuating age differences in cardiovascular drug use for prevention and treatment of coronary heart disease, 1998–2010. Eur Heart J 34:3198–3205
102. Peters SAE, Colantonio LD, Zhao H, Bittner V, Dai Y, Farkouh ME, Monda KL, Safford MM, Muntner P, Woodward M (2018) Sex Differences in High-Intensity Statin Use Following Myocardial Infarction in the United States. J Am Coll Cardiol 71(16):1729–1737
103. Wilson R (1966) Feminine forever. M. Evans and Company, Inc.

104. Houck JA (2003) 'What do these women want?': feminist responses to feminine forever, 1963–1980. Bull Hist Med 77:103–132+233
105. Ziel HK, Finkle WD (1975) Increased risk of endometrial carcinoma among users of conjugated estrogens. N Engl J Med 293:1167–1170
106. Smith DC, Prentice R, Thompson DJ, Herrmann WL (1975) Association of exogenous estrogen and endometrial carcinoma. N Engl J Med 293:1164–1167
107. Woodruff JD, Pickar JH (1994) Incidence of endometrial hyperplasia in postmenopausal women taking conjugated estrogens (Premarin) with medroxyprogesterone acetate or conjugated estrogens alone. The menopause study group. Am J Obstet Gynecol 170:1213–1223
108. Hulley S et al (1998) Randomized trial of estrogen plus progestin for secondary prevention of coronary heart disease in postmenopausal women. Heart and Estrogen/Progestin Replacement Study (HERS) Research Group. JAMA 280:605–613
109. Rossouw JE et al (2002) Risks and benefits of estrogen plus progestin in healthy postmenopausal women: principal results From the Women's Health Initiative randomized controlled trial. JAMA 288:321–333
110. Manson JAE et al (2019) Menopausal estrogen-alone therapy and health outcomes in women with and without bilateral oophorectomy: a randomized trial. Ann Intern Med 171:406–414
111. Boardman H, Hartley L, Eisinga A, Main C, Figuls MRI (2016) Cochrane corner: oral hormone therapy and cardiovascular outcomes in post-menopausal women. Heart 102:9–11
112. Hodis HN et al (2016) Vascular effects of early versus late postmenopausal treatment with estradiol. N Engl J Med 374:1221–1231
113. Boardman HM, Hartley L, Eisinga A, Main C, Roqué i Figuls M, Bonfill Cosp X, Gabriel Sanchez R, Knight B (2015) Hormone therapy for preventing cardiovascular disease in postmenopausal women. Cochrane Database Syst Rev (3):CD002229
114. Harper-Harrison G, Shanahan MM (2023) Hormone Replacement Therapy. In: StatPearls [Internet]. Treasure Island (FL): StatPearls Publishing
115. Pinkerton JAV et al (2017) The 2017 hormone therapy position statement of The North American Menopause Society. Menopause 24:728–753
116. Trémollieres FA et al (2022) Management of postmenopausal women: Collège National des Gynécologues et Obstétriciens Français (CNGOF) and Groupe d'Etude sur la Ménopause et le Vieillissement (GEMVi) Clinical Practice Guidelines. Maturitas 163:62–81
117. Lee SR et al (2020) The 2020 menopausal hormone therapy guidelines. J Menopausal Med 26:69
118. Arnal JF et al (2017) Membrane and nuclear estrogen receptor alpha actions: from tissue specificity to medical implications. Physiol Rev 97:1045–1087
119. Pare G et al (2002) Estrogen receptor-alpha mediates the protective effects of estrogen against vascular injury. Circ Res 90:1087–1092
120. Usselman CW, Stachenfeld NS, Bender JR (2016) The molecular actions of oestrogen in the regulation of vascular health. Exp Physiol 101:356–361
121. Störk S, Von Schacky C, Angerer P (2002) The effect of 17beta-estradiol on endothelial and inflammatory markers in postmenopausal women: a randomized, controlled trial. Atherosclerosis 165:301–307
122. Sumino H et al (2006) Different effects of oral conjugated estrogen and transdermal estradiol on arterial stiffness and vascular inflammatory markers in postmenopausal women. Atherosclerosis 189:436–442
123. Kip KE et al (2005) Global inflammation predicts cardiovascular risk in women: a report from the Women's Ischemia Syndrome Evaluation (WISE) study. Am Heart J 150:900–906
124. Novella S, Heras M, Hermenegildo C, Dantas AP (2012) Effects of estrogen on vascular inflammation: a matter of timing. Arterioscler Thromb Vasc Biol 32:2035–2042
125. Cutolo M et al (2006) Estrogens and autoimmune diseases. Ann NY Acad Sci 1089:538–547
126. Arnal JF et al (2004) Estrogens and atherosclerosis. Eur J Endocrinol 150:113–117
127. Shwaery GT, Vita JA, Keaney JF (1997) Antioxidant protection of LDL by physiological concentrations of 17 beta-estradiol. Requirement for estradiol modification. Circ 95:1378–1385

128. Hansson GK, Libby P (2006) The immune response in atherosclerosis: a double-edged sword. Nat Rev Immunol 6:508–519
129. Kuohung W, Shwaery GT, Keaney JF (2001) Tamoxifen, esterified estradiol, and physiologic concentrations of estradiol inhibit oxidation of low-density lipoprotein by endothelial cells. Am J Obstet Gynecol 184:1060–1063
130. Novensà L et al (2011) Aging negatively affects estrogens-mediated effects on nitric oxide bioavailability by shifting ERα/ERβ balance in female mice. PLoS One 6:e25335
131. Alm-Roijer C, Stagmo M, Udén G, Erhardt L (2004) Better knowledge improves adherence to lifestyle changes and medication in patients with coronary heart disease. Eur J Cardiovasc Nurs 3:321–330
132. Bairey Merz CN et al (2017) Knowledge, attitudes, and beliefs regarding cardiovascular disease in women: the women's heart alliance. J Am Coll Cardiol 70:123–132
133. McDonnell LA et al (2014) Perceived vs actual knowledge and risk of heart disease in women: findings from a Canadian survey on heart health awareness, attitudes, and lifestyle. Can J Cardiol 30:827–834
134. Hoare E, Stavreski B, Kingwell BA, Jennings GL (2017) Australian adults' behaviours, knowledge and perceptions of risk factors for heart disease: a cross-sectional study. Prev Med Rep 8:204–209
135. Miller VM et al (2013) Embedding concepts of sex and gender health differences into medical curricula. J Womens Health (Larchmt) 22:194–202
136. Anderson NN, Gagliardi AR (2021) Medical student exposure to women's health concepts and practices: a content analysis of curriculum at Canadian medical schools. BMC Med Educ 21
137. Rusiecki J, Rojas J, Oyler J, Pincavage A (2022) An expanded primary care-based women's health clinic to improve resident education and patient care in resident continuity clinic. J Gen Intern Med 37(9):2314–2317
138. Brooks HL, Pollow DP, Hoyer PB (2016) The VCD mouse model of menopause and perimenopause for the study of sex differences in cardiovascular disease and the metabolic syndrome. Physiol (Bethesda) 31:250–257
139. Konhilas JP et al (2020) Using 4-vinylcyclohexene diepoxide as a model of menopause for cardiovascular disease. Am J Physiol Heart Circ Physiol 318:H1461–H1473
140. Adreak N et al (2021) Incorporating a women's cardiovascular health curriculum into medical education. CJC Open 3:S187–S191

# Chapter 11
# Vascular Dysfunction in Women

**Danah S. Al-Hattab and Michael P. Czubryt**

**Abstract** Vascular dysfunction may arise when blood vessel function is impaired by alterations in the underlying cellular functions and responses to stimuli, changes in vascular composition, or damage. Vascular stiffness—in which compliance and elasticity of vessels become compromised—is a significant contributing factor to the development and adverse outcomes of cardiovascular diseases, including hypertension, atherosclerosis and ischemic heart disease. Growing evidence indicates that the development and impact of vascular stiffness is different in women versus men. While women have frequently been excluded from clinical studies of cardiovascular disease in the past, resulting in under-reporting and a lack of clarity of the disease burden faced by women, recent work has sought to address this major shortcoming, and has led to the identification of important sex-associated differences in cardiovascular disease risk, treatment and outcomes. In this chapter, we examine factors that contribute to vascular stiffness and dysfunction, with a focus on the differences between women and men, and consideration of how different life stresses impact women in unique ways.

## Introduction

Vascular stiffness (VS) occurs when the vessel wall has a reduced capacity to expand or contract normally in response to blood pressure changes within the vascular system. VS is caused either by cellular dysfunction within the vessel wall, including endothelial or smooth muscle cell dysfunction, or by an increase in vascular wall components, including extracellular matrix or cellularity. Measures of VS reflect the physical properties of arteries in the vascular system, and include distensibility,

---

D. S. Al-Hattab · M. P. Czubryt (✉)
Department of Physiology and Pathophysiology, Rady Faculty of Health Sciences, University of Manitoba, R4008 St. Boniface Hospital Albrechtsen Research Centre, 351 Tache Avenue, Winnipeg, MB R2H 2A6, Canada
e-mail: mczubryt@sbrc.ca

© The Author(s), under exclusive license to Springer Nature Switzerland AG 2023 
L. Kirshenbaum and I. Rabinovich-Nikitin (eds.), *Biology of Women's Heart Health*, Advances in Biochemistry in Health and Disease 26,
https://doi.org/10.1007/978-3-031-39928-2_11

compliance, and elasticity. Alterations of these properties can be a consequence of, or can exacerbate, numerous cardiovascular diseases (CVD) [1].

It is important to carefully consider the basic terminology describing the mechanical properties of blood vessels to help avoid mis-communication between basic sciences and clinical prognostic settings. The Elastic Modulus (EM) is defined as the ratio between stress, which is a measure of the force acting upon the vessel wall normalized to area, and strain, which is the deformation of the wall in response to force. Both terms are related to the resistance to, or related to the diameter change in response to, the force applied on the vascular wall, which is quantified as the change in vessel diameter when intravascular pressure increases. The EM is thus a measure of the stiffness of the vascular wall, and higher values of EM indicate stiffer vessels. The elasticity or deformability of the vessel wall is determined by the quantity and quality of vascular wall components including collagens, elastin fibers and the physical characteristics and number of vascular smooth muscle cells (VSMC). The relative proportions of these components impact the compliance and distensibility of the vessel wall, that is, its ability to accommodate pressure changes by changing shape [2].

## Arterial Stiffness and Cardiovascular Disease Risk

Arterial stiffness is an independent risk factor for all-cause mortality and for the development of CVD [3]. Women with arterial stiffness have a nearly threefold increase in mortality compared to men [4]. Furthermore, women tend to have increased pulse pressure and pulse wave velocity (PWV), which are correlated to stiffer arteries over time [5]. These two simple, non-invasive parameters are reliable gold standard techniques for measuring arterial stiffness in the clinic. Multi-factorial stressors throughout life may interfere with women's hormones which in turn increases susceptibility to cardiovascular diseases and vascular dysfunction [6]. According to Statistics Canada, CVD is the leading cause of premature death for Canadian women, with the top 3 CVD-related causes of death for women being ischemic heart disease (IHD), stroke, and heart failure (HF). Canadian women are more likely to develop angina as their first CVD manifestation and with HF, are at a 24% higher risk to develop atrial fibrillation than men [7]. Young Canadian women are more likely than men to die within 1 year after an acute MI [8]. Most of these diseases are initiated and aggravated by vascular dysfunction and/or vascular stiffness [9]. United Nations population data from 2000 to 2010 indicates that younger women (ages 25–49) have significantly higher rates versus men (6.31% vs. 2.89%) of peripheral vascular disease worldwide [10]. In 2019, the leading cause of death in women worldwide was CVD, with IHD accounting for 47% of the total, followed by stroke at 36% [11].

From an anatomical aspect, sex differences may contribute to the clinical manifestations and characteristics of certain CVD. For example, on average women have smaller hearts than men, and have smaller coronary arteries, making them more susceptible to atherosclerotic plaque build-up, which can contribute to IHD risk.

CVD is very common, responsible for approximately one-third of all female deaths globally, affecting nearly 48 million women in the United States, and in Canadian women is five times more likely to cause death than breast cancer [12, 13]. Differences in cardiomyocyte metabolism also exist between men and women, thus the differential impact of sex on cardiovascular risk may arise from a combination of factors unique to the heart versus the blood vessels, and it is important to consider all such contributors [14].

## Sex-Specific Factors Contributing to Vascular Stiffness

### *Sex Hormones Affecting Vascular Stiffness*

Sex-specific hormones affecting arterial stiffness in females have been investigated at different points of the life cycle. Studies from females at pre-puberty revealed stiffer large arteries which were indicated by an increase in carotid-femoral PWV measurements, and a higher pulse pressure when compared to males of similar age [15]. In contrast, after puberty, females exhibit increased arterial distensibility and attenuated stiffness compared to males. This is not only due to the effect of sex hormones on vascular stiffness, but also due to the potential role of overall body/aortic growth on arterial stiffness. Studies have linked this to the effect of the menstrual cycle on arterial compliance and endothelial function [15–17].

Many studies have investigated the protective role of estrogen in premenopausal women against the development of CVD when compared to men of similar age. Estrogen acts directly through its active receptors on endothelial cells and VSMC within the vessel wall. It promotes vasodilation by stimulating nitric oxide and prostacyclin synthesis and inhibits VSMC proliferation in part by altering its membrane ionic permeability by activating potassium efflux and by inhibiting calcium influx [18].

The Study of Women's Health Across the Nation (SWAN) examined the vascular alterations in midlife women transitioning through menopause. This study suggested that arterial stiffness is significantly increased within 1 year after the final menstrual period. This change is augmented with the increase in women's age, which is usually accompanied by multiple CVD risk factors including increases in blood pressure, obesity and lipid profile [19].

## Post-menopause, Hormone Replacement Therapies and Oral Contraceptives

Atherosclerosis is a significant contributor to vascular dysfunction in both men and women. It is characterised by arterial narrowing due to lipid accumulation, inflammation, VSMC migration and foam cell deposition [20]. Although women are less prone than men to develop atherosclerosis prior to menopause, they become more susceptible than men after menopause. In comparing menopausal women to men, the risk of hypercholesterolemia is increased, and is associated with increasing levels of total cholesterol and low-density lipoprotein by 10% and 14%, respectively, without further changes in high-density lipoprotein levels [21]. Post-menopausal women have a 3.4-fold increase in risk of developing atherosclerosis compared to pre-menopausal women [22, 23]. In addition, post-menopausal women with osteoporosis have increased arterial stiffness as measured by PWV, correlating significantly with an increased risk of developing coronary atherosclerosis [24]. Women with autoimmune diseases such as systemic lupus erythematosus or rheumatoid arthritis are also at greater risk for developing atherosclerosis compared to healthy women [25].

Hormonal fluctuations throughout a woman's life significantly determine vascular health resilience. Steroidal sex hormones and their receptors, including estrogen, progesterone and androgen receptors, are expressed in the vascular endothelium and VSMC [26]. During menopause, women have significantly reduced levels of estrogen, which can be treated by hormone replacement therapy (HRT). The most common HRT used during menopause is estrogen supplementation [27]. While some studies have reported that HRT suppresses atherosclerosis in the coronary arteries, others indicate that a major complication of HRT is development of clots and strokes [28]. Data from randomized trials suggest that standard hormone therapy increases stroke risk by 30% when using different types of HRT [29, 30]. Exogenous estrogen, as occurs in oral contraceptive (OC) use, may induce the development of thrombosis through increased coagulation factors, decreased platelet aggregation and altered lipid profile [31]. The prevalence of hypertension among females taking OC is significantly higher with increased duration of OC use [32]. This impact is highly augmented in the presence of factors such as arterial stiffness, endothelial dysfunction and oxidative stress [33, 34].

In addition, a role has been identified for the male sex hormone testosterone in inducing arterial stiffness in post-menopausal women [35]. Testosterone levels fall significantly with age, and loss of testosterone correlates highly with vascular impairment and endothelial dysfunction. However, additional studies are needed to explore its sex-specific effects on blood vessel function and elasticity [36, 37].

## Aging and Vascular Stiffness

Aging is an independent risk factor for increased vascular stiffness. Data from the 26-year Framingham study identified sex- and age-related changes to vascular stiffness and CVD risks, with risk increasing linearly in men, while women tend to have a more curved aging trend: a flatter curve in youth with increased steepness after menopause in both arterial stiffness and associated CVD risks [38]. Additional studies correlated the primary effect of aging with increasing central artery stiffness observed in healthy postmenopausal women. However, this increase in arterial stiffness could be a combined factor with age-related elevations in blood pressure in these women [39].

The Baltimore Longitudinal Study of Aging (BLSA) examined age-associated changes in arterial stiffness using aortic PWV and carotid applanation tonometry measurements in 50 healthy females (26–96 years). This study revealed that these measures are significantly increased with age, without showing any noticeable increase in their blood pressure [40]. This result suggests that such age-associated changes in VS may occur independently of age-related changes in blood pressure.

## Lifestyle (Exercise and Diet)

Additional studies have suggested that lifestyle changes such as exercise or diet could influence arterial elasticity and compliance in females independent of age. Post-menopausal women who are aerobically active have better vascular compliance, with reduced total cholesterol and LDL-cholesterol levels, in healthy females with different ages and physical activity status [39]. In contrast, female athletes showed an increase in arterial stiffness after their retirement from competition with high physical activity [41]. Additionally, post-menopausal women with elevated baseline blood pressure who restricted sodium from their diet showed a significant reduction in systolic blood pressure and pulse pressure [42]. These results indicate that vascular stiffness may be mediated in part by modifying diet or lifestyle.

## Pregnancy and Preeclampsia

Major events in the life cycle of women may also influence their overall health consequences. A 10-year study that assessed arterial function in pregnant vs. non-pregnant women of the same age showed that during pregnancy, arterial function is altered throughout the gestation period [43]. This study found that pregnant women exhibited statistically significant differences in systolic, diastolic and central blood pressure, and in pulse pressure values, compared with non-pregnant women, and that, in post-partum women, there is an increase in PWV and brachial augmentation index.

Parameters assessing arterial stiffness can be a predictor for the early diagnosis of pregnancy risks, complications or development of preeclampsia—an alarming medical condition that may occur midway through pregnancy (typically after 20 weeks). Women with preeclampsia experience a rise in blood pressure, protein in their urine, headaches and blurred vision. This condition could result in serious and potentially life-threatening outcomes for both the fetus and the mother and may be related to end-organ dysfunction and other CVD in the mother [44]. Higher arterial stiffness has been observed in women with prolonged postpartum preeclampsia. Findings from systemic meta-analyses including twenty-three related studies noted a significant increase in arterial stiffness measurements in women with preeclampsia vs. women with normotensive pregnancies [45]. Similarly, women with preeclampsia had higher arterial stiffness compared with women with gestational hypertension [45]. These factors may contribute to the development of long-term CVD as a result of persistent endothelial damage beyond the period of postpartum preeclampsia. It is important to note, however, that another study investigating vascular compliance one year after delivery showed no significant changes in arterial stiffness between women that had preeclampsia and women with normotensive pregnancies, suggesting that postpartum-related changes in arterial stiffness in women with preeclampsia are reversible [46]. Some studies have reported increased risk of IHD, cerebrovascular incidents and CVD mortality following preeclampsia. Furthermore, the risk for hypertension in women who experienced preeclampsia is increased 10 years after preeclamptic delivery [47]. Thus, additional studies in this area are required to reconcile these disparate results, and to gain clearer insight into the long-term risks of preeclampsia, including the role of underlying mechanisms such as short- or long-term alterations in VS.

## *Psychosocial Stresses*

It is notable that another life-related stressor in a woman's life may impact CVD risks. A study that included both males and females investigated the impact of childhood trauma and recent life stresses on increased central arterial stiffness in combination with depression and anxiety, which contribute to and exacerbate CVD [48]. Psychosocial stresses in women at mid-life, including family responsibilities, motherhood or caregiving, job stress, and perceived discrimination, can negatively impact women's cardiovascular health [49].

A systematic literature review involving more than a dozen studies has identified that the majority of these studies (83%) showed a significant connection between general stress exposures and increased later-life CVD risks. Most of these studies have used different indicators for stress-related CVD outcomes such as carotid intima-media thickness (cIMT), endothelial dysfunction, central arterial stiffness and presence of carotid plaque, including measurements of plaque numbers, thickness, area or 3-dimensional volume, which are all inclusive in predicting CV events [50]. For example, Mexican women who experienced more than five years of chronic stress or had experienced physical violence had thicker cIMT compared to non-stressed Mexican women [49]. Among these women, the longer they experienced physical violence, the greater the cIMT was measured. Another study from Denmark has shown that family and partnership breakups are associated with the development of myocardial infarction in middle-aged women [51]. Additional study has shown that highly demanding jobs are significantly associated with CVD risk factors amongst working women [52].

When examining the overall data correlating multi-factorial stresses and hormonal fluctuations in the different events in women's life, it is crucial to consider sex-specific influences on CVD risks and vascular health. An examination of the literature highlights the need for sex-specific research, since two-thirds of cardiovascular research to date has focused on males [13]. Sex-specific factors increase the probability of vascular disease development and CVD events, and there is an urgent need for studies exploring the interactions between sex, aging and cardiovascular health, and the specific role of VS.

## Conclusion and Future Directions

A host of factors related to lifestyle and physiology are important determinants of vascular stiffness, and in turn, vascular heath and development of cardiovascular diseases (Fig. 11.1). Recent work has clearly demonstrated how such factors may be unique to women, or may exert differential effects in women versus men. Identifying and characterizing these factors is critical for understanding cardiovascular disease risk, progression and treatment, and as importantly, for understanding how disease occurs differently in women and men.

Despite this significant progress, many questions remain open, and therefore, there is an urgent need to undertake cardiovascular disease research both in isolated populations of women, and in contrasting men and women in the same study to better assess the impact of sex on vascular health. Understanding how these factors change throughout the a women's life-cycle—particularly during key lifetime events such as puberty, parturition and menopause—is needed in order to better recognize how risk and treatments may differ in young versus older women, and how this may correlate or differ from men. Many funding agencies are now requiring considerations of sex- and gender-based analysis to be specifically noted in applications for grants, and these approaches are likely to lead to better study designs and greater translation of research to effective clinical practice. These studies will facilitate a deeper knowledge and understanding of potential therapeutic targets for intervention that may ultimately reduce cardiovascular disease risk and improve outcomes for both women and men.

**Fig. 11.1 Contributors to vascular stiffness in women.** This representative figure shows the numerous factors that may influence vascular elasticity and stiffness and that may be of greater significance in women than men, resulting in increased risk for cardiovascular disease or altered prognosis

**Acknowledgements** This work was supported by a Grant-in-Aid from the Heart and Stroke Foundation of Canada (G-18-0022111). We also would like to thank Rob Blaich (St. Boniface Hospital Research) for excellent assistance with the figure.

## References

1. DeLoach SS, Townsend RR (2008) Vascular stiffness: its measurement and significance for epidemiologic and outcome studies. Clin J Am Soc Nephrol 3(1):184–192
2. Lehmann ED (1999) Terminology for the definition of arterial elastic properties. Pathol Biol (Paris) 47(6):656–664
3. Laurent S, Boutouyrie P, Asmar R, Gautier I, Laloux B, Guize L et al (2001) Aortic stiffness is an independent predictor of all-cause and cardiovascular mortality in hypertensive patients. Hypertens 37(5):1236–1241
4. Regnault V, Thomas F, Safar ME, Osborne-Pellegrin M, Khalil RA, Pannier B et al (2012) Sex difference in cardiovascular risk: role of pulse pressure amplification. J Am Coll Cardiol 59(20):1771–1777
5. Mitchell GF, Gudnason V, Launer LJ, Aspelund T, Harris TB (2008) Hemodynamics of increased pulse pressure in older women in the community-based Age. Gene/Environ Susceptibility-Reykjavik Study Hypertens 51(4):1123–1128
6. Townsend RR (2017) Arterial stiffness: recommendations and standardization. Pulse (Basel) 4(Suppl 1):3–7
7. Pacheco C, Mullen KA, Coutinho T, Jaffer S, Parry M, Van Spall HGC et al (2022) The Canadian women's heart health alliance atlas on the epidemiology, diagnosis, and management of cardiovascular disease in women—Chapter 5: sex- and gender-unique manifestations of cardiovascular disease. CJC Open 4(3):243–62

8. Izadnegahdar M, Singer J, Lee MK, Gao M, Thompson CR, Kopec J et al (2014) Do younger women fare worse? Sex differences in acute myocardial infarction hospitalization and early mortality rates over ten years. J Womens Health (Larchmt) 23(1):10–17
9. Norris CM, Yip CYY, Nerenberg KA, Clavel MA, Pacheco C, Foulds HJA et al (2020) State of the science in women's cardiovascular disease: a Canadian perspective on the influence of sex and gender. J Am Heart Assoc 9(4):e015634
10. Fowkes FG, Rudan D, Rudan I, Aboyans V, Denenberg JO, McDermott MM et al (2013) Comparison of global estimates of prevalence and risk factors for peripheral artery disease in 2000 and 2010: a systematic review and analysis. Lancet 382(9901):1329–1340
11. Vogel B, Acevedo M, Appelman Y, Bairey Merz CN, Chieffo A, Figtree GA et al (2021) The lancet women and cardiovascular disease commission: reducing the global burden by 2030. Lancet 397(10292):2385–2438
12. Benjamin EJ, Blaha MJ, Chiuve SE, Cushman M, Das SR, Deo R et al (2017) Heart disease and stroke statistics-2017 update: a report from the American Heart Association. Circ 135(10):e146–e603
13. Heart and Stroke 2018 Heart Report (2018) Ms. Understood: Heart and stroke foundation of Canada. Available from https://www.heartandstroke.ca/-/media/pdf-files/canada/2018-heart-month/hs_2018-heart-report_en.ashx
14. Czubryt MP, Espira L, Lamoureux L, Abrenica B (2006) The role of sex in cardiac function and disease. Can J Physiol Pharmacol 84(1):93–109
15. Ahimastos AA, Formosa M, Dart AM, Kingwell BA (2003) Gender differences in large artery stiffness pre- and post puberty. J Clin Endocrinol Metab 88(11):5375–5380
16. Hayashi K, Miyachi M, Seno N, Takahashi K, Yamazaki K, Sugawara J et al (2006) Variations in carotid arterial compliance during the menstrual cycle in young women. Exp Physiol 91(2):465–472
17. Williams MR, Westerman RA, Kingwell BA, Paige J, Blombery PA, Sudhir K et al (2001) Variations in endothelial function and arterial compliance during the menstrual cycle. J Clin Endocrinol Metab 86(11):5389–5395
18. Tostes RC, Nigro D, Fortes ZB, Carvalho MH (2003) Effects of estrogen on the vascular system. Braz J Med Biol Res 36(9):1143–1158
19. Samargandy S, Matthews KA, Brooks MM, Barinas-Mitchell E, Magnani JW, Janssen I et al (2020) Arterial stiffness accelerates within 1 year of the final menstrual period: the SWAN heart study. Arterioscler Thromb Vasc Biol 40(4):1001–1008
20. Palombo C, Kozakova M (2016) Arterial stiffness, atherosclerosis and cardiovascular risk: pathophysiologic mechanisms and emerging clinical indications. Vascul Pharmacol 77:1–7
21. Maas AH, Appelman YE (2010) Gender differences in coronary heart disease. Neth Heart J 18(12):598–602
22. Witteman JC, Grobbee DE, Kok FJ, Hofman A, Valkenburg HA (1989) Increased risk of atherosclerosis in women after the menopause. BMJ 298(6674):642–644
23. Villablanca AC, Jayachandran M, Banka C (2010) Atherosclerosis and sex hormones: current concepts. Clin Sci (Lond) 119(12):493–513
24. Seo SK, Cho S, Kim HY, Choi YS, Park KH, Cho DJ et al (2009) Bone mineral density, arterial stiffness, and coronary atherosclerosis in healthy postmenopausal women. Menopause 16(5):937–943
25. Salmon JE, Roman MJ (2008) Subclinical atherosclerosis in rheumatoid arthritis and systemic lupus erythematosus. Am J Med 121(10 Suppl 1):S3–S8
26. Orshal JM, Khalil RA (2004) Gender, sex hormones, and vascular tone. Am J Physiol Regul Integr Comp Physiol 286(2):R233–R249
27. MacLennan A, Lester S, Moore V (2001) Oral estrogen replacement therapy versus placebo for hot flushes: a systematic review. Climacteric 4(1):58–74
28. Akhrass F, Evans AT, Wang Y, Rich S, Kannan CR, Fogelfeld L et al (2003) Hormone replacement therapy is associated with less coronary atherosclerosis in postmenopausal women. J Clin Endocrinol Metab 88(12):5611–5614

29. Henderson VW, Lobo RA (2012) Hormone therapy and the risk of stroke: perspectives 10 years after the Women's Health Initiative trials. Climacteric 15(3):229–234
30. Santen RJ, Allred DC, Ardoin SP, Archer DF, Boyd N, Braunstein GD, et al (2010) Postmenopausal hormone therapy: an endocrine society scientific statement. J Clin Endocrinol Metab 95(7 Suppl 1):s1–s66
31. Sitruk-Ware R (2016) Hormonal contraception and thrombosis. Fertil Steril 106(6):1289–1294
32. Liu H, Yao J, Wang W, Zhang D (2017) Association between duration of oral contraceptive use and risk of hypertension: a meta-analysis. J Clin Hypertens (Greenwich) 19(10):1032–1041
33. Hickson SS, Miles KL, McDonnell BJ, Yasmin, Cockcroft JR, Wilkinson IB et al (2011) Use of the oral contraceptive pill is associated with increased large artery stiffness in young women: the ENIGMA study. J Hypertens 29(6):1155–1159
34. Olatunji LA, Seok YM, Igunnu A, Kang SH, Kim IK (2016) Combined oral contraceptive-induced hypertension is accompanied by endothelial dysfunction and upregulated intrarenal angiotensin II type 1 receptor gene expression. Naunyn Schmiedebergs Arch Pharmacol 389(11):1147–1157
35. Creatsa M, Armeni E, Stamatelopoulos K, Rizos D, Georgiopoulos G, Kazani M et al (2012) Circulating androgen levels are associated with subclinical atherosclerosis and arterial stiffness in healthy recently menopausal women. Metabolism 61(2):193–201
36. DuPont JJ, Kenney RM, Patel AR, Jaffe IZ (2019) Sex differences in mechanisms of arterial stiffness. Br J Pharmacol 176(21):4208–4225
37. Moreau KL, Babcock MC, Hildreth KL (2020) Sex differences in vascular aging in response to testosterone. Biol Sex Differ 11(1):18
38. Lerner DJ, Kannel WB (1986) Patterns of coronary heart disease morbidity and mortality in the sexes: a 26-year follow-up of the Framingham population. Am Heart J 111(2):383–390
39. Tanaka H, DeSouza CA, Seals DR (1998) Absence of age-related increase in central arterial stiffness in physically active women. Arterioscler Thromb Vasc Biol 18(1):127–132
40. Vaitkevicius PV, Fleg JL, Engel JH, O'Connor FC, Wright JG, Lakatta LE et al (1993) Effects of age and aerobic capacity on arterial stiffness in healthy adults. Circ 88(4 Pt 1):1456–1462
41. Koshiba H, Maeshima E, Okumura Y (2017) The relationship between arterial stiffness and the lifestyle habits of female athletes after retiring from competitive sports: a prospective study. Clin Physiol Funct Imaging 37(5):474–480
42. Seals DR, Tanaka H, Clevenger CM, Monahan KD, Reiling MJ, Hiatt WR et al (2001) Blood pressure reductions with exercise and sodium restriction in postmenopausal women with elevated systolic pressure: role of arterial stiffness. J Am Coll Cardiol 38(2):506–513
43. Turi V, Dragan S, Iurciuc M, Moleriu L, Bungau S, Tit DM et al (2020) Arterial function in healthy pregnant women vs. non-pregnant women—a 10-year study. Diagn (Basel) 10(6):374
44. Robb AO, Mills NL, Din JN, Smith IB, Paterson F, Newby DE et al (2009) Influence of the menstrual cycle, pregnancy, and preeclampsia on arterial stiffness. Hypertens 53(6):952–958
45. Hausvater A, Giannone T, Sandoval YH, Doonan RJ, Antonopoulos CN, Matsoukis IL et al (2012) The association between preeclampsia and arterial stiffness. J Hypertens 30(1):17–33
46. Kim S, Lim HJ, Kim JR, Oh KJ, Hong JS, Suh JW (2020) Longitudinal change in arterial stiffness after delivery in women with preeclampsia and normotension: a prospective cohort study. BMC Pregnancy Childbirth 20(1):685
47. Christensen M, Kronborg CS, Eldrup N, Rossen NB, Knudsen UB (2016) Preeclampsia and cardiovascular disease risk assessment - Do arterial stiffness and atherosclerosis uncover increased risk ten years after delivery? Pregnancy Hypertens 6(2):110–114
48. Bomhof-Roordink H, Seldenrijk A, van Hout HP, van Marwijk HW, Diamant M, Penninx BW (2015) Associations between life stress and subclinical cardiovascular disease are partly mediated by depressive and anxiety symptoms. J Psychosom Res 78(4):332–339
49. Stewart AL, Kathawalla UK, Wolfe AG, Everson-Rose SA (2018) Women's heart health at mid-life: what is the role of psychosocial stress? Womens Midlife Health 4:11
50. Naqvi TZ, Lee MS (2014) Carotid intima-media thickness and plaque in cardiovascular risk assessment. JACC Cardiovasc Imaging 7(10):1025–1038

51. Kriegbaum M, Christensen U, Andersen PK, Osler M, Lund R (2013) Does the association between broken partnership and first time myocardial infarction vary with time after break-up? Int J Epidemiol 42(6):1811–1819
52. Slopen N, Glynn RJ, Buring JE, Lewis TT, Williams DR, Albert MA (2012) Job strain, job insecurity, and incident cardiovascular disease in the Women's Health Study: results from a 10-year prospective study. PLoS ONE 7(7):e40512

# Chapter 12
# Cardiometabolic Function in Women

**Jovana Joksimovic Jovic, Jovana Novakovic, Nevena Jeremic, and Jovana Bradic**

**Abstract** Metabolic syndrome and associated increased risk of cardiovascular diseases developed the new global concept of cardiometabolic risk. Insulin resistance, impaired glucose tolerance, dyslipidemia, hypertension, and central obesity are the major features of cardiometabolic syndrome, a collection of different metabolic dysfunctions. Cardiometabolic diseases represent the enormous burden of the modern population. Historically considered a 'man's disease', cardiometabolic diseases in women represent underdiagnosed, underestimated, and undertreated conditions. Various cardiovascular disease manifestations in women are often neglected, or do not require an extensive therapeutic regimen. Cardiovascular diseases in women are increased largely by risks such as diabetes, hypertension, dyslipidemia, and obesity. Although cardiovascular events incidence progresses with age in both genders, men have a higher prevalence of cardiovascular events until mid-age. However, after menopause, the women-to-men ratio of cardiovascular events increases. Biological explanations for gender differences in cardiometabolic diseases are complex. The main physiological features that underlie the gender differences in cardiometabolic diseases development could be defined as sex hormones and body composition disparities in men and women. Some cardiometabolic aspects are unique for women, such as pregnancy-related or menopause-related features, leading to exert adverse cardiometabolic features in present, or leave a deep imprint in later life.

**Keywords** Cardiovascular diseases · Metabolic diseases · Metabolic syndrome · Diabetes mellitus · Hypertension · Obesity · Dyslipidemias

---

J. J. Jovic
Department of Physiology, Faculty of Medical Science, University of Kragujevac, Kragujevac, Serbia

J. Novakovic (✉) · N. Jeremic · J. Bradic
Department of Pharmacy, Faculty of Medical Science, University of Kragujevac, Kragujevac, Serbia
e-mail: jovana.jeremic@medf.kg.ac.rs

© The Author(s), under exclusive license to Springer Nature Switzerland AG 2023 
L. Kirshenbaum and I. Rabinovich-Nikitin (eds.), *Biology of Women's Heart Health*, Advances in Biochemistry in Health and Disease 26,
https://doi.org/10.1007/978-3-031-39928-2_12

## Introduction

A class of disorders known as cardiometabolic diseases (CMD) includes heterogeneous group of clinical entities arising from endocrine, nutritional, and metabolic contributors. A group of metabolic dysregulations known as the metabolic syndrome, which includes central obesity, atherogenic dyslipidemia, and hypertension is strongly associated with an increased risk of diabetes mellitus and cardiovascular diseases (CVD) if untreated. Metabolic syndrome and associated increased risk of CVD morbidity developed the new global concept of cardiometabolic risk. Insulin resistance, impaired glucose tolerance, dyslipidemia, hypertension, and central obesity are the major features of the cardiometabolic syndrome (CMS), a collection of different metabolic dysfunctions. As a new disease entity defined by the World Health Organization (WHO) and the American Society of Endocrinology (ASE) [1], CMS represents a burden on all spheres of the health system.

The presence of metabolic syndrome is associated with markedly increased risk of CVD morbidity and mortality, which puts into light strong relationship between these two disorders. These common conditions are frequently preventable, although increasing prevalence of CMD in the last decades managed to promote CMD as a global health problem worldwide. People around the world are becoming more likely to develop one or more of these disorders at some point in their lives. The four main causes of the CMD prevalence include smoking, poor exercise habits, consuming large amounts of alcohol, and maintaining an unhealthy diet. Therefore, it is important to address these issues due to the significant socioeconomic cost to countries of all income levels. Importantly, CMD as a group of common conditions can be prevented with early diagnosis along with treatments and scientific efforts to find novel strategies for reducing the number of people experiencing one or more of these conditions worldwide.

The basic step in providing adequate treatment modalities refers to understanding the pathophysiological events that underlie CMD. In brief, the pathophysiology of CMD is related to insulin resistance, which can result in metabolic syndrome, prediabetes, and non-alcoholic fatty liver disease. Patients with metabolic syndrome, prediabetes, and/or insulin resistance have accelerated atherogenesis and a higher risk of CVD. Obesity can aggravate insulin resistance and hasten the development of CMD, although it can also exist independently of this condition. Interventions should be prompt and include prevention of metabolic and vascular outcomes and treatment of CMD. In order to prevent CVD, it is necessary to recognize the benefits of a proper diet and training regimen in time. Accordingly, controlling CVD risk factors will be narrowed and further diagnosis and treatment will have better outcome in confronting CMD.

## Gender Differences in Cardiovascular (CV) Risk Factors Between Men and Women

Despite having a significant impact on health, sex and gender are frequently ignored in medical care. Sex differences between men and women may relate to a combination of physiological, cultural, and behavioral contributors. Sex-related disparities in CV morbidity are related to different physiological and pathophysiological functions in women, which are most often associated with different physiological properties, such as hormonal milieu or body composition.

It is worth to mention that women were not involved in experimental studies until the latter part of the twentieth century, hence most of the knowledge we currently have on the major diseases affecting CVD originates from research performed predominantly in men, with its findings being applied to women.

CMD represent an enormous burden on the modern population. Historically considered a ‘man’s disease’, CMD in women represent an underdiagnosed, underestimated, and undertreated condition. However, CMD remains the leading cause of death in both genders [2]. Moreover, CVD in women is increased largely by risks such as diabetes, hypertension, dyslipidemia, and obesity. Socioeconomic and psychosocial factors also seem to have a higher impact on CVD in women [3]. Regarding metabolic syndrome, the prevalence could not be understood easily, because of numerous confounding factors, such as age, genetic background, diet, and exercise habits influencing the occurrence of metabolic syndrome and its components [4]. Although the prevalence of metabolic syndrome increased significantly from the 1980s to 2012 by 35%, the focus on gender difference in prevalence of metabolic syndrome should be put forward as demonstrated by the newest data from the NHANES study, showing that the prevalence for CMD declined by 24% in men and 22% in women [5].

It is known that CVD morbidity is more frequent in men compared to women, while women, on the other hand, exhibit a greater CVD mortality rate. Various CVD manifestations that occur in women are often neglected. When it comes to gender disparity in CVD, underestimating the cardiac risk and misinterpreting symptoms can lead to less referral for cardiac testing and inappropriate diagnosis and treatment in women compared to men. This underrepresentation of women in large clinical trials can also reflect this important issue. CVD have been largely underdiagnosed in women, or not required for extensive therapeutic regimens [6].

Both genders are affected by CV pathologies worldwide, with ischemic heart disease and stroke as leading causes of deaths in men and women. Although CV events incidence progresses with age in both genders, men have higher prevalence of cardiovascular events until mid-age [7]. After menopause, the women to men ratio of CV events increases.

## The Mechanistic Basis of Gender-Specific Contributors for CMD Development

Biological explanations for gender differences in CMDs are complex. However, the underlying mechanism of different cardiometabolic dysfunction between genders still remains unclear. Many studies have shown different risk factors for CVD and metabolic syndrome, different epidemiology, and clinical settings, but whether the observed gender differences regarding CVD pathologies originate from congenital sex-dependent characteristics, or derive from disproportional inclusion of both gender in clinical studies and analysis, remains unclear. The main physiological features underlying the gender differences in CMDs development could be defined as sex hormones and body composition disparities in men and women. It is more than clear that hormonal changes occurring in women during reproductive age and menopause affect CV and metabolic functions. Furthermore, differences in muscle and fat tissue distribution, quantity and function, also contribute to different metabolic regulation in men and women.

The physiological roles of estrogen and testosterone are well-known in regards to their role in the reproductive system, as well as metabolic regulatory roles. Estrogen could modulate different functions by activation of its intracellular alpha and beta receptors, as well as G-protein membrane-bound receptors. Activation of these receptors modulate feeding behavior, deposition of visceral fat tissue, insulin production and secretion, as well as glucose utilization [8]. Moreover, estrogen exerts a protective role on cardiomyocytes and vasculature, increase in bioavailability of nitric-oxide (NO), and thus decreasing arterial tone, arterial stiffness and vascular remodeling [9]. Besides the protective roles of estrogen in women of reproductive age, the role of androgens on cardiovascular system and metabolic functions is still controversial in both genders. Before menopause, women are considered to be protected from atherosclerosis to some extent in comparison to men. However, after the reproductive period, estrogen levels drop, and the risk of CVD is increases with age. It is well-known that after menopause, women have higher incidence of stroke, and hypertension and is more prevalent in women after 75 years old [10].

Testosterone inhibits lipid uptake, activity of lipoprotein lipase, decreases visceral fat tissue deposits and increases the lean body mass [11]. Androgen deficiency and excess androgen exert many adverse effects on cardiovascular health [12]. For example, it is known that lower levels of androgen in men lead to dyslipidemia, higher body mass index (BMI), hypertension, and CVD [13]. Similar situation is observed in post-menopausal women [14], however, sex steroid replacement therapy could ameliorate these changes [15]. On the other hand, high concentration of androgens in both genders exert adverse effects on CV system [16]. One of the representative entities of hyperandrogenic state in females is polycystic ovary syndrome (PCOS). In PCOS, cardiometabolic disturbances exist in extent to reproductive abnormalities. Frequently unrecognized, these changes could severely influence cardiovascular health and metabolic functions.

Regarding body composition, women have higher percentage of body fat tissue and lower percentage of lean body mass. Nevertheless, women have lower prevalence of metabolic syndrome compared to men, even after adjustment of BMI [17]. Fat tissue represents the endocrine tissue secreting adipokines, which exert numerous functions at local and systemic levels. It is known that fat tissue distribution differs between genders; women have high percentage of subcutaneous fat, which secretes adiponectin, a cytokine with anti-inflammatory action [18].

## Cardiometabolic Aspects in Women

### *Hypertension in Women*

Hypertension represents one of the major risk factors for development of CVD, with an increasing prevalence worldwide. In 2000, it was estimated that 26% of people in the general population have hypertension [19], however, these numbers increased to 31.1% by 2010 [20]. It is estimated that 29% of men and 29.5% of women population will have hypertension in 2025 [19]. In general, men have higher risk of CV and renal illness than premenopausal women of similar age, and day-time, ambulatory blood pressure measurements by 24 h monitoring revealed that normotensive men had higher blood pressure values than women [21]. In women, the prevalence of hypertension increases with age, ranging from 19% in premenopausal women up to 75% in postmenopausal (65–74 years), and 85% in women over 75 years old [22]. This marked discrepancy between younger and older females can be attributed, at least partially, to physiological hormonal changes following menopause. It is well-known that estrogen induces vasodilatation indirectly acting on secretion of vasodilatatory endothelial substances, however its direct effect on vascular smooth muscle cells (VSMC) was also reported [23]. The estrogen-induced vascular protection can provide an appropriate explanation for lower incidence of CVDs in premenopausal women, while on contrary, lack of estrogen during postmenopausal period may explain the increased hypertension rates in older women. Women can experience varying forms of hypertension that are associated with sex-specific differences, such as menopause, oral contraceptive pills (OCP) and pregnancy.

## Sex Specific Factors that Influence Hypertension

### *Hypertension and Menopause*

After losing the vasodilatory effects of endogenous estrogen during menopause, postmenopausal women experience greater systolic blood pressure compared to age-matched men. Moreover, these alterations are secondary to lose of nitric oxide production and increased expression of angiotensin II receptors [24]. Furthermore, estrogen decreases oxidative stress by producing more antioxidants and fewer reactive oxygen species, thus these mechanisms are also involved in complex pathophysiology of hypertension-related menopausal disorders. Notably, estrogen also decreases inflammation by activating neoangiogenesis and inhibiting pro-fibrotic genes. Therefore, losing the protective effects of estrogen increases women's risk for CV and developing adverse cardiac events [25]. However, studies have shown that exogenous estrogen administration by hormone replacement therapy, had no impact on hypertension in menopausal women, and further no beneficial effects on the CV system [26]. Moreover, obesity is present in 40% of postmenopausal women, which could further aggravate risk for hypertension [27].

## Hypertension and Oral Contraceptive Pills Usage

OCP usage in young women is associated with risk for hypertension and CV events, while discontinuation OCP decreases and reverses the risk [28]. Interestingly, women who used OCP for more than 24 months had a 1.96-fold increased risk of hypertension [29]. Moreover, it was shown that the risk for developing hypertension was almost 50% higher at the highest OCP doses compared to the lowest ones [30]. It is also known that smoking prevalence is high in women, with the pooled prevalence of ever cigarette smoking in adolescent girls/students of the school, adult women, pregnant women, and women with the disease was 23, 27, 32, and 38%, respectively [31], and the risk involved with smoking is augmented by usage of OCP in younger women. The decreased concentration of ethinyl estradiol in third generation OCP decreased risk of hypertension development [32]. Moreover, data to date indicate that progesterone-only contraceptive pills do not raise the risk of hypertension or short-term cardiometabolic consequences [33]. The mechanisms involved in OCP-induced hypertension include progesterone actions, estrogen-induced fluid retention, renal dysfunction driven by sodium-lithium counter transport, and intra-renal vascular lesions [34].

## Hypertension and Pregnancy

Hypertension during pregnancy could be represented as preeclampsia/eclampsia (preeclampsia with seizures), gestational hypertension, chronic hypertension of any cause, and chronic hypertension with superimposed preeclampsia. The risk of hypertension and metabolic syndrome is increased in women experiencing pregnancy-related hypertensive disorders compared to normal blood pressure pregnancies [35]. Moreover, hypertensive disorders of pregnancy are associated with increased risk of CVD 20–30 years later while the risk of hypertension in women after pregnancy is 2.4-fold increased ten years postpartum [36].

During pregnancy, physiological hemodynamic changes in the women organism include plasma volume and cardiac output increases, systemic vascular resistance decreases, and renal blood flow increases. In contrast to real hypovolemia and underfilling, symptoms including suppressed plasma renin activity, elevated blood pressure, lower glomerular filtration rate, and frequent edema development are more indicative of an overfilled, vasoconstricted circulation in preeclampsia settings [37]. In these patients, cardiometabolic risk is more pronounced [38], while state of hypercoagulability is further exaggerated [39].

Using a lower diagnostic threshold for hypertension diagnosis in pregnancy may lead to better diagnosis of women at risk for developing preeclampsia and pregnancies at risk for adverse fetal/neonatal outcomes. According to the American College of Cardiology, lower threshold is recommended to diagnose hypertension in pregnancy: systolic blood pressure of 130 mm Hg or diastolic blood pressure of 80 mm Hg, or higher [40].

## Hypertension and PCOS

Although various clinical studies have reported an increased hypertension risk among PCOS patients [41], there is still lots of inconsistency regarding this topic. However, some studies also report increased hypertension prevalence among PCOS women, independently of BMI [42]. In one study, a 40% increased risk for elevated blood pressure in PCOS women was reported, independently of age, BMI, diabetes or dyslipidemia [43]. According to one research, this risk was threefold higher in PCOS women (mean age around 30 years) than in age-matched controls [44]. Furthermore, the large Danish study, enrolling dominantly premenopausal PCOS women, reported elevated blood pressure (130/85 mmHg or above) in 30% of patients compared to controls [45]. On the other hand, some studies reported no difference in 24-h BP measurements among PCOS and non-PCOS women [46, 47], others demonstrated a modest elevation of systolic blood pressure during daytime, as well as mean arterial pressure in young PCOS patients compared to controls [48]. Moreover, rodent studies demonstrated that rats spontaneously hypertensive rats expressed similar ovarian morphological features, hormonal status, and blood pressure values as normotensive rats with induced PCOS [49]. Furthermore, there are evidence from preclinical studies that PCOS modeling increases left ventricle wall thickness and produces cardiomyocytes hypertrophy [50].

## Insulin Resistance and Diabetes Mellitus in Women

Diabetes represents the fastest growing chronic disease worldwide with number of patients projected to increase to 592 million by 2035. Numerous data support the fact that gender influences the pathophysiology of insulin resistance, diabetes, relative risk of diabetes-associated cardiometabolic complications and response to therapy [51, 52]. It has been reported that complications of diabetes are more frequent and particularly detrimental in women, however, understanding the precise influence and contribution of sex hormones on occurrence of CV complications is challenging [51, 53]. Complex factors affect the development and outcomes of CV system complications in diabetes, such as sex hormones and other reproductive factors among women including age of menarche and menopause, levels of sex hormones, history of hormonal medicines intake, and history of childbearing [54, 55].

According to meta-analyses, diabetic women had a 27% higher relative risk for stroke and 50% higher relative risk of CHD than men [51, 56]. Evidence is accruing that the diabetes-linked CHD mortality was significantly higher in women compared to men [53]. According to literature, the major causes of morbidity and mortality among diabetic patients include peripheral arterial disease and myocardial infarction (MI) [52]. Epidemiological data suggest that mortality rate in patients who experience MI is seven times higher in woman and four times higher in man with diabetes compared to non-diabetic individuals.

It has not been completely known what is causing women with diabetes to have higher relative risk of CV morbidity and mortality. Importantly, occurrence of diabetes type 2 reverses the beneficial effects of female sex hormones in cardioprotection, thus resulting in greater burden of ischemic complications in diabetic woman [57]. Complex pathophysiological mechanisms have been proposed to contribute to adverse CV outcome in women. One of them is referring to the fact that women are exposed to metabolic stage of prediabetes for a longer period of time than man [51, 53, 58]. Also, treatment modalities are less efficient in diabetic women especially referring to achieving an optimal treatment that will decrease mortality risk. Late diagnosis and implementation of adequate treatment approach in women than men are one of the explanations for worse clinical outcome in women with diabetes. Generally viewed, women suffering from diabetes type 1 (DM1) are characterized with overall worse metabolic control and greater exposure to hyperglycemia during life [59–61].

Another factor that also increases the risk of diabetes-related CV complications is BMI for which men have lower tolerance threshold before developing diabetes compared to women [62, 63]. Previously conducted clinical studies revealed that BMI in female with diabetes type 2 was 1.79 kg/m$^2$ higher than in man [51, 64]. Gender-specific differences in fat storage have been reflected in greater subcutaneous fat storage in women and visceral and ectopic fat storage in man, while only visceral and ectopic fat storage are responsible for the increased risk for insulin resistance, thus explaining the necessity for greater amount of weight gain in women before diabetes onset [60].

An increase in androgen levels appears to be beneficial against diabetes in man. Nevertheless, it markedly increases risk for development of CVD complications in women with diabetes. Recent findings suggest that women with diabetes may more likely experience dyslipidemia, hypercoagulability, and endothelium-dependent vasodilation impairments [8]. In fact, significantly elevated factor VII, PAI-1 activity as well as adiponectin were revealed in women, which may explain higher cardiovascular risk and mortality compared to men [65, 66].

## Insulin Resistance, Diabetes Mellitus and Women-Specific Factors

### *Insulin Resistance, Diabetes Mellitus and Menopause*

A growing body of evidence suggests a strong relation between diabetes, menopause and CV risk. Numerous reports highlight that postmenopausal woman with metabolic disorders such as T2DM are prone to elevated risk for CVD. Importantly, these women are three times more likely to experience CV complications and stroke in comparison to non-diabetic women, while early menopause further increases the risk of impaired CV function [67]. The majority of experimental and clinical data indicate that T1DM and T2DM accelerate reproductive ageing through increase in LH and FSH levels and decrease in estradiol levels. It's important to emphasize that T2DM triples the risk for menopause, and ovarian ageing [68, 69]. Importantly, estrogen deficiency and a rise in cortisol during menopause further disturb glycemic control and have important CV health implications in women [70]. Alterations in the sex hormone levels in postmenopausal women with diabetes lead to activation of the renin–angiotensin–aldosterone system, which results in impaired endothelial function [71]. Since coexistence of diabetes and menopause represent predictors of CV morbidity and mortality, their early recognition and identification are of crucial importance in order to provide adequate treatment modality to the patients and decrease the burden of CV disease [72].

## Insulin Resistance, Diabetes Mellitus and Pregnancy

Gestational diabetes mellitus (GDM) is characterized by glucose intolerance during pregnancy. GDM is one of the most serious health problems that may not only affect pregnancy outcomes, but also elevate the risk for establishment of diabetes 2 postpartum by sevenfold [58]. It has been considered that secretion of placental hormones such as estrogen, progesterone, PGH, hPL [73] as well as cortisol can lead to insulin resistance by blocking insulin effects. Women with history of GDM are associated with markedly higher CV risk factors such as hypertension [73], dyslipidemia, hypoadiponectinemia and metabolic syndrome [74] compared to their peers. Unfortunately, women who suffered from pregnancy-induced diabetes are two times more likely to experience future CV events in comparison to women without history of diabetes during pregnancy [74]. The Coronary Artery Risk Development in Young Adults study suggested that GDM is associated with impaired cardiac function such as altered left ventricular relaxation and compromised left ventricular systolic capacity years after postpartum [75]. Furthermore, GDM strongly impairs vascular function and structure, therefore resulting in future atherosclerosis. Nevertheless, the mechanisms that underlie GDM pathology, endothelial injury and CVD remain uncertain [73]. Pathophysiological events that have been proposed to contribute to endothelial dysfunction include subclinical vascular inflammation and suppressed apelin, while hypoadiponectinemia and elevation of PAI-1 [76] represent interesting biomarkers for CMD diagnosis in women with history of GDM [77]. Interestingly, it has been discovered that certain cardiometabolic alterations could be present before pregnancy and diagnosis of GDM [78]. Therefore, it is of great significance to follow up with women following GDM in order to prevent, alleviate and properly manage future adverse CV outcomes.

## Insulin Resistance, Diabetes Mellitus and PCOS

According to estimates, 65–70% of PCOS women have insulin resistance and compensatory hyperinsulinemia, which appear to be the root cause of many of the endocrine symptoms of PCOS [79]. In recent years lean PCOS patients are also considered as high risk for hyperlipidemia and hyperinsulinemia, which contributes to increased risk of diabetes mellitus type 2 (DMT2). Genetic factors have a great influence on PCOS occurrence; women are more prone to have PCOS if their mother or sister has DMT2 or PCOS [80]. Insulin plays a central role in metabolic dysfunction, hyperandrogenism and reproductive dysfunction of PCOS. Hyperinsulinemia could aggravate hyperandrogenemia in PCOS individuals, both obese and lean [81]; compensatory hyperinsulinemia predisposes granulosa cells to be more sensitive to lueinizing hormone (LH) actions, which in turn causes theca cells to release more androgen while producing less sex hormone-binding globulin in the liver. A dominant follicle cannot be matured because greater insulin levels switch follicle stimulating hormone (FSH) to LH [82], leading to anovulation or oligo-ovulation. In addition, hyperinsulinemia can reduce aromatase activity, which leads to hyperandrogenemia. In parallel, high insulin levels raise free insulin-like growth factor (IGF), which enhances androgenesis and ovarian theca cell proliferation, while decreasing IGFBP-1 [83]. A link between PCOS and insulin resistance was also demonstrated by the therapeutic potential of insulin-sensitizing substances such as myo- and D-chiro-inositol [84], metformin, and delivery of oral contraceptives [85], together with lifestyle changes including diet and exercise [86]. Reducing the reproductive and endocrine effects of PCOS, is important in reducing CV risk factors simultaneously, considering that PCOS has a negative impact on quality of life.

## Obesity in Women

Obesity is considered a major risk factor for cardiometabolic disorders. Although obesity affects both genders, the prevalence of obesity is increased by 50% in women compared to men, to a total of 300 million women worldwide [87, 88]. By 2025, the prevalence of obesity is predicted to rise to 18% of men and 21% of women [89]. According to 2017–2018 data from the NHANES of America, the percentage of adult women who are severely obese (BMI of 40 kg/m$^2$) is higher (11.5%) than the percentage of adult men who are severely obese (6.9%). Interestingly, the novel aspect of "normal weight obesity" is important to be considered when obesity as a risk factor for CMS development is debatable in these patients which may also be at a higher risk [90]. Differences between urban and rural obesity prevalence should also be considered for management of overweight and obesity, since greater socioeconomic status is associated with greater risk of obesity in urban regions [91].

## Obesity and Women-Specific Factors

### *Obesity and Menopause*

Obesity rates are increasing in pre-pregnancy, childbearing-age women [92], and in menopause [93]. Postmenopausal women have 4.88-fold higher risk for obesity development compared to premenopausal women [94]. When estrogen levels are low, relative hyperandrogenemia occurs, which contributes to the development of obesity with physiologically unfavorable fat redistribution from gynoid to abdominal regions, leading to central type of obesity. Sex-hormone binding globulin is produced to a lower degree by the liver, which raises androgen bioavailability during the development of central obesity. In adipose tissue, testosterone converts to estrogen via aromatization process. However, although obese postmenopausal women have higher levels of estrogens, they do not exert protective and beneficial roles and higher rates of CV diseases in postmenopausal women are still observed. Hormone replacement treatment is still debatable because there may be advantages for women who are in the early stages of menopause.

Furthermore, when discussing obesity in menopause, controlling appetite by estrogen should be considered. Decline in estrogen levels could be responsible for enhancing neuropeptide Y (NPY) effects, since NPY is inhibited in the presence of estrogen [95]. The absence of its inhibition leads to increases food intake, particularly carbohydrate rich food, which could contribute to obesity development.

## Obesity and Pregnancy

Pregnancy-related obesity increases risks for both the mother and the fetus. One of the most frequent problems linked to pregnancy fat is gestational diabetes mellitus (GDM), which develops suddenly in pregnant women without a history of diabetes. Later in life, GDM becomes a risk factor for CVD [96].

Obese pregnant women have markedly different cardiac functions, as evidenced by their higher heart rates, lower cardiac and stroke volume indices, and increased systemic vascular resistance [97]. It was observed that overweight and obese pregnant women was are at higher risk for congenital heart defects in their fetuses [98]. The effects of maternal obesity during pregnancy can include higher risk for metabolic syndrome and obesity in children as well as other negative health outcomes. Pregnancy-related increases in fat mass, particularly in the lower body, causes metabolic dysregulation, hyperglycemia, lipotoxicity, and inflammation, which may affect endothelial function, placental development, and outcomes of pregnancy [99]. Postpartum weight retention increases future cardiometabolic risks and pre-pregnancy obesity in subsequent pregnancies because 50–60% of overweight or obese women gain even more weight later on [100]. Leptin and adiponectin are essential adipokines for the proper development of the placenta. Recent findings show that placentas of obese women had lower levels of the two adipokines, while epigenetic modification of their promoters (DNA methylation) found to be highlighted as the molecular mechanisms involved in the placental adaptation to harmful maternal environment [101].

Pre-pregnancy BMI and gestational weight gain are independent risk factors for preeclampsia occurrence [102]. Preeclampsia intensifies pregnancy's "physiological" symptoms, such as insulin resistance, hyperlipidemia, inflammation, and hypercoagulability, which may result in metabolic syndrome in pregnant women [103]. Women with preeclamptic have higher risk for hypertension, ischemic heart disease (IHD), and stroke compared to the general population later in life [104].

Increased adiposity is undoubtedly linked to CVD. It's important to note that women who are obese and insulin resistant have higher chances of developing CVD, particularly ischemic heart disease and heart failure with preserved ejection fraction (HFpEF).

After menopause, ischemic heart disease becomes more common and is the leading cause of mortality in women, demonstrating the negative effects of lower sex hormones on metabolic profile and fat redistribution [105]. Angiographic presentations of ischemic heart disease in women differ from those in men, with lesser obstructive lesions and more coronary vasospasm. Higher mortality rates have been seen in women with non-obstructive lesions, and dyslipidemia, hypertension, and type 2 diabetes, suggesting a significant predictor of the overall mortality rates [106].

Higher BMI levels were associated with a lifetime risk of developing heart failure (HF). However, according to Framingham Heart Study, the rise of one unit BMI (1 kg/m$^2$) on the risk of HF is sex dependent, with 5% increased risk for men and 7% in women [107]. The differential risk of HFpEF among obese individuals seems particularly profound among women and may underlie sex differences in HF subtypes. In other words, when obesity is combined with female sex, HFpEF stands out as a special phenomenon, contrary to men, which develop both HFpEF and heart failure with reduced ejection fraction (HFrEF) in the presence of the same risk factors [108]. Estrogenic vasodilatory signaling and the differential use of myocardial energy substrates seem to contribute to these differences [109]. The entity called "obesity paradox" describes a phenomenon where overweight and obese individuals have better prognosis for certain recognized CVD symptoms than those who are lean, notably for heart failure [110]. Higher level of arterial stiffening is present in obese women and it has a greater negative impact on diastolic function compared to men [111].

Some of the hypothesized causes for enhanced CVD risk include increased aldosterone and mineralocorticoid receptor activation, abnormal estrogenic signaling, and higher levels of androgens [112].

## Obesity and PCOS

Cardiometabolic dysfunction is associated with the majority of obesity-related comorbidities. It was shown that 38–80% women with PCOS are obese [113]. When compared to women who were not obese, women with obesity had an odds ratio of 2.77 for developing PCOS, according to a meta-analysis of pertinent research published [114]. PCOS represents obesity-related condition; higher body weight aggravated PCOS symptoms and reducing the body weight, even modestly by 5%, improve reproductive, metabolic and other PCOS manifestations, including risk of CV events [115]. Regardless of the method of conception, obese women have lower reproductive outcomes, while having a higher BMI is linked to worse fertility prognosis [116]. Obesity and weight gain are linked to PCOS because of the impact of weight increase on insulin resistance and hyperinsulinemia, as well as the dysmetabolic and steroidogenic effects of defective PI3-kinase and intact MAP kinase post-receptor insulin pathways. Insulin sensitivity and cardiometabolic risk are influenced by several adipokines. Moreover, it indicates that secretion of adipokines by visceral and subcutaneous fat affects metabolic processes [117]. For example, visfatin has roles in metabolic, inflammatory, and insulin sensitivity pathways [118], and its levels are increased in PCOS patients [119]. In girls and women who are genetically susceptible to PCOS, obesity plays a role in insulin resistance, which promotes clinical presentation of PCOS. The underlying reason why women with PCOS are obese is that obesity increases the susceptibility for PCOS, as recent study of Barber explained. In this study, factors such as epidemiology, genetic and epigenetic aspects of PCOS, including the underlying role of insulin resistance should be considered crucial contributors in weight gain during PCOS development. Obesity, acting through enhanced insulin resistance, promotes the clinical manifestation of PCOS in those girls and women who are genetically predisposed. Therefore, obesity increases the propensity for PCOS, and this is the true explanation for why women with PCOS are obese [120]. The interrelationships between genetic factors, environmental contributors, insulin resistance, obesity, metabolic and reproductive dysfunctions should be considered in estimation of CV risk factors in PCOS patients.

## Dyslipidemia in Women

An increase in serum total cholesterol (TC), low-density lipoprotein (LDL) or tryglicerides (TG), and a decrease in serum high-density lipoprotein (HDL) concentration are signs of dyslipidemia. Dyslipidemia is one of the most significant risk factors for atherosclerosis and associated CVD.

Interestingly, 50% of those who experience a heart attack have normal TC levels. In addition to standard ways of biochemical analyses of lipid status to evaluate cardiometabolic risks, nowadays there are advanced lipid assays, markers of oxidative stress and inflammation, hormone levels, and body composition, which all provide a better chance for early therapies that are safer and more successful in prevention and treatment of CMD. In order to develop strategies for blood lipid management, which is a crucial step in the primary prevention of CVD, basic data on the prevalence and factors linked to various lipid indicators is necessary.

The prevalence of metabolic dyslipidemia among people with diabetes is 40%, while independently of other CVD risk factors, low HDL-C and metabolic dyslipidemia were separately linked to an elevated risk of coronary artery disease events and contribute to the end-point of cardiovascular death, MI, stroke, and angina [121]. Although lowering LDL with statins has decreased the incidence of atherosclerotic CVD events, people with T2DM still have an elevated residual CVD risk [122]. The condition known as metabolic dyslipidemia is characterized by aberrant lipid profiles with excessive TG and/or low HDL [123]. Compared to men, women with T2DM are known to have higher mean LDL, TC, and TG levels, and receive lowering lipid treatment in lesser times [124].

Serum levels of LDL, TC, and TG were positively associated with CVD risk in women. However, after adjusting confounding factors, such as diet, the significance of that correlation was diminished, while still exists among men, especially for MI occurrence and TC levels [125].

The study of Peters and coauthors investigated sex-difference in the prevalence, trends, and CV risk factors in the United States from 2001 to 2016. A greater percentage of men with CVD had dyslipidemia compared to women, whereas in those without CVD, rates showed similar values between men and women [126]. According to this study, older men were more likely to receive treatment and have controlled dyslipidemia, contrary to women, which were less likely to have adequate control of dyslipidemia. However, younger women often have better lipid profiles than men, but after menopause, cholesterol levels rise to levels greater than in men [127].

Wang and coworkers investigated the gender differences in middle-age rural China and concluded that females had much higher prevalence of dyslipidemia, with so-called "scissor shaped" age-gender trend between the sexes after the age of 50 to 55 [128]. Moreover, in this study, the risk of dyslipidemia was associated with obesity in males, and with hypertension in females. Similar "scissor shaped" observations were also found between age and gender in other studies [129], while the underlying mechanisms referred to estrogen levels decline during menopause [130].

## Dyslipidemia and Women-Specific Factors

### *Dyslipidemia and Pregnancy*

Pregnancy is related to altered lipid metabolism, which is important for fetal growth. Hormonal alterations such as increase in insulin and progesterone levels lead to increased lipogenesis and decreased lipolysis, while lipids are transferred via placenta to fetal tissues providing development and growth of fetus [131]. Both TG and TC are rising in their levels as pregnancy progresses. However, these alterations seem to be not atherogenic, although their levels are 3–4 times higher in the third trimester [132] and normalized after the delivery.

Pregnant women who have greater levels of small dense LDL fractions are more likely to have cardiovascular disease later in life. The smaller and denser LDL particles fraction increases in pregnancy, while such a particle is thought to be more atherogenic [133].

Atherogenic dyslipidemia, which is characterized by high TG, small dense LDL, and low HDL levels, is showing increased evidence for adverse pregnancy outcomes and CV risk later in life. Moreover, pregnancy complications, such as preeclampsia, gestational diabetes mellitus (GDM) may highlight these detrimental alterations in lipid patterns and clinical results, particularly because the children of these mothers are more likely to have fatty streaks and have a higher chance of developing progressive atherosclerosis [134]. It is suggested that dyslipidemia observed in early pregnancy (< 20 weeks) could be associated with increased risk for preeclampsia [135].

In predisposed women or those with familiar forms of hyperlipidemia, increase of lipids carries an increased risk for maternal–fetal complications.

## Dyslipidemia and Menopause

Changes in cardiometabolic parameters, such as dyslipidemia, have higher influence on CVD risk after menopause. Menopausal status is linked to elevated TC, LDL, apolipoproteins, and TG as well as decreased HDL cholesterol (predominantly in the HDL2 sub-fraction [136]), even independently of age. Additionally, recent studies indicate that HDL cardioprotective qualities are lost after menopause. However, although hormone replacement therapy exerts beneficial effects on lipid status, it should not be recommended for CVD prevention because improving CVD outcomes is omitted [137].

Women with PCOS, post-menopausa, premature menopause, early menopause, premature ovarian insufficiency, and familial hypercholesterolemia had higher risk of dyslipidemia [137, 138].

Physiological estrogen decline during menopause plays a major role in altered lipid metabolism such as increase in LDL concentration. Menopausal women have high frequency of dyslipidemia, yet research has shown that this population has a very low degree of disease knowledge. Therefore, early diagnosis of dyslipidemia in menopausal and young women who have not yet experienced menopause is necessary to lower the risks of CVD and the mortality rate [139].

## Dyslipidemia and PCOS

Nearly 70% of PCOS women suffer of dyslipidemia [140], while the prevalence of metabolic syndrome varies between 34 and 46% [141–144]. Dyslipidemia, concomitantly with higher androgen levels and body weight increase, particularly in early adulthood, is important for evaluation of PCOS symptoms and diagnosis. One meta-analysis study of PCOS demonstrated significant difference in lipid status among PCOS and non-PCOS women. Levels of TG were 26 mg/dL higher and HDL-C was 6 mg/dL lower in PCOS-affected women than in control women. Additionally, the values of LDL and non-HDL were 12 and 19 mg/dL higher, respectively. LDL and non-HDL were still increased in PCOS participants even after BMI was matched [145]. These data showed that dyslipidemia existence in newly recognized cardiometabolic aspects of reproductive disorders such as PCOS, has significant influence on augmentation of further complications regarding CV events. According to Talbott et al. total HDL and HDL2 levels were considerably lower in women with PCOS than controls even after adjusting for age and BMI [146]. This study compared

206 women with PCOS and 206 age-matched controls. Moreover, lean women with PCOS have shown lower serum HDL and HDL2 sub-fraction levels compared to control subjects [147]. Complete lipid testing should be done on all PCOS patients, as first line of treatment for all PCOS patients—and especially for those with dyslipidemia in order to reduce risk of further CV events and cardiometabolic consequences. These features could pinpoint a critical window of time when weight gain plays a significant role in the development of PCOS and indicates the need for preventative measures against metabolic and cardiovascular illnesses.

## Conclusion

CMD in women are increased largely by risks such as diabetes, hypertension, dyslipidemia, and obesity. Although cardiovascular events incidence progresses with age in both genders, men have a higher prevalence of cardiovascular events until mid-age. However, after menopause, the women-to-men ratio of cardiovascular events increases. The basic step in providing adequate treatment modalities refers to understanding the pathophysiological events that underlie CMD. Biological explanations for gender differences in cardiometabolic diseases are complex. The main physiological features that underlie the gender differences in CMD development could be defined as sex hormones and body composition disparities in men and women. Some cardiometabolic aspects are unique for women, such as pregnancy-related or menopause-related features, leading to exert adverse cardiometabolic features in present, or leave a deep imprint later in life (Fig. 12.1).

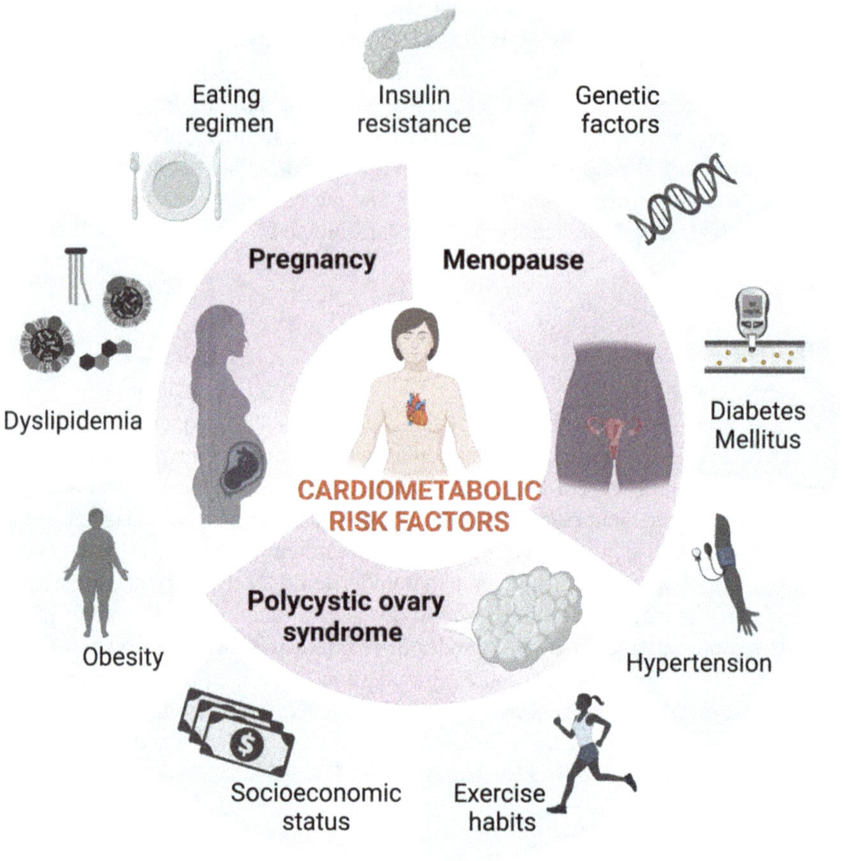


**Fig. 12.1** Schematic representation summaries risk factors for cardiometabolic disease in women

# References


1. Srivastava AK (2012) Challenges in the treatment of cardiometabolic syndrome. Indian J Pharmacol 44(2):155–156. https://doi.org/10.4103/0253-7613.93579
2. Zhao M, Woodward M, Vaartjes I, Millett ERC, Klipstein-Grobusch K, Hyun K, Carcel C, Peters SAE (2020) Sex differences in cardiovascular medication prescription in primary care: a systematic review and meta-analysis. J Am Heart Assoc 9(11):e014742. https://doi.org/10.1161/JAHA.119.014742. Epub 2020 May 20
3. Möller-Leimkühler AM (2007) Gender differences in cardiovascular disease and comorbid depression. Dialogues Clin Neurosci 9(1):71–83. https://doi.org/10.31887/DCNS.2007.9.1/ammoeller
4. Pitsavos C, Panagiotakos D, Weinem M, Stefanadis C (2006) Diet, exercise and the metabolic syndrome. Rev Diabet Stud 3(3):118–26. https://doi.org/10.1900/RDS.2006.3.118. Epub 2006 Nov 10
5. Swarup S, Goyal A, Grigorova Y, Zeltser R (2022) Metabolic syndrome. In: StatPearls [Internet]. Treasure Island (FL). StatPearls Publishing

6. Di Giosia P, Passacquale G, Petrarca M, Giorgini P, Marra AM, Ferro A (2017) Gender differences in cardiovascular prophylaxis: focus on antiplatelet treatment. Pharmacol Res 119:36–47. https://doi.org/10.1016/j.phrs.2017.01.025. Epub 2017 Jan 25
7. Humphries KH, Izadnegahdar M, Sedlak T, Saw J, Johnston N, Schenck-Gustafsson K, Shah RU, Regitz-Zagrosek V, Grewal J, Vaccarino V, Wei J, Bairey Merz CN (2017) Sex differences in cardiovascular disease—impact on care and outcomes. Front Neuroendocrinol 46:46–70. https://doi.org/10.1016/j.yfrne.2017.04.001. Epub 2017 Apr 18
8. Pradhan AD (2014) Sex differences in the metabolic syndrome: implications for cardiovascular health in women. Clin Chem 60(1):44–52. https://doi.org/10.1373/clinchem.2013.202549. Epub 2013 Nov 19
9. Coutinho T (2014) Arterial stiffness and its clinical implications in women. Can J Cardiol 30(7):756–764. https://doi.org/10.1016/j.cjca.2014.03.020. Epub 2014 Mar 20
10. Stoberock K, Debus ES, Atlihan G, Daum G, Larena-Avellaneda A, Eifert S, Wipper S (2016) Gender differences in patients with carotid stenosis. Vasa 45(1):11–16. https://doi.org/10.1024/0301-1526/a000490
11. Dušková M, Pospíšilová H (2011) The role of non-aromatizable testosterone metabolite in metabolic pathways. Physiol Res 60(2):253–261. https://doi.org/10.33549/physiolres.932080. Epub 2010 Nov 29
12. Soisson V, Brailly-Tabard S, Helmer C, Rouaud O, Ancelin ML, Zerhouni C, Guiochon-Mantel A, Scarabin PY (2013) A J-shaped association between plasma testosterone and risk of ischemic arterial event in elderly men: the French 3C cohort study. Maturitas 75(3):282–288. https://doi.org/10.1016/j.maturitas.2013.04.012. Epub 2013 May 22
13. Srinath R, Hill Golden S, Carson KA, Dobs A (2015) Endogenous testosterone and its relationship to preclinical and clinical measures of cardiovascular disease in the atherosclerosis risk in communities study. J Clin Endocrinol Metab 100(4):1602–1608. https://doi.org/10.1210/jc.2014-3934. Epub 2015 Jan 13
14. Ouyang P, Vaidya D, Dobs A, Golden SH, Szklo M, Heckbert SR, Kopp P, Gapstur SM (2009) Sex hormone levels and subclinical atherosclerosis in postmenopausal women: the multi-ethnic study of Atherosclerosis. Atherosclerosis 204(1):255–261. https://doi.org/10.1016/j.atherosclerosis.2008.08.037. Epub 2008 Sep 6
15. Britto R, Araújo L, Barbosa I, Silva L (2012) Improvement of the lipid profile in post menopausal women who use estradiol and testosterone implants. Gynecol Endocrinol 28(10):767–769. https://doi.org/10.3109/09513590.2012.664191. Epub 2012 Mar 8
16. Xu L, Freeman G, Cowling BJ, Schooling CM (2013) Testosterone therapy and cardiovascular events among men: a systematic review and meta-analysis of placebo-controlled randomized trials. BMC Med 18(11):108. https://doi.org/10.1186/1741-7015-11-108
17. Slagter SN, van Waateringe RP, van Beek AP, van der Klauw MM, Wolffenbuttel BHR, van Vliet-Ostaptchouk JV (2017) Sex, BMI and age differences in metabolic syndrome: the Dutch Lifelines Cohort Study. Endocr Connect 6(4):278–288. https://doi.org/10.1530/EC-17-0011. Epub 2017 Apr 18
18. Blüher M (2020) Metabolically healthy obesity. Endocr Rev 41(3):bnaa004. https://doi.org/10.1210/endrev/bnaa004
19. Kearney PM, Whelton M, Reynolds K, Muntner P, Whelton PK, He J (2005) Global burden of hypertension: analysis of worldwide data. Lancet 365(9455):217–23. https://doi.org/10.1016/S0140-6736(05)17741-1
20. Mills KT, Bundy JD, Kelly TN, Reed JE, Kearney PM, Reynolds K, Chen J, He J (2016) Global disparities of hypertension prevalence and control: a systematic analysis of population-based studies from 90 countries. Circ 134(6):441–450. https://doi.org/10.1161/CIRCULATIONAHA.115.018912
21. Wiinberg N, Høegholm A, Christensen HR, Bang LE, Mikkelsen KL, Nielsen PE, Svendsen TL, Kampmann JP, Madsen NH, Bentzon MW (1995) 24-h ambulatory blood pressure in 352 normal Danish subjects, related to age and gender. Am J Hypertens 8(10 Pt 1):978–986. https://doi.org/10.1016/0895-7061(95)00216-2

22. Brahmbhatt Y, Gupta M, Hamrahian S (2019) Hypertension in premenopausal and postmenopausal women. Curr Hypertens Rep 21(10):74. https://doi.org/10.1007/s11906-019-0979-y
23. White RE (2002) Estrogen and vascular function. Vascul Pharmacol 38(2):73–80. https://doi.org/10.1016/s0306-3623(02)00129-5
24. Izumi Y, Matsumoto K, Ozawa Y, Kasamaki Y, Shinndo A, Ohta M, Jumabay M, Nakayama T, Yokoyama E, Shimabukuro H, Kawamura H, Cheng Z, Ma Y, Mahmut M (2007) Effect of age at menopause on blood pressure in postmenopausal women. Am J Hypertens 20(10):1045–1050. https://doi.org/10.1016/j.amjhyper.2007.04.019
25. Iorga A, Cunningham CM, Moazeni S, Ruffenach G, Umar S, Eghbali M (2017) The protective role of estrogen and estrogen receptors in cardiovascular disease and the controversial use of estrogen therapy. Biol Sex Differ 8(1):33. https://doi.org/10.1186/s13293-017-0152-8
26. Wenger NK, Ferdinand KC, Bairey Merz CN, Walsh MN, Gulati M, Pepine CJ (2016) American College of Cardiology Cardiovascular Disease in Women Committee. Women, hypertension, and the systolic blood pressure intervention trial. Am J Med 129(10):1030–1036. https://doi.org/10.1016/j.amjmed.2016.06.022. Epub 2016 Jul 15
27. Lobo RA (2008) Metabolic syndrome after menopause and the role of hormones. Maturitas 60(1):10–18. https://doi.org/10.1016/j.maturitas.2008.02.008. Epub 2008 Apr 14
28. Abramson BL, Melvin RG (2014) Cardiovascular risk in women: focus on hypertension. Can J Cardiol 30(5):553–559. https://doi.org/10.1016/j.cjca.2014.02.014. Epub 2014 Feb 26
29. Park H, Kim K (2013) Associations between oral contraceptive use and risks of hypertension and prehypertension in a cross-sectional study of Korean women. BMC Womens Health 21(13):39. https://doi.org/10.1186/1472-6874-13-39
30. Liu H, Yao J, Wang W, Zhang D (2017) Association between duration of oral contraceptive use and risk of hypertension: a meta-analysis. J Clin Hypertens (Greenwich) 19(10):1032–1041. https://doi.org/10.1111/jch.13042. Epub 2017 Jun 13
31. Jafari A, Rajabi A, Gholian-Aval M, Peyman N, Mahdizadeh M, Tehrani H (2021) National, regional, and global prevalence of cigarette smoking among women/females in the general population: a systematic review and meta-analysis. Environ Health Prev Med 26(1):5. https://doi.org/10.1186/s12199-020-00924-y
32. Bonnema RA, McNamara MC, Spencer AL (2010) Contraception choices in women with underlying medical conditions. Am Fam Physician 82(6):621–628
33. Glisic M, Shahzad S, Tsoli S, Chadni M, Asllanaj E, Rojas LZ, Brown E, Chowdhury R, Muka T, Franco OH (2018) Association between progestin-only contraceptive use and cardiometabolic outcomes: a systematic review and meta-analysis. Eur J Prev Cardiol 25(10):1042–1052. https://doi.org/10.1177/2047487318774847. Epub 2018 May 10
34. Woods JW (1988) Oral contraceptives and hypertension. Hypertens 11(3 Pt 2):II11– II15. https://doi.org/10.1161/01.hyp.11.3_pt_2.ii11
35. Behrens I, Basit S, Lykke JA, Ranthe MF, Wohlfahrt J, Bundgaard H, Melbye M, Boyd HA (2016) Association between hypertensive disorders of pregnancy and later risk of cardiomyopathy. JAMA 315(10):1026–1033. https://doi.org/10.1001/jama.2016.1869
36. Levine LD, Ky B, Chirinos JA, Koshinksi J, Arany Z, Riis V, Elovitz MA, Koelper N, Lewey J (2022) Prospective evaluation of cardiovascular risk 10 years after a hypertensive disorder of pregnancy. J Am Coll Cardiol 79(24):2401–2411. https://doi.org/10.1016/j.jacc.2022.03.383
37. Lindheimer MD, August P (2009) Aldosterone, maternal volume status and healthy pregnancies: a cycle of differing views. Nephrol Dial Transplant 24(6):1712–1714. https://doi.org/10.1093/ndt/gfp093. Epub 2009 Mar 3
38. Spracklen CN, Smith CJ, Saftlas AF, Robinson JG, Ryckman KK (2014) Maternal hyperlipidemia and the risk of preeclampsia: a meta-analysis. Am J Epidemiol 180(4):346–58. https://doi.org/10.1093/aje/kwu145. Epub 2014 Jul 2
39. Hellgren M (2003) Hemostasis during normal pregnancy and puerperium. Semin Thromb Hemost 29(2):125–130. https://doi.org/10.1055/s-2003-38897
40. Bello NA, Zhou H, Cheetham TC, Miller E, Getahun DT, Fassett MJ, Reynolds K (2021) Prevalence of hypertension among pregnant women when using the 2017 American College

of Cardiology/American Heart Association blood pressure guidelines and association with maternal and fetal outcomes. JAMA Netw Open 4(3):e213808. Erratum in: JAMA Netw Open. 2021 Apr 1;4(4):e2112000. https://doi.org/10.1001/jamanetworkopen.2021.3808
41. Vrbíková J, Cífková R, Jirkovská A, Lánská V, Platilová H, Zamrazil V, Stárka L (2003) Cardiovascular risk factors in young Czech females with polycystic ovary syndrome. Hum Reprod 18(5):980–984. https://doi.org/10.1093/humrep/deg218
42. Joham AE, Boyle JA, Zoungas S, Teede HJ (2015) Hypertension in reproductive-aged women with polycystic ovary syndrome and association with obesity. Am J Hypertens 28(7):847–851. https://doi.org/10.1093/ajh/hpu251. Epub 2014 Dec 26
43. Lo JC, Feigenbaum SL, Yang J, Pressman AR, Selby JV, Go AS (2006) Epidemiology and adverse cardiovascular risk profile of diagnosed polycystic ovary syndrome. J Clin Endocrinol Metab 91(4):1357–1363. https://doi.org/10.1210/jc.2005-2430. Epub 2006 Jan 24
44. Glintborg D, Hass Rubin K, Nybo M, Abrahamsen B, Andersen M (2015) Morbidity and medicine prescriptions in a nationwide Danish population of patients diagnosed with polycystic ovary syndrome. Eur J Endocrinol 172(5):627–638. https://doi.org/10.1530/EJE-14-1108. Epub 2015 Feb 5
45. Glintborg D, Rubin KH, Nybo M, Abrahamsen B, Andersen M (2018) Cardiovascular disease in a nationwide population of Danish women with polycystic ovary syndrome. Cardiovasc Diabetol 17(1):37. https://doi.org/10.1186/s12933-018-0680-5
46. Meyer C, McGrath BP, Teede HJ (2005) Overweight women with polycystic ovary syndrome have evidence of subclinical cardiovascular disease. J Clin Endocrinol Metab 90(10):5711–5716. https://doi.org/10.1210/jc.2005-0011. Epub 2005 Jul 26
47. Zimmermann S, Phillips RA, Dunaif A, Finegood DT, Wilkenfeld C, Ardeljan M, Gorlin R, Krakoff LR (1992) Polycystic ovary syndrome: lack of hypertension despite profound insulin resistance. J Clin Endocrinol Metab 75(2):508–513. https://doi.org/10.1210/jcem.75.2.1639952
48. Holte J, Gennarelli G, Berne C, Bergh T, Lithell H (1996) Elevated ambulatory day-time blood pressure in women with polycystic ovary syndrome: a sign of a pre-hypertensive state? Hum Reprod 11(1):23–28. https://doi.org/10.1093/oxfordjournals.humrep.a019028
49. Joksimovic Jovic J, Jovic N, Sretenovic J, Zivkovic V, Nikolic M, Rudic J, Milošević V, Ristić N, Andric K, Dimkic Tomic T, Milicic B, Jakovljevic V (2021) Normotensive rats with PCOS exhibit the hypertensive pattern: focus on oxidative stress. Reprod 163(1):11–21. https://doi.org/10.1530/REP-21-0219
50. Joksimovic Jovic J, Sretenovic J, Jovic N, Rudic J, Zivkovic V, Srejovic I, Mihajlovic K, Draginic N, Andjic M, Milinkovic M, Milosavljevic Z, Jakovljevic V (2021) Cardiovascular properties of the androgen-induced PCOS model in rats: the role of oxidative stress. Oxid Med Cell Longev 31(2021):8862878. https://doi.org/10.1155/2021/8862878
51. de Jong M, Oskam MJ, Sep SJS, Ozcan B, Rutters F, Sijbrands EJG, Elders PJM, Siegelaar SE, DeVries JH, Tack CJ, Schroijen M, de Valk HW, Abbink EJ, Stehouwer CDA, Jazet I, Wolffenbuttel BHR, Peters SAE, Schram MT (2020) Diabetes Pearl from the Parelsnoer Initiative. Sex differences in cardiometabolic risk factors, pharmacological treatment and risk factor control in type 2 diabetes: findings from the Dutch Diabetes Pearl cohort. BMJ Open Diabetes Res Care 8(1):e001365. https://doi.org/10.1136/bmjdrc-2020-001365
52. Campesi I, Franconi F, Seghieri G, Meloni M (2017) Sex-gender-related therapeutic approaches for cardiovascular complications associated with diabetes. Pharmacol Res 119:195–207. https://doi.org/10.1016/j.phrs.2017.01.023. Epub 2017 Feb 9
53. Xu G, You D, Wong L, Duan D, Kong F, Zhang X, Zhao J, Xing W, Han L, Li L (2019) Risk of all-cause and CHD mortality in women versus men with type 2 diabetes: a systematic review and meta-analysis. Eur J Endocrinol 180(4):243–255. https://doi.org/10.1530/EJE-18-0792
54. Peters SA, Huxley RR, Woodward M (2016) Women's reproductive health factors and body adiposity: findings from the UK Biobank. Int J Obes (Lond) 40(5):803–808. https://doi.org/10.1038/ijo.2015.254. Epub 2015 Dec 24
55. Yang L, Li L, Millwood IY, Lewington S, Guo Y, Sherliker P, Peters SA, Bian Z, Wu X, Yu M, Liu H, Wang H, Mao E, Chen J, Woodward M, Peto R, Chen Z (2017) China Kadoorie

Biobank study collaborative group (members listed at end of report). Adiposity in relation to age at menarche and other reproductive factors among 300,000 Chinese women: findings from China Kadoorie Biobank study. Int J Epidemiol 46(2):502–512. https://doi.org/10.1093/ije/dyw165

56. Peters SA, Huxley RR, Woodward M (2014) Diabetes as risk factor for incident coronary heart disease in women compared with men: a systematic review and meta-analysis of 64 cohorts including 858,507 individuals and 28,203 coronary events. Diabetologia 57(8):1542–1551. https://doi.org/10.1007/s00125-014-3260-6. Epub 2014 May 25
57. Seghieri G, Policardo L, Anichini R, Franconi F, Campesi I, Cherchi S, Tonolo G (2017) The effect of sex and gender on diabetic complications. Curr Diabetes Rev 13(2):148–160. https://doi.org/10.2174/1573399812666160517115756
58. Green JB (2021) Cardiovascular consequences of gestational diabetes. Circ 143(10):988–990. https://doi.org/10.1161/CIRCULATIONAHA.120.052995. Epub 2021 Mar 8
59. Manicardi V, Russo G, Napoli A, Torlone E, Li Volsi P, Giorda CB, Musacchio N, Nicolucci A, Suraci C, Lucisano G, Rossi MC (2016) AMD Annals Study Group. Gender-disparities in adults with type 1 diabetes: more than a quality of care issue. A cross-sectional observational study from the AMD annals initiative. PLoS One 11(10):e0162960. https://doi.org/10.1371/journal.pone.0162960
60. Kwon SK (2014) Women are diagnosed with type 2 diabetes at higher body mass indices and older ages than men: Korea national health and nutrition examination survey 2007–2010. Diabetes Metab J 38(1):74–80. https://doi.org/10.4093/dmj.2014.38.1.74. Epub 2014 Feb 19
61. Huxley R, Barzi F, Woodward M (2006) Excess risk of fatal coronary heart disease associated with diabetes in men and women: meta-analysis of 37 prospective cohort studies. BMJ 332(7533):73–8. https://doi.org/10.1136/bmj.38678.389583.7C. Epub 2005 Dec 21
62. Peters SAE, Woodward M (2018) Sex differences in the burden and complications of diabetes. Curr Diab Rep 18(6):33. https://doi.org/10.1007/s11892-018-1005-5
63. Al-Salameh A, Chanson P, Bucher S, Ringa V, Becquemont L (2019) Cardiovascular disease in type 2 diabetes: a review of sex-related differences in predisposition and prevention. Mayo Clin Proc 94(2):287–308. https://doi.org/10.1016/j.mayocp.2018.08.007
64. Rossi MC, Cristofaro MR, Gentile S, Lucisano G, Manicardi V, Mulas MF, Napoli A, Nicolucci A, Pellegrini F, Suraci C, Giorda C (2013) AMD Annals Study Group. Sex disparities in the quality of diabetes care: biological and cultural factors may play a different role for different outcomes: a cross-sectional observational study from the AMD Annals initiative. Diabetes Care 36(10):3162–3168. https://doi.org/10.2337/dc13-0184. Epub 2013 Jul 8
65. Singer JR, Palmas W, Teresi J, Weinstock R, Shea S, Luchsinger JA (2012) Adiponectin and all-cause mortality in elderly people with type 2 diabetes. Diabetes Care 35(9):1858–1863. https://doi.org/10.2337/dc11-2215. Epub 2012 Jul 6
66. Donahue RP, Rejman K, Rafalson LB, Dmochowski J, Stranges S, Trevisan M (2007) Sex differences in endothelial function markers before conversion to pre-diabetes: does the clock start ticking earlier among women? The Western New York Study. Diabetes Care 30(2):354–359. https://doi.org/10.2337/dc06-1772
67. Yoshida Y, Chen Z, Baudier RL, Krousel-Wood M, Anderson AH, Fonseca VA, Mauvais-Jarvis F (2021) Early menopause and cardiovascular disease risk in women with or without type 2 diabetes: a pooled analysis of 9,374 postmenopausal women. Diabetes Care 44(11):2564–2572. https://doi.org/10.2337/dc21-1107. Epub 2021 Sep 2
68. Brand JS, Onland-Moret NC, Eijkemans MJ, Tjønneland A, Roswall N, Overvad K, Fagherazzi G, Clavel-Chapelon F, Dossus L, Lukanova A, Grote V, Bergmann MM, Boeing H, Trichopoulou A, Tzivoglou M, Trichopoulos D, Grioni S, Mattiello A, Masala G, Tumino R, Vineis P, Bueno-de-Mesquita HB, Weiderpass E, Redondo ML, Sánchez MJ, Castaño JM, Arriola L, Ardanaz E, Duell EJ, Rolandsson O, Franks PW, Butt S, Nilsson P, Khaw KT, Wareham N, Travis R, Romieu I, Gunter MJ, Riboli E, van der Schouw YT (2015) Diabetes and onset of natural menopause: results from the European Prospective Investigation into Cancer and Nutrition. Hum Reprod 30(6):1491–1498. https://doi.org/10.1093/humrep/dev054. Epub 2015 Mar 15

69. Sekhar TV, Medarametla S, Rahman A, Adapa SS (2015) Early menopause in type 2 diabetes—a study from a South Indian Tertiary Care Centre. J Clin Diagn Res 9(10):OC08–OC10. https://doi.org/10.7860/JCDR/2015/14181.6628. Epub 2015 Oct 1
70. Stachowiak G, Pertyński T, Pertyńska-Marczewska M (2015) Metabolic disorders in menopause. Prz Menopauzalny 14(1):59–64. https://doi.org/10.5114/pm.2015.50000. Epub 2015 Mar 25
71. Matthews KA, Gibson CJ, El Khoudary SR, Thurston RC (2013) Changes in cardiovascular risk factors by hysterectomy status with and without oophorectomy: study of Women's Health Across the Nation. J Am Coll Cardiol 62(3):191–200. https://doi.org/10.1016/j.jacc.2013.04.042. Epub 2013 May 15
72. Kim C (2021) Management of cardiovascular risk in perimenopausal women with diabetes. Diabetes Metab J 45(4):492–501. https://doi.org/10.4093/dmj.2020.0262. Epub 2021 Jul 30
73. Burlina S, Dalfrà MG, Chilelli NC, Lapolla A (2016) Gestational diabetes mellitus and future cardiovascular risk: an update. Int J Endocrinol 2016:2070926
74. Retnakaran R, Shah BR (2017) Role of type 2 diabetes in determining retinal, renal, and cardiovascular outcomes in women with previous gestational diabetes mellitus. Diabetes Care 40(1):101–108. https://doi.org/10.2337/dc16-1400. Epub 2016 Nov 7
75. Appiah D, Schreiner PJ, Gunderson EP, Konety SH, Jacobs DR Jr, Nwabuo CC, Ebong IA, Whitham HK, Goff DC Jr, Lima JA, Ku IA, Gidding SS (2016) Association of gestational diabetes mellitus with left ventricular structure and function: the CARDIA study. Diabetes Care 39(3):400–407. https://doi.org/10.2337/dc15-1759. Epub 2016 Jan 6
76. Mehmood S, Ye C, Connelly PW, Hanley AJ, Zinman B, Retnakaran R (2018) Rising plasminogen activator inhibitor-1 and hypoadiponectinemia characterize the cardiometabolic biomarker profile of women with recent gestational diabetes. Cardiovasc Diabetol 17(1):133. https://doi.org/10.1186/s12933-018-0776-y
77. Akinci B, Celtik A, Tunali S, Genc S, Yuksel F, Secil M, Ozcan MA, Bayraktar F (2014) Circulating apelin levels are associated with cardiometabolic risk factors in women with previous gestational diabetes. Arch Gynecol Obstet 289(4):787–793. https://doi.org/10.1007/s00404-013-3070-y. Epub 2013 Nov 2
78. Gunderson EP, Quesenberry CP Jr, Jacobs DR Jr, Feng J, Lewis CE, Sidney S (2010) Longitudinal study of prepregnancy cardiometabolic risk factors and subsequent risk of gestational diabetes mellitus: the CARDIA study. Am J Epidemiol 172(10):1131–1143. https://doi.org/10.1093/aje/kwq267. Epub 2010 Oct 7
79. Mathur R, Alexander CJ, Yano J, Trivax B, Azziz R (2008) Use of metformin in polycystic ovary syndrome. Am J Obstet Gynecol 199(6):596–609. https://doi.org/10.1016/j.ajog.2008.09.010
80. Khan MJ, Ullah A, Basit S (2019) Genetic basis of polycystic ovary syndrome (PCOS): current perspectives. Appl Clin Genet 24(12):249–260. https://doi.org/10.2147/TACG.S200341
81. Marshall JC, Dunaif A (2012) Should all women with PCOS be treated for insulin resistance? Fertil Steril 97(1):18–22. https://doi.org/10.1016/j.fertnstert.2011.11.036
82. Kalra B, Kalra S, Sharma JB (2016) The inositols and polycystic ovary syndrome. Indian J Endocrinol Metab 20(5):720–724. https://doi.org/10.4103/2230-8210.189231
83. Scicchitano P, Dentamaro I, Carbonara R, Bulzis G, Dachille A, Caputo P, Riccardi R, Locorotondo M, Mandurino C, Matteo Ciccone M (2012) Cardiovascular risk in women With PCOS. Int J Endocrinol Metab 10(4):611–618. https://doi.org/10.5812/ijem.4020. Epub 2012 Sep 30
84. Unfer V, Facchinetti F, Orrù B, Giordani B, Nestler J (2017) Myo-inositol effects in women with PCOS: a meta-analysis of randomized controlled trials. Endocr Connect 6(8):647–658. https://doi.org/10.1530/EC-17-0243.
85. Aversa A, La Vignera S, Rago R, Gambineri A, Nappi RE, Calogero AE, Ferlin A (2020) Fundamental concepts and novel aspects of polycystic ovarian syndrome: expert consensus resolutions. Front Endocrinol (Lausanne) 11(11):516. https://doi.org/10.3389/fendo.2020.00516
86. Szczuko M, Kikut J, Szczuko U, Szydłowska I, Nawrocka-Rutkowska J, Ziętek M, Verbanac D, Saso L (2021) Nutrition strategy and life style in polycystic ovary syndrome-narrative review. Nutrients 13(7):2452. https://doi.org/10.3390/nu13072452

87. Finucane MM, Stevens GA, Cowan MJ, Danaei G, Lin JK, Paciorek CJ, Singh GM, Gutierrez HR, Lu Y, Bahalim AN, Farzadfar F, Riley LM, Ezzati M (2011) Global burden of metabolic risk factors of chronic diseases collaborating group (body mass index). National, regional, and global trends in body-mass index since 1980: systematic analysis of health examination surveys and epidemiological studies with 960 country-years and 9·1 million participants. Lancet 377(9765):557–67. https://doi.org/10.1016/S0140-6736(10)62037-5. Epub 2011 Feb 3
88. Templeton A (2014) Obesity and women's health. Facts Views Vis Obgyn 6(4):175–176
89. NCD Risk Factor Collaboration (NCD-RisC) (2016) Trends in adult body-mass index in 200 countries from 1975 to 2014: a pooled analysis of 1698 population-based measurement studies with 19.2 million participants. Lancet 387(10026):1377–1396. https://doi.org/10.1016/S0140-6736(16)30054-X. Erratum in: Lancet. 2016 May 14;387(10032):1998
90. Mohammadian Khonsari N, Khashayar P, Shahrestanaki E, Kelishadi R, Mohammadpoor Nami S, Heidari-Beni M, Esmaeili Abdar Z, Tabatabaei-Malazy O, Qorbani M (2022) Normal weight obesity and cardiometabolic risk factors: a systematic review and meta-analysis. Front Endocrinol (Lausanne) 24(13):857930. https://doi.org/10.3389/fendo.2022.857930
91. Thapa R, Dahl C, Aung WP, Bjertness E (2021) Urban-rural differences in overweight and obesity among 25–64 years old Myanmar residents: a cross-sectional, nationwide survey. BMJ Open 11(3):e042561. https://doi.org/10.1136/bmjopen-2020-042561
92. Driscoll AK, Gregory ECW (2020) Increases in prepregnancy obesity: United States, 2016–2019. NCHS Data Brief (392):1–8
93. Kozakowski J, Gietka-Czernel M, Leszczyńska D, Majos A (2017) Obesity in menopause—our negligence or an unfortunate inevitability? Prz Menopauzalny 16(2):61–65. https://doi.org/10.5114/pm.2017.68594. Epub 2017 Jun 30
94. Donato GB, Fuchs SC, Oppermann K, Bastos C, Spritzer PM (2006) Association between menopause status and central adiposity measured at different cutoffs of waist circumference and waist-to-hip ratio. Menopause 13(2):280–285. https://doi.org/10.1097/01.gme.0000177907.32634.ae
95. Rebouças EC, Leal S, Sá SI (2016) Regulation of NPY and α-MSH expression by estradiol in the arcuate nucleus of Wistar female rats: a stereological study. Neurol Res 38(8):740–747. https://doi.org/10.1080/01616412.2016.1203124. Epub 2016 Jun 30
96. Vrachnis N, Augoulea A, Iliodromiti Z, Lambrinoudaki I, Sifakis S, Creatsas G (2012) Previous gestational diabetes mellitus and markers of cardiovascular risk. Int J Endocrinol 2012:458610. https://doi.org/10.1155/2012/458610. Epub 2012 Mar 18
97. Vinayagam D, Gutierrez J, Binder J, Mantovani E, Thilaganathan B, Khalil A (2017) Impaired maternal hemodynamics in morbidly obese women: a case-control study. Ultrasound Obstet Gynecol 50(6):761–765. https://doi.org/10.1002/uog.17428. Epub 2017 Nov 8
98. Zheng Z, Yang T, Chen L, Wang L, Zhang S, Wang T, Zhao L, Ye Z, Chen L, Qin J (2018) Increased maternal Body Mass Index is associated with congenital heart defects: an updated meta-analysis of observational studies. Int J Cardiol 15(273):112–120. https://doi.org/10.1016/j.ijcard.2018.09.116. Epub 2018 Oct 1
99. Jarvie E, Hauguel-de-Mouzon S, Nelson SM, Sattar N, Catalano PM, Freeman DJ (2010) Lipotoxicity in obese pregnancy and its potential role in adverse pregnancy outcome and obesity in the offspring. Clin Sci (Lond) 119(3):123–129. https://doi.org/10.1042/CS20090640
100. Catalano PM, Shankar K (2017) Obesity and pregnancy: mechanisms of short term and long term adverse consequences for mother and child. BMJ 8(356):j1. https://doi.org/10.1136/bmj.j1
101. Nogues P, Dos Santos E, Jammes H, Berveiller P, Arnould L, Vialard F, Dieudonné MN (2019) Maternal obesity influences expression and DNA methylation of the adiponectin and leptin systems in human third-trimester placenta. Clin Epigenetics 11(1):20. https://doi.org/10.1186/s13148-019-0612-6
102. Shao Y, Qiu J, Huang H, Mao B, Dai W, He X, Cui H, Lin X, Lv L, Wang D, Tang Z, Xu S, Zhao N, Zhou M, Xu X, Qiu W, Liu Q, Zhang Y (2017) Pre-pregnancy BMI, gestational

weight gain and risk of preeclampsia: a birth cohort study in Lanzhou, China. BMC Pregnancy Childbirth 17(1):400. https://doi.org/10.1186/s12884-017-1567-2
103. Rodie VA, Freeman DJ, Sattar N, Greer IA (2004) Pre-eclampsia and cardiovascular disease: metabolic syndrome of pregnancy? Atherosclerosis 175(2):189–202. https://doi.org/10.1016/j.atherosclerosis.2004.01.038
104. Bellamy L, Casas JP, Hingorani AD, Williams DJ (2007) Pre-eclampsia and risk of cardiovascular disease and cancer in later life: systematic review and meta-analysis. BMJ 335(7627):974. https://doi.org/10.1136/bmj.39335.385301.BE. Epub 2007 Nov 1
105. Anand SS, Islam S, Rosengren A, Franzosi MG, Steyn K, Yusufali AH, Keltai M, Diaz R, Rangarajan S, Yusuf S (2008) INTERHEART investigators. Risk factors for myocardial infarction in women and men: insights from the INTERHEART study. Eur Heart J 29(7):932–940. https://doi.org/10.1093/eurheartj/ehn018. Epub 2008 Mar 10
106. Sharaf B, Wood T, Shaw L, Johnson BD, Kelsey S, Anderson RD, Pepine CJ, Bairey Merz CN (2013) Adverse outcomes among women presenting with signs and symptoms of ischemia and no obstructive coronary artery disease: findings from the National Heart, Lung, and Blood Institute-sponsored Women's Ischemia Syndrome Evaluation (WISE) angiographic core laboratory. Am Heart J 166(1):134–141. https://doi.org/10.1016/j.ahj.2013.04.002. Epub 2013 May 2
107. Kenchaiah S, Evans JC, Levy D, Wilson PW, Benjamin EJ, Larson MG, Kannel WB, Vasan RS (2002) Obesity and the risk of heart failure. N Engl J Med 347(5):305–313. https://doi.org/10.1056/NEJMoa020245
108. Savji N, Meijers WC, Bartz TM, Bhambhani V, Cushman M, Nayor M, Kizer JR, Sarma A, Blaha MJ, Gansevoort RT, Gardin JM, Hillege HL, Ji F, Kop WJ, Lau ES, Lee DS, Sadreyev R, van Gilst WH, Wang TJ, Zanni MV, Vasan RS, Allen NB, Psaty BM, van der Harst P, Levy D, Larson M, Shah SJ, de Boer RA, Gottdiener JS, Ho JE (2018) The Association of Obesity and Cardiometabolic Traits With Incident HFpEF and HFrEF. JACC Heart Fail 6(8):701–709. https://doi.org/10.1016/j.jchf.2018.05.018. Epub 2018 Jul 11
109. Crescioli C (2021) The role of estrogens and vitamin D in cardiomyocyte protection: a female perspective. Biomol 11(12):1815. https://doi.org/10.3390/biom11121815
110. Schmidt D, Salahudeen A (2007) The obesity-survival paradox in hemodialysis patients: why do overweight hemodialysis patients live longer? Nutr Clin Pract 22(1):11–15. https://doi.org/10.1177/011542650702200111
111. Redfield MM, Jacobsen SJ, Borlaug BA, Rodeheffer RJ, Kass DA (2005) Age- and gender-related ventricular-vascular stiffening: a community-based study. Circulation 112(15):2254–2262. https://doi.org/10.1161/CIRCULATIONAHA.105.541078. Epub 2005 Oct 3
112. Manrique-Acevedo C, Chinnakotla B, Padilla J, Martinez-Lemus LA, Gozal D (2020) Obesity and cardiovascular disease in women. Int J Obes (Lond) 44(6):1210–1226. https://doi.org/10.1038/s41366-020-0548-0. Epub 2020 Feb 17
113. Barber TM, McCarthy MI, Wass JA, Franks S (2006) Obesity and polycystic ovary syndrome. Clin Endocrinol (Oxf) 65(2):137–145
114. Lim SS, Davies MJ, Norman RJ, Moran LJ (2012) Overweight, obesity and central obesity in women with polycystic ovary syndrome: a systematic review and meta-analysis. Hum Reprod Update 18(6):618–637
115. Holte J, Bergh T, Berne C, Wide L, Lithell H (1995) Restored insulin sensitivity but persistently increased early insulin secretion after weight loss in obese women with polycystic ovary syndrome. J Clin Endocrinol Metab 80(9):2586–2593. https://doi.org/10.1210/jcem.80.9.7673399
116. Talmor A, Dunphy B (2015) Female obesity and infertility. Best Pract Res Clin Obstet Gynaecol 29(4):498–506. https://doi.org/10.1016/j.bpobgyn.2014.10.014. Epub 2014 Nov 7
117. Barber TM, Hanson P, Weickert MO, Franks S (2019) Obesity and polycystic ovary syndrome: implications for pathogenesis and novel management strategies. Clin Med Insights Reprod Health 9(13):1179558119874042. https://doi.org/10.1177/1179558119874042
118. Barber TM, Franks S (2012) Adipocyte biology in polycystic ovary syndrome. Mol Cell Endocrinol 373:68–76

119. Jongwutiwes T, Lertvikool S, Leelaphiwat S, Rattanasiri S, Jultanmas R, Weerakiet S (2009) Serum visfatin in Asian women with polycystic ovary syndrome. Gynecol Endocrinol 25(8):536–542
120. Barber TM (2022) Why are women with polycystic ovary syndrome obese? Br Med Bull 143(1):4–15. https://doi.org/10.1093/bmb/ldac007
121. Kaze AD, Santhanam P, Musani SK, Ahima R, Echouffo-Tcheugui JB (2021) Metabolic dyslipidemia and cardiovascular outcomes in type 2 diabetes mellitus: findings from the look AHEAD study. J Am Heart Assoc 10(7):e016947. https://doi.org/10.1161/JAHA.120.016947. Epub 2021 Mar 17. Erratum in: J Am Heart Assoc. 2021 Jul 20;10(14):e020749
122. Pyŏrälä K, Pedersen TR, Kjekshus J, Faergeman O, Olsson AG, Thorgeirsson G (1997) Cholesterol lowering with simvastatin improves prognosis of diabetic patients with coronary heart disease. A subgroup analysis of the Scandinavian Simvastatin Survival Study (4S). Diabetes Care 20(4):614–620. https://doi.org/10.2337/diacare.20.4.614. Erratum in: Diabetes Care 20(6):1048
123. Taskinen MR (2003) Diabetic dyslipidaemia: from basic research to clinical practice. Diabetologia 46(6):733–749. https://doi.org/10.1007/s00125-003-1111-y. Epub 2003 May 28
124. Zhang X, Ji L, Ran X, Su B, Ji Q, Hu D (2017) Gender disparities in lipid goal attainment among type 2 diabetes outpatients with coronary heart disease: results from the CCMR-3B Study. Sci Rep 7(1):12648. https://doi.org/10.1038/s41598-017-13066-z
125. Hedayatnia M, Asadi Z, Zare-Feyzabadi R, Yaghooti-Khorasani M, Ghazizadeh H, Ghaffarian-Zirak R, Nosrati-Tirkani A, Mohammadi-Bajgiran M, Rohban M, Sadabadi F, Rahimi HR, Ghalandari M, Ghaffari MS, Yousefi A, Pouresmaeili E, Besharatlou MR, Moohebati M, Ferns GA, Esmaily H, Ghayour-Mobarhan M (2020) Dyslipidemia and cardiovascular disease risk among the MASHAD study population. Lipids Health Dis 19(1):42. https://doi.org/10.1186/s12944-020-01204-y
126. Peters SAE, Muntner P, Woodward M (2019) Sex differences in the prevalence of, and trends in, cardiovascular risk factors, treatment, and control in the United States, 2001 to 2016. Circ 139(8):1025–1035. https://doi.org/10.1161/CIRCULATIONAHA.118.035550
127. Wang X, Magkos F, Mittendorfer B (2011) Sex differences in lipid and lipoprotein metabolism: it's not just about sex hormones. J Clin Endocrinol Metab 96(4):885–893. https://doi.org/10.1210/jc.2010-2061
128. Wang M, Liu M, Li F, Guo C, Liu Z, Pan Y, Liu Y, Liu F, Cai H, Wu Y, He Z, Ke Y (2020) Gender heterogeneity in dyslipidemia prevalence, trends with age and associated factors in middle age rural Chinese. Lipids Health Dis 19(1):135. https://doi.org/10.1186/s12944-020-01313-8
129. Boo S, Yoon YJ, Oh H (2018) Evaluating the prevalence, awareness, and control of hypertension, diabetes, and dyslipidemia in Korea using the NHIS-NSC database: A cross-sectional analysis. Medicine (Baltimore) 97(51):e13713. https://doi.org/10.1097/MD.0000000000013713
130. Lizcano F, Guzmán G (2014) Estrogen deficiency and the origin of obesity during menopause. Biomed Res Int 2014:757461. https://doi.org/10.1155/2014/757461. Epub 2014 Mar 6
131. Vrijkotte TG, Krukziener N, Hutten BA, Vollebregt KC, van Eijsden M, Twickler MB (2012) Maternal lipid profile during early pregnancy and pregnancy complications and outcomes: the ABCD study. J Clin Endocrinol Metab 97(11):3917–3925. https://doi.org/10.1210/jc.2012-1295. Epub 2012 Aug 29
132. Hadden DR, McLaughlin C (2009) Normal and abnormal maternal metabolism during pregnancy. Semin Fetal Neonatal Med 14(2):66–71. https://doi.org/10.1016/j.siny.2008.09.004. Epub 2008 Nov 4. Erratum in: Semin Fetal Neonatal Med. 2009 Dec;14(6):401
133. Brizzi P, Tonolo G, Esposito F, Puddu L, Dessole S, Maioli M, Milia S (1999) Lipoprotein metabolism during normal pregnancy. Am J Obstet Gynecol 181(2):430–434. https://doi.org/10.1016/s0002-9378(99)70574-0
134. Nasioudis D, Doulaveris G, Kanninen TT (2019) Dyslipidemia in pregnancy and maternal-fetal outcome. Minerva Ginecol 71(2):155–162. https://doi.org/10.23736/S0026-4784.18.04330-7. Epub 2018 Oct 11

135. Chaudhary S, Hiranwal M, Chaudhary D et al (2020) Hyperlipidemia of pregnancy: normal or predictor of preeclampsia. J South Asian Feder Obst Gynae 12(1):31–33
136. Cífková R, Krajčoviechová A (2015) Dyslipidemia and cardiovascular disease in women. Curr Cardiol Rep 17(7):609. https://doi.org/10.1007/s11886-015-0609-5
137. Torosyan N, Visrodia P, Torbati T, Minissian MB, Shufelt CL (2022) Dyslipidemia in midlife women: approach and considerations during the menopausal transition. Maturitas 166:14–20. https://doi.org/10.1016/j.maturitas.2022.08.001. Epub ahead of print
138. Echiburú B, Crisosto N, Maliqueo M, Pérez-Bravo F, de Guevara AL, Hernández P, Cavada G, Rivas C, Clavel A, Sir-Petermann T (2016) Metabolic profile in women with polycystic ovary syndrome across adult life. Metabolism 65(5):776–782. https://doi.org/10.1016/j.metabol.2016.01.006. Epub 2016 Jan 16
139. Jeong J, Kim M (2022) Awareness and related factors of dyslipidemia in menopausal women in Korea. Healthcare (Basel). 10(1):112. https://doi.org/10.3390/healthcare10010112
140. Legro RS, Kunselman AR, Dunaif A (2001) Prevalence and predictors of dyslipidemia in women with polycystic ovary syndrome. Am J Med 111:607–613
141. Barber TM, McCarthy MI, Franks S et al (2007) Metabolic syndrome in polycystic ovary syndrome. Endokrynol Pol 58:34–41
142. Barber TM, Franks S (2021) Obesity and polycystic ovary syndrome. Clin Endocrinol (Oxf) 95:531–541
143. Barber TM, Dimitriadis GK, Andreou A et al (2015) Polycystic ovary syndrome: insight into pathogenesis and a common association with insulin resistance. Clin Med (Lond) 15:s72–s76
144. Joharatnam J, Barber TM, Webber L et al (2011) Determinants of dyslipidaemia in probands with polycystic ovary syndrome and their sisters. Clin Endocrinol (Oxf) 74:714–719
145. Wild RA, Rizzo M, Clifton S, Carmina E (2011) Lipid levels in polycystic ovary syndrome: systematic review and meta-analysis. Fertil Steril 95:1073–1079
146. Talbott E, Guzick D, Clerici A, Berga S, Detre K, Weimer K et al (1995) Coronary heart disease risk factors in women with polycystic ovary syndrome. Arterioscler Thromb Vasc Biol 15:821–826
147. Conway GS, Agrawal R, Betteridge DJ, Jacobs HS (1992) Risk factors for coronary artery disease in lean and obese women with the polycystic ovary syndrome. Clin Endocrinol (Oxf) 37(2):119–125. https://doi.org/10.1111/j.1365-2265.1992.tb02295.x

# Chapter 13
# Molecular Basis of the Circadian Mechanism in Women

**Molly Crandall, Inna Rabinovich-Nikitin, and Lorrie A. Kirshenbaum**

**Abstract** Circadian rhythm is present in all living life forms and influences daily major physiological processes through synchronized molecular interactions within a 24-h cycle. The circadian mechanism is a coordinated pathway instigated via receival of various triggers, causing a cascade of molecular signals acting to achieve a shift in gene regulation. Interestingly, there are apparent sex differences within the circadian mechanism. Modulation of the circadian rhythm in sex-dependent manner leads to physiological variations in homeostatic functions that are specific to women. Finally, recent evidence in the past 20 years has indicated that the disruption of circadian rhythm can influence the development of many different pathologies. Moreover, there is increasing information that certain diseases linked to the female sex may have a link to the circadian function. The cardiovascular system is significantly impacted by the circadian clock, and we are beginning to understand the effects of disrupted clock upon the heart. Notably, research in women's heart health has proven existence of sex-dependent characteristics of cardiovascular disease (CVD). This novel intersection of the cardiac disease and misaligned internal clock presents isolated risks that can be specific for women. Herein, we will discuss how dysregulation of circadian rhythms impact women's heart health.

**Keywords** Circadian rhythm · Clock · Light/dark cycles · Circadian misalignment · Chronotherapy

M. Crandall · I. Rabinovich-Nikitin · L. A. Kirshenbaum (✉)
Department of Physiology and Pathophysiology, The Institute of Cardiovascular Sciences, St. Boniface Hospital Albrechtsen Research Centre, Winnipeg, MB, Canada
e-mail: lkirshenbaum@sbrc.ca

L. A. Kirshenbaum
Department of Pharmacology and Therapeutics, Rady College of Medicine, Max Rady Faculty of Health Sciences, University of Manitoba, Winnipeg, MB, Canada

© The Author(s), under exclusive license to Springer Nature Switzerland AG 2023 
L. Kirshenbaum and I. Rabinovich-Nikitin (eds.), *Biology of Women's Heart Health*, Advances in Biochemistry in Health and Disease 26,
https://doi.org/10.1007/978-3-031-39928-2_13

## Overview of Circadian Rhythm

The circadian mechanism is represented by diurnal oscillations of multiple regulatory components that modulate daily physiological functions. Internal and external signals, also known as zeitgebers, can trigger molecular responses creating a cascade of signals throughout the body [1]. Light–dark cycles are a principal factor in the modulation of circadian machinery. Photoreceptors within the retina of the eye receive light, initiating a response from the suprachiasmatic nucleus (SCN) in the hypothalamus of the brain [2]. The SCN is a core component of circadian mechanism as it instigates organ specific responses via neural and humoral signals. Furthermore, peripheral cells contain their own circadian machinery, known as peripheral clocks, that can be stimulated through messages from the SCN or via inter- and intracellular zeitgebers. Both the central and peripheral clocks regulate the expression of circadian genes to maintain daily oscillations of key cellular events through positive and negative feedback loops [3]. Transcription factors, CLOCK (circadian locomotor output cycles kaput) and BMAL1 (brain and muscle Arnt-like protein-1), mark the beginning of the positive arm through their heterodimerization within the nucleus. The CLOCK:BMAL1 dimer bind to E-box elements present within promoter regions of circadian-dependent genes. Specifically, CLOCK:BMAL1 will induce the transcription of Period (*PER1, PER2 and PER3*) and Cryptochrome (*CRY1* and *CRY2*) genes [4]. Upon their translation within the cytoplasm and subsequence heterodimerization, PER and CRY will inhibit the transcriptional activity of CLOCK:BMAL1. Accordingly, PER and CRY are fundamental aspects within the negative feedback loop to regulate the transcription of core clock genes (Fig. 13.1). Furthermore, there are additional components that concern the regulation of core clock genes [5]. In this regard, the heterodimer CLOCK:BMAL1 regulates the transcription of the orphan nuclear-receptor, RORα/β and REV-ERBα/β, which provide further transcriptional control of the circadian machinery. Particularly, ROR proteins function to initiate the transcription of *Bmal1* whereas the REV-ERB isoforms will inhibit this transcriptional activity. Thus, circadian rhythm is a multifaceted process that must act in turn to modulate the timely expression of target genes [1].

In the past years, it has become increasingly apparent that there are sex differences regarding circadian rhythm. In general, women tend to align with a morning chronotype and report to have increased mental performance at an earlier time of day when compared to their male counterparts [6, 7]. Furthermore, women have circadian periods shorter than 24 h as demonstrated with earlier entrained phase core body temperature (CBT) and melatonin rhythms [8]. The role of estrogen within the context of the circadian clock has been suggested as a possible reason for a reduced intrinsic circadian period [9]. It is well understood that estrogen is involved in circadian pathways. Estrogen receptor 2 (*ESR2*) encodes estrogen-receptor β (ERb) and contains an E-box sequence within the promotor region indicating possible regulation via CLOCK:BMAL1. When *Bmal1* is knocked down, there is disruption of timely ERb transcription [10]. In addition, PER1 and PER2 can differentially be modulated through estrogen, implying that estrogen can control the inhibition of

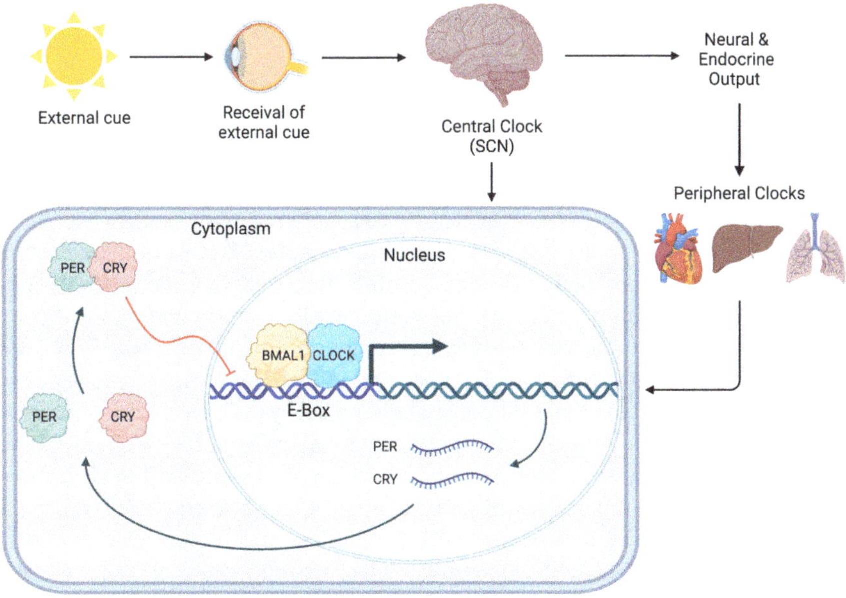


**Fig. 13.1 Molecular mechanism of the circadian clock.** Light is an external cue received through the retina of the eye and relays this information to the circadian master clock within the superchiasmatic nucleus (SCN) of the brain. This results in a cascade of neural and endocrine signals throughout the body to reach peripheral clocks. Within these tissues, the cell modulates the molecular mechanism of circadian clocks through activation of the positive or negative transcriptional-translational feedback loops in the nucleus. The positive arm begins with the expression of two transcription factors, CLOCK (circadian locomotor output cycles kaput) and BMAL1 (brain and muscle Arnt-like protein-1), who heterodimerize in the nucleus to bind to E-box promotor sequences in DNA. Together, they activate two genes of the negative arm of the circadian mechanism, PERIOD (PER1, PER2, PER3) and CRYPTOCHROME (CRY1 AND CRY2). Upon translation in the cytoplasm, PER and CRY heterodimerize to inhibit the transcriptional activity of CLOCK:BMAL1. Furthermore, BMAL1 has another layer of regulation through ROR and REV-ERB. Specifically, while ROR proteins function to initiate the transcription of *Bmal1,* REV-ERB isoforms inhibit this transcriptional activity

the primary transcription factors of the circadian clock [11]. When blind hamsters were subjected to an estradiol implant, their circadian locomotor period was reduced [12] and, ovariectomized C57BL/6 J mice demonstrated shortened circadian activity [9]. This is indicative of a species-dependent effect of estrogen upon the circadian period. Additionally, the menstrual cycle follows an ultradian rhythm and unequivocally, provides a distinct interrelation between chronobiology and women. Clearly, the unique impacts of estrogen upon the circadian clock demonstrates the complicated relationship between the female sex and circadian rhythmicity.

## Consequences of Circadian Misalignment in the Female Sex

Within humans, it has been demonstrated that most reproductive hormones follow daily rhythmicity of their levels, such as follicle stimulating hormone (FSH), luteinizing hormone (LH) and progesterone. Notably, the SCN initiates release of gonadotropin-releasing hormone (GnRH) to drive LH secretion to cause follicular rupture and oocyte release. While GnRH may be an autonomous circadian oscillator, this information leads to the rise of questions surrounding the role of the circadian mechanism in the reproductive system, specifically concerning the repercussions of a disrupted circadian rhythm [13]. Furthermore, ultradian rhythm of distal body temperature and heart rate variability can predict the surge of LH commonly seen during the pre-ovulatory stage of the menstrual cycle. The fertile window within women follows ultradian timing, demonstrating control by the internal clock of reproductive potential [14]. Women often fulfill caregiving roles in the family dynamic and participate in professions with irregular scheduling, known as shift work, leading to disrupted circadian rhythm that disproportionally affect women. Misalignment of circadian rhythm has been associated with a multitude of pathologies across many physiological systems. Notably, emerging evidence strongly indicates that women are uniquely affected by this disruption as effects of the circadian clock can be seen within different diseases.

Polycystic ovary syndrome (PCOS) is an endocrine disorder that affects 4% to 10% of women worldwide and is characterized by hyperandrogenism, polycystic ovaries and ovulatory dysfunction [15]. While better known as a reproductive disorder, it has been evident since the late 1990s that PCOS is associated with increased risk for the development of metabolic and cardiovascular disorders [16]. Recently, night shift work had a significant correlation with PCOS that is exacerbated with prolonged participation in this work schedule which prompted further exploration into the relationship between circadian mechanism and PCOS. It was found that core circadian genes were reduced in amplitude in human granulosa cells from individuals with PCOS. Furthermore, the same study deduced through KEGG analysis that genes involved in metabolic disorders, including the TCA cycle and carbohydrate acid metabolism have altered cyclic patterns of expression between PCOS and control groups [17]. In further inquiry into how night shift work impacts women, female rats exposed to continuous light exhibited PCOS-like symptoms such as prolonged and disordered estrous cycles, oligo/anovulation and atretic cyst-like follicles. Additionally, the 24-h light exposure resulted in abnormal glucose metabolism in the liver, adipose and muscle tissue. Metagenomic analysis also revealed that rats subjected to consistent light harbored gut microbiota containing reduced amounts of reproductive and metabolic-related genes [18]. Markedly, there is a potential underlying circadian role within the pathology of PCOS and further research is necessary to fully understand the impact of circadian disruption upon the reproductive systems in women.

Globally, CVD is the leading cause of mortality for women worldwide, although it remains vastly misdiagnosed and understudied [19, 20]. With new research, it is becoming increasingly apparent that CVD appears differently in women than in men and that women are more susceptible to specific types of heart disease. In men, the main physical symptom of acute myocardial infarction (AMI) is typical chest pain or discomfort that is described as pressure, tightness, and squeezing. Whereas in women, AMI presents as atypical chest pain, which is noted as sharp, burning, and sore combined with other symptoms such as fatigue, dyspnea, nausea/vomiting, and anxiety [21]. Sex-dependent differences in clinical presentation are a plausible cause for disparities in CVD mortality in women albeit, there are separate risk factors associated with the female sex. Certain cardiomyopathies have been shown to have a direct link to the circadian dysregulation in women. Furthermore, depression, which affects millions of women worldwide has been implicated in the development of CVD and may contribute partially to circadian disruption. Furthermore, professions employing shift work and at-home childcare are predominately performed by women, leading to circadian disruption which impacts the health of the cardiovascular system. Thus, we shall investigate the correlation between cardiovascular events and circadian rhythm and unique repercussions that women face from this association.

Historically, there has always been an awareness that the cardiovascular system is influenced by circadian oscillators, indicating a long withstanding link between the internal clock and cardiomyopathies [22]. Foremost, heart rate (HR) and blood pressure (BP) are known to decrease during the evening and upon entering sleep stages. Respectively, these cardiovascular parameters will increase during the day [23, 24]. In fact, the circadian phase corresponding to 9:00–11:00 a.m. has a marked peak in heartbeat scaling exponent $\alpha$ which corresponds to a time frame of cardiac vulnerability [25] Cardiac events, such as MI, are more likely to be experienced during the early hours of the day as well [26]. Moreover, some cardiomyopathies are manipulated by the circadian clock in a sex-dependent manner. Acute stress-induced cardiomyopathy, such as Takotsubo syndrome (TTS) has a similar presentation and prognosis rate of acute coronary syndrome but is a heart failure disease [27, 28]. Characteristics of TTS include unobstructed coronary arteries and ballooning of the left ventricle with normal or near-normal ejection fraction [29]. Although previous preconceptions have noted otherwise, it is becoming increasingly apparent that TTS causes long-term heart failure phenotype and global severe edema of both ventricular myocardium [29, 30]. This heart failure disorder primarily affects middle-aged or elderly women and occurrence follows a circadian rhythm [31]. Chronobiological variation of TTS follows a wave-like pattern of presentation with majority of events occurring between noon and 6 p.m. with fewer incidents happening between midnight and 6 a.m. Interestingly, this deviates from the standard understanding that adverse cardiac event occurs during the early morning hours. Furthermore, most individuals are diagnosed specifically within October and upon Wednesdays. Therefore, TTS not only demonstrates daily chronological occurrence but also weekly and monthly rhythmicity [32]. Additionally, spontaneous coronary artery dissection (SCAD) has shown to have chronobiological significance related to the menstruation. Sudden cardiac death and MI are linked to the appearance of SCAD within women, and they present with recurring chest pain despite lacking ischemia within the heart or coronary obstruction [33, 34]. A normal menstrual cycle follows an ultradian rhythm and relies on the cyclical regulation of reproductive hormones [35]. Interestingly, SCAD related chest pain has been shown to worsen with upcoming menstruation in a predictable manner [33]. This is suggestive that adverse cardiac events associated with SCAD may follow an ultradian cycle and be associated with hormonal fluctuations. Moreover, SCAD has also been noted to frequently occur within the first month postpartum and were more likely to have undergone infertility treatments, furthering the notion that SCAD is associated with hormonal regulation [36]. Further research is necessary into the implications of circadian and sex related cardiomyopathies as there is clear evidence that belonging to the female sex is an important risk factor for circadian-related cardiovascular events.

As research into mental health increases, there is significant implication that circadian misalignment has negative impacts upon mental health while mental illnesses are also considered a risk for CVD with apparent gender dispositions. Depression disorder (DD) is a common, although serious, mental health disorder that typically presents as combinations of various somatic and cognitive symptoms with the most predominant quality being anhedonia; the inability to experience pleasure [37, 38]. In addition, women are at greater risk of developing DD, particularly during mid-puberty and adult years due to the hormonal changes that women will enter throughout various stages in their life [39]. For instance, post-partum depression is one of the most common complications associated with pregnancy and childbirth. The occurrence of post-partum depression affects 13–15% of pregnant individuals although the risk is dependent on multiple factors such as cultural aspects, socioeconomic status, and social support [40]. Increasing evidence has recently demonstrated that depression has a relationship with a disrupted circadian rhythm. Individuals with major depressive disorder (MDD) had low levels of motor activity with increases only present upon antidepressant therapy [41]. Additionally, depressed individuals had clear circadian variation deviations in body temperature with reduced amplitude at the diurnal peak [42]. Genotyping and selection of 248 single-nucleotide polymorphisms (SNPs) at 19 circadian genes within a population study of individuals with unipolar MDD, and bipolar disorder (BD) revealed significant associations between circadian genes and mental illness. The top genes nominally associated with the group combining these mood disorders were *CRY1*, *NPAS2* (paralogous to *CLOCK*), and Vasoactive intestinal peptide receptor 2 (G-protein coupled receptor implicated in circadian mechanism). In particular, the strongest correlation in only the MDD group was found at a SNP, rs2287161, located at the 3' end close to *CRY1* [43]. In a separate analysis of university students, evening chronotype females were at most risk for depression and anxiety. Furthermore, via computer modelling of five circadian polymorphisms, individuals with allelic or genotypic variants of genes regarding circadian period length or amplitude were more likely to have depression. This association was particularly significant in the population of depressed individuals with the *PER3* variable tandem repeat length polymorphism allele, rs57875989 [44]. Hence, not only is it apparent at a societal level that depression is correlated with circadian disruption in women, but it can also be found on a genome-wide level.

Women with depression have an increased risk of coronary artery disease (CAD) and other CVDs and since women are twice as likely to experience depression, there is pronounced reasoning for investigation into this relationship [20, 45]. These individuals are more likely to develop acute myocardial infarction (AMI), heart failure (HF) or stroke that is associated with depression. In fact, ischemic heart disease mortality had 3.70 (95% confidence interval [CI] 1.32–10.35) for depression and 7.12 (95% CI 2.67–18.98) for history of attempted suicide with the strongest associations among women [39]. Regarding coronary heart disease (CHD), women had improved long-term clinical outcomes with darapladib therapy for stabilization of atherosclerotic plaque. This benefit was lost when these women demonstrated high frequency of depressive symptoms [46]. Additionally, following MI, 71.4% of female patients reported developing a mood disorder or emotional mental health issue after diagnosis [47]. The relationship between depression with CVDs and circadian misalignment in women requires further studies as this overlooked risk factor has substantial implications in heart related pathologies.

With increasing demand for individuals to participate in shift work, even from a young age, there is a societal elevated risk in the development of cardiovascular and circadian disorders. This results in a unique situation where women are independently affected by disruption of the circadian mechanism and are subjected to increased incidences of having their circadian rhythm disturbed resulting in an elevated risk of cardiovascular disease. In fact, shift work disorder (SWD), a circadian rhythm sleep disorder associated with an abnormal work schedule, has psychological and physical consequences. For instance, individuals with SWD describe symptoms such as insomnia, inappropriately timed and excessive sleepiness, irritability, and depression [48]. In a 2019 meta-analysis study, results indicated that shift work was associated with a general increased risk of poor mental health outcomes. This study identified that female shift workers had a 33% higher risk of depression when compared to female non-shift workers [49]. In a cross-sectional study surrounding SWD and nurses, women in this profession were more likely to be under constant stress than men and this stress was exacerbated when in a situation of low social support. Moreover, women were more likely to experience insomnia and excessive daytime sleepiness with no significant association with high job strain [50]. Shift work has also been associated with a heightened risk of adverse cardiac events and diseases. In U.S. cohort of female nurses working rotating night shift work, it was also shown that participation in this kind of shift work increases all-cause and CVD-related mortality [51]. Through Nurses' Health Studies with 24 years of follow-up, rotating night shift work was associated with an elevated risk for CHD in a length-dependent manner where risk declined over time upon leaving night shift work [52]. Night shift work was also found to be linked to CHD and atrial fibrillation in a duration-dependent manner within a different study although the sex differences were not reported, while a different study did not state the gender statistics within their study population [53]. Therefore, not only does shift work elevate the risk of CVD but also shows that increased incidence of night shift work worsens these risks. In female nurses, those that had worked 4 or less night shifts in two weeks had lower HDL cholesterol levels and nurses with a history of rotating night shift work had higher C-reactive protein levels. These two biomarkers have been previously associated with an increased probability of CVD [54, 55]. Other biomarkers associated with a prevalent risk of CVD are inflammatory agents, such as tumour necrosis factor alpha (TNF$\alpha$), resisting and interleukin-6 (IL-6). These three markers were elevated among individuals subjected to circadian misalignment [56]. While the cohort was not separated by sex, this further demonstrates that shift work has harmful consequences. Additionally, shift work and rotating shift work was associated with adverse pregnancy outcomes with some evidence of elevated risk for preeclampsia, gestational hypertension, and gestational diabetes mellitus [57]. These pregnancy related diseases have previously been shown to be associated with higher probability of subsequent CVD [58]. With this knowledge, we can see that women in shift work associated professions have increased likelihood of circadian-related disorders, cardiovascular events, and adverse maternal outcomes during pregnancy.

## Potential Therapeutics and Conclusions

Over the years, the relationship between CVD and circadian rhythm is gradually being elucidated, however, the effects of circadian disruption and CVD within women is not well understood. With the apparent gender differences in circadian rhythm and CVD, it is crucial to further our knowledge in this area to develop more clinically relevant diagnostic criteria and medical therapeutics. At present, there are few treatments available for those with a disrupted circadian rhythm and the most known is relatively simple: bright light therapy. In circadian rhythm desynchronization during chemotherapy for breast cancer, daily exposure that mimics sunlight without ultraviolet light showed significant improvements to circadian variables such as amplitude and mean of activity levels, particularly during recovery weeks following their fourth cycle of chemotherapy [59]. In corroboration of the previous findings, bright light improved daytime sleepiness, depression, sleep onset latency and sleep efficiency in women receiving chemotherapy for breast cancer [60]. Bright light therapy has also been previously reported in improving seasonal and non-seasonal depression and mood disorders such as bipolar disorder which could act to decrease mood disorder related cardiovascular risks [61]. In fact, exposure to intense light in mice and humans lead to increased glycolysis through upregulation of a Per2 mimicking miRNA, therefore suggesting bright light therapy as a potential therapeutic for MI [62, 63] Regarding cardiovascular diseases, chronopharmacological circadian-timed drug administration has shown improvements of hypertension. Ingestion of hypertension medications at bedtime resulted in lower mean asleep BP, greater decline of BP relative to time of sleep, and were strongly correlated to lowered CVD risk [64, 65]. For women specifically, there have been many studies concerning estrogen replacement treatment (ERT) due to the theory of cardioprotective estrogen effects, although the results are controversial. ERT in postmenopausal women increased the 24-h amplitude of serum cortisol levels and lowered the 24-h mean temperature demonstrating its abilities to modulate the biological clock [66]. In opposition, postmenopausal women subjected to estrogen and progestin replacement therapy were found to have increased incidences of MI, stroke, venous thromboembolisms,

CHD, and breast cancer [67]. Largely, there are no specific therapeutics targeted towards CVD and circadian disruption, other than appropriately timed drug administration, ways to reduce cardiovascular risks and traditional CVD treatments. There are distinctive risks for women concerning circadian misalignment and CVDs as they are prone to shift work, familial caregiver roles and other circadian- or cardiovascular-related comorbidities. Further research is required to fully elucidate the molecular relationship women have to circadian rhythm disorders and CVDs. The relationship between cardiovascular disease and circadian rhythm in women is a novel field that few are familiar with, but it is imperative to comprehend the common link to identify potential interventions and therapeutics.

## References


1. Manella G, Aviram R, Bolshette N, Muvkadi S, Golik M, Smith DF et al (2020) Hypoxia induces a time- and tissue-specific response that elicits intertissue circadian clock misalignment. Proc Natl Acad Sci 117(1):779–786
2. Vandewalle G, Maquet P, Dijk DJ (2009) Light as a modulator of cognitive brain function. Trends Cogn Sci 13(10):429–438
3. Rabinovich-Nikitin I, Lieberman B, Martino TA, Kirshenbaum LA (2019) Circadian-regulated cell death in cardiovascular diseases. Circ 139(7):965–980
4. Reinke H, Asher G (2019) Crosstalk between metabolism and circadian clocks. Nat Rev Mol Cell Biol 20(4):227–41. Available from: https://doi.org/10.1038/s41580-018-0096-9
5. Portaluppi F, Tiseo R, Smolensky MH, Hermida RC, Ayala DE, Fabbian F (2012) Circadian rhythms and cardiovascular health. Sleep Med Rev 16(2):151–66. Available from: https://doi.org/10.1016/j.smrv.2011.04.003
6. Jones SE, Lane JM, Wood AR, van Hees VT, Tyrrell J, Beaumont RN et al (2019) Genome-wide association analyses of chronotype in 697,828 individuals provides insights into circadian rhythms. Nat Commun 10(1). Available from: https://doi.org/10.1038/s41467-018-08259-7
7. Adan A, Natale V (2002) Gender differences in morningness-eveningness preference. Chronobiol Int 19(4):709–720
8. Duffy JF, Cain SW, Chang AM, Phillips AJK, Muñch MY, Gronfier C et al (2011) Sex difference in the near-24-hour intrinsic period of the human circadian timing system. Proc Natl Acad Sci USA 108(Suppl 3):15602–15608
9. Blattner MS, Mahoney MM (2014) Estrogen receptor 1 modulates circadian rhythms in adult female mice. Chronobiol Int 31(5):637–644
10. Cai W, Rambaud J, Teboul M, Masse I, Benoit G, Gustafsson JÅ et al (2022) Expression levels of estrogen receptor β are modulated by components of the molecular clock. Mol Cell Biol 28(2):784. Available from: https://doi.org/10.1128/MCB.00233-07
11. Nakamura TJ, Moriya T, Inoue S, Shimazoe T, Watanabe S, Ebihara S et al (2005) Estrogen differentially regulates expression of Per1 and Per2 genes between central and peripheral clocks and between reproductive and nonreproductive tissues in female rats. J Neurosci Res 82(5):622–630
12. Takahashi JS, Menaker M (1980) Interaction of estradiol and progesterone: effects on circadian locomotor rhythm of female golden hamsters. Am J Physiol 239(5):R497–504
13. Bungum L, Jacobsson AK, Rosn F, Becker C, Yding Andersen C, Gner N et al (2011) Circadian variation in concentration of anti-Mllerian hormone in regularly menstruating females: relation to age, gonadotrophin and sex steroid levels. Hum Reprod 26(3):678–684
14. Grant AD, Newman M, Kriegsfeld LJ (2020) Ultradian rhythms in heart rate variability and distal body temperature anticipate onset of the luteinizing hormone surge. Sci Rep 10(1). Available from: https://doi.org/10.1038/s41598-020-76236-6

15. Bozdag G, Mumusoglu S, Zengin D, Karabulut E, Yildiz BO (2016) The prevalence and phenotypic features of polycystic ovary syndrome: a systematic review and meta-analysis. Hum Reprod 31(12):2841–2855. Available from: https://academic.oup.com/humrep/article/31/12/2841/2730240
16. Sirmans SM, Pate KA (2013) Epidemiology, diagnosis, and management of polycystic ovary syndrome. Clin Epidemiol 6(1):1–13
17. Wang F, Xie N, Wu Y, Zhang Q, Zhu Y, Dai M et al (2021) Association between circadian rhythm disruption and polycystic ovary syndrome. Fertil Steril 115(3):771–781
18. Chu W, Zhai J, Xu J, Li S, Li W, Chen ZJ et al (2020) Continuous light-induced PCOS-like changes in reproduction, metabolism, and gut microbiota in Sprague-Dawley rats. Front Microbiol 10. Available from: https://doi.org/10.3389/fmicb.2019.03145
19. Wild RA, Hovey KM, Andrews C, Robinson JG, Kaunitz AM, Manson JAE et al (2021) Cardiovascular disease (CVD) risk scores, age, or years since menopause to predict cardiovascular disease in the Women's Health Initiative. Menopause 28(6):610–618
20. Saeed A, Kampangkaew J, Nambi V (2017) Prevention of cardiovascular disease in women. Prev Cardiovasc Dis Women Anum 13(4):183–184
21. Mehta LS, Beckie TM, DeVon HA, Grines CL, Krumholz HM, Johnson MN, et al (2016) Acute myocardial infarction in women: a scientific statement from the American Heart Association. Circ 133:916–947
22. Kümmell HC, van Leeuwen P, Heckmann C, Engelke P, Resting G, Kremer H et al (1993) Quality of life and circadian variation of heart rate and heart rate variability in short-term survivors and nonsurvivors after acute myocardial infarction. Clin Cardiol 16(11):776–782
23. Haghayegh S, Smolensky MH, Khoshnevis S, Hermida RC, Castriotta RJ, Diller KR (2021) The circadian rhythm of thermoregulation modulates both the sleep/wake cycle and 24 h pattern of arterial blood pressure. Compr Physiol 11(4):2645–2658
24. Boudreau P, Yeh WH, Dumont GA, Boivin DB (2013) Circadian Variation of Heart Rate Variability Across Sleep Stages. Sleep 36(12):1919. Available from: https://doi.org/10.5665/sleep.3230
25. Hu K, Ivanov PC, Hilton MF, Chen Z, Ayers RT, Stanley HE et al (2004) Endogenous circadian rhythm in an index of cardiac vulnerability independent of changes in behavior. Proc Natl Acad Sci USA 101(52):18223–18227
26. Muller JE, Stone PH, Turi ZG, Rutherford JD, Czeisler CA, Parker C et al (2010) Circadian variation in the frequency of onset of acute myocardial infarction. 313(21):1315–1322. http://dx.doi.org.uml.idm.oclc.org/101056/NEJM198511213132103. Available from: https://www.nejm-org.uml.idm.oclc.org/https://doi.org/10.1056/NEJM198511213132103
27. Redfors B, Vedad R, Angerås O, Råmunddal T, Petursson P, Haraldsson I et al (2015) Mortality in takotsubo syndrome is similar to mortality in myocardial infarction - A report from the SWEDEHEART. Int J Cardiol 15(185):282–289
28. Manfredini R, Manfredini F, Fabbian F, Salmi R, Gallerani M, Bossone E et al (2016) Chronobiology of takotsubo syndrome and myocardial infarction: analogies and differences. Heart Fail Clin 12(4):531–542. Available from: https://doi.org/10.1016/j.hfc.2016.06.004
29. Eitel I, von Knobelsdorff-Brenkenhoff F, Bernhardt P, Carbone I, Muellerleile K, Aldrovandi A et al (2011) Clinical characteristics and cardiovascular magnetic resonance findings in stress (Takotsubo) cardiomyopathy. JAMA 306(3):277–286. Available from: https://jamanetwork-com.uml.idm.oclc.org/journals/jama/fullarticle/1104115
30. Scally C, Ahearn T, Rudd A, Neil CJ, Srivanasan J, Jagpal B et al (2016) Right ventricular involvement and recovery after acute stress-induced (Tako-tsubo) cardiomyopathy. Am J Cardiol 117(5):775–780
31. Scally C, Abbas H, Ahearn T, Srinivasan J, Mezincescu A, Rudd A et al (2019) Myocardial and systemic inflammation in acute stress-induced (Takotsubo) cardiomyopathy. Circulation 139(13):1581–1592
32. El-Battrawy I, Aweimer A, Lang S, Ansari U, Gietzen T, Ullrich N et al (2021) Impact of chronobiological variation in takotsubo syndrome: prognosis and outcome. Front Cardiovasc Med 26:8

33. Tweet MS, Codsi E, Best PJM, Gulati R, Rose CH, Hayes SN (2017) Menstrual chest pain in women with history of spontaneous coronary artery dissection. Physiol Behav 176(5):139–148
34. Saw J, Co-Chair F, Adlam D, Arslanian-Engoren C, Economy KE, Ganesh SK et al (2018) Spontaneous coronary artery dissection: current state of the science: on behalf of the American Heart Association Council on peripheral vascular disease; Council on Clinical Cardiology; Council on Cardiovascular and Stroke Nursing; Council on Genomic. Heart Association, vol 137, pp 523–557
35. Baker FC, Driver HS (2007) Circadian rhythms, sleep, and the menstrual cycle. Sleep Med 8(6):613–622
36. Tweet MS, Hayes SN, Codsi E, Gulati R, Rose CH, Best PJM (2017) Spontaneous coronary artery dissection associated with pregnancy. J Am Coll Cardiol 70(4):426–435
37. Bucciarelli V, Caterino AL, Bianco F, Caputi CG, Salerni S, Sciomer S et al (2020) Depression and cardiovascular disease: the deep blue sea of women's heart. Trends Cardiovasc Med 30(3):170–176
38. Tomfohr LM, Martin TM, Miller GE (2008) Symptoms of depression and impaired endothelial function in healthy adolescent women. J Behav Med 31(2):137–143
39. Goldstein BI, Carnethon MR, Matthews KA, McIntyre RS, Miller GE, Raghuveer G et al (2015) Major depressive disorder and bipolar disorder predispose youth to accelerated atherosclerosis and early cardiovascular disease: a scientific statement from the American Heart Association. Circ 132(10):965–986
40. Pearlstein T, Howard M, Salisbury A, Zlotnick C (2009) Postpartum depression. Am J Obstet Gynecol 200(4):357–364. Available from: https://doi.org/10.1016/j.ajog.2008.11.033
41. Raoux N, Benoit O, Dantchev N, Denise P, Franc B, Alliale JF et al (1994) Circadian pattern of motor activity in major depressed patients undergoing antidepressant therapy: relationship between actigraphic measures and clinical course. Psychiatry Res 52(1):85–98
42. Souêtre E, Salvati E, Belugou JL, Pringuey D, Candito M, Krebs B et al (1989) Circadian rhythms in depression and recovery: evidence for blunted amplitude as the main chronobiological abnormality. Psychiatry Res 28(3):263–278
43. Soria V, Martínez-Amorós È, Escaramís G, Valero J, Pérez-Egea R, García C et al (2010) Differential association of circadian genes with mood disorders: CRY1 and NPAS2 are associated with unipolar major depression and CLOCK and VIP with bipolar disorder. Neuropsychopharmacology 35(6):1279–1289. Available from: https://www.nature.com/articles/npp2009230
44. Liberman AR, Halitjaha L, Ay A, Ingram KK (2018) Modeling strengthens molecular link between circadian polymorphisms and major mood disorders. J Biol Rhythms 33(3):318–336. Available from: https://www.journals.sagepub.com/https://doi.org/10.1177/0748730418764540
45. Ferketich AK, Schwartzbaum JA, Frid DJ, Moeschberger ML (2000) Depression as an antecedent to heart disease among women and men in the NHANES I Study. Arch Intern Med 160(9):1261–1268. Available from: https://jamanetwork-com.uml.idm.oclc.org/journals/jamainternalmedicine/fullarticle/485308
46. H PO, Granger CB, Stebbins A, Chiswell K, Held C, Hochman JS et al (2017) Sex differences in clinical characteristics, psychosocial factors, and outcomes among patients with stable coronary heart disease: Insights from the STABILITY (Stabilization of Atherosclerotic Plaque by Initiation of Darapladib Therapy) trial. J Am Heart Assoc 6(9). Available from: http://jaha.ahajournals.org/content/6/9/e006695/DC1/embed/inline-supplementary-material-1.pdf
47. Serpytis P, Navickas P, Lukaviciute L, Navickas A, Aranauskas R, Serpytis R et al (2018) Gender-based differences in anxiety and depression following acute myocardial infarction. Arq Bras Cardiol 111(5):676. Available from: https://doi.org/10.5935/abc.20180161
48. Kervezee L, Shechter A, Boivin DB (2018) Impact of shift work on the circadian timing system and health in women. Sleep Med Clin 13(3):295–306. Available from: https://doi.org/10.1016/j.jsmc.2018.04.003
49. Torquati L, Mielke GI, Brown WJ, Burton NW, Kolbe-Alexander TL (2019) Shift work and poor mental health: a meta-analysis of longitudinal studies. Am J Public Health 109(11):E13-20

50. D'ettorre G, Pellicani V, Caroli A, Greco M (2020) Shift work sleep disorder and job stress in shift nurses: implications for preventive interventions. Med Lav 111(3):195–202
51. Gu F, Han J, Laden F, Pan A, Caporaso NE, Stampfer MJ et al (2015) Total and cause-specific mortality of U.S. nurses working rotating night shifts. Am J Prev Med 48(3):241–252. Available from: https://doi.org/10.1016/j.amepre.2014.10.018
52. Vetter C, Devore EE, Wegrzyn LR, Massa J, Speizer FE, Kawachi I et al (2016) Association between rotating night shift work and risk of coronary heart disease among women. JAMA 315(16):1726. Available from: https://doi.org/10.1001/jama.2016.4454
53. Wang N, Sun Y, Zhang H, Wang B, Chen C, Wang Y et al (2021) Long-term night shift work is associated with the risk of atrial fibrillation and coronary heart disease. Eur Heart J 42(40):4180–4188. Available from: https://academic.oup.com/eurheartj/article/42/40/4180/6347324
54. Jia C, Anderson JLC, Gruppen EG, Lei Y, Bakker SJL, Dullaart RPF et al (2021) High-density lipoprotein anti-inflammatory capacity and incident cardiovascular events. Circ 143(20):1935–1945. Available from: https://www.ahajournals-org.uml.idm.oclc.org/https://doi.org/10.1161/CIRCULATIONAHA.120.050808
55. González-Cantero A, Ortega-Quijano D, Álvarez-Díaz N, Ballester MA, Jimenez-Gomez N, Jaen P et al (2021) Impact of biological agents on imaging and biomarkers of cardiovascular disease in patients with psoriasis: a systematic review and meta-analysis of randomized placebo-controlled trials. J Investig Dermatol 141(10):2402–2411
56. Morris CJ, Purvis TE, Hu K, Scheer FAJL (2016) Circadian misalignment increases cardiovascular disease risk factors in humans. Proc Natl Acad Sci USA 113(10):E1402–E1411. Available from: https://doi.org/10.1073/pnas.1516953113
57. Cai C, Vandermeer B, Khurana R, Nerenberg K, Featherstone R, Sebastianski M et al (2019) The impact of occupational shift work and working hours during pregnancy on health outcomes: a systematic review and meta-analysis. Am J Obstet Gynecol 221(6):563–576
58. Parikh NI, Gonzalez JM, Anderson CAM, Judd SE, Rexrode KM, Hlatky MA et al (2021) Adverse pregnancy outcomes and cardiovascular disease risk: unique opportunities for cardiovascular disease prevention in women: a scientific statement from the American Heart Association. Circ 143:E902–E916
59. Neikrug AB, Rissling M, Trofimenko V, Liu L, Natarajan L, Lawton S et al (2012) Bright light therapy protects women from circadian rhythm desynchronization during chemotherapy for breast cancer. https://doi-org.uml.idm.oclc.org/101080/154020022011634940 10(3):202–16. Available from: https://www.tandfonline-com.uml.idm.oclc.org/doi/abs/https://doi.org/10.1080/15402002.2011.634940
60. Wu HS, Davis JE, Chen L (2021) Bright light shows promise in improving sleep, depression, and quality of life in women with breast cancer during chemotherapy: findings of a pilot study. https://doi-org.uml.idm.oclc.org/101080/0742052820211871914 38(5):694–704. Available from: https://www.tandfonline-com.uml.idm.oclc.org/doi/abs/https://doi.org/10.1080/07420528.2021.1871914
61. Wirz-Justice A, Benedetti F, Berger M, Lam RW, Martiny K, Terman M et al (2005) Chronotherapeutics (light and wake therapy) in affective disorders. Psychol Med 35(7):939–944
62. Bartman CM, Oyama Y, Brodsky K, Khailova L, Walker L, Koeppen M et al (2017) Intense light-elicited upregulation of miR-21 facilitates glycolysis and cardioprotection through Per2-dependent mechanisms. PLoS One 12(4). Available from: https://doi.org/10.1371/journal.pone.0176243
63. Brainard J, Gobel M, Scott B, Koeppen M, Eckle T (2015) Health implications of disrupted circadian rhythms and the potential for daylight as therapy. Anesthesiology 122(5):1170. Available from: https://doi.org/10.1097/ALN.0000000000000596
64. Hermida RC, Ayala DE, Mojón A, Fernández JR (2010) Influence of circadian time of hypertension treatment on cardiovascular risk: results of the MAPEC study. Chronobiol Int 27(8):1629–1651

65. Hermida RC, Smolensky MH, Balan H, Castriotta RJ, Crespo JJ, Dagan Y et al (2021) Guidelines for the design and conduct of human clinical trials on ingestion-time differences—chronopharmacology and chronotherapy—of hypertension medications. Chronobiol Int 38(1):1. Available from: https://doi.org/10.1080/07420528.2020.1850468
66. Gudmundsson A, Goodman B, Lent S, Barczi S, Grace A, Boyle L et al (1999) Effects of estrogen replacement therapy on the circadian rhythms of serum cortisol and body temperature in postmenopausal women. Exp Gerontol 34(6):809–818
67. Rossouw JE, Anderson GL, Prentice RL, LaCroix AZ, Kooperberg C, Stefanick ML et al (2002) Risks and benefits of estrogen plus progestin in healthy postmenopausal women: principal results from the women's health initiative randomized controlled trial. JAMA 288(3):321–333. Available from: https://jamanetwork-com.uml.idm.oclc.org/journals/jama/fullarticle/195120

# Chapter 14
# Cardio-Rheumatology and Women's Hearts

**Shadi Akhtari and Paula Harvey**

**Abstract** Autoimmune rheumatic disorders (ARD) such as systemic lupus erythematosus (SLE) and rheumatoid arthritis (RA) affect 8% of the population, of which approximately 80% are women (Mavrogeni Heart Fail Rev 24:489–498, 2019 [1]). The presence of chronic systemic inflammation increases the risk of a variety of cardiovascular disorders, making cardiovascular disease (CVD) a leading cause of morbidity and mortality in this patient population. Moreover, cardiac symptoms in these women can be atypical, clinically silent or be misinterpreted as being related to their underlying ARD (Mulvagh et al., CJC Open 4(2):115–13, 2022 [2]). Many are affected from an early age not only by debilitating physical symptoms from their underlying autoimmune disease but also by psychological aspects including a disrupted sense of identity, loss of independence, and societal stigma (Sutanto Arthritis Care Res (Hoboken) 65:1752–1765, 2013 [3]). These factors, combined with sociocultural factors and ethnicity play an additional important role in cardiovascular event rates and outcomes. Finally, women with ARDs require special attention with regards to reproductive health issues including contraception, fertility, pregnancy, lactation, and menopause, particularly with regards to safety of therapies used for their ARD at various stages of their reproductive life. In this chapter, we aim to provide a general overview of cardiovascular considerations in women with ARDs.

**Keywords** Cardio-rheumatology · Autoimmune rheumatic disorders (ARD) · Systemic lupus erythematosus (SLE) · Rheumatoid arthritis (RA) · Inflammation

S. Akhtari (✉) · P. Harvey
Department of Medicine, Women's College Hospital, University of Toronto, 76 Grenville Street, Toronto, ON M5S 1B2, Canada
e-mail: Shadi.Akhtari@wchospital.ca

© The Author(s), under exclusive license to Springer Nature Switzerland AG 2023 
L. Kirshenbaum and I. Rabinovich-Nikitin (eds.), *Biology of Women's Heart Health*, Advances in Biochemistry in Health and Disease 26,
https://doi.org/10.1007/978-3-031-39928-2_14

## Cardiovascular Manifestations of Rheumatic Diseases

Patients with ARDs can present with a broad spectrum of cardiovascular manifestations including premature atherosclerotic cardiovascular disease (ASCVD) (Table 14.1). The mechanism for development of ASCVD in the presence of systemic inflammatory disease has been linked not only to a higher prevalence of traditional cardiovascular risk factors, but importantly to chronic inflammation contributing to impaired endothelial and microvascular function, leading to early atherosclerosis [4–6]. Other than a predisposition to premature ASCVD, patients with systemic inflammatory disorders are at high risk for a wide range of other cardiac and vascular problems, including myocarditis, ischemic as well as non-ischemic cardiomyopathy, pericarditis, pericardial effusions, valvulitis, cardiac arrhythmias, pulmonary hypertension, and vasculitis. A hypercoagulable state may also ensue, predisposing patients to thrombotic events including vascular thrombosis and nonbacterial thrombotic endocarditis.

**Table 14.1** Range of cardiovascular problems in autoimmune rheumatic disease

| | |
|---|---|
| Vascular | • Accelerated atherosclerosis<br>• Myocardial infarction<br>• Coronary microvascular dysfunction<br>• Vasculitis including aortitis and coronary arteritis<br>• Aortic aneurysms and dissections<br>• Arterial stiffness<br>• Pulmonary hypertension<br>• Thrombosis<br>• Cerebrovascular disease |
| Myocardial | • Myocarditis ± myocardial fibrosis<br>• CHF (HFrEF and HFpEF)<br>• Infiltrative cardiomyopathy |
| Pericardial | • Pericarditis<br>• Pericardial effusion<br>• Pericardial nodules<br>• Pericardial constriction |
| Valvular | • Valvulitis/fibrosis<br>• Valvular regurgitation<br>• Libman-Sacks endocarditis |
| Conduction Disease and Arrhythmias | • Atrial arrhythmias<br>• Ventricular arrhythmias<br>• Conduction abnormalities and heart block<br>• Sudden cardiac death |
| Related to use of medication | • Hypertension<br>• Dyslipidemia<br>• Heart failure<br>• Increased risk of myocardial infarction<br>• Antimalarial-induced Cardiomyopathy |
| Other | • Autonomic insufficiency |

Therapies used in the treatment of these systemic inflammatory disorders such as glucocorticoids, non-steroidal anti-inflammatory drugs (NSAIDs), and some disease-modifying anti-rheumatic drugs (DMARDs) may also contribute to an increased cardiovascular risk. For example, glucocorticoid use is associated with a variety of adverse cardiovascular effects including hypertension, fluid retention, premature atherosclerosis, risk of myocardial infarction, arrhythmias, and heart failure (HF) [7]. Use of NSAIDs is also associated with an increased risk of adverse cardiovascular events, including hypertension, myocardial infarction, stroke, and HF [8]. Tofacitinib, a targeted synthetic DMARD that inhibits Janus Kinase (JAK) has been linked to higher risk of major adverse cardiovascular events (MACE) compared to anti-TNF therapy [9]. Conversely, other therapies such as methotrexate and anti-TNF therapies may reduce this risk [10–12]. A number of anti-inflammatory medications including colchicine and canakinumab (an anti-IL 1b) have been shown in randomized controlled trials [13, 14] to be effective in secondary prevention of CVD events in the general population but evidence for efficacy and safety on CVD outcomes in ARD is currently lacking. Other medications such as the antimalarial drugs chloroquine and hydroxychloroquine can have both important cardioprotective as well as rare, but potentially life-threatening cardiotoxic side effects such as HF and arrhythmias in the small number of patients who develop cardiomyopathy with long-term use [15].

## ASCVD Risk Assessment

Cardiovascular risk assessment in the presence of chronic systemic inflammatory disease is challenging. The commonly used traditional cardiovascular risk calculators such as the Framingham Risk Score (FRS) and the Pooled Cohort Equations (PCE) usually underestimate and sometimes overestimate CV risk in patients with ARDs [16]. Traditional CV risk factors such as elevated cholesterol and blood pressure (BP) can vary markedly over time in patients with a chronic relapsing and remitting disease partly due to variable disease activity and treatment-related factors such as use of glucocorticoids, NSAIDs and DMARDs [17]. Cumulative exposure over time to risk factors such as elevated cholesterol and hypertension are better able to quantify ASCVD risk [18], yet these dynamic risk factors are difficult to capture in a single-point-in-time model. To further complicate assessment, some risk factors behave paradoxically in the inflammatory milieu; for example, lipid levels may appear falsely low in presence of active inflammation and start rising once inflammation is better controlled, the so-called 'lipid paradox'[19]. Efforts to include non-traditional risk factors, disease-specific parameters, multipliers, and biomarkers have not yet been successful at improving risk estimate in this population [16]. In addressing this issue, the European League Against Rheumatism (EULAR) recommended a multiplier of 1.5 to the calculated score using traditional risk models for RA in 2016 [20]. Despite this recommendation, validation studies have shown that the multiplication factor does not significantly improve risk prediction [21, 22]. In 2021, EULAR did not endorse use of any particular risk calculators for SLE due to lack of validated SLE-specific tools, but instead recommended thorough assessment of traditional and disease-specific factors to guide prevention efforts [23]. For other ARDs, use of prediction tools for the general population was recommended due to lack of validated disease-specific tools [23]. Newer approaches incorporating non-invasive imaging of subclinical atherosclerosis such as carotid ultrasound or coronary artery calcium scoring are promising in more accurate ASCVD risk stratification in this population [24–26]. The key factors in prevention of cardiovascular complications include aggressive management of traditional risk factors as well as achievement of effective inflammatory control.

## Select Conditions Disproportionately Affecting Women

As shown in Table 14.2, women are at greater risk of developing ARDs compared to men and therefore more likely to be affected by the accompanying cardiovascular manifestations.

### *Rheumatoid Arthritis (RA)*

Rheumatoid arthritis (RA) is the most common form of inflammatory arthritis, affecting approximately 1% of the general population and is twice as common in women compared with men [27, 28]. Common cardiovascular complications in RA include an elevated risk of ASCVD, HF, arrhythmias, and strokes [29]. Clinically apparent myocarditis and pericarditis are rare in RA but may occur. When compared to healthy individuals, RA portends a similar risk to type 2 diabetes of developing CV events [30]. The observed increased risk of ASCVD is thought to be not only due to under-recognition and under-treatment of traditional CV risk factors but importantly, due to chronic systemic inflammation. Additionally, some medications may increase this risk (NSAIDs, glucocorticoids, and JAK inhibitors) while others, such as methotrexate, may decrease it. A two-fold increased risk for development of HF

**Table 14.2** Female predominance of various autoimmune rheumatic conditions

| Rheumatic condition | Female: male ratio |
|---|---|
| RA [28] | >2:1 |
| SLE [34] | 9:1 |
| Systemic sclerosis [41] | 3:1 |
| Giant cell arteritis [46] | 2–3:1[a] |
| Takayasu arteritis [52] | 6–10:1[a] |

[a] Ranges reflect different age-groups and populations studied

has been reported in patients with RA compared to those without RA, a risk that persists after adjusting for traditional cardiovascular risk factors and ischemic heart disease [31]. The pathogenesis of HF in RA is not fully understood; an increased prevalence of established risk factors contributes, but RA related chronic production of inflammatory cytokines also appears to be an independent risk factor [31]. Risk of conduction disturbances, cardiac arrhythmias and sudden cardiac death has been shown to be higher in RA patients compared to controls [32]. The key elements in prevention of cardiac complications are aggressive treatment of traditional risk factors and optimization of anti-inflammatory and immunomodulatory therapies.

## *Systemic Lupus Erythematosus (SLE)*

SLE is an autoimmune disease that can affect multiple organs including the heart [33]. There is a female preponderance (9F:1 M) with symptoms usually presenting in the second or third decades of life [34]. CVD is a leading cause of morbidity and mortality in patients with SLE and includes coronary artery disease, pericarditis, valvular disease, myocarditis, and cardiomyopathy. Young women 35–44 years of age were found to be over 50 times more likely to have a myocardial infarction than age-matched controls in the Framingham Offspring Study [35]. Older age at lupus diagnosis, longer lupus disease duration, longer duration of glucocorticoid use, hypercholesterolemia, and post-menopausal status were more common in those who had CV events than those who did not [35]. Pericarditis is the most common cardiac manifestation in lupus with autopsy series reporting a prevalence of 43–83% [36]. Valvular heart disease is often asymptomatic, involving left-sided valves, presenting as thickening of the valves due to underlying inflammation and fibrosis. Libman-sacks endocarditis is associated with more severe valvular dysfunction and risk of thromboembolic events due to fibrin and platelet emboli and requires lifelong anticoagulation therapy. Myocarditis in SLE was previously considered uncommon but prevalence appears to be higher than previously thought with the use of more sophisticated diagnostic imaging modalities such as cardiovascular magnetic resonance (CMR) imaging. Antimalarial drugs, chloroquine and hydroxychloroquine (HCQ), are important in treatment of rheumatic conditions including SLE and RA. HCQ has been shown to improve survival, prevent disease progression, reduce the need for steroid use, improve lipid levels, and decrease thrombotic as well as cardiovascular events [18, 37]. However, serious toxicity from antimalarials can occur, including retinal toxicity, skin hyperpigmentation, neuromyopathy and cardiotoxicity. The cardiotoxic effects of antimalarials are collectively referred to as antimalarial induced cardiomyopathy (AMIC), which can present either as HF due to left ventricular (LV) systolic and diastolic dysfunction, with conduction abnormalities, or with arrhythmias. AMIC is a rare, under-recognized, complication of prolonged antimalarial treatment [38]. Early recognition and drug withdrawal are critical to avoid poor patient outcomes, with a survival rate of almost 55% [15]. The best approach for early identification of AMIC is currently not well-defined and remains an area of active research, focusing on the use of serum biomarkers such as cardiac troponin and B-type natriuretic peptides (BNP), serial ECG monitoring, cardiac imaging including echocardiographic techniques as well as assessment of myocardium by CMR imaging [39].

## *Systemic Sclerosis (SSc)*

Systemic sclerosis is a rare, chronic, multisystem disease characterized by vascular dysfunction and fibrosis of the skin and internal organs. The majority of patients with SSc are female [40, 41]. Cardiac involvement in SSc is common and can be asymptomatic but overt cardiac involvement can occur and is a poor prognostic factor [42]. The dysregulated immune system and endothelial dysfunction in SSc result in diffuse microvascular disease and recurrent ischemia–reperfusion injury, leading to inflammation and eventually tissue fibrosis and adverse remodeling. This can manifest as microvascular ischemia, LV systolic and diastolic dysfunction, right ventricular (RV) dysfunction, pulmonary hypertension, or acute/chronic myocarditis [40]. Myocardial fibrosis can also lead to electrical abnormalities, including atrial and ventricular tachyarrhythmias, as well as bradyarrhythmias with sinus node dysfunction and bundle branch blocks [32]. Pericardial effusions are often asymptomatic but pericarditis and cardiac tamponade have been described [43]. Autonomic insufficiency is frequently seen in SSc [44]. Lack of heart rate variability and resting tachycardia have been associated with increased mortality in SSc [44]. Given high prevalence of cardiac involvement in SSc with associated increase in morbidity and mortality, annual cardiac screening with biomarkers such as troponins and BNP, ECG, echocardiogram is performed regardless of symptoms, with additional more advanced testing as guided by symptoms. There are currently no SSc-specific therapies for most cardiovascular complications and management is generally the same as for patients without SSc. Beta blocker use should be avoided due to effects on Raynaud's phenomenon[23]. For biopsy-proven myocarditis, treatment with immunosuppressive therapy in addition to the typical cardiomyopathy treatment is recommended [42].

## *Giant Cell Arteritis (GCA)*

Giant Cell Arteritis is an idiopathic systemic vasculitis of large- and medium-sized vessels, twice as frequent in women compared to men [45, 46]. Vascular involvement can be widespread, leading to stenoses and aneurysm formation of the affected vascular beds. Vision loss is the most feared complication in patients with GCA, though cardiovascular disease risk is another important consideration. A significantly higher risk of cardiovascular mortality and morbidity including higher rate of strokes, thoracic aortic aneurysms and dissection, ischemic heart disease and peripheral vascular disease has been identified in patients with GCA [45]. The association with increased CVD morbidity and mortality may be due to prolonged duration of therapy with glucocorticoids or due to ongoing inflammation due to lack of complete cessation of inflammation. With more recent development of novel biologic therapies including tocilizumab, patients now have a steroid-free option to achieve remission [47]. Aspirin [48] and statins [49, 50] have been evaluated as adjunct therapies to glucocorticoids in treatment of GCA, but results have been mixed. Important clinical questions remain including whether the use of statins post-clinical remission in patients with GCA decreases cardiovascular events in these patients and whether it allows a longer course of disease remission compared to having the patient without any therapy after clinical remission is achieved.

## *Takayasu Arteritis (TA)*

TA is an uncommon chronic idiopathic large-vessel vasculitis, primarily affecting the aorta, its primary branches, and the pulmonary arteries. Women are affected in 80–90% of cases with age of onset usually between 20 and 40 years, with the highest prevalence in Asia [51, 52]. The inflammatory process within the arteries can lead to narrowing, occlusion, or dilatation of the involved portion of the artery [53]. TA can present with constitutional symptoms, hypertension, arterial bruits, or absent/diminished pulses. Limb claudication, discrepant blood pressure between arms, hypertension, angina due to coronary artery ostial narrowing from aortitis or coronary arteritis, aortic regurgitation due to aortic dilatation, pericardial disease and myocardial infarction are some of the more specific potential cardiovascular manifestations [53]. Accelerated atherosclerosis is well known and is thought to be due to arterial injury and endothelial dysfunction. Traditional CVD risk factors have also been shown to be more prevalent in patients with TA than controls, which can be contributory [54]. Patients may present with myocardial infarction, angina, or sudden death. Revascularization procedures cannot be performed until inflammation is controlled. When revascularization is indicated, coronary artery bypass graft (CABG) is preferred to percutaneous coronary intervention (PCI), with superior outcomes with regards to patency rates as well as mortality [55]. CHF can be related to significant valvular dysfunction, mainly aortic regurgitation, severe hypertension, pulmonary arterial involvement, or rarely inflammatory myocarditis [53]. Aortic regurgitation is the most common valvular dysfunction in patients with TA, caused primarily by annular dilatation and dilatation of the ascending aorta. Conduction disease can be seen with frequent or complex ventricular arrhythmias, prolonged QT dispersion, and complete AV block [56, 57].

### *Osteoarthritis (OA)*

Though not an autoimmune condition, osteoarthritis cannot be excluded from this discussion. OA is the most common type of arthritis, affecting approximately 10% of the adult population, disproportionately affecting women (> 70%). It is the fourth most common predictor of health problems worldwide in women [58]. Obesity and the associated cardiovascular risk factors such as hypertension, diabetes, and dyslipidemia are commonly observed in individuals with OA [59]. Use of therapies such as NSAIDs is associated with additional cardiac risk. On the other hand, inadequately treated pain from OA can be an important limiting factor, leading to avoidance of regular physical activity, weight gain, poorer cardiovascular fitness, and physical disability. Among those with OA of the hip and/or knee, severity of OA disability has been shown to be associated with a significant and independent increase in all-cause mortality and cardiovascular events [59]. There are data to suggest that receipt of an elective hip or knee total joint arthroplasty is associated with improved survival and reduced risk for CVD events [59]. This apparent protective effect may be related to more physical activity, pain relief and reduction in NSAID use, though more research is needed to elucidate the explanation for this observed relationship [59].

## Other Important Considerations

### *Ethnicity*

Significant ethnic and sociocultural variation in CVD event rates have been observed in patients with ARD. SLE is better studied with this regard. Black women with SLE, for example, have more SLE-related morbidity and mortality compared to White women [60] with a reported 14% increased risk of CVD. Risk of MI has been observed to be lower among Hispanic and Asian women with SLE compared to White women, and risk of stroke observed to be higher among Black and Hispanic compared to White women [61]. Furthermore, significant ethnic disparities with regards to age at the time of hospitalization for CVD events has also been observed. Among different groups with SLE, Black women were the youngest to be admitted with CVD and have a CVD-associated in-hospital death, on average 19.8 years younger than ethnicity—and sex-matched controls at the time of CVD-associated death [62]. There is a paucity of data on interaction of ethnicity and ASCVD outcomes in RA and other ARDs [63]. Further research to better understand the role of ethnicity among various groups as it relates to CVD risk in this population is needed.

## Reproductive Health Issues

Autoimmune conditions can be associated with increased prevalence of adverse pregnancy outcomes. The risk of pregnancy complications depends on factors including diagnosis, disease activity and organ damage, medications, and presence of antiphospholipid (aPL) antibodies. For example, women with SLE are at significantly higher risk of developing pre-eclampsia, preterm live birth, low birth weight, spontaneous abortions, and stillbirth compared to controls [60]. Pulmonary arterial hypertension related to some ARDs is another example of a situation associated with high risk of maternal mortality, estimated at up to 20% [64]. Pre-pregnancy counseling and tailored safe and effective contraception is therefore an important part of patient management. Choice of contraception itself can be complicated in women with ARD; for example, the combined progesterone and estrogen oral contraceptive pill is not recommended in those with positive aPL antibodies or those who have more severe SLE [65]. Patients who are positive for aPL experience additional challenges with regards to fertility and higher risk of thrombosis and spontaneous abortions once pregnancy is achieved [65]. Therapies used in patients with ARD are not all safe preconception, or during pregnancy and lactation and therefore, careful review of medications is essential if pregnancy is desired or contemplated. As an example, cyclophosphamide, a potent immunosuppressant used for severe SLE complications and systemic sclerosis, among other indications, is highly teratogenic and can lead to increased risk of infertility and premature menopause [65]. Patients of childbearing age should be counseled prior to starting such agents and highly effective contraception should be initiated. In general, data is limited on medication effects on fertility, pregnancy, or lactation as women in these stages are typically excluded from studies. Additionally, many women with ARDs avoid or postpone pregnancy due to anxiety about risks of pregnancy complications, miscarriage, and genetic transmission to their offspring [3]. They may also feel anxious about caring for a child given their potentially debilitating disease [3]. Asking patients about their desire for pregnancy early and often can help keep an open dialogue. Post-menopausal issues are also not uncommon in patients with ARD. HRT remains an important consideration in those with disabling vasomotor symptoms, with current evidence supporting the use of HRT in ARD patients in absence of aPL, including those with SLE [66].

## Societal Stigma

Despite significant improvements in treatment options and patient survival, ARDs can have a significant impact on patients' sense of independence, self-esteem, and quality of life. Many patients suffer from debilitating pain, fatigue and cutaneous manifestations which can limit ability to work and participate in desired social activities. Patients with SLE often perceive that SLE is misunderstood and stigmatized by their family, friends, and physicians, leading to a sense of isolation [3]. The significant physical disability can cause feelings of anxiety about their future. Similarly, patients with scleroderma have been reported to experience difficulties with regards to physical, emotional, and social limitations [67]. Studies in RA have shown that patients often experience physical restriction, social restriction and anxiety about the future and loss of independence [68]. Further research on interventions to promote positive coping strategies and mental resilience can help improve quality of life and overall health outcomes, including cardiovascular health in these patients.

## Future Directions

The field of cardio-rheumatology is emerging, appreciating the need for prioritizing cardiovascular care in patients with ARD. As the field evolves, it is important to acknowledge that many ARDs disproportionately affect women. Framing the importance of ARDs as a women's health issue can be helpful in the design and delivery of care that is consistent with the needs of this growing population.

## References

1. Mavrogeni SI et al (2019) Pathophysiology and imaging of heart failure in women with autoimmune rheumatic diseases. Heart Fail Rev 24(4):489–498
2. Mulvagh SL et al (2022) The Canadian women's heart health alliance atlas on the epidemiology, diagnosis, and management of cardiovascular disease in women—Chapter 4: sex- and gender-unique disparities: CVD across the lifespan of a woman. CJC Open 4(2):115–132
3. Sutanto B et al (2013) Experiences and perspectives of adults living with systemic lupus erythematosus: thematic synthesis of qualitative studies. Arthritis Care Res (Hoboken) 65(11):1752–1765
4. Ciftci O et al (2008) Impaired coronary microvascular function and increased intima-media thickness in rheumatoid arthritis. Atherosclerosis 198(2):332–337
5. Liao KP et al (2021) Coronary microvascular dysfunction in rheumatoid arthritis compared to diabetes mellitus and association with all-cause mortality. Arthritis Care Res (Hoboken) 73(2):159–165
6. Ikonomidis I et al (2008) Inhibition of interleukin-1 by anakinra improves vascular and left ventricular function in patients with rheumatoid arthritis. Circ 117(20):2662–2669
7. Avina-Zubieta JA et al (2008) Risk of cardiovascular mortality in patients with rheumatoid arthritis: a meta-analysis of observational studies. Arthritis Rheum 59(12):1690–1697
8. Bally M et al (2018) Risk of acute myocardial infarction with real-world NSAIDs depends on dose and timing of exposure. Pharmacoepidemiol Drug Saf 27(1):69–77
9. Ytterberg SR, Bhatt DL, Connell CA (2022) Cardiovascular and cancer risk with tofacitinib in Rheumatoid arthritis. N Engl J Med 386(18):1768
10. Choi HK et al (2002) Methotrexate and mortality in patients with rheumatoid arthritis: a prospective study. Lancet 359(9313):1173–1177
11. Westlake SL et al (2010) The effect of methotrexate on cardiovascular disease in patients with rheumatoid arthritis: a systematic literature review. Rheumatology (Oxford) 49(2):295–307

12. Guin A et al (2013) Effects of disease modifying anti-rheumatic drugs on subclinical atherosclerosis and endothelial dysfunction which has been detected in early rheumatoid arthritis: 1-year follow-up study. Semin Arthritis Rheum 43(1):48–54
13. Tardif JC et al (2019) Efficacy and safety of low-dose colchicine after myocardial infarction. N Engl J Med 381(26):2497–2505
14. Ridker PM et al (2017) Antiinflammatory therapy with canakinumab for atherosclerotic disease. N Engl J Med 377(12):1119–1131
15. Tselios K et al (2018) Antimalarial-induced cardiomyopathy: a systematic review of the literature. Lupus 27(4):591–599
16. Colaco K et al (2020) Predictive utility of cardiovascular risk prediction algorithms in inflammatory rheumatic diseases: a systematic review. J Rheumatol 47(6):928–938
17. Nikpour M et al (2010) Variability over time and correlates of cholesterol and blood pressure in systemic lupus erythematosus: a longitudinal cohort study. Arthritis Res Ther 12(3):R125
18. Nikpour M et al (2011) Importance of cumulative exposure to elevated cholesterol and blood pressure in development of atherosclerotic coronary artery disease in systemic lupus erythematosus: a prospective proof-of-concept cohort study. Arthritis Res Ther 13(5):R156
19. Gonzalez-Gay MA, Gonzalez-Juanatey C (2014) Inflammation and lipid profile in rheumatoid arthritis: bridging an apparent paradox. Ann Rheum Dis 73(7):1281–1283
20. Agca R et al (2017) EULAR recommendations for cardiovascular disease risk management in patients with rheumatoid arthritis and other forms of inflammatory joint disorders: 2015/2016 update. Ann Rheum Dis 76(1):17–28
21. Navarini L et al (2018) Performances of five risk algorithms in predicting cardiovascular events in patients with Psoriatic Arthritis: an Italian bicentric study. PLoS ONE 13(10):e0205506
22. Arts EE et al (2016) Prediction of cardiovascular risk in rheumatoid arthritis: performance of original and adapted SCORE algorithms. Ann Rheum Dis 75(4):674–680
23. Drosos GC et al (2022) EULAR recommendations for cardiovascular risk management in rheumatic and musculoskeletal diseases, including systemic lupus erythematosus and antiphospholipid syndrome. Ann Rheum Dis 81(6):768–779
24. Corrales A et al (2014) Carotid ultrasound is useful for the cardiovascular risk stratification of patients with rheumatoid arthritis: results of a population-based study. Ann Rheum Dis 73(4):722–727
25. Corrales A et al (2013) Cardiovascular risk stratification in rheumatic diseases: carotid ultrasound is more sensitive than coronary artery calcification score to detect subclinical atherosclerosis in patients with rheumatoid arthritis. Ann Rheum Dis 72(11):1764–1770
26. Evans MR et al (2011) Carotid atherosclerosis predicts incident acute coronary syndromes in rheumatoid arthritis. Arthritis Rheum 63(5):1211–1220
27. Crowson CS et al (2011) The lifetime risk of adult-onset rheumatoid arthritis and other inflammatory autoimmune rheumatic diseases. Arthritis Rheum 63(3):633–639
28. Linos A et al (1980) The epidemiology of rheumatoid arthritis in Rochester, Minnesota: a study of incidence, prevalence, and mortality. Am J Epidemiol 111(1):87–98
29. Lindhardsen J et al (2012) Risk of atrial fibrillation and stroke in rheumatoid arthritis: Danish nationwide cohort study. BMJ 344:e1257
30. Peters MJ et al (2009) Does rheumatoid arthritis equal diabetes mellitus as an independent risk factor for cardiovascular disease? A prospective study. Arthritis Rheum 61(11):1571–1579
31. Nicola PJ et al (2005) The risk of congestive heart failure in rheumatoid arthritis: a population-based study over 46 years. Arthritis Rheum 52(2):412–420
32. Seferovic PM et al (2006) Cardiac arrhythmias and conduction disturbances in autoimmune rheumatic diseases. Rheumatology (Oxford) 45(Suppl 4):iv39– iv42
33. Tan EM et al (1982) The 1982 revised criteria for the classification of systemic lupus erythematosus. Arthritis Rheum 25(11):1271–1277
34. Fessel WJ (1974) Systemic lupus erythematosus in the community. Incidence, prevalence, outcome, and first symptoms; the high prevalence in black women. Arch Intern Med 134(6):1027–1035

35. Manzi S et al (1997) Age-specific incidence rates of myocardial infarction and angina in women with systemic lupus erythematosus: comparison with the Framingham Study. Am J Epidemiol 145(5):408–415
36. Jain D, Halushka MK (2009) Cardiac pathology of systemic lupus erythematosus. J Clin Pathol 62(7):584–592
37. Magder LS, Petri M (2012) Incidence of and risk factors for adverse cardiovascular events among patients with systemic lupus erythematosus. Am J Epidemiol 176(8):708–719
38. Tselios K et al (2019) Antimalarial-induced cardiomyopathy in systemic lupus erythematosus: as rare as considered? J Rheumatol 46(4):391–396
39. Hanneman K et al (2020) Antimalarial-induced cardiomyopathy resembles fabry disease on cardiac MRI. JACC Cardiovasc Imaging 13(3):879–881
40. Champion HC (2008) The heart in scleroderma. Rheum Dis Clin North Am 34(1):181–190
41. van den Hoogen F et al (2013) 2013 classification criteria for systemic sclerosis: an American College of Rheumatology/European League against Rheumatism collaborative initiative. Arthritis Rheum 65(11):2737–2747
42. Ioannidis JP et al (2005) Mortality in systemic sclerosis: an international meta-analysis of individual patient data. Am J Med 118(1):2–10
43. Hosoya H, Derk CT (2018) Clinically symptomatic pericardial effusions in hospitalized systemic sclerosis patients: demographics and management. Biomed Res Int 2018:6812082
44. Othman KM et al (2010) Autonomic dysfunction predicts early cardiac affection in patients with systemic sclerosis. Clin Med Insights Arthritis Musculoskelet Disord 3:43–54
45. Lee YH, Song GG (2018) Overall and cause-specific mortality in giant cell arteritis: a meta-analysis. Z Rheumatol 77(10):946–951
46. Gonzalez-Gay MA et al (2009) Epidemiology of giant cell arteritis and polymyalgia rheumatica. Arthritis Rheum 61(10):1454–1461
47. Stone JH et al (2017) Trial of tocilizumab in giant-cell arteritis. N Engl J Med 377(4):317–328
48. Martinez-Taboada VM et al (2014) Effect of antiplatelet/anticoagulant therapy on severe ischemic complications in patients with giant cell arteritis: a cumulative meta-analysis. Autoimmun Rev 13(8):788–794
49. Narvaez J et al (2007) Statin therapy does not seem to benefit giant cell arteritis. Semin Arthritis Rheum 36(5):322–327
50. Pugnet G et al (2016) Predictors of cardiovascular hospitalization in giant cell arteritis: effect of statin exposure. A French population-based study. J Rheumatol 43(12):2162–2170
51. Kerr GS et al (1994) Takayasu arteritis. Ann Intern Med 120(11):919–929
52. Lupi-Herrera E et al (1977) Takayasu's arteritis. Clinical study of 107 cases. Am Heart J 93(1):94–103
53. Talwar KK et al (1991) Cardiac involvement in nonspecific aortoarteritis (Takayasu's arteritis). Am Heart J 122(6):1666–1670
54. Tervaert JW (2009) Translational mini-review series on immunology of vascular disease: accelerated atherosclerosis in vasculitis. Clin Exp Immunol 156(3):377–385
55. Liang P, Hoffman GS (2005) Advances in the medical and surgical treatment of Takayasu arteritis. Curr Opin Rheumatol 17(1):16–24
56. Siburian G, Hashimoto Y, Numano F (1993) Ventricular arrhythmias in Takayasu arteritis. Int J Cardiol 40(3):243–249
57. Kato T et al (2000) QT dispersion in patients with Takayasu arteritis. Angiology 51(9):751–756
58. Murray CJ, Lopez AD (1996) Evidence-based health policy–lessons from the global burden of disease study. Science 274(5288):740–743
59. Hawker GA et al (2014) All-cause mortality and serious cardiovascular events in people with hip and knee osteoarthritis: a population based cohort study. PLoS ONE 9(3):e91286
60. Barnado A et al (2014) Pregnancy outcomes among African-American patients with systemic lupus erythematosus compared with controls. Lupus Sci Med 1(1):e000020
61. Barbhaiya M et al (2017) Race/ethnicity and cardiovascular events among patients with systemic lupus erythematosus. Arthritis Rheumatol 69(9):1823–1831

62. Scalzi LV, Hollenbeak CS, Wang L (2010) Racial disparities in age at time of cardiovascular events and cardiovascular-related death in patients with systemic lupus erythematosus. Arthritis Rheum 62(9):2767–2775
63. Daniel CM et al (2020) Ethnic disparities in atherosclerotic cardiovascular disease incidence and prevalence among rheumatoid arthritis patients in the United States: a systematic review. ACR Open Rheumatol 2(9):525–532
64. Meng ML et al (2017) Pulmonary hypertension in pregnancy: a report of 49 cases at four tertiary North American sites. Obstet Gynecol 129(3):511–520
65. Sammaritano LR et al (2020) 2020 American College of Rheumatology Guideline for the management of reproductive health in rheumatic and musculoskeletal diseases. Arthritis Rheumatol 72(4):529–556
66. Buyon JP et al (2005) The effect of combined estrogen and progesterone hormone replacement therapy on disease activity in systemic lupus erythematosus: a randomized trial. Ann Intern Med 142(12 Pt 1):953–962
67. Cinar FI et al (2012) Living with scleroderma: patients' perspectives, a phenomenological study. Rheumatol Int 32(11):3573–3579
68. Whalley D et al (1997) Quality of life in rheumatoid arthritis. Br J Rheumatol 36(8):884–888

# Chapter 15
# Cardiovascular Adverse Effects of Breast Cancer Chemotherapy

**Haojun Huang and Liam R. Brunham**

**Abstract** Breast cancer is the most commonly diagnosed cancer globally, with more than 2.3 million cases per year and resulting in the death of 685,000 women [1]. As a result of population growth and aging, the International Agency for Research on Cancer estimates that the burden of breast cancer will increase to over 3 million new cases and 1 million deaths per year by 2040 [1]. With the introduction of early diagnosis, prevention measures, and effective treatment approaches, the mortality of breast cancer has decreased by 42% from 1989 to 2022 [2]. Various treatments have been developed based on the stage of breast cancer and the presence or absence of molecular markers such as estrogen or progesterone receptors and human epidermal growth factor 2. However, with improved survival and outcomes, the long-term side effects of breast cancer treatment are a growing concern. Cardiovascular complications are among the most frequent adverse effects of chemotherapies, which increases the concern of premature morbidity and mortality among cancer survivors [3]. In this chapter, we will focus on the adverse cardiovascular effects caused by chemotherapeutic agents used in the treatment of breast cancer.

**Keywords** Breast cancer · Chemotherapy · Doxorubicin · Cardiomyopathy · Topoisomerase II

---

H. Huang · L. R. Brunham (✉)
Centre for Heart Lung Innovation, University of British Columbia, Vancouver, BC V6Z 1Y6, Canada
e-mail: liam.brunham@ubc.ca

L. R. Brunham
Department of Medicine, University of British Columbia, Vancouver, BC V6T 1Z3, Canada

Department of Medical Genetics, University of British Columbia, Vancouver, BC V6H 3N1, Canada

© The Author(s), under exclusive license to Springer Nature Switzerland AG 2023 
L. Kirshenbaum and I. Rabinovich-Nikitin (eds.), *Biology of Women's Heart Health*, Advances in Biochemistry in Health and Disease 26,
https://doi.org/10.1007/978-3-031-39928-2_15

## Cardiovascular Complications of Chemotherapies

The era of chemotherapy began in 1946 when researchers from Memorial Sloan Kettering Cancer Center reported that the chemical warfare agents, nitrogen mustards, are effective against certain forms of cancer by chemically modifying the DNA of cancer cells [4, 5]. Soon after the discovery of the first anthracycline drug, daunorubicin, in the 1950s, doxorubicin was isolated from a mutant variant of *Streptomyces peucetius* (var. *caesius*) and found to be highly effective against a wide range of solid tumours and hematologic malignancies [6, 7]. Other analogues such as epirubicin and idarubicin were also approved for clinical use [8]. Today, a broad range of chemotherapy drugs (such as alkylating agents, antimetabolites, and anti-tumour antibiotics) have been developed to target different cancer types.

Cardiovascular complications such as arrhythmias, myocardial dysfunction, and heart failure are the most severe side effects of the cancer therapeutics [9]. For example, doxorubicin-induced cardiac damage is irreversible and cumulative dose-dependent. The incidence of developing impaired myocardial function with doxorubicin, based on the decline in left ventricular ejection fraction (LVEF), is estimated to be 3–5% at a cumulative dose of 400 mg/m$^2$ of doxorubicin, 7–26% at a dose of 550 mg/m$^2$, and up to 48% at 700 mg/m$^2$ [10]. Other chemotherapy drugs such as monoclonal antibodies (trastuzumab, bevacizumab and pertuzumab) and tyrosine kinase inhibitors (sunitinib and sorafenib) may also induce cardiotoxicity but tend to be reversible [11–13].

## Known Mechanisms of Doxorubicin-Induced Cardiotoxicity

### *Doxorubicin Interacts with DNA and Topoisomerase II*

The primary mode of action of doxorubicin appears to be intercalation with nucleotide base, causing deformation of DNA in tumour cells [14, 15]. Doxorubicin is also known to intercalate with topoisomerase II-DNA complex, inhibiting replication fork progression and leading to double-strand DNA breaks (DSB) [16]. Topoisomerase II (TOP2) is an ATP-dependent nuclear enzyme that induces transient DNA breaks to release the torsional stress that occurs during DNA replication, transcription, and other nuclear processes. Both human topoisomerase II isoforms, TOP2A and TOP2B, are targeted by doxorubicin [17]. TOP2A is required during mitosis and is highly expressed in proliferating cells [18], while TOP2B is predominant in non-dividing cells such as cardiomyocytes [19]. Cardiomyocyte-specific deletion of *Top2b* in mice results in protection from doxorubicin-induced cardiotoxicity (DIC) [20]. In addition to nuclear DNA damage, doxorubicin also targets mitochondrial DNA, forming high levels of adducts and cross-links and leading to breakage of the mitochondrial DNA helix [21, 22]. Mitochondrial DNA lesions and free radical-associated mitochondrial dysfunction are reported after doxorubicin exposure [23].

## Doxorubicin Increases Cellular Oxidative Stress

Another important mechanism of DIC is increased oxidative stress [24]. Reactive oxygen species (ROS) are natural by-products of cellular activity, while disruption of antioxidant functions or overproduction of ROS leads to oxidative stress. Mitochondria supply 95% of the ATP through fatty acid and glucose oxidation as well as producing the principal amount of ROS in cardiomyocytes [25]. Assisted by NADH dehydrogenase on complex I of the mitochondrial electron transport chain, the quinone moiety of doxorubicin is reduced to a semiquinone radical, leading to the generation of superoxide anion $O_2^-$ and highly toxic hydroxyl radical OH• in the presence of iron [26, 27]. The formation of free radicals during doxorubicin metabolism increases mitochondrial oxidative stress and membrane lipid peroxidation, eventually causing mitochondrial dysfunction and lowering ATP synthesis [28].

Besides mitochondria, cytoplasmic oxidative stress also increases in response to doxorubicin. Firstly, doxorubicin can directly interact with iron and form a DOX-Fe complex, which promotes ROS production through Fe(II) and Fe(III) cycling [29]. Nitric oxide (NO) is also a source of DOX-induced oxidative stress by catalyzation of nitric oxide synthase (NOS). Similar to NADH dehydrogenase, doxorubicin can also bind to a reductase enzyme, endothelial nitric oxide synthase (eNOS), and induce semiquinone radical and superoxide $O_2^-$ formation. Subsequentially, the imbalance between eNOS and nitric oxide leads eNOS to generate the potent oxidants peroxynitrite and hydrogen peroxide [30]. Furthermore, doxorubicin-induced down-regulation of endogenous antioxidant enzymes (e.g., peroxidase, catalase, and superoxide dismutase) limits the ability of ROS elimination and enhances the oxidative stress in cardiomyocytes [31, 32].

## Doxorubicin Induces Cell Death Pathways

Doxorubicin-induced apoptosis has been extensively studied. DNA damage and oxidative stress induced by doxorubicin activate tumour suppressor p53 and Bcl-2-associated X (Bax) proteins, c-Jun N-terminal kinase (JNK) and p38 mitogen-activated protein kinase (MAPK) signaling and down-regulation of GATA Binding Protein 4 (GATA4) as well as PI-3K/Akt survival pathway [33–35]. Activation of the extrinsic apoptotic pathway has also been studied in DIC. Oxidative stress up-regulates the expression of transcription factor NF-κB and nuclear factor-activated T cell-4 (NFAT4), leading to increased Fas/FasL and p53 [36, 37]. Additionally, increased ROS production stimulates endoplasmic/sarcoplasmic reticulum (ER/SR) stress, triggering the apoptotic pathway by activating caspase 12 [38, 39].

In addition to apoptosis, other cell death pathways such as autophagy, ferroptosis and necrotic cell death have also been studied [40]. Autophagy is a cell defence mechanism differentially regulated in doxorubicin-treated cardiomyocytes to help the degradation of cellular components [41]. The control of autophagy level is essential to maintain cellular homeostasis, while excessive activation of autophagic flux can lead to cardiotoxicity and cell death [42, 43]. Interestingly, the critical regulator of ferroptosis, glutathione peroxidase 4 (GPx4), is downregulated by doxorubicin, resulting in excessive mitochondrial lipid peroxidation and mitochondria-dependent ferroptosis through the formation of DOX-Fe complex [44]. Treatment with the ferroptosis inhibitor ferrostatin-1 protects against DIC [44, 45], suggesting that modulation of the ferroptotic pathway may prove to be a successful method to mitigate cardiotoxicity. Additionally, doxorubicin increases necrotic cell death through BH3-only protein Bcl-2-like 19 kDa-interacting protein 3 (Bnip3)-mediated mitochondrial dysfunction [46]. Further study reveals that the elevated level of necrotic cell death is mediated by doxorubicin-induced proteasomal degradation of the ubiquitin E3-ligase TNF receptor associated factor 2 (TRAF2), leading to impaired tumor necrosis factor-α (TNFα)-mediated nuclear factor-κB (NF-κB) activation and increased cardiotoxicity [47].

## Prediction and Prevention of Doxorubicin-Induced Cardiotoxicity

### *Current Preventive Strategies for Doxorubicin-Induced Cardiotoxicity*

For patients deemed to be at highest risk of cardiotoxicity, for example due to pre-existing cardiovascular diseases, additional prevention management can be considered before doxorubicin treatment. These include limiting the total cumulative doxorubicin doses, altered delivery methods with lower cardiotoxicity (e.g., continuous infusions and liposomal doxorubicin) and co-treatment with cardioprotective drugs (e.g., dexrazoxane, angiotensin-converting enzyme inhibitors, beta-blockers or statins) [10].

The iron-chelator dexrazoxane binds free iron and decreases mitochondrial iron levels [29, 48]. Dexrazoxane also leads to TOP2B depletion and a reduction in DNA double-strand breaks were observed [49, 50]. Recent studies also found that dexrazoxane inhibits both apoptosis and necroptosis in doxorubicin-treated cardiomyocytes [51, 52]. Usage of dexrazoxane is recommended in a European Society of Cardiology (ESC) guideline for patients with advanced or metastatic breast cancer receiving a cumulative dose of > 300 mg/m$^2$ doxorubicin [10]. Dexrazoxane treatment in children with high-risk acute lymphoblastic leukemia or solid tumours improves left ventricular structure and function compared to those who receive doxorubicin treatment only [53, 54]. However, high-quality evidence and further randomized trials of dexrazoxane treatment DIC are still needed [55].

In recent years, multiple clinical trials have provided compelling evidence of a cardioprotective effect of sodium-glucose transport protein 2 (SGLT2) inhibitors

(SGLT2i) in multiple forms of heart failure [56–60], which raises the intriguing possibility that this class of medication could also have cardioprotective effects for DIC. SGLT2i is a well-tolerated and effective glucose-lowering medication for treating type 2 diabetes mellitus (T2DM) [61]. Four SGLT2i (empagliflozin, dapagliflozin, canagliflozin, and ertugliflozin) have been approved for the T2DM treatment [62]. Empagliflozin has been found to prevent DIC in mice [63–69] and lower the rate of cardiac events in anthracycline-treated patients [70], but its exact mechanism is still under evaluation.

## *Advances in Approaches for Cardiovascular Research*

Although there is high interest in developing new therapeutic compounds for cardiotoxicity and other cardiovascular diseases, it is challenging to access human primary cardiac cells and maintain them in cell culture during the preclinical research [71]. In vivo studies using laboratory animals, such as mice or rats, are commonly used, but display significant differences in cardiac anatomy and physiology compared to humans, limiting their ability to accurately reflect pathological mechanisms [72–74]. A major breakthrough came in 2007 with the description of human pluripotent stem cells (iPSCs) by Shinya Yamanaka [75] and the development of directed differentiation methods to generate cardiomyocytes from iPSCs (iPSC-CMs) [76].

Human iPSCs have been shown great potential in differentiation to board range of cell types, including cardiomyocytes, hepatocytes, and neuronal cells. In cardiovascular research, iPSC-CMs derived from patients with specific phenotypes and genetic backgrounds have proven to be excellent models for studying cellular mechanisms and mimicking drug responses. An advantage of this model is the suitability to high-throughput drug screening platform for multiple disease-specific measurements, including multielectrode array, calcium handling and toxicity measurements. Moreover, combined with other cutting-edge techniques, such as genome editing, next-generation sequencing, and three-dimension cell culture/heart-on-a-chip technology, iPSC-CMs offer an efficient and scalable approaches in cardiotoxicity disease modelling and regenerative medicine.

## *Pharmacogenomics and Precision Medicine*

Although chemotherapeutic agents are among the most cardiotoxic of all medicines, the risk factors for developing DIC range from patient-associated (age, sex, pre-existing heart disease, etc.) to therapy-associated (drug formulations, intravenous injection administration and history of irradiation, etc.) [10]. In some cases, patients suffer from DIC at low doses of doxorubicin, suggesting genetic diversity effects in susceptibility [77]. Genetic influences have been proposed to explain up to 95% of the interpatient variability in drug disposition and therapeutic effects in general

[78]. One large-scale study on over 44,000 participants found that over 99% of those individuals assessed had a genotype associated with a higher risk of at least one medication [79]. Unlike other risk factors, genetic susceptibility exists throughout a patient's lifetime [80].

Pharmacogenomics research aims to minimize adverse drug reactions. Precision medicine involves using pharmacogenetics, biomarker, and phenotypic testing to help physicians select the appropriate drug and dosage for the patient [81]. Over 200 drugs have label information regarding pharmacogenomic biomarkers [82].

Nineteen published genetic association studies have identified 28 genetic variants significantly associated with DIC [83]. These variants are located in genes related to drug transporter (*ABCC1*, *ABCC2*, *ABCC5* and *SLC28A3*), antioxidants (*CAT*, *GSTP1* and *HAS3*), drug metabolism (*CBR3*, *NOS3*, *POR* and *UGT1A6*), DNA repair (*ERCC2*), NAD(P)H oxidase multienzyme complex (*NCF4*, *RAC2* and CYBA), iron metabolism (*HFE*), sarcomere structure and function (*CELF4* and *MYH7*), and transcription factor (*RARG*).

The rs2229774 variant, located in the transcription factor retinoic acid receptor gamma (*RARG*), was the first DIC-associated variant validated in a human iPSC-CM model. Christidi et al. compared the effects of the rs2229774 variant in multiple patient-specific iPSC-CM lines and found that the presence of the *RARG* variant was necessary and sufficient to increase susceptibility to doxorubicin [84]. Further studies using genome-edited isogenic iPSC-CMs pairs differing only by the presence of rs2229774 showed that the presence of this variant leads to reduced activation of RARG target genes, and consequently attenuated DNA repair activity in response to doxorubicin [85]. Another group also found rs2229774 increases sensitivity to doxorubicin through activating the cardioprotective extracellular regulated kinase (ERK) pathway and suppression of topoisomerase 2b, while co-treatment with RARG agonist CD1530 attenuates DIC [86].

Another DIC-related gene identified by human genetics, *SLC28A3*, encodes solute carrier family 28 member 3, and has also been studied in iPSC-CMs. Two variants rs7853758 and rs885004 located in *SLC28A3* have been associated with a lower risk of DIC [87, 88]. However, since rs7853758 is a synonymous variant and rs885004 is located in the intron region, it is unclear how they change the expression and function of *SLC28A3*. Magdy et al. first focused on the impact of *SLC28A3* in DIC and found that overexpression of *SLC28A3* significantly increased iPSC-CMs sensitivity to doxorubicin, while knocking out *SLC28A3* showed higher cell viability compared to the wild-type iPSC-CMs in DIC. Using nanopore-based fine-mapping and base-editing, they further identified a novel variant, rs11140490, within the overlap region of *SLC28A3* and an antisense long noncoding RNA (SLC28A3-AS1) that appeared to be the causal cardioprotective variant. The cardioprotective action of rs11140490 exerted in regulating the transcription of SLC28A3-AS1, which resulted in down-regulating the expression of *SLC28A3* and protected against DIC in patient-specific iPSC-CMs [89].

These studies of DIC-associated variants have illustrated the utility of iPSC-CMs in validating the functional effects of genetic variants and identifying novel pharmacogenetic mechanisms that underlie the pathogenesis of DIC. This type of functional validation supports the clinical implementation of genetic testing to predict risk of chemotherapy-induced cardiotoxicity prior to treatment. Furthermore, as the genetic background of each patient is unique and complex, generating iPSC-CMs from breast cancer patients and testing drug susceptibility prior to chemotherapy could lead to a personalized medicine approach to predict and prevent DIC.

## Conclusion

Despite substantial improvements in the treatment of breast cancer, major concern remains about the cardiovascular complication of breast cancer and its treatment. The anthracyclines are among the most cardiotoxic chemotherapies. Advances in the identification of genetic variants associated with chemotherapy-associated cardiotoxicity, as well as the development of iPSC-CMs as physiologically relevant model systems which validate the functional effects of these genetic variants, has moved us closer to personalizing treatment in order to predict and prevent cardiotoxicity (Fig. 15.1).

Chemotherapy is an effective method in treating breast cancer but may also induce toxicity to the heart. Progress in using patient-specific induced pluripotent stem cell-derived cardiomyocytes (iPSC-CMs) modelling and pharmacogenomic studies help to predict and prevent this side effect. Cardiovascular risk assessment pre- and post-chemotherapy as well as cardioprotective medications are also used for monitoring

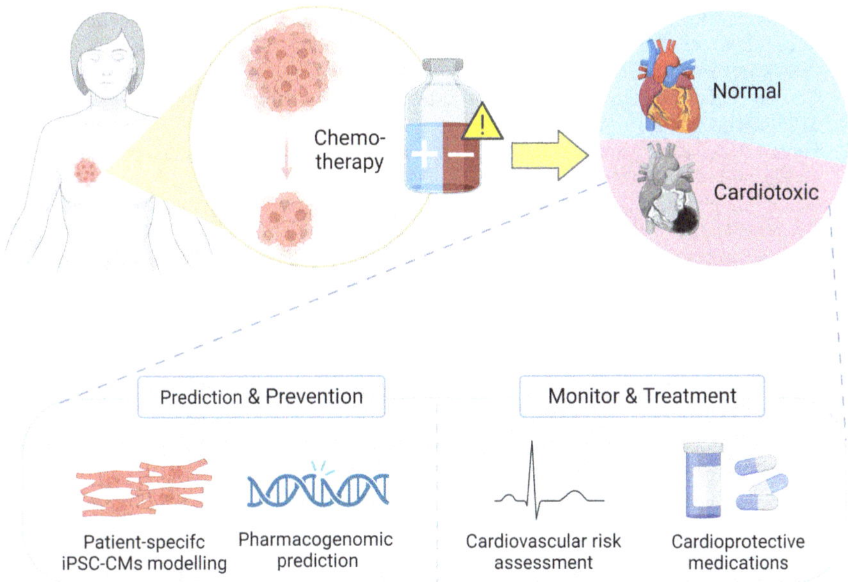


**Fig. 15.1** Advanced research of chemotherapy-induced cardiotoxicity in breast cancer patients

and treating breast cancer patients at high risk of developing cardiotoxicity. Figure was created with BioRender.

## References

1. Arnold M et al (2022) Current and future burden of breast cancer: global statistics for 2020 and 2040. Breast 66:15–23
2. Siegel RL et al (2022) Cancer statistics, 2022. CA Cancer J Clin 72(1):7–33
3. Ewer MS, Ewer SM (2015) Cardiotoxicity of anticancer treatments. Nat Rev Cardiol 12(9):547–558
4. Goodman LS et al (1946) Nitrogen mustard therapy—use of methyl-bis(beta-chloroethyl)amine hydrochloride and tris(beta-chloroethyl)amine hydrochloride for hodgkins disease, lymphosarcoma, leukemia and certain allied and miscellaneous disorders. JAMA-J Am Med Assoc 132(3):126–132
5. Kohn KW, Hartley JA, Mattes WB (1987) Mechanisms of DNA sequence selective alkylation of guanine-N7 positions by nitrogen mustards. Nucleic Acids Res 15(24):10531–10549
6. Arcamone F et al (1969) Adriamycin, 14-hydroxydaunomycin, a new antitumor antibiotic from S. peucetius var. caesius. Biotechnol Bioeng 11(6):1101–1110
7. Blum RH, Carter SK (1974) Adriamycin. A new anticancer drug with significant clinical activity. Ann Intern Med 80(2):249–259
8. Weiss RB (1992) The anthracyclines: will we ever find a better doxorubicin? Semin Oncol 19(6):670–686
9. Minotti G et al (2004) Anthracyclines: molecular advances and pharmacologic developments in antitumor activity and cardiotoxicity. Pharmacol Rev 56(2):185–229
10. Zamorano JL et al (2016) 2016 ESC Position Paper on cancer treatments and cardiovascular toxicity developed under the auspices of the ESC Committee for practice guidelines: the task force for cancer treatments and cardiovascular toxicity of the European Society of Cardiology (ESC). Eur Heart J 37(36):2768–2801
11. Chu TF et al (2007) Cardiotoxicity associated with tyrosine kinase inhibitor sunitinib. Lancet 370(9604):2011–2019
12. Force T, Krause DS, Van Etten RA (2007) Molecular mechanisms of cardiotoxicity of tyrosine kinase inhibition. Nat Rev Cancer 7(5):332–344
13. Hansel TT et al (2010) The safety and side effects of monoclonal antibodies. Nat Rev Drug Discov 9(4):325–338
14. Chen KS, Gresh N, Pullman B (1986) A theoretical investigation on the sequence selective binding of adriamycin to double-stranded polynucleotides. Nucleic Acids Res 14(5):2251–2267
15. Hortobagyi GN (1997) Anthracyclines in the treatment of cancer. An overview. Drugs 54(Suppl 4):1–7
16. Yang F et al (2014) Doxorubicin, DNA torsion, and chromatin dynamics. Biochim Biophys Acta 1845(1):84–89
17. Pommier Y et al (2010) DNA topoisomerases and their poisoning by anticancer and antibacterial drugs. Chem Biol 17(5):421–433
18. Taagepera S et al (1993) DNA topoisomerase II alpha is the major chromosome protein recognized by the mitotic phosphoprotein antibody MPM-2. Proc Natl Acad Sci USA 90(18):8407–8411
19. Capranico G et al (1992) Different patterns of gene expression of topoisomerase II isoforms in differentiated tissues during murine development. Biochim Biophys Acta 1132(1):43–48
20. Zhang S et al (2012) Identification of the molecular basis of doxorubicin-induced cardiotoxicity. Nat Med 18(11):1639–1642

21. Ashley N, Poulton J (2009) Mitochondrial DNA is a direct target of anti-cancer anthracycline drugs. Biochem Biophys Res Commun 378(3):450–455
22. Cullinane C et al (2000) Interstrand cross-linking by adriamycin in nuclear and mitochondrial DNA of MCF-7 cells. Nucleic Acids Res 28(4):1019–1025
23. Lebrecht D et al (2005) Tissue-specific mtDNA lesions and radical-associated mitochondrial dysfunction in human hearts exposed to doxorubicin. J Pathol 207(4):436–444
24. Songbo M et al (2019) Oxidative stress injury in doxorubicin-induced cardiotoxicity. Toxicol Lett 307:41–48
25. Zhou B, Tian R (2018) Mitochondrial dysfunction in pathophysiology of heart failure. J Clin Invest 128(9):3716–3726
26. Davies KJ, Doroshow JH (1986) Redox cycling of anthracyclines by cardiac mitochondria. I. Anthracycline radical formation by NADH dehydrogenase. J Biol Chem 261(7):3060–3067
27. Thomas CE, Aust SD (1986) Release of Iron from ferritin by cardiotoxic anthracycline antibiotics. Arch Biochem Biophys 248(2):684–689
28. Hortobágyi GN (1997) Anthracyclines in the treatment of cancer. Drugs 54:1–7
29. Ichikawa Y et al (2014) Cardiotoxicity of doxorubicin is mediated through mitochondrial iron accumulation. J Clin Investig 124(2):617–630
30. Vasquez-Vivar J et al (1997) Endothelial nitric oxide synthase-dependent superoxide generation from adriamycin. Biochem 36(38):11293–11297
31. Sangomla S et al (2018) Nanoceria ameliorates doxorubicin induced cardiotoxicity: possible mitigation via reduction of oxidative stress and inflammation. J Trace Elem Med Biol 47:53–62
32. Pai VB, Nahata MC (2000) Cardiotoxicity of chemotherapeutic agents—Incidence, treatment and prevention. Drug Saf 22(4):263–302
33. Zhu WQ et al (2009) Acute doxorubicin cardiotoxicity is associated with p53-induced inhibition of the mammalian target of rapamycin pathway. Circ 119(1):99–U195
34. Aries A et al (2004) Essential role of GATA-4 in cell survival and drug-induced cardiotoxicity. Proc Natl Acad Sci USA 101(18):6975–6980
35. Shi J, Abdelwahid E, Wei L (2011) Apoptosis in anthracycline cardiomyopathy. Curr Pediatr Rev 7(4):329–336
36. Kalivendi SV et al (2005) Doxorubicin activates nuclear factor of activated T-lymphocytes and Fas ligand transcription: role of mitochondrial reactive oxygen species and calcium. Biochem J 389:527–539
37. Wang S et al (2002) Activation of nuclear factor-kappaB during doxorubicin-induced apoptosis in endothelial cells and myocytes is pro-apoptotic: the role of hydrogen peroxide. Biochem J 367(Pt 3):729–740
38. Jang YM et al (2004) Doxorubicin treatment in vivo activates caspase-12 mediated cardiac apoptosis in both male and female rats. FEBS Lett 577(3):483–490
39. Glembotski CC (2007) Endoplasmic reticulum stress in the heart. Circ Res 101(10):975–984
40. Christidi E, Brunham LR (2021) Regulated cell death pathways in doxorubicin-induced cardiotoxicity. Cell Death Dis 12(4):339
41. Xiao B, Hong L, Cai X, Mei S, Zhang P, Shao L (2019) The true colors of autophagy in doxorubicin-induced cardiotoxicity. Oncol Lett 18(3):2165–2172
42. Thorburn A (2014) Autophagy and its effects: making sense of double-edged swords. PLoS Biol 12(10):e1001967
43. Ryter SW, Mizumura K, Choi AM (2014) The impact of autophagy on cell death modalities. Int J Cell Biol 2014:502676
44. Tadokoro T et al (2020) Mitochondria-dependent ferroptosis plays a pivotal role in doxorubicin cardiotoxicity. JCI Insight 5(9)
45. Fang X et al (2019) Ferroptosis as a target for protection against cardiomyopathy. Proc Natl Acad Sci 116:2672–2680
46. Dhingra R et al (2014) Bnip3 mediates doxorubicin-induced cardiac myocyte necrosis and mortality through changes in mitochondrial signaling. Proc Natl Acad Sci USA 111(51):E5537–E5544

47. Dhingra R et al (2022) Proteasomal degradation of TRAF2 mediates mitochondrial dysfunction in doxorubicin-cardiomyopathy. Circ 146(12):934–954
48. Wexler LH et al (1996) Randomized trial of the cardioprotective agent ICRF-187 in pediatric sarcoma patients treated with doxorubicin. J Clin Oncol 14(2):362–372
49. Deng SW et al (2014) Dexrazoxane may prevent doxorubicin-induced DNA damage via depleting both Topoisomerase II isoforms. BMC Cancer 14
50. Attia SM et al (2017) Dexrazoxane averts idarubicin-evoked genomic damage by regulating gene expression profiling associated with the DNA damage-signaling pathway in BALB/c mice. Toxicol Sci 160(1):161–172
51. Yu XX et al (2020) Dexrazoxane ameliorates doxorubicin-induced cardiotoxicity by inhibiting both apoptosis and necroptosis in cardiomyocytes. Biochem Biophys Res Commun 523(1):140–146
52. Yu X, Ruan Y, Shen T, Qiu Q, Yan M, Sun S, Dou L, Huang X, Wang Q, Zhang X, Man Y, Tang W, Jin Z, Li J (2020) Dexrazoxane protects cardiomyocyte from doxorubicin-Induced apoptosis by modulating miR-17-5p. Biomed Res Int 2020:5107193
53. Liesse K et al (2018) Dexrazoxane significantly reduces anthracycline-induced cardiotoxicity in pediatric solid tumor patients: a systematic review. J Pediatr Hematol Oncol 40(6):417–425
54. Cvetkokic RS, Scott LJ (2005) Dexrazoxane—a review of its use for cardioprotection during anthracycline chemotherapy. Drugs 65(7):1005–1024
55. Macedo AVS et al (2019) Efficacy of dexrazoxane in preventing anthracycline cardiotoxicity in breast cancer. JACC Cardio Oncol 1(1):68–79
56. Zannad F et al (2020) SGLT2 inhibitors in patients with heart failure with reduced ejection fraction: a meta-analysis of the EMPEROR-Reduced and DAPA-HF trials. Lancet 396(10254):819–829
57. Cardoso R et al (2021) SGLT2 inhibitors decrease cardiovascular death and heart failure hospitalizations in patients with heart failure: a systematic review and meta-analysis. EClinicalMedicine 36:100933
58. McDonagh TA et al (2021) 2021 ESC Guidelines for the diagnosis and treatment of acute and chronic heart failure. Eur Heart J 42(36):3599–3726
59. McDonald M et al (2021) CCS/CHFS heart failure guidelines update: defining a new pharmacologic standard of care for heart failure with reduced ejection fraction. Can J Cardiol 37(4):531–546
60. Heidenreich PA et al (2022) 2022 AHA/ACC/HFSA guideline for the management of heart failure: executive summary: a report of the American College of Cardiology/American Heart Association Joint Committee on clinical practice guidelines. J Am Coll Cardiol 79(17):1757–1780
61. Rizzo MR et al (2022) Cognitive impairment and type 2 diabetes mellitus: focus of SGLT2 inhibitors treatment. Pharmacol Res 176:106062
62. Hsia DS, Grove O, Cefalu WT (2017) An update on sodium-glucose co-transporter-2 inhibitors for the treatment of diabetes mellitus. Curr Opin Endocrinol Diabetes Obes 24(1):73–79
63. Baris VO et al (2021) Empagliflozin significantly prevents the doxorubicin-induced acute cardiotoxicity via non-antioxidant pathways. Cardiovasc Toxicol 21(9):747–758
64. Sabatino J et al (2020) Empagliflozin prevents doxorubicin-induced myocardial dysfunction. Cardiovasc Diabetol 19(1):66
65. Quagliariello V et al (2021) The SGLT-2 inhibitor empagliflozin improves myocardial strain, reduces cardiac fibrosis and pro-inflammatory cytokines in non-diabetic mice treated with doxorubicin. Cardiovasc Diabetol 20(1):150
66. Yang CC et al (2019) Early administration of empagliflozin preserved heart function in cardiorenal syndrome in rat. Biomed Pharmacother 109:658–670
67. Oh CM et al (2019) Cardioprotective potential of an SGLT2 inhibitor against doxorubicin-induced heart failure. Korean Circ J 49(12):1183–1195
68. Chang WT et al (2022) Dapagliflozin protects against doxorubicin-induced cardiotoxicity by restoring STAT3. Arch Toxicol 96(7):2021–2032

69. Wang CY et al (2020) TLR9 binding to Beclin 1 and mitochondrial SIRT3 by a sodium-glucose co-transporter 2 inhibitor protects the heart from doxorubicin toxicity. Biology (Basel) 9(11)
70. Gongora CA et al (2022) Sodium-glucose co-transporter-2 inhibitors and cardiac outcomes among patients treated with anthracyclines. JACC Heart Fail 10(8):559–567
71. Lundy DJ, Lee DS, Hsieh PCH (2017) Solving the puzzle of pluripotent stem cell-derived cardiomyocyte maturation: piece by piece. Ann Transl Med 5(6)
72. Yang X, Pabon L, Murry CE (2014) Engineering adolescence: maturation of human pluripotent stem cell-derived cardiomyocytes. Circ Res 114(3):511–523
73. Peter AK, Bjerke MA, Leinwand LA (2016) Biology of the cardiac myocyte in heart disease. Mol Biol Cell 27(14):2149–2160
74. van der Worp HB et al (2010) Can animal models of disease reliably inform human studies? Plos Med 7(3)
75. Takahashi K et al (2007) Induction of pluripotent stem cells from adult human fibroblasts by defined factors. Cell 131(5):861–872
76. Lian XJ et al (2013) Directed cardiomyocyte differentiation from human pluripotent stem cells by modulating Wnt/beta-catenin signaling under fully defined conditions. Nat Protoc 8(1):162–175
77. Lipshultz SE et al (2005) Chronic progressive cardiac dysfunction years after doxorubicin therapy for childhood acute lymphoblastic leukemia. J Clin Oncol 23(12):2629–2636
78. Tang WKB-K, Endrenyi L (1998) Hypothesis: comparisons of inter- and intra-individual variations can substitute for twin studies in drug research. Pharmacogenet Genomics 8(4):283–289
79. Reisberg S et al (2019) Translating genotype data of 44,000 biobank participants into clinical pharmacogenetic recommendations: challenges and solutions. Genet Med 21(6):1345–1354
80. Evans WE, McLeod HL (2003) Pharmacogenomics—drug disposition, drug targets, and side effects. N Engl J Med 348(6):538–549
81. Epstein RS et al (2010) Warfarin genotyping reduces hospitalization rates results from the MM-WES (Medco-Mayo Warfarin effectiveness study). J Am Coll Cardiol 55(25):2804–2812
82. U.S. Food and Drug Administration (2022) Table of pharmacogenomic biomarkers in drug labeling. Available from: https://www.fda.gov/drugs/science-and-research-drugs/table-pharmacogenomic-biomarkers-drug-labeling
83. Linschoten M, Teske AJ, Cramer MJ, van der Wall E, Asselbergs FW (2018) Chemotherapy-related cardiac dysfunction: a systematic review of genetic variants modulating individual risk. Circ Genom Precis Med 11(1):e001753
84. Christidi E, Huang H, Shafaattalab S, Maillet A, Lin E, Huang K, Laksman Z, Davis MK, Tibbits GF, Brunham LR (2020) Variation in RARG increases susceptibility to doxorubicin-induced cardiotoxicity in patient specific induced pluripotent stem cell-derived cardiomyocytes. Sci Rep 10(1):10363
85. Huang H et al (2022) RARG S427L attenuates the DNA repair response to doxorubicin in induced pluripotent stem cell-derived cardiomyocytes. Stem Cell Reports 17(4):756–765
86. Magdy T, Jiang Z, Jouni M, Fonoudi H, Lyra-Leite D, Jung G, Romero-Tejeda M, Kuo HH, Fetterman KA, Gharib M, Burmeister BT, Zhao M, Sapkota Y, Ross CJ, Carleton BC, Bernstein D, Burridge PW (2021). RARG variant predictive of doxorubicin-induced cardiotoxicity identifies a cardioprotective therapy. Cell Stem Cell 28(12):2076–2089.e7
87. Visscher H et al (2012) Pharmacogenomic prediction of anthracycline-induced cardiotoxicity in children. J Clin Oncol 30(13):1422–1428
88. Visscher H et al (2013) Validation of variants in SLC28A3 and UGT1A6 as genetic markers predictive of anthracycline-induced cardiotoxicity in children. Pediatr Blood Cancer 60(8):1375–1381
89. Magdy T et al (2022) Identification of drug transporter genomic variants and inhibitors that protect against doxorubicin-induced cardiotoxicity. Circ 145(4):279–294

# Chapter 16
# Sex-Dependent Differences in the Diagnosis, Treatment and Causes of Heart Failure

**Jessica A. M. McBride and Jeffrey T. Wigle**

**Abstract** The impact of heart failure on women's health is only now being fully realized. There has previously been a mistaken belief that heart failure was predominantly a man's condition. However, research has demonstrated that the causes and treatments of heart failure differ between males and females. Females are exposed to unique stresses on their cardiovascular system (hypertensive disorders of pregnancy, gestational diabetes and chemotherapy for breast cancer) that can ultimately lead to increased risk of heart failure. As well, risk factors such as diabetes can have a larger impact in females as compared to males. Effective treatment for female heart failure patients has been hindered by challenges in accurately diagnosing the disease, the lack of relevant preclinical models, a historic underrepresentation of females in clinical trials and a lack of treatments for heart failure with preserved ejection fraction, the most common form of heart failure in females. Biological differences between males and females can account for some of the differences observed in heart failure outcomes but sex dependent impacts of social determinants of health can also account for some of the adverse outcomes for females with heart failure. In this review we will briefly describe the two major subtypes of heart failure and the use of new biomarkers to more effectively diagnose heart failure. We also discuss four specific risk factors for heart failure that predominantly affect heart failure incidence in females, the impacts of social determinants of health on heart failure risk, and how women may be more adversely impacted by them. Finally, we will propose approaches to narrow the gap in heart failure treatment for females in the future.

**Keywords** Heart failure · HFpEF · HFrEF · HFmrEF · Ejection fraction

---

J. A. M. McBride
Institute of Cardiovascular Sciences, St. Boniface Hospital Albrechtsen Research Centre, Winnipeg, MB, Canada
e-mail: mcbridej@myumanitoba.ca

J. T. Wigle (✉)
Department of Biochemistry and Medical Genetics, Max Rady College of Medicine, Rady Faculty of Health Sciences, University of Manitoba, Winnipeg, MB, Canada
e-mail: JWigle@sbrc.ca

© The Author(s), under exclusive license to Springer Nature Switzerland AG 2023 
L. Kirshenbaum and I. Rabinovich-Nikitin (eds.), *Biology of Women's Heart Health*, Advances in Biochemistry in Health and Disease 26,
https://doi.org/10.1007/978-3-031-39928-2_16

## Introduction

Heart failure (HF) is the end-stage manifestation of various cardiovascular diseases that occur when the heart is unable to meet the metabolic demands of peripheral tissues [23, 73]. HF affects a variety of organs in the body and is typically accompanied by multiple co-morbidities thereby complicating its' prevention and treatment. In 2020, approximately 64 million people worldwide were living with HF with an incidence in high-income countries of around 1–9 per 1000 people [25]. Currently, HF incidence is higher in males but HF prevalence is equal between the sexes, as females have lower mortality from the disease [40]. In Canada, HF prevalence is between 1.2 and 3.6% of the population, although this is likely an underestimate of the true prevalence because of difficulties in definitively diagnosing HF [25, 57]. The lack of data from patient populations outside of Europe and North America means that the worldwide burden of HF is also unclear [25].

HF pathology is complex because there are multiple etiologies that can lead to HF and only a limited understanding of how these risk factors work together to produce the disease [49]. Chronic inflammation is a common feature between many risk factors and is thought to play a large role in HF pathology, contributing to pulmonary hypertension, extracellular matrix deposition, and eventually cardiac remodeling and impairment of heart function [49]. An elevated level of neurohormonal activation creates a positive feedback loop of continued remodeling and hemodynamic alterations [32]. Symptoms of HF worsen with the stage of disease, from mild exercise intolerance, fatigue, and weakness to zero exercise tolerance and trouble breathing while seated [23, 73]. There are many co-morbidities that put individuals at risk for heart failure—ischemic heart disease, coronary artery disease (CAD), and atrial fibrillation are major factors for males whereas hypertension, diabetes mellitus, and obesity are major co-morbidities for females [8, 62, 73].

Functional and symptomatic classifications of HF describe the severity of HF, like the Framingham or New York Heart Association criteria. Classifying individuals by ventricular ejection fraction (LVEF) status is commonly used because individuals with reduced ejection fraction (HFrEF) (EF $<$ 40%) typically have different disease etiologies and treatment plans than individuals who have HF with preserved ejection fraction (HFpEF) (EF $>$ 50%). Those with an EF between 40 and 50% have recently been classified as having HF with mid-range EF (HFmrEF), although this classification has little clinical significance because there are currently no prevention or treatment guidelines specific to individuals with HFmrEF [57].

This chapter will focus on HF in the female sex. The history of HF diagnosis and treatments, the gaps in research on HF in females, and female-specific considerations for HF will be discussed. Suggestions will also be made on future work to improve the outcomes for females with HF. In this chapter, use of the words 'female' or 'male' indicates a person's biological sex, while 'woman' or 'man' will indicate their gender. The complex interplay of sex and gender and how it affects HF will be discussed briefly.

## History

In the late twentieth century HF was declared an epidemic due to its' large burden (with respect to both patient quality of life and financial costs) on Western populations and the lack of effective interventions to decrease morbidity and mortality [57]. Incidence of HF steadily decreased until two decades ago when mortality and incidence rates both plateaued [77]. The nature of HF has shifted as well—HFrEF was the predominant form of HF before declining to less than half of diagnosed cases in the twenty-first century [8]. This change in type of HF is attributed to: (1) the discovery of multiple pharmaceuticals and other interventions to treat HFrEF, (2) better prevention and treatment for MI and other ischemic heart disease that leads to HFrEF, and (3) the ability to accurately diagnose HF [25]. Two additional reasons that HFpEF prevalence has increased are the improved diagnosis of HFpEF and recognition that HF affects males and females equally [64].

The wide-spread belief of HF as a 'men's disease' has greatly affected all aspects of the HF field. Pre-clinical and clinical studies researching diagnosis, prognosis, or treatments for HF were conducted primarily in male animal or cell models, or male patients [7, 12]. In addition, the diversity of male samples was low regarding patient socioeconomic status, race, and ethnicity [80]. This has resulted in clinical guidelines being implemented for populations that were not represented in the clinical trials. Western research is still lagging in including more diverse groups of people in clinical studies [45]. This is an important gap in the HF field since females and males have very different disease profiles. A sex-specific approach to HF is required and will lead to better treatment of cardiovascular disease.

## HFpEF

An in-depth examination of HFpEF is necessary when examining the effects HF has on females since females tend to be diagnosed with HFpEF rather than HFrEF [57, 77]. While HFpEF and HFrEF are two classifications of HF, they are distinct diseases. HFpEF typically presents in individuals with chronic disorders like hypertension or diabetes mellitus and is associated with age-related comorbidities like arterial stiffness [5, 19, 71]. These diverse etiologies are thought to produce multiple phenotypes of HFpEF that often include interstitial fibrosis and concentric remodeling of the heart [5, 19, 71]. In contrast, HFrEF commonly follows ischemic or acute damage such as that from myocardial infarction or valve disease which lead to cardiomyocyte death, replacement fibrosis, and eccentric remodeling of the heart [5, 19, 71]. HFpEF is harder to diagnose: approximately 76% of undiagnosed HF cases are later classified as HFpEF [25] and there is no unifying diagnostic criteria for HFpEF [50]. There also remain no effective treatments for HFpEF while various pharmaceuticals and interventions are known to be effective in lowering HFrEF morbidity, as described below [37]. However, HFrEF incidence continues to fall while that of HFpEF rises

[8], providing a grim picture for treating HF in the future without effective treatments for patients.

The higher percentage of females with HFpEF, and a lower percentage with HFrEF, has led to the assumption that females are biologically predisposed to HFpEF and protected from HFrEF [5, 16]. Fundamental differences between females and males have been used to explain this notion: for example females have higher estrogen levels, higher amounts of inflammation, different associated comorbidities (diabetes, preeclampsia, hypertension), smaller average LV size, and differential gene expression [5]. Fundamental differences do affect cardiac metabolism, myocardial structure, and vascular dysfunction, and may predispose females to HFpEF. However, recent studies are challenging the notion that HFpEF is more prevalent in females. Pooled longitudinal cohort studies, preclinical studies, and clinical trials have found HFpEF to have equal incidence in females and males [37, 65, 76]. It is likely that the increased prevalence of HFpEF in females is a result of the underlying risk factors themselves having sex differences in their distribution [50]. This debate is ongoing, but no matter the outcome, understanding the fundamental differences between females and males will improve the treatment of females with HF in a sex-specific manner.

## Current Treatments

There are currently no effective treatments for HF, and barring a heart transplant, the outcome of HF is ultimately death. Pharmacological therapies and interventions can manage HF symptoms, improve quality of life, and prolong lifespan, but the availability of treatments depends on the etiology of HF. For patients with HFrEF, therapeutic regimes are well established and include Angiotensin-Converting Enzyme (ACE) inhibitors, angiotensin receptor blockers, aldosterone antagonists, beta-blockers, diuretics, and hyper polarization blockers [45]. Surgical interventions like coronary revascularization or valve replacement are offered to those HF patients with CAD or valve disease, respectively [37].

Therapy guidelines for patients with HFpEF are much less established. ACE inhibitors and aldosterone antagonists are commonly prescribed although there is little evidence that these inhibitors improve outcomes of HFpEF patients [37]. Spironolactone has recently been added to clinical guidelines for managing HF and modestly improves HFpEF symptoms and outcomes, with the best results seen in females [37, 74]. Clinical trials have provided evidence that sodium glucose co-transportor-2 (SGLT2) inhibitors decrease major adverse cardiac events in both females and males [54]. Ongoing studies suggest that these molecules may become another option for HFpEF patients in the future [69].

The dearth of treatment guidelines for HFpEF patients is due to many factors, including a lack of knowledge on the cellular and molecular pathobiology of HFpEF [46], and a historic lack of an animal model for HFpEF [46]. Recently, the laboratory of Dr. Hill has generated a two-hit model of HFpEF, where mice are fed a high-fat diet and then treated with N-nitro-1-arginine methyl ester (L-NAME), that better models human HFpEF [61]. Additionally, poor historical clinical trial design (inadequate treatments, poorly selected treatment target, and/or improper patient selection) have also hampered the development of treatment guidelines [74]. Targets for future investigation are loop or thiazide diuretics, SGLT-2 inhibitors, anti-inflammatory drugs, inhibitors of cardiac fibrosis, and drugs targeting cardiometabolic abnormalities [37, 68].

Investments have been made into investigating sex-specific HF treatments as females tend to respond to some therapies differently than do males. Exercise and dietary modifications have benefitted diastolic function, EF, and blood volume in HF trials where females were well represented [74]. Much more work is needed for these treatments to be accepted and routinely used in clinical practice, and more research into other sex-specific therapies will benefit females with HF.

## Female-Specific Considerations for HF (Fig. 16.1)

(A) Breast Cancer

Breast cancer (BC) is the most common cancer diagnosed in women although cases do occur rarely in males [21]. It is treated with a combination of cytoreductive surgery, targeted endocrine or molecular therapy, radiation treatment, and chemotherapy [21]. Cardiotoxicity is a major concern in BC patients with up to ~ 1/3 of BC patients developing treatment-related adverse cardiovascular events [58]. Chemotherapeutics like anthracyclines are associated specifically with HFrEF development while radiation is associated with development of HFpEF [56].

Much research has been put into preventing and alleviating chemotherapy-induced cardiotoxicity. Physical activity in any form is recommended for BC patients, as it has been shown to preserve long-term cardiac function in BC survivors [48]. Co-therapy with angiotensin receptor blockers or ACE inhibitors is also recommended for patients receiving either anthracycline or trastuzumab chemotherapy [39] to preserve LVEF. A modified, conserved non-coding RNA (Circ-INSR) shows promise for protecting against doxorubicin-induced cardiac dysfunction in preclinical studies [41]. Anthracycline-based BC treatment is common, so methods focusing on protecting against anthracycline-induced cardiotoxicity are important to improve treatment of BC patients.

While the emphasis to treat chemotherapy-induced cardiotoxicity has benefitted those who develop HFrEF, HFpEF is more common in breast cancer survivors [56]. BC patients face similar comorbidities associated with HFpEF development, like obesity, hypertension, a sedentary lifestyle, and older age [27]. Older females are especially vulnerable to HFpEF after receiving radiotherapy with or without previous cardiovascular disease [60]. Unfortunately, there is a large gap in research on BC patients who develop HFpEF. A challenge in researching this patient population is the patients themselves, as following older populations with higher mortality rates for long time periods is quite challenging. New, large cohort studies specifically on

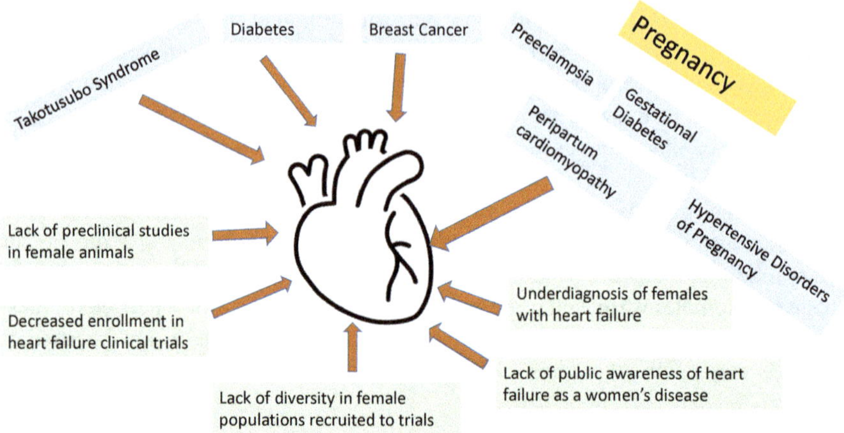


**Fig. 16.1** Female-specific risk factors for heart failure. Risk factors for heart failure that have a greater effect in female populations are shown in blue. The pregnancy associated risk factors and breast cancer are uniquely female specific, whereas diabetes and Takotusubo syndrome are more common in females with heart failure. Systemic barriers to improving treatment for women with heart failure are shown in green. The lack of preclinical models (animals or cells), especially for heart failure with preserved ejection fraction has hampered translation of results into the clinic. Clinical trials for heart failure have historically had poor recruitment of female subjects. There is a need to continue to educate clinicians and the public that heart failure is not only a male disease but is found in females as well

females with BC could ameliorate the lack of understanding the HF field has on BC patients with HFpEF.

The detection of cardiotoxicity relies mostly on measurement of LVEF function by echocardiography, but cardiac MRI can also be used [4]. These methods are effective but are not able to detect early, reversible stages of cardiotoxicity with high sensitivity [78]. The measurement of blood procollagen-derived type I C-terminal propeptide (PICP) levels is being investigated as an early minimally invasive marker of cardiotoxicity. Increases in PICP levels 3 months after beginning anthracycline-based chemotherapy predicted cardiotoxicity in BC patients 1 year after the patients completed therapy [14]. This early detection allows patients and health care teams to take interventions to prevent cardiotoxicity and related cardiovascular events, improving quality of life in women and decreasing costs to the healthcare system.

(B) Pregnancy

Pregnancy can lead to many health complications in women such as preeclampsia, gestational diabetes, and hypertensive disorders of pregnancy (HDP). Cardiovascular diseases were recently reported to complicate 10% of pregnancies and cause 15% of maternal deaths globally [55] When women with cardiovascular disease become pregnant their chance of having long-term adverse cardiovascular events (including heart failure), increased onset of diabetes or new hypertensive disease increases [66]. Tools that evaluate pregnancy risk for these patients were also able to identify those patients at the highest risk for long-term adverse outcomes. Early life-style intervention, monitoring and treatments may reduce the burden of cardiovascular disease for these women. In the Women Health Initiative study, disorders of pregnancy were associated with increased prevalence of diabetes, hypertension and coronary artery disease [26]. However, only hypertensive disorders of pregnancy were significantly associated with heart failure incidence with a 1.8 fold increase in risk [26]. Furthermore, pre-eclampsia for even one child was associated in another study with a doubling of the incidence of heart failure in the affected women versus women who did not have preeclampsia during pregnancy [29]. The cardiovascular strains that result from pregnancy result in long-term complications for women especially those with pre-existing cardiovascular disease [66]. Peripartum cardiomyopathy is a form of heart failure with reduced ejection fraction that develops late in pregnancy or within 5 months post-delivery [28, 30]. This disease occurs without an obvious cause and is more prevalent in African women [28, 30] Around 80% of women with peripartum cardiomyopathy recover function but the long-term outcomes for these patients is not clear currently [28, 30]. Pregnancy is a uniquely female risk factor for the development of heart failure and the ability to identify at risk women in the future should improve long-term outcomes.

(C) Diabetes

45% of individuals with HFpEF also have diabetes mellitus [49]. Diabetes mellitus combined with HF produces worse mortality and morbidity than either disease alone [49]. Diabetes mellitus is associated with increased incidence of heart failure in both men and women but the effect is 2.5 times more pronounced in females [31]. Non-diabetic women with HF have an increased survival rate as compared to non-diabetic men with heart failure but this survival advantage is not seen in diabetic populations [43]. A study of Asian HFrEF patients showed that diabetes is prevalent in this population and female HF patients had increased rates of diabetes even at lower body mass index [11]. Outcomes for diabetic females were shown to be worse than outcomes for diabetic men in this study, although diabetes was associated with poorer outcome for both sexes as compared to non-diabetic patients [11]. The effect of diabetes on heart failure could result more from ischemic heart disease in men as opposed to an increase in HFpEF in women [36]. In a Canadian cohort study, women who have gestational diabetes during pregnancy have been shown to have an increased rate of HF as compared to women who did not have gestational diabetes during pregnancy [20]. Maternal diabetes was shown to be associated with an earlier onset of cardiovascular disease in their children, including a 1.45 fold increase rate of heart failure [82].

(D) Takotsubo Syndrome

Takotsubo syndrome (TTS), or acute stress-induced cardiomyopathy, is a form of acute heart failure recognized ~ 25 years ago that is characterized by a temporary and reversible ballooning of the apical (75%) or mid-apical (25%) area of the LV in absence of pre-existing CAD, myocardial infarction, or myocarditis [1, 13]. It occurs in 1–2.5% of the population and has a clear sex-bias with 90% of cases occurring in females and is frequently underdiagnosed because its presentation is similar to that of myocardial infarction [1, 13]. TTS may be triggered by emotional or physical stress, or arise from no known triggers, that result in a large catecholamine release. CNS activation may cause calcium-overload cardiotoxicity and microvascular dysfunction seen in TTS [1] or it may be associated with the stress trigger [13]. A recent debate has emerged over whether TTS is caused by catecholamine release or if this release is an epiphenomenon of TTS pathology [3].

Studying Takotsubo syndrome has many challenges, one being a lack of an animal model for the syndrome as the cause is still not understood and another being the difficulty diagnosing these patients [13]. Despite these challenges, the prognosis of TTS is favourable: TTS can be easily reversed, and ~ 80% of patients fully recover within a few months [22]. There is a lack of clinically proven, effective treatments for TTS [22, 42], but patients receive beta-blockers, ACE inhibitors, angiotensin receptor blockers, diuretics to speed the recovery of the LV [1]. TTS is a significant HF etiology that females are more susceptible to whose pathology and treatment is poorly understood.

## Diagnosing HF and Biomarkers

Diagnosing HF with high sensitivity and specificity has remained a challenge, even with tools like echocardiography and radionuclide angiography [23]. The current clinical gold standard of HF diagnosis is right heart catheterization, an invasive method to measure cardiac output and chamber pressures. Despite these methods, there are still individuals who do not present with detectable symptoms of HF but nonetheless experience deteriorating cardiovascular function. In addition, the diagnostic cutoffs for these methods are based on data gathered from males despite females' thinner average left ventricular wall thickness and lower ejection fraction, two measures typically used to diagnose HF [6, 64]. Less invasive, more sensitive, and sex-specific methods are required to diagnose HF more accurately.

Biomarkers are a diagnostic method that are much less invasive than right heart catheterization and are being increasingly used clinically. Biomarkers are molecules in the blood including proteins, hormones, and RNA transcripts that are produced throughout the body from normal or disease-state physiological processes [71]. Levels of biomarkers are measured from blood samples taken from patients, and if levels meet or exceed a set threshold, a patient is recommended for more testing from the high likelihood of heart failure present. The amino terminal of the precursor of natriuretic peptide B (NT-proBNP) and cardiac troponins (cTns) are commonly used in clinical diagnosis of HF as they indicate stretching of the myocardium from volume overload [71, 72] and non-specific myocardial damage [9, 72] that occur during HF. More molecules are being investigated to improve the diagnostic value of biomarkers like soluble interleukin-1 receptor-like 1 (sST2) [75]. sST2 binds IL33, preventing the cardioprotective effects that IL33 induces, and indicates that myocardial damage is occurring [9]. Inflammatory biomarkers like C-reactive protein (CRP) and IL6 may be useful in supporting the diagnosis of HF as they indicate general inflammation and decreased myocardial function respectively [65, 67].

Currently, a universal cutoff is used in most clinical settings when diagnosing HF using biomarkers. Sex-specific cutoff levels are being advocated because multiple biomarkers have been identified to have different baseline levels in females as compared to males: levels of natriuretic peptides are higher in females while levels of cTns, CRP, and IL6 are lower in females [67, 72]. Females may be missed when screening for HF using universal biomarker cutoff levels, and since biomarker levels are typically the first diagnostic measure taken of individuals suspected of having HF, this has large implications for females with HFpEF who will not present with reduced EF [9, 67].

Researchers have been studying if biomarkers can be used to differentiate HFpEF and HFrEF. Toma et al. [75] found that HFpEF and HFrEF have different transcriptomic signatures which also differ between sexes, although they do require validation with larger sample sizes [75]. Duprez et al. [19] found that the biomarker procollagen type III N-terminal propeptide (PIIINP), a marker of collagen synthesis in the heart, was able to differentiate patients with HFpEF from healthy patients. Moving this research to the clinic has the potential to improve the diagnosis of HFpEF, especially in females with HFpEF who are often missed in earlier HF screening.

There are limits to the diagnostic use of biomarkers. For one, biomarker levels can be greatly affected by different physiological states like obesity and therefore biomarker results should account for conditions such as obesity [71]. In addition, there are reports that while biomarker levels differ between sexes at baseline, they do not differ between sexes after incident HF [9, 53]. Future work will determine the ability to use biomarkers as a sex-specific indicator of HF, although the prognostic ability of biomarkers, including NT-proBNP, cTns, and sST2, remains relevant in both sexes [9].

## Equity, Diversity and Inclusion (EDI)

Western health care systems were designed under a colonial mindset and inherently were built to disadvantage some groups of people more than others. Structural racism leads to poor treatment of people of colour and increases their overall morbidity and mortality, including in the HF field. Sexism is also present in the healthcare system and historically studies were conducted on males and through a patriarchal lens. Non-heterosexual, non-cisgendered, and disabled people also endure discrimination and face poorer health outcomes within colonial health care systems than their heterosexual, cisgendered, or able-bodied counterparts. Discrimination leads to real consequences in health outcomes, which will be outlined below as it relates to HF.

As of 2020, females were still underrepresented in clinical studies on HF [64] despite the continued recommendations by scientists in the field to recruit more females to studies. As HF occurs equally between the sexes, the reason females are underrepresented is not a lack of females with HF. Saeed et al. [ 59] propose that (1) reduced referrals to cardiovascular specialists due to atypical HF presentation and underestimation of cardiovascular disease risk in females, (2) an age-bias preventing female recruitment to studies, as females develop HF later in life than males, (3) fear and mistrust in the research system decreasing the likelihood of females to join clinical studies, (4) study interference with work or family responsibilities, and (5) financial costs all can contribute to the underrepresentation of females in HF clinical studies [59]. Physicians have been found to be less capable of identifying female patients with cardiovascular disease [6, 59], which would decrease the number of referrals of females to specialists. Evidence-based drugs are also used less often to treat females with HF than males with HF [6, 38], and especially Black or Hispanic females [44], which may cause distrust in females towards the health care and research systems. Promoting research into females with HF, changing the education healthcare workers receive on HF in females, and getting input from females with HF on how to improve study accessibility are three approaches that may increase representation of females in clinical studies on HF.

Social determinants of health (SDOH) must be considered when seeking to reduce HF disease risk. Income is known to be the SDOH that impacts health the greatest and is a risk factor comparable to age, myocardial infarction, or diabetes in developing HF [79, 80]. Since women are overrepresented in those living in poverty, the effects of income as an SDOH are compounded by sex and further disadvantage females [63]. Low socioeconomic status has been linked to poor cardiovascular outcomes specifically [63], highlighting the importance of considering this SDOH when attempting to reduce cardiovascular disease prevalence and incidence. Income level, employment status, education and neighbourhood socioeconomic factors have been associated with cardiovascular health in high income countries [63]. For example, a higher income level is correlated with a decreased burden of CVD disease even when confounders such as smoking levels and alcohol consumption are accounted for [35, 47]. All these factors may result in patients receiving a reduced standard of care resulting in increased mortality [17, 70].

Race and ethnicity are another health determinant that directly impact the health of individuals. The place of race within medicine has been debated, since race is a social construct with no biological basis, but we argue that race is correlated with ~ 15% of genetic variation between individuals [51] and race changes health outcomes expected by the average person, so it must be evaluated in medical research. Individuals who are Black, Indigenous, or are people of colour (BIPOC) are subjected to more frequent risk factors for HF including hypertension, diabetes, and obesity, and bear an increased burden of HF [52, 57, 79–81]. In the US specifically, Black individuals are more likely to develop HF after myocardial infarction than their White counterparts, have increased incidence of HF (especially in younger age groups [57], and have higher rates of hospitalization from HF [52, 81]. Hispanic Americans also face worse outcomes after treatment and higher morbidity than White individuals [81]. Knowledge gaps exist on how structural racism impacts other non-White populations living in North America, especially South Asian persons [57]. Despite the large burden of HF on BIPOC individuals, and the knowledge that BIPOC individuals may have worse responses to medications than White individuals, they are severely underrepresented in clinical trials [79, 80]. There needs to be proper representation in HF research to provide equitable treatment guidelines and improve the outcomes of people with HF, and a willingness for research staff and healthcare workers to engage with BIPOC communities to translate this research effectively to the clinic.

## Next Steps and Suggestions

(A) Study Design

It has been well established how the design of studies investigating HF, especially regarding patient recruitment, is inadequate when translating research to national populations in North America. Female, black, and elderly patients are not included in cardiovascular disease clinical studies to the degree they make up cardiovascular cases [2, 25, 33, 34]. Elderly patients (> 85 years) are typically not included in clinical studies and are therefore underrepresented in data [34]. It is imperative to increase the diversity in HF studies so that all persons can be treated equitably for cardiovascular diseases. Requirements by grant-giving organizations and high-impact journals to have diversity in patient recruitment would be impactful to change these statistics. Strikingly, clinical studies led by women were more likely to include BIPOC study participants and report race/ethnicity [79, 80]. National funding agencies have implemented guidelines for researchers to clearly outline how they will include both female and male animal or cell models in their research [12], which will have an impact on pre-clinical studies.

Increasing diversity in clinical studies is more difficult, as studies can depend on volunteers for trials to occur. Volunteers are more likely to be of higher socioeconomic status and have disposable time and resources to participate in the trial, have their identity represented in the researchers undertaking the study, and have had favourable experiences with the healthcare system in the past. Unfortunately, structural racism means BIPOC individuals tend to be of lower socioeconomic status, face barriers to higher education and therefore becoming a clinical researcher, and experience discrimination within the healthcare system that leads to distrust. In addition, females are more likely to have a higher burden of unpaid work, like child rearing, preparing food, and taking care of family members. Clinical trials should recognize these barriers faced by BIPOC and female individuals and make efforts to ameliorate them, like providing financial compensation, reevaluating hiring criteria, or increasing the flexibility for participants to participate in trials. Recommendations for building back trust with BIPOC individuals after centuries of colonialism and racism include: rebuilding trust between the healthcare system and research teams with BIPOC communities, implementing cultural competency training for health care workers and researchers involved in HF research, and properly communicating with communities affected by colonialism and racism [2].

(B) Treatment Considerations

Females are chronically undertreated with available therapies for HFpEF and HFrEF [24, 81]. As of 2019, females were still underprescribed with diuretics, were less frequently enrolled in exercise prescriptions and disease management programs than males, and cardiac resynchronization therapy was underutilized in females [15, 64]. This is troubling as evidence has shown these therapies might benefit females more than males [15, 16] Health care providers and the public need to be continually educated regarding the presentation and prevalence of HF in females.

The goals of treatment in females and males should be reevaluated. Females experience a worse quality of life, decreased functional capacity, and worse prognosis as compared to men [15, 80], although they do have a lower mortality rate. This differs greatly from males who tend to experience higher rates of sudden death, quality of life and functional capacity. Goals for males with HF trend towards improving underlying (ischemic) causes of HF and prolonging life, but as women tend not to have ischemic disease and live longer lives with HF, we suggest attention be paid to how quality of life and functional capacity can be increased with treatment. Females are also readmitted to hospital more often and may benefit from more frequent, personalized healthcare [39]. In the future, the development of new biomarker assays that can robustly detect HF in females would help with both earlier treatment of the disease itself and allow for ongoing monitoring of response to treatment.

(C) Changing Public Perception

The traditional view of cardiovascular disease as a 'men's disease' led to the widespread belief in Western countries that females are protected from HF, possibly because females more often experience HFpEF instead of HFrEF and went under-diagnosed for decades. Many studies since have proven this view to be incorrect [64], but public perception still views females to be at low risk for cardiovascular disease. This means that females seek treatment later than males would for similar severity of symptoms, are less capable of identifying cardiovascular disease in females [59], and adhere less strictly to modifying cardiovascular disease risk factors than males.

Females experience different modifiable risk factors to cardiovascular disease than males do [10], so efforts to reduce cardiovascular disease should be sex-specific. Obesity is a risk factor that should be targeted in females, as more than half of females with HFpEF are also obese [50]. Heart failure can also result from altered mental health states as exampled by TTS. Addressing mental health and mental health-affecting SDOHs with a systems-wide approach will have far reaching effects on HF, risk factors for HF, and other cardiovascular diseases. It will also increase the quality of life in females with HFpEF, as depression is common in these patients [50].

## Conclusion

Heart failure is a complex disease and it has been historically underdiagnosed in females (Fig. 16.1). This underdiagnosis results from an interplay of different factors such as preclinical models being predominantly male, presentation of females with heart failure is different than that for males, unique risk factors that trigger heart failure in females, and under recruitment of females for heart failure clinical trials. Future studies that address these issues will aid in the identification of approaches to both better diagnose and treat heart failure in females.

## References

1. Amin HZ, Amin LZ, Pradipta A (2020) Takotsubo cardiomyopathy: a brief review. J Med Life 13(1):3–7. https://doi.org/10.25122/jml-2018-0067
2. Anaba U, Ishola A, Alabre A, Bui A, Prince M, Okafor H, Kola-Kehinde O, Joseph JJ, Mitchell D, Odei BC, Uzendu A, Williams KP, Capers Q, Addison D (2022) Diversity in modern heart failure trials: where are we, and where are we going. Int J Cardiol 348:95–101. https://doi.org/10.1016/j.ijcard.2021.12.018
3. Angelini P, Uribe C, Tobis JM (2021) Pathophysiology of takotsubo cardiomyopathy: reopened debate. Texas Heart Inst J 48(3):1–7. https://doi.org/10.14503/THIJ-20-7490
4. Awadallaa M, Hassana MZO, Alvia RM, Neilan TG (2018) Advanced imaging modalities to detect cardiotoxicity Magid. Curr Probl Cancer 42(4):386–396. https://doi.org/10.1016/j.currproblcancer.2018.05.005.Advanced
5. Beale AL, Meyer PMD, Marwick TH, Lam CSP, Kaye DM (2018) Sex differences in cardiovascular pathophysiology why women are overrepresented in heart failure with preserved ejection fraction. Circ 138(2):198–205. https://doi.org/10.1161/CIRCULATIONAHA.118.034271
6. Bhatia RT, Papadakis M (2022) Female sex and persistent inequalities in the care of patients with hypertrophic obstructive cardiomyopathy: a call to action. Eur J Prev Cardiol 29(11):1542–1544
7. Blenck CL, Harvey PA, Reckelhoff JF, Leinwand LA (2016) The importance of biological sex and estrogen in rodent models of cardiovascular health and disease. Circ Res 118(8):1294–1312
8. Bui AL, Horwich TB, Fonarow GC (2011) Epidemiology and risk profile of heart failure. In Nature reviews cardiology, 8(1):30–41. https://doi.org/10.1038/nrcardio.2010.165
9. Cediel G, Codina P, Spitaleri G, Domingo M, Santiago-Vacas E, Lupón J, Bayes-Genis A (2021) Gender-related differences in heart failure biomarkers. Front Cardiovasc Med 7(January):1–10. https://doi.org/10.3389/fcvm.2020.617705
10. Cesaroni G, Mureddu GF, Agabiti N, Mayer F, Stafoggia M, Forastiere F, Latini R, Masson S, Davoli M, Boccanelli A, Boccanelli A, Cacciatore G, Mureddu GF, Rizzello V, Agabiti N, Cesaroni G, Forastiere F, Perucci CA, Davoli M et al (2021) Sex differences in factors associated with heart failure and diastolic left ventricular dysfunction: a cross-sectional population-based study. BMC Public Health 21(1):1–13. https://doi.org/10.1186/s12889-021-10442-3
11. Chandramouli C, Teng THK, Tay WT, Yap J, MacDonald MR, Tromp J, Yan L, Siswanto B, Reyes EB, Ngarmukos T, Yu CM, Hung CL, Anand I, Richards AM, Ling LH, Regensteiner JG, Lam CSP, Anand I, Hung CL et al (2019) Impact of diabetes and sex in heart failure with reduced ejection fraction patients from the ASIAN-HF registry. Eur J Heart Fail 21(3). https://doi.org/10.1002/ejhf.1358
12. Clayton JA, Collins FS (2014). NIH to balance sex in cell and animal studies. Nat 509(7500). https://doi.org/10.1038/509282a
13. Dawson DK (2018) Acute stress-induced (takotsubo) cardiomyopathy. Heart 104(2):96–102. https://doi.org/10.1136/heartjnl-2017-311579

14. de la Fuente A, Santisteban M, Lupón J, Aramendía JM, Díaz A, Santaballa A, Hernándiz A, Sepúlveda P, Cediel G, López B, Picazo JML, Mazo MM, Rábago G, Gavira JJ, García-Bolao I, Díez J, González A, Bayés-Genís A, Ravassa S (2022) A fibrosis biomarker early predicts cardiotoxicity due to anthracycline-based breast cancer chemotherapy. Cancers 14(12):2941. https://doi.org/10.3390/cancers14122941
15. Dewan P, Rørth R, Jhund PS, Shen L, Raparelli V, Petrie MC, Abraham WT, Desai AS, Dickstein K, Køber L, Mogensen UM, Packer M, Rouleau JL, Solomon SD, Swedberg K, Zile MR, McMurray JJV (2019) Differential impact of heart failure with reduced ejection fraction on men and women. J Am Coll Cardiol 73(1):29–40. https://doi.org/10.1016/j.jacc.2018.09.081
16. Dewan P, Rørth R, Raparelli V, Campbell RT, Shen L, Jhund PS, Petrie MC, Anand IS, Carson PE, Desai AS, Granger CB, Køber L, Komajda M, McKelvie RS, O'Meara E, Pfeffer MA, Pitt B, Solomon SD, Swedberg K, Zile MR, McMurray JJV (2019) Sex-related differences in heart failure with preserved ejection fraction. Circ Heart Fail 12(12):e006539
17. Dunlay SM, Pack QR, Thomas RJ, Killian JM, Roger VL (2014) Participation in cardiac rehabilitation, readmissions, and death after acute myocardial infarction. Am J Med 127(6). https://doi.org/10.1016/j.amjmed.2014.02.008
18. Dunlay SM, Roger VL, Redfield MM (2017) Epidemiology of heart failure with preserved ejection fraction. Nat Rev Cardiol 14(10):591–602. https://doi.org/10.1038/nrcardio.2017.65
19. Duprez DA, Gross MD, Kizer JR, Ix JH, Hundley WG, Jacobs DR (2018) Predictive value of collagen biomarkers for heart failure with and without preserved ejection fraction: MESA (Multi-Ethnic Study of Atherosclerosis). J Am Heart Assoc 7(5):13–16. https://doi.org/10.1161/JAHA.117.007885
20. Echouffo-Tcheugui JB, Guan J, Retnakaran R, Shah BR (2021) Gestational diabetes and incident heart failure: a cohort study. Diabetes Care 44(10). https://doi.org/10.2337/dc21-0552
21. Fahad Ullah M (2019) Breast cancer: current perspectives on the disease status. Adv Exp Med Biol 1152:51–64. https://doi.org/10.1007/978-3-030-20301-6_4
22. Farris C, McEnroe-Petittee D, Kanayama T (2014) Takotsubo cardiomyopathy: can hearts really break? Eur Cardiol Rev 32(7):410–414. https://doi.org/10.15420/ecr.2015.10.01.25
23. Gaggin HK, Januzzi JL (2013) Biomarkers and diagnostics in heart failure. Biochimica et Biophysica Acta—Mol Basis Dis 1832(12)–2442–2450. https://doi.org/10.1016/j.bbadis.2012.12.014
24. Ghali JK, Krause-Steinrauf HJ, Adams KF, Khan SS, Rosenberg YD, Yancy CW, Young JB, Goldman S, Peberdy MA, Lindenfeld J (2003) Gender differences in advanced heart failure: insights from the BEST study. J Am Coll Cardiol 42(12):2128–2134. https://doi.org/10.1016/j.jacc.2003.05.012
25. Groenewegen A, Rutten FH, Mosterd A, Hoes AW (2020) Epidemiology of heart failure. Eur J Heart Fail 22(8):1342–1356. https://doi.org/10.1002/ejhf.1858
26. Hansen AL, Søndergaard MM, Hlatky MA, Vittinghof E, Nah G, Stefanick ML, Manson JAE, Farland, Lv, Wells GL, Mongraw-Chaffin M, Gunderson EP, van Horn L, Wild RA Liu B, Shadyab AH, Allison MA, Liu S, Eaton CB, Honigberg MC, Parikh NI (2021) Adverse pregnancy outcomes and incident heart failure in the women's health initiative. JAMA Netw Open 4(12):e2138071. https://doi.org/10.1001/jamanetworkopen.2021.38071
27. Haykowsky MJ, Beaudry R, Brothers RM, Nelson MD, Sarma S, La Gerche A (2016) Pathophysiology of exercise intolerance in breast cancer survivors with preserved left ventricular ejection fraction. Clin Sci 130(24):2239–2244. https://doi.org/10.1042/CS20160479
28. Honigberg MC, Givertz MM (2019) Peripartum cardiomyopathy. BMJ 364:k5287
29. Honigberg MC, Riise HKR, Daltveit AK, Tell GS, Sulo G, Igland J, Klungsøyr K, Scott NS, Wood MJ, Natarajan P, Rich-Edwards JW (2020). Heart failure in women with hypertensive disorders of pregnancy: insights from the cardiovascular disease in Norway project. Hypertens 76(5):1506–1513
30. Iorgoveanu C, Zaghloul A, Ashwath M (2021) Peripartum cardiomyopathy: a review. Heart Fail Rev 26(6):1287–1296. https://doi.org/10.1007/s10741-020-10061-x

31. Kannel WB, McGee DL (1979) Diabetes and cardiovascular disease: the Framingham study. JAMA: J Am Med Assoc 241(19):2035. https://doi.org/10.1001/jama.1979.03290450033020
32. Kemp CD, Conte JV (2012) The pathophysiology of heart failure. Cardiovasc Pathol 21(5):365–371. https://doi.org/10.1016/j.carpath.2011.11.007
33. Khan MS, Shahid I, Siddiqi TJ, Khan SU, Warraich HJ, Greene SJ, Butler J, Michos ED (2020) Ten-year trends in enrollment of women and minorities in pivotal trials supporting recent us food and drug administration approval of novel cardiometabolic drugs. J Am Heart Assoc 9(11). https://doi.org/10.1161/JAHA.119.015594
34. Khan SU, Khan MZ, Raghu Subramanian C, Riaz H, Khan MU, Lone AN, Khan MS, Benson EM, Alkhouli M, Blaha MJ, Blumenthal RS, Gulati M, Michos ED (2020) Participation of women and older participants in randomized clinical trials of lipid-lowering therapies: a systematic review. JAMA Netw Open 3(5):e205202. https://doi.org/10.1001/jamanetworkopen.2020.5202
35. Kucharska-Newton AM, Harald K, Rosamond WD, Rose KM, Rea TD, Salomaa V (2011) Socioeconomic indicators and the risk of acute coronary heart disease events: comparison of population-based data from the United States and Finland. Ann Epidemiol 21(8). https://doi.org/10.1016/j.annepidem.2011.04.006
36. Kwak S, Hwang IC, Park JJ, Park JH, Park JB, Cho GY (2021) Sex-specific impact of diabetes mellitus on left ventricular systolic function and prognosis in heart failure. Sci Rep 11(1). https://doi.org/10.1038/s41598-021-91170-x
37. Lam CSP, Voors AA, De Boer RA, Solomon SD, Van Veldhuisen DJ (2018) Heart failure with preserved ejection fraction: from mechanisms to therapies. Eur Heart J 39(30):2780–2792. https://doi.org/10.1093/eurheartj/ehy301
38. Lenzen MJ, Rosengren A, Scholte op Reimer WJ, Follath F, Boersma E, Simoons ML, Cleland JG, Komajda M (2008) Management of patients with heart failure in clinical practice: differences between men and women. Heart (Br Cardiac Soc) 94(3):1–5. https://doi.org/10.1136/hrt.2006.099523
39. Lewinter C, Nielsen TH, Edfors LR, Linde C, Bland JM, LeWinter M, Cleland JGF, Køber L, Braunschweig F, Mansson-Broberg A (2022) A systematic review and meta-analysis of beta-blockers and renin-angiotensin system inhibitors for preventing left ventricular dysfunction due to anthracyclines or trastuzumab in patients with breast cancer. Eur Heart J 43(27):2562–2569
40. López-Vilella R, Marqués-Sulé E, Laymito Quispe R, del P, Sánchez-Lázaro I, Donoso Trenado V, Martínez Dolz L, Almenar Bonet L (2021) The female sex confers different prognosis in heart failure: same mortality but more readmissions. Front Cardiovasc Med 8(Mar):1–8. https://doi.org/10.3389/fcvm.2021.618398
41. Lu D, Chatterjee S, Xiao K, Riedel I, Huang CK, Costa A, Cushman S, Neufeldt D, Rode L, Schmidt A, Juchem M, Leonardy J, Büchler G, Blume J, Gern OL, Kalinke U, Wen Tan WL, Foo R, Vink A, van Laake LW, van der Meer P, Bär C, Thum T (2022) A circular RNA derived from the insulin receptor locus protects against doxorubicin-induced cardiotoxicity. Eur Heart J 43(42):4496–4511
42. Lyon AR, Citro R, Schneider B, Morel O, Ghadri JR, Templin C, Omerovic E (2021) Pathophysiology of Takotsubo syndrome: JACC state-of-the-art review. J Am Coll Cardiol 77(7):902–921. https://doi.org/10.1016/j.jacc.2020.10.060
43. Martínez-Sellés M, Doughty RN, Poppe K, Whalley GA, Earle N, Tribouilloy C, McMurray JJV, Swedberg K, Køber L, Berry C, Squire I (2012) Gender and survival in patients with heart failure: interactions with diabetes and aetiology. Results from the MAGGIC individual patient meta-analysis. Eur J Heart Fail 14(5). https://doi.org/10.1093/eurjhf/hfs026
44. Mensah GA (2021) The implementation frontier: impact on cardiovascular health in racial and ethnic minority populations, pp 35–45. https://doi.org/10.1007/978-3-030-81034-4_5
45. Mentzer G, Hsich EM (2019) Heart failure with reduced ejection fraction in women. Heart Fail Clin 15(1):19–27. https://doi.org/10.1016/j.hfc.2018.08.003
46. Mishra S, Kass DA (2015) Cellular and molecular pathobiology of heart failure with preserved ejection fraction. Nat Rev Cardiol 18(6):400–423. https://doi.org/10.1038/s41569-020-00480-6.Cellular

47. Mosquera PA, San Sebastian M, Waenerlund AK, Ivarsson A, Weinehall L, Gustafsson PE (2016) Income-related inequalities in cardiovascular disease from mid-life to old age in a Northern Swedish cohort: a decomposition analysis. Soc Sci Med 149:135–144
48. Naaktgeboren WR, Groen WG, Jacobse JN, Steggink LC, Walenkamp AME, van Harten WH, Stuiver MM, Aaronson NK, Aleman BMP, van der Meer P, Schaapveld M, Sonke GS, Gietema JA, van Leeuwen FE, May AM (2022) Physical activity and cardiac function in long-term breast cancer survivors: a cross-sectional study. JACC: Cardio Oncol 4(2):183–191. https://doi.org/10.1016/j.jaccao.2022.02.007
49. Nair N (2020) Epidemiology and pathogenesis of heart failure with preserved ejection fraction. Rev Cardiovasc Med 21(4):531–540. https://doi.org/10.31083/J.RCM.2020.04.154
50. O'Kelly AC, Lau ES (2020) Sex differences in HFpEF. Curr Treat Options Cardiovasc Med 22(12). https://doi.org/10.1007/s11936-020-00856-4
51. Oni-Orisan A, Mavura Y, Banda Y, Thornton TA, Sebro R (2021) Embracing genetic diversity to improve black health. N Engl J Med 384(12):1163–1167
52. Powell-Wiley TM, Ngwa J, Kebede S, Lu D, Schulte PJ, Bhatt DL, Yancy C, Fonarow GC, Albert MA (2018) Impact of body mass index on heart failure by race/ethnicity from the get with the guidelines–heart failure (GWTG–HF) registry. JACC: Heart Failure 6(3):233–242. https://doi.org/10.1016/j.jchf.2017.11.011
53. Raafs A, Verdonschot J, Ferreira JP, Wang P, Collier T, Henkens M, Björkman J, Boccanelli A, Clark AL, Delles C, Diez J, González A, Girerd N, Jukema JW, Pinet F, Rossignol P, Thum T, Vodovar N, de Boer RA et al (2021) Identification of sex-specific biomarkers predicting new-onset heart failure. ESC Heart Failure 8(5):3512–3520. https://doi.org/10.1002/ehf2.13476
54. Rådholm K, Zhou Z, Clemens K, Neal B, Woodward M (2019) Effects of SGLT2 inhibitors in type 2 diabetes, comparing women to men. Diabetes Obes Metab 46(May):1–4
55. Ramlakhan KP, Johnson MR, Roos-Hesselink JW (2020) Pregnancy and cardiovascular disease. Nat Rev Cardiol 17(11). https://doi.org/10.1038/s41569-020-0390-z
56. Reding KW, Cheng RK, Barac A, Vasbinder A, Hovsepyan G, Stefanick M, Simon MS (2022). Toward a better understanding of the differential impact of heart failure phenotypes after breast cancer. J Clin Oncol 40(32):3688–3691
57. Roger VL (2021) Epidemiology of heart failure: a contemporary perspective. Circ Res 128(10):1421–1434. https://doi.org/10.1161/CIRCRESAHA.121.318172
58. Rosenkaimer S, Sieburg T, Winter L, Mavratzas A, Hofmann WK, Hofheinz RD, Akin I, Duerschmied D, Hohneck A (2022) Adverse cardiovascular effects of anti-tumor therapies in patients with breast cancer: a single-center cross-sectional analysis. Anticancer Res 42(6):3075–3084. https://doi.org/10.21873/anticanres.15795
59. Saeed A, Kampangkaew J, Nambi V (2017) Prevention of cardiovascular disease in women. Methodist DeBakey Cardiovasc J 13(4):185–192. https://doi.org/10.14797/mdcj-13-4-185
60. Saiki H, Petersen IA, Scott CG, Bailey KR, Dunlay SM, Finley RR, Ruddy KJ, Yan E, Redfield MM (2018) Women after contemporary radiotherapy for breast cancer. 135(15):1388–1396. https://doi.org/10.1161/CIRCULATIONAHA.116.025434.Risk
61. Schiattarella GG, Altamirano F, Tong D, French KM, Villalobos E, Kim SY, Luo X, Jiang N, May HI, Wang Zv, Hill TM, Mammen PPA, Huang J, Lee DI, Hahn VS, Sharma K, Kass DA, Lavandero S, Gillette TG, Hill JA (2019) Nitrosative stress drives heart failure with preserved ejection fraction. Nat 568(7752):351–356. https://doi.org/10.1038/s41586-019-1100-z
62. Schreuder MM, Schuurman A, Akkerhuis KM, Constantinescu AA, Caliskan K, van Ramshorst J, Germans T, Umans VA, Boersma E, Roeters van Lennep JE, Kardys I (2021) Sex-specific temporal evolution of circulating biomarkers in patients with chronic heart failure with reduced ejection fraction. Int J Cardiol 334:126–134. https://doi.org/10.1016/j.ijcard.2021.04.061
63. Schultz WM, Kelli HM, Lisko JC, Varghese T, Shen J, Sandesara P, Quyyumi AA, Taylor HA, Gulati M, Harold JG, Mieres JH, Ferdinand KC, Mensah GA, Sperling LS (2018) Socioeconomic status and cardiovascular outcomes: challenges and interventions. Circ 137(20):2166–2178. https://doi.org/10.1161/CIRCULATIONAHA.117.029652
64. Sciomer S, Moscucci F, Salvioni E, Marchese G, Bussotti M, Corrà U, Piepoli MF (2020). Role of gender, age and BMI in prognosis of heart failure. Eur J Prev Cardiol 27(2_suppl):46–51. https://doi.org/10.1177/2047487320961980

65. Silverman MG, Patel B, Blankstein R, Lima JAC, Blumenthal RS, Nasir K, Blaha MJ (2016) Impact of race, ethnicity, and multimodality biomarkers on the incidence of new-onset heart failure with preserved ejection fraction (from the multi-ethnic study of atherosclerosis). Am J Cardiol 117(9):1474–1481. https://doi.org/10.1016/j.amjcard.2016.02.017
66. Siu SC, Lee DS, Rashid M, Fang J, Austin PC, Silversides CK (2021) Long-term cardiovascular outcomes after pregnancy in women with heart disease. J Am Heart Assoc 10(11). https://doi.org/10.1161/JAHA.120.020584
67. Sobhani K, Nieves Castro DK, Fu Q, Gottlieb RA, Van Eyk JE, Noel Bairey Merz C (2018) Sex differences in ischemic heart disease and heart failure biomarkers. Biol Sex Differ 9(1):1–13. https://doi.org/10.1186/s13293-018-0201-y
68. Solomon SD, de Boer RA, DeMets D, Hernandez AF, Inzucchi SE, Kosiborod MN, Lam CSP, Martinez F, Shah SJ, Lindholm D, Wilderäng U, Öhrn F, Claggett B, Langkilde AM, Petersson M, McMurray JJV (2021) Dapagliflozin in heart failure with preserved and mildly reduced ejection fraction: rationale and design of the DELIVER trial. Eur J Heart Fail 23(7). https://doi.org/10.1002/ejhf.2249
69. Solomon SD, McMurray JJV, Claggett B, de Boer RA, DeMets D, Hernandez AF, Inzucchi SE, Kosiborod MN, Lam CSP, Martinez F, Shah SJ, Desai AS, Jhund PS, Belohlavek J, Chiang C-E, Borleffs CJW, Comin-Colet J, Dobreanu D, Drozdz J et al (2022) Dapagliflozin in heart failure with mildly reduced or preserved ejection fraction. N Engl J Med 387(12):1089–1098. https://doi.org/10.1056/NEJMoa2206286
70. Stirbu I, Looman C, Nijhof GJ, Reulings PG, Mackenbach JP (2012) Income inequalities in case death of ischaemic heart disease in the Netherlands: a national record-linked study. J Epidemiol Community Health 66(12). https://doi.org/10.1136/jech-2011-200924
71. Suthahar N, Lau ES, Blaha MJ, Paniagua SM, Larson MG, Psaty BM, Benjamin EJ, Allison MA, Bartz TM, Januzzi JL, Levy D, Meems LMG, Bakker SJL, Lima JAC, Cushman M, Lee DS, Wang TJ, deFilippi CR, Herrington DM et al (2020) Sex-specific associations of cardiovascular risk factors and biomarkers with incident heart failure. J Am Coll Cardiol 76(12):1455–1465. https://doi.org/10.1016/j.jacc.2020.07.044
72. Suthahar N, Meems LMG, Ho JE, de Boer RA (2020) Sex-related differences in contemporary biomarkers for heart failure: a review. Eur J Heart Fail 22(5):775–788. https://doi.org/10.1002/ejhf.1771
73. Tanai E, Frantz S (2016) Pathophysiology of heart failure. Compr Physiol 6(1):187–214. https://doi.org/10.1002/cphy.c140055
74. Tibrewala A, Yancy CW (2019) Heart failure with preserved ejection fraction in women. Heart Fail Clin 15(1):9–18. https://doi.org/10.1016/j.hfc.2018.08.002
75. Toma M, Mak GJ, Chen V, Hollander Z, Shannon CP, Lam KKY, Ng RT, Tebbutt SJ, Wilson-McManus JE, Ignaszewski A, Anderson T, Dyck JRB, Howlett J, Ezekowitz J, McManus BM, Oudit GY (2017) Differentiating heart failure phenotypes using sex-specific transcriptomic and proteomic biomarker panels. ESC Heart Fail 4(3):301–311. https://doi.org/10.1002/ehf2.12136
76. Tong D, Schiattarella GG, Jiang N, May HI, Lavandero S, Gillette TG, Hill JA (2019) Female sex is protective in a preclinical model of heart failure with preserved ejection fraction. Circ 140(21):1769–1771. Lippincott Williams and Wilkins. https://doi.org/10.1161/CIRCULATIONAHA.119.042267
77. Tsao CW, Lyass A, Enserro D, Larson MG, Ho JE, Kizer JR, Gottdiener JS, Psaty BM, Vasan RS (2018) Temporal trends in the incidence of and mortality associated with heart failure with preserved and reduced ejection fraction. JACC: Heart Fail 6(8):678–685. https://doi.org/10.1016/j.jchf.2018.03.006
78. Varghese SS, Johnston WJ, Eekhoudt CR, Keats MR, Jassal DS, Grandy SA (2021) Exercise to reduce anthracycline-mediated cardiovascular complications in breast cancer survivors. Curr Oncol 28(5):4139–4156. https://doi.org/10.3390/curroncol28050351
79. Wei S, Le N, Zhu JW, Breathett K, Greene SJ, Mamas MA, Zannad F, Van Spall HGC (2022) Factors associated with racial and ethnic diversity among heart failure trial participants: a systematic bibliometric review. Circul Heart Fail 15(3):E008685. https://doi.org/10.1161/CIRCHEARTFAILURE.121.008685

80. Ma Y, Shi Y, Ma W, Yang D, Hu Z, Wang M, Cao X, Zhang C, Luo X, He S, Zhang M, Duan Y, Cai H (2022) A prospective study on sex differences in functional capacity, quality of life and prognosis in patients with heart failure. Medicine (Baltimore) 101(26):e29795
81. Youmans QR, Okwuosa IS, Yancy CW (2021) Heart failure in African Americans and Hispanic Americans: a persistent and disproportionate burden in underrepresented minorities. In: Cardiovascular disease in racial and ethnic minority populations, pp 55–74
82. Yu Y, Arah OA, Liew Z, Cnattingius S, Olsen J, Sørensen HT, Qin G, Li J (2019) Maternal diabetes during pregnancy and early onset of cardiovascular disease in offspring: population based cohort study with 40 years of follow-up. BMJ 367:l6398

# Chapter 17
# Sex-Dependent Cardiac Fibrosis After Myocardial Infarction: A Function of Differential Periostin Signaling?

**Besher M. Abual'anaz, Sunil G. Rattan, and Ian M. C. Dixon**

**Abstract** Sex-related differences among male and female patients are well known but much less information is available with respect to sex-specific pathogenesis of cardiac fibrosis in hearts of these patients. A common form of cardiac fibrosis occurs after chronic myocardial infarction (MI). As a major etiology of heart failure (HF) is ischemic heart disease with attendant MI and as HF is characterized by remodeling of the extracellular matrix (ECM), this chapter will highlight the literature that features differences between males and females in this pathology. Normal ECM scaffolding may lose normal configuration in acute MI (including a loss of scaffold structure and its protein components) or in the excessive deposition of matrix components seen in chronic wound healing in the heart following MI. Overt cardiac fibrosis is now known to contribute to the loss of normal cardiac function via increased ventricular wall stiffness which is in turn manifest as the incidence of heart failure and cardiac fibrosis per se and is known to be a primary contributor to the occurrence of heart failure. In this chapter, we seek to explore possible signaling differences among males and females that might contribute to sex-specific progression of cardiac fibrosis in the aftermath of acute MI. In this case, secreted periostin, known to be a marker for the activation of quiescent cardiac fibroblasts to hypersecretory myofibroblasts, may be crucial for the early onset of cardiac wound healing. Periostin is secreted by myofibroblasts and is a profibrotic matricellular protein that is known to be re-expressed in the adult heart after pathological insult. We suggest that sex differences in periostin-dependent mechanisms to promote i) acute wound healing and ii) chronic cardiac fibrosis that may underpin the differential occurrence of heart failure in males and females following MI.

---

B. M. Abual'anaz · S. G. Rattan · I. M. C. Dixon
Institute of Cardiovascular Sciences, Albrechtsen Research Centre, Winnipeg, Canada

B. M. Abual'anaz · S. G. Rattan · I. M. C. Dixon (✉)
Department of Physiology and Pathophysiology, St. Boniface Hospital Albrechtsen Research Centre, University of Manitoba, R3010 - 351 Taché Avenue, Winnipeg, MB R2H 2A6, Canada
e-mail: idixon@sbrc.ca; ian.dixon@umanitoba.ca

Rady Faculty of Health Sciences, Max Rady College of Medicine, University of Manitoba, Winnipeg, Canada

© The Author(s), under exclusive license to Springer Nature Switzerland AG 2023 
L. Kirshenbaum and I. Rabinovich-Nikitin (eds.), *Biology of Women's Heart Health*, Advances in Biochemistry in Health and Disease 26,
https://doi.org/10.1007/978-3-031-39928-2_17

**Keywords** Sex-dependent fibrosis · Cardiac fibroblast · Myofibroblast · Cardiac fibrosis · Extracellular matrix · Periostin · Myocardial infarction

## Introduction

Sex-related differences among male and female patients with common etiologies of cardiovascular diseases are well-known, however very little work has been published with respect to specific mechanisms of cardiovascular disease progression in the context of women's heart heath [1, 2]. Periostin was first described as bone specific adhesion protein and was thus named as osteoblast specific factor 2 (OSF-2) in 1993, however the role of periostin since its discovery has evolved to be included in the remodeling of ECM scaffolds in heart and vasculature, as well as dental tissues and ligaments [3, 4]. Further, periostin is now known to be involved in the remodeling of ECM in response to different pathological stimuli including vascular and cardiac muscle injury [4].

Normal ECM scaffolding may lose normal configuration in acute myocardial infarction (loss of structure and protein components) or in excessive deposition of matrix components seen in chronic wound healing/cardiac fibrosis which may develop in the heart following myocardial infarction [5]. Overt cardiac fibrosis is now known to contribute to the loss of normal cardiac function via increased ventricular wall stiffness which in turn manifests as an incidence of heart failure [6, 7]. Cardiac fibrosis per se is known to be a primary contributor to the occurrence of heart failure [8, 9]. Further, secreted periostin is also known as a marker for the activation of quiescent cardiac fibroblasts to hypersecretory myofibroblasts [10], and that periostin may be crucial for the early onset of cardiac wound healing [11]. As the literature is replete with data to underscore the marked re-expression of periostin secretion after myocardial injury, we used coronary artery ligation to reproduce a common means of induction of periostin expression [12–16].

## Myocardial Infarction and Cardiac Fibrosis in Males and Females

The myocardium is extraordinarily sensitive to oxygen availability and each cardiomyocyte contains a relatively large number of active mitochondria (compared to mitochondrial numbers in smooth muscle or skeletal muscle) producing ATP for cellular contraction. Myocardial infarction (MI) is defined as a pathological state wherein cardiomyocyte death ensues as a result of prolonged myocardial ischemia. The causes of cardiac ischemia can be divided into atherothrombotic ischemia as a result of a thrombus that fully or partially occludes the coronary artery, which leads to acute coronary syndrome, whereas non-thrombotic ischemia results from an oxygen supply–demand imbalance [17]. MI may be diagnosed from a clinical assessment of the patient (showing obvious chest pain and discomfort, unexplained weakness, nausea, dyspnea or a combination of these symptoms), biomarker levels (leaking cardiac troponin I or T to the plasma from ruptured cardiac myocytes), recordings of aberrant electrical activity of the heart from electrocardiogram (ECG) readings, and echocardiographic imaging showing disturbances in M-mode readouts [17, 18]. Based on generation of cardiac ECG depolarization and repolarization patterns, acute myocardial infarction can be classified into two categories—and these are characterized by elevation of ST segments into two main patterns. In the first instance one may consider ST segment elevation acute myocardial infarction (STEMI) and in the second case may manifest as non-ST segment elevation acute myocardial infarction (non-STEMI) [19]. Both STEMI and Non-STEMI are major causes of cardiac emergency.

Cardiac fibrosis is defined as a pathological disorder that affects the heart as a result of excessive production of extracellular matrix (ECM) proteins and deposition in the cardiac interstitium. The cardiac interstitial space is defined as the space between and among cardiomyocytes and may contain structural proteins such as fibrillar collagens, fibronectins, laminins, other types of collagens (non-fibrillar), as well as non-structural matrikines, and nascent signaling peptides (matricryptic signal proteins and known cytokines including latent $TGF\beta_1$).

The main structural ECM proteins are fibrillar collagens, especially types I and type III, which are produced mainly from the relatively quiescent cardiac fibroblast (CF—these cells are evolved and present for low levels of output of ECM proteins in normal myocardium that is suited for slow matrix turnover) in the myocardial interstitium. Any activation of the normal fibroblast to a hypersynthetic myofibroblast represents a primary switch, and perhaps is a singular event in the evolution of cardiac fibrosis—heralded by an increase of ECM deposition in the myocardial interstitial space. A net increase in ECM proteins will usually result in an increased myocardial stiffness which will have a negative impact on cardiac function such as increased incidence of arrythmia, or development of systolic dysfunction as well as diastolic dysfunction which eventually lead to HF [20–23]. Cardiac fibrosis can be classified based on histopathological criteria into three subtypes—and they are replacement fibrosis, interstitial fibrosis as well as perivascular fibrosis [23].

Cardiac fibroblasts are mesenchymal-derived cells that normally exist throughout the heart and are located in interstitial spaces between cardiomyocytes. In healthy hearts, fibroblasts are responsible for the maintenance of the homeostasis between the secreted ECM components which is the result of a balance in secretion of tissue inhibitors of metalloproteinases (TIMPs) and matrix metalloproteinases (MMPs) [24] (Fig. 17.1). After any type of heart injury, this cell undergoes activation to transition phenotypes to that of a myofibroblast. This cell is hypersecretory in injured hearts, and it contributes to the development of cardiac fibrosis by secreting collagen and other ECM components [25, 26] (Fig. 17.2). Resident cardiac fibroblasts and their activated phenotype, i.e., cardiac myofibroblasts, are the main contributors in the development of cardiac fibrosis [10, 27–29], however their distribution following MI in hearts of male and female patients has not been rigorously compared.

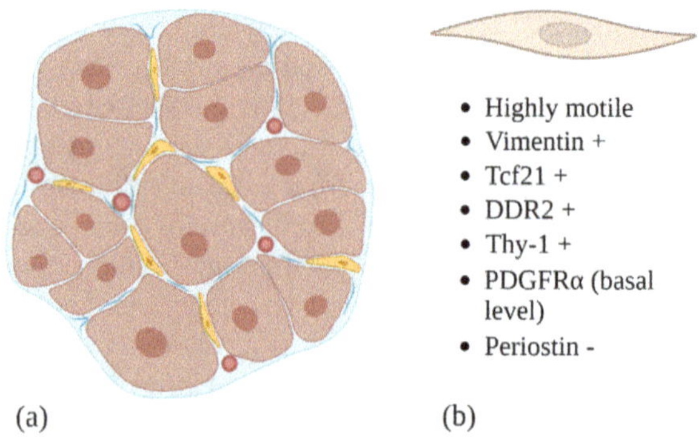


**Fig. 17.1** *Normal cardiac interstitium.* **Panel A** depicts resting cardiac fibroblasts, with balanced deposition and turnover of ECM components. **Panel B** provides a listing of resting cardiac fibroblast properties with emphasis on a catalogue of expressed proteins. The cardiac fibroblast is characterized with relatively high motility, while secreting vimentin, transcription factor 21(Tcf21), discoidin domain receptor 2 (DDR2) and thymus cell antigen 1 (Thy1). It is also characterized with the basal expression of platelet-derived growth factor receptor type α (PDGFRα) and expresses vimentin with little to no periostin expression

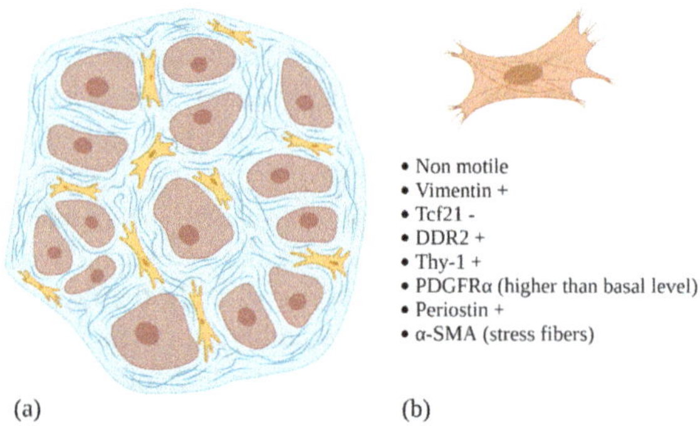


**Fig. 17.2** *Fibrosed cardiac interstitium.* **Panel A** depicts the fibrosed cardiac interstitium, and shows activated cardiac fibroblasts (e.g., myofibroblasts) with the attendant hypersecretory phenotype. Prolonged population of the interstitial space with myofibroblasts results in the enhanced deposition of ECM components mainly collagen type I and type III. **Panel B**: A depiction of an activated cardiac myofibroblast. The cardiac myofibroblast is characterized by being relatively non-motile, while secreting periostin. It is also characterized by the incorporation of α- SMA proteins incorporated into stress fibers, and these cells are known to express PDGFRα above the basal level along with vimentin [26]

## Biology of Periostin

Periostin was first identified in 1993, and given the name osteoblast specific factor 2 (OSF-2), it was first discovered in bone and lung. Periostin was characterized as adhesion protein for preosteoblasts cells and its expression was increased by TGF-β [3,

30]. Since 1993, periostin has been recognized as a player in the ECM remodeling in response to different pathologies including vascular and muscle injuries [4]. Periostin expression is detected in the developing hearts but not in healthy adult hearts. [31]. Following any type of cardiac injury, periostin protein is re-expressed exclusively from cardiac myofibroblasts [10]. The periostin gene in human and mouse has been cloned, in mouse it is located on chromosome number 3, and it has 23 exons with genomic footprint of approximately 30 kilobases. The terminal exons are protein coding as shown in Fig. 17.3.

The full length of mouse periostin protein is 838 amino acids (AA) which is encoded by 23 exons with a molecular weight of approximately 93 kDa [32]. The protein structure of periostin consists of four distinct regions as shown in Fig. 17.3. The first region of the periostin gene -signal peptide- is a relatively short signal peptide sequence (AA sequence from 1 to 24) included entirely on the first exon. It is highly conserved and is responsible for periostin protein secretion and regulation of cell function through integrin binding [3, 30, 33]. The second region is the EMI domain, named after the same domain in the EMILLIN family of proteins (the AA sequence includes amino acids from 25 to 108). This region is contained within the coding region of exons two and three and it is responsible for the formation of multimers which are important in protein–protein interactions (PPIs) [33, 34]. Through the EMI domain, periostin can interact with fibronectin [35], collagen type I [36] and Notch1 [37]. The third structural region of periostin consists of four fasciclin (FAS1) domains which are highly conserved - each domain has approximately 150 AA (AA sequence from 109 to 634) included between exons three and fourteen. The second and fourth FAS1 domains have an integrin-binding motif that is responsible for cell adhesion [32, 33, 38]. The fourth region is the C-terminal region (the AA sequence extends from amino acids 635–838) and is encoded between exons fifteen and twenty-three. It is also called the alternatively spliced region that is responsible for the formation of the periostin protein variants or isoforms, resulting in variation in molecular weight of the secreted periostin between 86 and 93 kDa [4, 33, 38–40]. Moreover, this region contains proteolytic cleavage sites which are important in regulation of the cell matrix

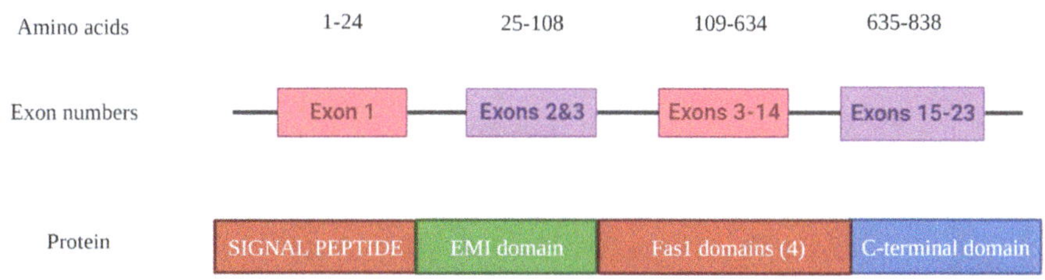


**Fig. 17.3** *Periostin gene structure.* The amino acid sequence that encodes each protein domain is labelled above (first line). The second line shows the exact exon number that encodes each protein domain. The third line shows the four characteristic periostin protein domains. The signal peptide is encoded in exon 1, the EMI domain is encoded in exon 2 and 3, and the FAS1 domains are encoded by exons 3 to 14 that have an integrin-binding motif that is responsible for cell adhesion. Finally, the C-terminal domain is encoded by exons 15–23

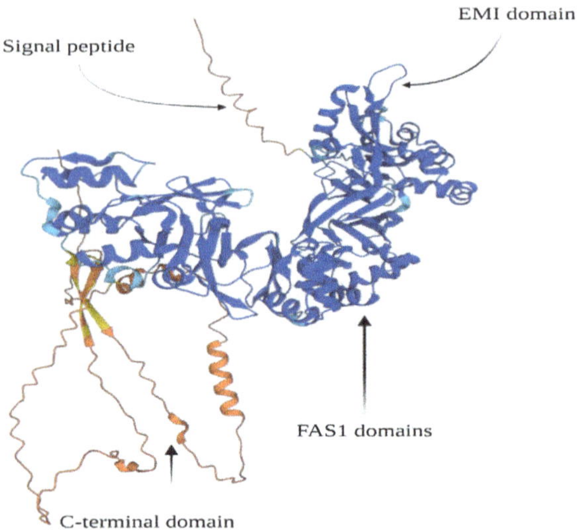


**Fig. 17.4** *The 3D structure of mouse periostin protein.* This figure was created with the complete folded prediction of structure obtained from databases in the SWISS-MODEL Repository [42]. Herein we provide the N-terminal which contains the signal peptide (periostin secretion), followed by EMI domain (fibronectin binding) and four fasciclin domains (FAS1 domains, Integrins, BMP-1 and Tenascin-C binding) and end with the C-terminal domain (periostin protein variants). The variants formation is induced by TGF-β1, their expression appears to be developmentally regulated and contribute differently to ECM fibrillogenesis [41]

organization [41]. The mouse periostin protein structure in three dimension (3D) is shown in Fig. 17.4. The figure was created with the complete folded prediction of structure obtained from databases in the so-called SWISS-MODEL Repository [42].

## Periostin in Post MI Cardiac Interstitial Remodeling

Periostin functions as a matricellular protein in heart, and is produced as extracellular secreted protein in interstitial space [43]. In healthy adult heart, periostin is expressed at very low basal levels. Moreover, it is expressed in the heart in vascular smooth muscle cells (VSMCs) in vasculature [44], valve interstitial cells (VIC) [45] and in cardiac fibroblasts within myocardium [4]. However, after any type of cardiac injury, periostin is highly expressed from the activated cardiac myofibroblast [5]. Periostin promotes collagen fibrogenesis which may explain why it is expressed in different cell types that populate matrix-rich tissues [46]. Periostin interacts with several proteins in the myocardial interstitial space, including other extracellular matrix proteins, enzymes, matricellular proteins and receptors [47]. Fibronectin and tenascin-C both are ECM proteins that interact with periostin via direct interaction through EMI and FAS1 domains respectively [35, 48]. Collagen type I and Laminin $\gamma 2$ are also ECM component proteins that interact with periostin indirectly via unknown mechanisms [48–51]. Periostin can bind to collagen type I and induce collagen fibrogenesis by promoting collagen crosslinking [48].

Periostin can also bind directly and indirectly to two different enzymes, and in the first instance, by direct binding via the FAS1 domain to bone morphogenetic protein-1 (BMP-1). This binding results in enhanced deposition of BMP-1 in ECM. Secondly, periostin may indirectly bind via EMI domain to lysyl oxidase (LOX) enzyme [52, 53]. BMP-1 is one of the procollagen C proteinases that activates LOX enzyme because LOX is produced in an inactive form (pro-LOX) [54]. The active form of LOX is responsible for collagen stabilization in ECM by forming the inter- and intramolecular cross linking of fibrillar collagens between lysine residues [55]. It is also reported that LOX colocalized and interacts with fibronectin [56].

Periostin may also interact with cell receptors like integrins (integrin $\alpha V\beta 3$, integrin $\alpha V\beta 5$ and integrin $\alpha 6\beta 4$) and Notch-1 by unknown mechanisms. This interaction may lead to cell migration, proliferation and alteration in ECM properties [57–61]. In cardiac fibroblasts periostin can bind to integrin $\alpha V\beta 3$ and integrin $\alpha V\beta 5$ and this mode of binding induces cell migration through focal adhesion kinase (FAK) phosphorylation [15]. After MI, mesenchymal cells including cardiac fibroblasts may alter their phenotype and become activated [62]. Activation happens due to response to mechanical stressors and to binding of $TGF\beta_1$ [63–65]. Activated fibroblasts are hypersecretory myofibroblasts which make a major contribution to accumulation of ECM protein production and in the generation of the fibrotic scar as well as cardiac fibrosis. This scar is also called replacement fibrosis [4].

Previous studies performed with periostin gene knock out mice ($Postn^{-/-}$) using an MI heart failure model showed that periostin is essential for the development and establishment of fibrotic scar. $Postn^{-/-}$ animals were highly susceptible to cardiac rupture (CR) in the infarcted area within two weeks after inducing MI [14]. However, in periostin overexpressing transgenic mice CR was absent following MI—nonetheless these mice developed cardiac hypertrophy and spontaneous interstitial cardiac fibrosis [14]. It is now becoming clear that different pathologies may (in common) be linked to elevated levels of periostin release in myocardium—these etiologies include MI, hypertension, diabetic cardiomyopathy, dilated cardiomyopathy, cardiac hypertrophy and aging [15, 66–69]. Nonetheless, a pathologic stimulus seems to be required for its release in humans, as periostin is strongly expressed in failing hearts but not in healthy hearts [15, 70, 71]. Further, whether differences in how the myocardium (and its constituent cells) respond to periostin signaling in male vs. female patients is largely unexplored and represents a significant knowledge gap.

## Periostin's Role in Male and Female Post-MI Patients

Information dealing with the periostin role in acute cardiac wound healing is lacking in female patients due to an overwhelming presence of data generated using male subjects. Moving forward, one would need to include female subjects in any meaningful experimental design. In acute MI, a major determinant of death in both males and females is CR which again reflects the same causality of death in male mice from the Oka et al. study results [14]. However, in the detailed study performed by Gao et al. group the cause of death in wild type C57B/6 J mice was due to HF and CR [72]. On the other hand, very little information is available on parallel outcomes in female mice.

It has been reported that estrogen has the ability to downregulate proinflammatory molecules including adhesion molecules and cytokines which ultimately lead to suppression of vascular inflammation [73]. We suggest that the lack of this effect in males may contribute to the marked mortality at 3 days after MI in male animals, as male mice do not access this anti-inflammatory effect borne by estrogen, simply due to their intrinsic lower levels of circulating estrogen when compared to female animals. We expect that this difference contributes to a relatively higher inflammatory response in male heart after MI versus age-matched female hearts post—MI [74]. Moreover, there is a growing body of evidence to indicate that peak contractility (velocity of cardiac muscle contraction) in intact female hearts are slower compared to male contractility [75]. Thus, we may hypothesize that several causal factors may lead to a putative accentuation of peak mortality in males following MI compared to late mortality in post MI female subjects, and that these include the role of periostin itself in acute wound healing in males, the lack of estrogen protective effects on hearts and finally the higher native contractility in male hearts vs female hearts.

## Cardiovascular Disease Progression is Distinct in Male and Female Patients

In adulthood, male patients are more susceptible to morbidity and mortality than women with the onset of cardiovascular disease (CVDs) and development of HF [2]. Other sex-dependent trends in the pathogenesis of post-MI heart failure are also apparent. In particular, epidemiological data have shown consistently that women are more prone to develop heart failure with preserved ejection fraction (HFpEF) [76, 77] when compared to men. However, men are much more prone than their female compatriots to develop HFrEF [78, 79]. Hearts from males and females in both human patients and preclinical animal models exposed to pressure overload respond differently when compared to each other [1]. Male hearts respond to pressure overload by developing eccentric myocardial hypertrophy and fibrosis which has a negative effect on systolic function. Female hearts respond mainly by developing concentric myocardial hypertrophy which is marked by significantly less cardiac fibrosis [1]. Female hearts in preclinical models of MI (cardiac ischemia) develop significantly smaller infarct scars when compared to aged matched infarcted hearts from the male cohort [1]. As females' hearts following MI develop smaller infarcts, it follows then that they exhibit lower incidence of cardiac rupture compared to hearts from males after MI. The difference in infarct size is a result of delayed myocardial healing because males' hearts after MI exhibit greater activation and expression of MMPs as well as a more robust inflammatory response compared to females [80, 81]. Thus, while there is no significant difference of expression of αSMA or in collagen type IIIαI among males and females, we suggest that the sex-specificity for wound healing mechanisms in heart may exist. Moreover, the work of Molkentin et al. provides results that stipulate that POSTN KO mouse heart is characterized by the significant gene alteration of 449 specific genes when compared to the basal expression of wild type hearts [14]. Groups of genes within the 449 significantly altered genes can be sub-classified into different groups that involve either cell adhesion, fibrosis, fibroblast function or ECM [14]. These findings point to a broad and significant diversity of the response of protein expression patterns in heart and raise the possibility of putative sex-dependent differences.

Prior work by Norris et al. [48], has provided data to indicate that periostin is responsible for collagen I fibrogenesis and crosslinking as in the periostin KO mice, they reveal a reduction in collagen fibril diameter and decreased denaturation temperature of collagen [48]. These changes reflect a putative qualitative change in the "toughness" of fibrillar collagen type I when deprived of normal periostin exposure in tissues. Furthermore, a recent study by Angelini et al. addressed an important knowledge gap in our understanding of the regulating mechanisms of cardiac remodeling and fibrosis in the aged male and female hearts [2]. More specifically, this group has shown that sex differences exist in ECM scaffold components of the aged male hearts and the aged female hearts in response to 5-aminoimidazole-4-carboxamide riboside (AICAR) treatment by decreasing the ECM deposition while this treatment has no effect in males. AICAR treatment itself is responsible for the attenuation of periostin signal intensity in aged female hearts but not the aged male hearts having the same treatment regimen [2]. AICAR treatment provided the advantageous effect associated with a significant increase of adenosine monophosphate-activated protein kinase (AMPK) phosphorylation. Interestingly, AMPK phosphorylation was significantly higher in AICAR treated female derived fibroblasts compared to AICAR treated male heart derived fibroblasts. Moreover, these findings reveal that fibronectin is significantly different in aged male ECM scaffolds compared to aged female ECM scaffolds [2].

## Summary

As periostin release in the myocardium is required for acute cardiac wound healing and maintenance of the integrity of the left ventricular wall after MI, it provides an important target for future studies addressing sex-dependency in the pathogenesis of cardiac fibrosis in hearts of female and male subjects. Whether or not there is a significantly higher dependency of periostin expression in males than in females in the context of acute wound healing is a knowledge gap, and is currently unaddressed in the literature. Thus, the prioritization and implementation of experimental designs to interrogate the hypothesis that differential operation of periostin in hearts of males and females stimulates acute cardiac wound healing and also chronic cardiac fibrosis is required to shed light on these sex-dependent mechanisms following myocardial infarction. Data from these specific future studies will inform accurate treatment strategies for both female and male patients who have suffered myocardial infarction.

**Acknowledgements** The work relevant to the generation of this manuscript was supported with funding from the Canadian Institutes of Health Research (CIHR) operating grant and the CAN/USA joint funding initiative organized by the St. Boniface General Hospital (Winnipeg, MB, Canada) and the Mayo Clinic (Rochester, MN, USA) to IMCD. BA was supported with a studentship from a Bank of Montreal (BMO) studentship and the GETS program (University of Manitoba). SGR was supported with funding from the CIHR and our laboratory is grateful for ongoing support from the St. Boniface General Hospital Foundation. We thank Drs. Jeff Wigle and Andrew Halayko for their careful reading and critique of the penultimate draft.

**Funding** Supported by an operating grant from the Canadian Institutes for Health Research to IMCD.

## References

1. Regitz-Zagrosek V, Kararigas G (2017) Mechanistic pathways of sex differences in cardiovascular disease. Physiol Rev 97(1):1–37
2. Angelini A, Ortiz-Urbina J, Trial J, Reddy AK, Malovannaya A, Jain A, Entman ML, Taffet GE, Cieslik KA (2022) Sex-specific phenotypes in the aging mouse heart and consequences for chronic fibrosis. Am J Physiol Heart Circ Physiol 323(2):H285–H300
3. Takeshita S, Kikuno R, Tezuka K, Amann E (1993) Osteoblast-specific factor 2: cloning of a putative bone adhesion protein with homology with the insect protein fasciclin I. Biochem J 294(Pt 1):271–278
4. Landry NM, Cohen S, Dixon IMC (2018) Periostin in cardiovascular disease and development: a tale of two distinct roles. Basic Res Cardiol 113(1):1
5. Dixon IMC, Landry NM, Rattan SG (2019) Periostin reexpression in heart disease contributes to cardiac interstitial remodeling by supporting the cardiac myofibroblast phenotype. Adv Exp Med Biol 1132:35–41
6. Martos R, Baugh J, Ledwidge M, O'Loughlin C, Conlon C, Patle A, Donnelly SC, McDonald K (2007) Diastolic heart failure: evidence of increased myocardial collagen turnover linked to diastolic dysfunction. Circulation 115(7):888–895
7. Martos R, Baugh J, Ledwidge M, O'Loughlin C, Murphy NF, Conlon C, Patle A, Donnelly SC, McDonald K (2009) Diagnosis of heart failure with preserved ejection fraction: improved accuracy with the use of markers of collagen turnover. Eur J Heart Fail 11(2):191–197
8. Thum T, Gross C, Fiedler J, Fischer T, Kissler S, Bussen M, Galuppo P, Just S, Rottbauer W, Frantz S, Castoldi M, Soutschek J, Koteliansky V, Rosenwald A, Basson MA, Licht JD, Pena JT, Rouhanifard SH, Muckenthaler MU, Tuschl T, Martin GR, Bauersachs J, Engelhardt S (2008) MicroRNA-21 contributes to myocardial disease by stimulating MAP kinase signalling in fibroblasts. Nature 456(7224):980–984

9. Adam O, Lohfelm B, Thum T, Gupta SK, Puhl SL, Schafers HJ, Bohm M, Laufs U (2012) Role of miR-21 in the pathogenesis of atrial fibrosis. Basic Res Cardiol 107(5):278
10. Kanisicak O, Khalil H, Ivey MJ, Karch J, Maliken BD, Correll RN, Brody MJ, SC JL, Aronow BJ, Tallquist MD, Molkentin JD (2016) Genetic lineage tracing defines myofibroblast origin and function in the injured heart. Nat Commun 7:12260
11. Walker JT, McLeod K, Kim S, Conway SJ, Hamilton DW (2016) Periostin as a multifunctional modulator of the wound healing response. Cell Tissue Res 365(3):453–465
12. Snider P, Hinton RB, Moreno-Rodriguez RA, Wang J, Rogers R, Lindsley A, Li F, Ingram DA, Menick D, Field L, Firulli AB, Molkentin JD, Markwald R, Conway SJ (2008) Periostin is required for maturation and extracellular matrix stabilization of noncardiomyocyte lineages of the heart. Circ Res 102(7):752–760
13. Hakuno D, Kimura N, Yoshioka M, Mukai M, Kimura T, Okada Y, Yozu R, Shukunami C, Hiraki Y, Kudo A, Ogawa S, Fukuda K (2010) Periostin advances atherosclerotic and rheumatic cardiac valve degeneration by inducing angiogenesis and MMP production in humans and rodents. J Clin Invest 120(7):2292–2306
14. Oka T, Xu J, Kaiser RA, Melendez J, Hambleton M, Sargent MA, Lorts A, Brunskill EW, Dorn GW, Conway SJ, Aronow BJ, Robbins J, Molkentin JD (2007) Genetic manipulation of periostin expression reveals a role in cardiac hypertrophy and ventricular remodeling. Circ Res 101(3):313–321
15. Shimazaki M, Nakamura K, Kii I, Kashima T, Amizuka N, Li M, Saito M, Fukuda K, Nishiyama T, Kitajima S, Saga Y, Fukayama M, Sata M, Kudo A (2008) Periostin is essential for cardiac healing after acute myocardial infarction. J Exp Med 205(2):295–303
16. Minicucci MF, Santos PP, Rafacho BP, Goncalves AF, Ardisson LP, Batista DF, Azevedo PS, Polegato BF, Okoshi K, Pereira EJ, Paiva SA, Zornoff LA (2013) Periostin as a modulator of chronic cardiac remodeling after myocardial infarction. Clinics (Sao Paulo) 68(10):1344–1349
17. Anderson JL, Morrow DA (2017) Acute Myocardial Infarction. N Engl J Med 376(21):2053–2064
18. Saleh M, Ambrose JA (2018) Understanding myocardial infarction. F1000Res. 7:F1000 Faculty Rev-1378
19. Thygesen K, Alpert JS, Jaffe AS, Simoons ML, Chaitman BR, White HD; Joint ESC/ACCF/AHA/WHF Task Force for Universal Definition of Myocardial Infarction; Authors/Task Force Members Chairpersons; Thygesen K, Alpert JS, White HD; Biomarker Subcommittee; Jaffe AS, Katus HA, Apple FS, Lindahl B, Morrow DA; ECG Subcommittee; Chaitman BR, Clemmensen PM, Johanson P, Hod H; Imaging Subcommittee; Underwood R, Bax JJ, Bonow JJ, Pinto F, Gibbons RJ; Classification Subcommittee; Fox KA, Atar D, Newby LK, Galvani M, Hamm CW; Intervention Subcommittee; Uretsky BF, Steg PG, Wijns W, Bassand JP, Menasche P, Ravkilde J; Trials & Registries Subcommittee; Ohman EM, Antman EM, Wallentin LC, Armstrong PW, Simoons ML; Trials & Registries Subcommittee; Januzzi JL, Nieminen MS, Gheorghiade M, Filippatos G; Trials & Registries Subcommittee; Luepker RV, Fortmann SP, Rosamond WD, Levy D, Wood D; Trials & Registries Subcommittee; Smith SC, Hu D, Lopez-Sendon JL, Robertson RM, Weaver D, Tendera M, Bove AA, Parkhomenko AN, Vasilieva EJ, Mendis S; ESC Committee for Practice Guidelines (CPG); Bax JJ, Baumgartner H, Ceconi C, Dean V, Deaton C, Fagard R, Funck-Brentano C, Hasdai D, Hoes A, Kirchhof P, Knuuti J, Kolh P, McDonagh T, Moulin C, Popescu BA, Reiner Z, Sechtem U, Sirnes PA, Tendera M, Torbicki A, Vahanian A, Windecker S; Document Reviewers; Morais J, Aguiar C, Almahmeed W, Arnar DO, Barili F, Bloch KD, Bolger AF, Botker HE, Bozkurt B, Bugiardini R, Cannon C, de Lemos J, Eberli FR, Escobar E, Hlatky M, James S, Kern KB, Moliterno DJ, Mueller C, Neskovic AN, Pieske BM, Schulman SP, Storey RF, Taubert KA, Vranckx P, Wagner DR (2012). Third universal definition of myocardial infarction. J Am Coll Cardiol 60(16):1581–1598
20. Jellis C, Martin J, Narula J, Marwick TH (2010) Assessment of nonischemic myocardial fibrosis. J Am Coll Cardiol 56(2):89–97
21. Schelbert EB, Fridman Y, Wong TC, Abu Daya H, Piehler KM, Kadakkal A, Miller CA, Ugander M, Maanja M, Kellman P, Shah DJ, Abebe KZ, Simon MA, Quarta G, Senni M, Butler J, Diez J, Redfield MM, Gheorghiade M (2017) Temporal relation between myocardial

fibrosis and heart failure with preserved ejection fraction: association with baseline disease severity and subsequent outcome. JAMA Cardiol 2(9):995–1006
22. Butler J, Fonarow GC, Zile MR, Lam CS, Roessig L, Schelbert EB, Shah SJ, Ahmed A, Bonow RO, Cleland JG, Cody RJ, Chioncel O, Collins SP, Dunnmon P, Filippatos G, Lefkowitz MP, Marti CN, McMurray JJ, Misselwitz F, Nodari S, O'Connor C, Pfeffer MA, Pieske B, Pitt B, Rosano G, Sabbah HN, Senni M, Solomon SD, Stockbridge N, Teerlink JR, Georgiopoulou VV, Gheorghiade M (2014) Developing therapies for heart failure with preserved ejection fraction: current state and future directions. JACC Heart Fail 2(2):97–112
23. Frangogiannis NG (2019) Cardiac fibrosis: cell biological mechanisms, molecular pathways and therapeutic opportunities. Mol Aspects Med 65:70–99
24. Ahmed SH, Clark LL, Pennington WR, Webb CS, Bonnema DD, Leonardi AH, McClure CD, Spinale FG, Zile MR (2006) Matrix metalloproteinases/tissue inhibitors of metalloproteinases: relationship between changes in proteolytic determinants of matrix composition and structural, functional, and clinical manifestations of hypertensive heart disease. Circ 113(17):2089–2096
25. Tallquist MD (2020) Cardiac fibroblast diversity. Annu Rev Physiol 82:63–78
26. de Oliveira CR, Abual'anaz B, Rattan SG, Filomeno KL, Dixon IMC (2021) Novel factors that activate and deactivate cardiac fibroblasts: a new perspective for treatment of cardiac fibrosis. Wound Repair Regen 29(4):667–677
27. Moore-Morris T, Cattaneo P, Guimaraes-Camboa N, Bogomolovas J, Cedenilla M, Banerjee I, Ricote M, Kisseleva T, Zhang L, Gu Y, Dalton ND, Peterson KL, Chen J, Puceat M, Evans SM (2018) Infarct fibroblasts do not derive from bone marrow lineages. Circ Res 122(4):583–590
28. Li Y, Lui KO, Zhou B (2018) Reassessing endothelial-to-mesenchymal transition in cardiovascular diseases. Nat Rev Cardiol 15(8):445–456
29. Valiente-Alandi I, Schafer AE, Blaxall BC (2016) Extracellular matrix-mediated cellular communication in the heart. J Mol Cell Cardiol 91:228–237
30. Horiuchi K, Amizuka N, Takeshita S, Takamatsu H, Katsuura M, Ozawa H, Toyama Y, Bonewald LF, Kudo A (1999) Identification and characterization of a novel protein, periostin, with restricted expression to periosteum and periodontal ligament and increased expression by transforming growth factor beta. J Bone Miner Res 14(7):1239–1249
31. Lorts A, Schwanekamp JA, Elrod JW, Sargent MA, Molkentin JD (2009) Genetic manipulation of periostin expression in the heart does not affect myocyte content, cell cycle activity, or cardiac repair. Circ Res 104(1):e1–7
32. Hoersch S, Andrade-Navarro MA (2010) Periostin shows increased evolutionary plasticity in its alternatively spliced region. BMC Evol Biol 10:30
33. Kudo A (2017) Introductory review: periostin-gene and protein structure. Cell Mol Life Sci 74(23):4259–4268
34. Doliana R, Bot S, Bonaldo P, Colombatti A (2000) EMI, a novel cysteine-rich domain of EMILINs and other extracellular proteins, interacts with the gC1q domains and participates in multimerization. FEBS Lett 484(2):164–168
35. Kii I, Nishiyama T, Li M, Matsumoto K, Saito M, Amizuka N, Kudo A (2010) Incorporation of tenascin-C into the extracellular matrix by periostin underlies an extracellular meshwork architecture. J Biol Chem 285(3):2028–2039
36. Norris RA, Moreno-Rodriguez RA, Sugi Y, Hoffman S, Amos J, Hart MM, Potts JD, Goodwin RL, Markwald RR (2008) Periostin regulates atrioventricular valve maturation. Dev Biol 316(2):200–213
37. Tanabe H, Takayama I, Nishiyama T, Shimazaki M, Kii I, Li M, Amizuka N, Katsube K, Kudo A (2010) Periostin associates with Notch1 precursor to maintain Notch1 expression under a stress condition in mouse cells. PLoS ONE 5(8):e12234
38. Litvin J, Selim AH, Montgomery MO, Lehmann K, Rico MC, Devlin H, Bednarik DP, Safadi FF (2004) Expression and function of periostin-isoforms in bone. J Cell Biochem 92(5):1044–1061
39. Clout NJ, Tisi D, Hohenester E (2003) Novel fold revealed by the structure of a FAS1 domain pair from the insect cell adhesion molecule fasciclin I. Structure 11(2):197–203
40. Litvin J, Zhu S, Norris R, Markwald R (2005) Periostin family of proteins: therapeutic targets for heart disease. Anat Rec A Discov Mol Cell Evol Biol 287(2):1205–1212

41. Nuzzo PV, Buzzatti G, Ricci F, Rubagotti A, Argellati F, Zinoli L, Boccardo F (2014) Periostin: a novel prognostic and therapeutic target for genitourinary cancer? Clin Genitourin Cancer 12(5):301–311
42. Kiefer F, Arnold K, Kunzli M, Bordoli L, Schwede T (2009) The SWISS-MODEL Repository and associated resources. Nucleic Acids Res 37(Database issue):D387–392
43. Li L, Zhao Q, Kong W (2018) Extracellular matrix remodeling and cardiac fibrosis. Matrix Biol 68–69:490–506
44. Li G, Oparil S, Sanders JM, Zhang L, Dai M, Chen LB, Conway SJ, McNamara CA, Sarembock IJ (2006) Phosphatidylinositol-3-kinase signaling mediates vascular smooth muscle cell expression of periostin in vivo and in vitro. Atherosclerosis 188(2):292–300
45. Horne TE, VandeKopple M, Sauls K, Koenig SN, Anstine LJ, Garg V, Norris RA, Lincoln J (2015) Dynamic heterogeneity of the heart valve interstitial cell population in mitral valve health and disease. J Cardiovasc Dev Dis 2(3):214–232
46. Tkatchenko TV, Moreno-Rodriguez RA, Conway SJ, Molkentin JD, Markwald RR, Tkatchenko AV (2009) Lack of periostin leads to suppression of Notch1 signaling and calcific aortic valve disease. Physiol Genomics 39(3):160–168
47. Kii I (2019) Periostin functions as a scaffold for assembly of extracellular proteins. Adv Exp Med Biol 1132:23–32
48. Norris RA, Damon B, Mironov V, Kasyanov V, Ramamurthi A, Moreno-Rodriguez R, Trusk T, Potts JD, Goodwin RL, Davis J, Hoffman S, Wen X, Sugi Y, Kern CB, Mjaatvedt CH, Turner DK, Oka T, Conway SJ, Molkentin JD, Forgacs G, Markwald RR (2007) Periostin regulates collagen fibrillogenesis and the biomechanical properties of connective tissues. J Cell Biochem 101(3):695–711
49. Kii I, Amizuka N, Minqi L, Kitajima S, Saga Y, Kudo A (2006) Periostin is an extracellular matrix protein required for eruption of incisors in mice. Biochem Biophys Res Commun 342(3):766–772
50. Takayama G, Arima K, Kanaji T, Toda S, Tanaka H, Shoji S, McKenzie AN, Nagai H, Hotokebuchi T, Izuhara K (2006) Periostin: a novel component of subepithelial fibrosis of bronchial asthma downstream of IL-4 and IL-13 signals. J Allergy Clin Immunol 118(1):98–104
51. Nishiyama T, Kii I, Kashima TG, Kikuchi Y, Ohazama A, Shimazaki M, Fukayama M, Kudo A (2011) Delayed re-epithelialization in periostin-deficient mice during cutaneous wound healing. PLoS ONE 6(4):e18410
52. Hwang EY, Jeong MS, Park EK, Kim JH, Jang SB (2014) Structural characterization and interaction of periostin and bone morphogenetic protein for regulation of collagen cross-linking. Biochem Biophys Res Commun 449(4):425–431
53. Maruhashi T, Kii I, Saito M, Kudo A (2010) Interaction between periostin and BMP-1 promotes proteolytic activation of lysyl oxidase. J Biol Chem 285(17):13294–13303
54. Uzel MI, Scott IC, Babakhanlou-Chase H, Palamakumbura AH, Pappano WN, Hong HH, Greenspan DS, Trackman PC (2001) Multiple bone morphogenetic protein 1-related mammalian metalloproteinases process pro-lysyl oxidase at the correct physiological site and control lysyl oxidase activation in mouse embryo fibroblast cultures. J Biol Chem 276(25):22537–22543
55. Banse X, Sims TJ, Bailey AJ (2002) Mechanical properties of adult vertebral cancellous bone: correlation with collagen intermolecular cross-links. J Bone Miner Res 17(9):1621–1628
56. Fogelgren B, Polgar N, Szauter KM, Ujfaludi Z, Laczko R, Fong KS, Csiszar K (2005) Cellular fibronectin binds to lysyl oxidase with high affinity and is critical for its proteolytic activation. J Biol Chem 280(26):24690–24697
57. Bao S, Ouyang G, Bai X, Huang Z, Ma C, Liu M, Shao R, Anderson RM, Rich JN, Wang XF (2004) Periostin potently promotes metastatic growth of colon cancer by augmenting cell survival via the Akt/PKB pathway. Cancer Cell 5(4):329–339
58. Ghatak S, Misra S, Norris RA, Moreno-Rodriguez RA, Hoffman S, Levine RA, Hascall VC, Markwald RR (2014) Periostin induces intracellular cross-talk between kinases and hyaluronan in atrioventricular valvulogenesis. J Biol Chem 289(12):8545–8561

59. Vadon-Le Goff S, Hulmes DJ, Moali C (2015) BMP-1/tolloid-like proteinases synchronize matrix assembly with growth factor activation to promote morphogenesis and tissue remodeling. Matrix Biol 44–46:14–23
60. Sugiyama A, Kanno K, Nishimichi N, Ohta S, Ono J, Conway SJ, Izuhara K, Yokosaki Y, Tazuma S (2016) Periostin promotes hepatic fibrosis in mice by modulating hepatic stellate cell activation via alphav integrin interaction. J Gastroenterol 51(12):1161–1174
61. Shao R, Bao S, Bai X, Blanchette C, Anderson RM, Dang T, Gishizky ML, Marks JR, Wang XF (2004) Acquired expression of periostin by human breast cancers promotes tumor angiogenesis through up-regulation of vascular endothelial growth factor receptor 2 expression. Mol Cell Biol 24(9):3992–4003
62. Hinz B (2007) Formation and function of the myofibroblast during tissue repair. J Invest Dermatol 127(3):526–537
63. Landry NM, Rattan SG, Dixon IMC (2019) An improved method of maintaining primary murine cardiac fibroblasts in two-dimensional cell culture. Sci Rep 9(1):12889
64. Santiago JJ, Dangerfield AL, Rattan SG, Bathe KL, Cunnington RH, Raizman JE, Bedosky KM, Freed DH, Kardami E, Dixon IM (2010) Cardiac fibroblast to myofibroblast differentiation in vivo and in vitro: expression of focal adhesion components in neonatal and adult rat ventricular myofibroblasts. Dev Dyn 239(6):1573–1584
65. Talman V, Ruskoaho H (2016) Cardiac fibrosis in myocardial infarction-from repair and remodeling to regeneration. Cell Tissue Res 365(3):563–581
66. Guan J, Liu WQ, Xing MQ, Shi Y, Tan XY, Jiang CQ, Dai HY (2015) Elevated expression of periostin in diabetic cardiomyopathy and the effect of valsartan. BMC Cardiovasc Disord 15:90
67. Iekushi K, Taniyama Y, Azuma J, Katsuragi N, Dosaka N, Sanada F, Koibuchi N, Nagao K, Ogihara T, Morishita R (2007) Novel mechanisms of valsartan on the treatment of acute myocardial infarction through inhibition of the antiadhesion molecule periostin. Hypertens 49(6):1409–1414
68. Li Q, Liu X, Wei J (2014) Ageing related periostin expression increase from cardiac fibroblasts promotes cardiomyocytes senescent. Biochem Biophys Res Commun 452(3):497–502
69. Katsuragi N, Morishita R, Nakamura N, Ochiai T, Taniyama Y, Hasegawa Y, Kawashima K, Kaneda Y, Ogihara T, Sugimura K (2004) Periostin as a novel factor responsible for ventricular dilation. Circ 110(13):1806–1813
70. Zhao S, Wu H, Xia W, Chen X, Zhu S, Zhang S, Shao Y, Ma W, Yang D, Zhang J (2014) Periostin expression is upregulated and associated with myocardial fibrosis in human failing hearts. J Cardiol 63(5):373–378
71. Gupta S, Halushka MK, Hilton GM, Arking DE (2012) Postmortem cardiac tissue maintains gene expression profile even after late harvesting. BMC Genomics 13:26
72. Gao XM, Xu Q, Kiriazis H, Dart AM, Du XJ (2005) Mouse model of post-infarct ventricular rupture: time course, strain- and gender-dependency, tensile strength, and histopathology. Cardiovasc Res 65(2):469–477
73. Novella S, Heras M, Hermenegildo C, Dantas AP (2012) Effects of estrogen on vascular inflammation: a matter of timing. Arterioscler Thromb Vasc Biol 32(8):2035–2042
74. Prabhavathi K, Selvi KT, Poornima KN, Sarvanan A (2014) Role of biological sex in normal cardiac function and in its disease outcome—a review. J Clin Diagn Res 8(8):BE01–04
75. Parks RJ, Howlett SE (2013) Sex differences in mechanisms of cardiac excitation-contraction coupling. Pflugers Arch 465(5):747–763
76. Beale AL, Meyer P, Marwick TH, Lam CSP, Kaye DM (2018) Sex differences in cardiovascular pathophysiology: why women are overrepresented in heart failure with preserved ejection fraction. Circ 138(2):198–205
77. Dunlay SM, Roger VL, Redfield MM (2017) Epidemiology of heart failure with preserved ejection fraction. Nat Rev Cardiol 14(10):591–602

78. Dewan P, Jackson A, Jhund PS, Shen L, Ferreira JP, Petrie MC, Abraham WT, Desai AS, Dickstein K, Kober L, Packer M, Rouleau JL, Solomon SD, Swedberg K, Zile MR, McMurray JJV (2020) The prevalence and importance of frailty in heart failure with reduced ejection fraction - an analysis of PARADIGM-HF and ATMOSPHERE. Eur J Heart Fail 22(11):2123–2133
79. Chen X, Savarese G, Dahlstrom U, Lund LH, Fu M (2019) Age-dependent differences in clinical phenotype and prognosis in heart failure with mid-range ejection compared with heart failure with reduced or preserved ejection fraction. Clin Res Cardiol 108(12):1394–1405
80. Cavasin MA, Tao Z, Menon S, Yang XP (2004) Gender differences in cardiac function during early remodeling after acute myocardial infarction in mice. Life Sci 75(18):2181–2192
81. Fang L, Gao XM, Moore XL, Kiriazis H, Su Y, Ming Z, Lim YL, Dart AM, Du XJ (2007) Differences in inflammation, MMP activation and collagen damage account for gender difference in murine cardiac rupture following myocardial infarction. J Mol Cell Cardiol 43(5):535–544

# Chapter 18
# A Narrative Review of the Vascular Consequences of Estrogen Deficiency: A Focus on Female Athletes with Functional Hypothalamic Amenorrhea

**Emma O'Donnell and Paula J. Harvey**

**Abstract** Estrogen deficiency is linked with accelerated progression of atherosclerosis. Premenopausal women presenting with ovarian disruption due to functional hypothalamic amenorrhea (FHA) are characterised by estrogen deficiency. One common and reversible form of FHA in association with energy deficiency is exercise-associated amenorrhea (EAA). Despite being young, otherwise healthy and participating in regular exercise training, women with EAA demonstrate paradoxical changes in cardiovascular function, including impaired endothelial function, a known permissive factor for the progression and development of atherosclerosis. Such alterations suggest a profound effect of estrogen deficiency on vascular function. The long-term cardiovascular consequences of impaired vascular function in response to ovarian disruption in women with EAA remain to be determined. However, retrospective studies indicate premature development and progression of coronary artery disease in older premenopausal women reporting a history of hypothalamic ovarian disruption. Importantly, in women with EAA, estrogen therapy and resumption of menses restores endothelial function. In this review we focus on the influence of estrogen deficiency in association with energy deficiency in mediating changes in cardiovascular function in women with EAA, including endothelial function, regional blood flow, arterial stiffness and intima media thickness. The influence of energy deficiency and exercise training are also considered.

**Keywords** Arterial stiffness · Endothelial function · Energy deficiency · Estrogen · Exercise · Intima media thickness

---

E. O'Donnell (✉)
School of Sport and Exercise Health Sciences, National Centre of Sports and Exercise Medicine, Loughborough University, Loughborough, UK
e-mail: E.ODonnell@lboro.ac.uk

P. J. Harvey
Women's College Hospital, University of Toronto, Toronto, ON, Canada

© The Author(s), under exclusive license to Springer Nature Switzerland AG 2023 
L. Kirshenbaum and I. Rabinovich-Nikitin (eds.), *Biology of Women's Heart Health*, Advances in Biochemistry in Health and Disease 26,
https://doi.org/10.1007/978-3-031-39928-2_18

## Introduction

During their reproductive years, women are at lower risk of cardiovascular disease (CVD) than age matched men [1]. After menopause, this cardiovascular advantage is lost, such that by the 7th decade of life CVD risk in women surpasses that observed in age-matched men [2]. The loss of cardioprotection in women after menopause has been attributed, in part, to the beneficial effects of endogenous estrogen. Studies indicate that 17β-estradiol confers cardioprotection via numerous multifaceted mechanisms that act upon the myocardium and vascular system [3, 4]. For example, estrogen deficiency is associated with decreased production of endothleium-derived nitric oxide (NO), increased oxidative stress and changes in regulation of factors involved in inflammatory and coagulation/fibrinolytic cascades [3], each of which are independently linked with impaired endothelial function, increased arterial stiffness and carotid intima media thickness [5–11]. Accordingly, estrogen deficiency due to ovarian disruption during the premenopausal years is thought to contribute to premature development and accelerated progression of atherosclerosis, possibly via one or more of these mechanisms [12]. This notion is supported in part by evidence linking a history of menstrual irregularities due to either organic or functional hypothalamic origin to increased future cardiovascular disease risk in women [13, 14].

In premenopausal women, functional hypothalamic amenorrhea (FHA) is a reversible cause of non-organic ovarian disruption [15]. It is characterized by the absence of menses and marked estrogen deficiency due to suppression of the hypothalamic–pituitary–ovarian (HPO) axis [16]. Three commonly inter-related types of FHA are observed: stress-related amenorrhea, weight loss-related amenorrhea and exercise-associated amenorrhea (EAA) [16]. The current review focuses on women with EAA. EAA has been reported in occupationally, recreationally and competitively physically active women, and has been causally linked with energy deficiency due to insufficient caloric intake to meet high energy expenditure levels (e.g. exercise) [17]. In contrast to the well-documented adverse cardiovascular consequences of FHA in association with clinical eating disorders [18] or psychosocial stress [19], the cardiovascular sequelae of estrogen deficiency due to EAA in women who are weight-stable and otherwise healthy are less well-described. In these women, the absence of co-morbidities, organic endocrine dysfunction, chronic malnutrition, and ageing, provides a unique model to investigate the cardiovascular effects of endogenous estrogen, and estrogen deficiency in premenopausal women.

At the level of the vasculature, endogenous estrogen plays a vital role in both the maintenance of vessel integrity and vascular function. This role is integral to the development and progression of atherosclerosis, with impaired vascular function evident decades prior to overt coronary artery disease (CAD) [20]. Given that estrogen deficiency results in loss of the vasculo-protective effects of estrogen, and that premenopausal women are typically overlooked in terms of their CVD risk [21], understanding the possible cardiovascular consequences of estrogen deficiency on vascular health in otherwise healthy, young, estrogen deficient exercising women with EAA is of importance. Thus, the aims of this chapter will be to focus on the vascular effects of hypoestrogenemia due to ovulatory disruption in women with EAA. Due to the etiology of EAA in this group, the effects of aerobic exercise training and energy deficiency on cardiovascular function are also considered. For the purpose of this chapter, the term estrogen will refer to endogenous 17 beta-estradiol, and the term energy deficiency will refer to caloric deficit, due to either insufficient caloric intake to meet the energy demands of the individual (e.g., low energy availability), or due to caloric restriction per se. The sources, physiological actions, metabolism, and transport of estrogen are well characterized and are not reviewed here [22–25]. The cardiovascular consequences of psychogenic stress (e.g., anxiety), eating disorders (e.g., anorexia nervosa), and polycystic ovarian syndrome (PCOS), all of which are independently associated with ovulatory disruption, have also been described in detail elsewhere [14, 18, 26, 27].

## Ovarian Disruption in Exercising Premenopausal Women

Perturbations to the hypothalamic–pituitary–ovarian (HPO) axis elicits menstrual disturbances (MD). MD exist across a spectrum, from subtle and asymptomatic (luteal phase defect and anovulation) to severe and symptomatic (oligomenorrhea and amenorrhea). Not all women who participate in physical activity will experience MD. However, hypothalamic MD are more common among physically active women, including women with physically demanding occupations versus sedentary women [28–33]. For example, studies report that across 166 menstrual cycles from 87 eumenorrheic women, asymptomatic MD were identified in only 5% of cycles in sedentary women compared with 48% of cycles in physically active women [29]. The estimated prevalence of the most severe menstrual disturbance, amenorrhea (i.e., cessation of menses for at least 90 consecutive days [34]), is also markedly higher among physically active versus inactive women, ranging between 2 and 46% among competitive and recreational athletes [29, 32, 33, 35], as compared with 2–5% in non-exercising women [29, 34, 36]. It is noteworthy that the prevalence of amenorrhea among competitive athletes and recreationally physically active women are reported to be similar [29, 34, 36], identifying that this severe MD is not exclusive to competitive athletes.

MD in exercising women have been causally linked to energy deficiency, or low energy availability, in association with a high energy expenditure (i.e., exercise) frequently combined with restrictive eating patterns without obvious caloric deficits [17, 37, 38]. In women with EAA, metabolic and hormonal perturbations include decreased resting metabolic rate, and low circulating levels of estrogen, triiodothyronine, insulin, glucose, IGF-1 and leptin, and elevated levels of cortisol, ghrelin, PYY, and adiponectin [39–43]. The physiological and health consequences associated with these energy deficiency related perturbations in exercising women include MD and cardiovascular alterations [29, 44]. They also include, impaired bone health [30,

42], with possible detrimental effects to the immune, gastro-intestinal, and musculoskeletal systems [45]. Decrements in exercise performance and increased risk for injury are also postulated [45].

The suppression of reproductive function in response to energy deficiency is initiated through inhibition of the hypothalamic gonadotropin-releasing-hormone (GnRH) pulse generator, causing altered amplitude and frequency of the pituitary release of luteinising hormone, causing ovarian disruption. A genetic predisposition to susceptibility to developing FHA has also been identified [16]. Importantly, exercise per se does not disrupt menstrual function when energy balance (i.e., a eucaloric state) is maintained in exercising women [37]. Furthermore, no set weight point or critical fat mass has been identified that enables menstruation [46].

## Cardiovascular Consequences of Ovarian Dusruption in Premenopausal Women

CVD remains one of the leading causes of premature death in women in the United States [47] and Canada [48], yet the pathophysiology of CVD in women, particularly premenopausal women, remains surprisingly under-investigated. Given that regular cyclic 'normal' ovarian function promotes cardioprotection, ovarian disruption may conversely impair cardiovascular health in young premenopausal women. Despite the paucity of data, this notion is supported by studies reporting adverse cardiovascular effects of both subclinical ovarian disruptions and altered ovarian hormone levels. For example, studies examining the potential influence of fluctuations in endogenous estrogen levels across the menstrual cycle on cardiac events, such as unstable angina or myocardial infarction, demonstrate that they are more likely to occur during the early follicular (low estrogen) compared with all other phases of the menstrual cycle [49, 50]. Most [13, 51–53] but not all [54] studies also report that decreased estrogen exposure due to irregular or decreased number of menstrual cycles per year, is identified as an independent contributing factor to the development and progression of premenopausal CAD. It is pertinent to note that in women with irregular menstrual cycles of organic origin, such as PCOS, estrogen deficiency is more likely to co-exist with other co-morbidities, such as obesity, metabolic syndrome, or hyperinsulinemia [55], making it difficult to distinguish the independent cardiovascular effects of estrogen deficiency in these women. Among the largest prospective cohort studies in women, the Nurses Health Study, identified that a history of irregular menstrual cycles during the reproductive years is linked with a 53% increased risk for fatal or nonfatal CAD compared with women reporting a history of regular cycles [13]. A smaller case-controlled retrospective study of 202 women similarly reported that in premenopausal women under the age of 55 years with confirmed myocardial infarction, a higher lifelong incidence of irregular menstrual cycles were reported compared with controls [52].

The potential deleterious effects of hypothalamic hypoestrogenemia due to FHA on the cardiovascular system was investigated in the Women's Ischemia Syndrome Evaluation (WISE) [53]. This study examined the relationship between circulating estrogen levels and angiographically confirmed CAD in a small group (n = 95) of untrained premenopausal women [53]. The investigators observed that independent of other risk factors, estrogen deficiency, due to psychogenic stress-related FHA, was identified as a risk factor for angiographically documented CAD [53]. Taken together, these findings support an important role for 'normal' ovulatory cycling and maintenance of cardiovascular health in premenopausal women. Prospective larger studies to examine the cardiovascular effects of ovulatory disruption in premenopausal women are clearly warranted.

## Nitric Oxide and the Endothelium: Effects of Estrogen, Exercise and Caloric Restriction

As the mono-layer of cells lining the vasculature, the endothelium participates in all aspects of vascular health. This is achieved in part by the balanced synthesis and release of vasoactive substances from the endothelial cells, including NO, a potent vasodilator. In addition to modulating vascular tone, NO also exerts vasculo-protective effects through multiple mechanisms, including inhibition of inflammation and platelet aggregation, and suppression of smooth muscle cell proliferation [3, 56]. Thus, NO modulates both arterial function and structure and plays a central role in maintaining vascular health, and is considered to be "atheroprotective". Accordingly, decreased endothelium-derived NO production and/or bioavailability is both a characteristic of, and contributor to, disruption of vascular function and structure, including: (i) endothelial dysfunction, a recognized permissive factor for the development and progression of atherosclerosis [56]; (ii) increased arterial stiffness, an independent risk factor for the development of hypertension, left ventricular hypertrophy and heart failure [57], and (iii) increased carotid intima media thickness, a surrogate marker of early CVD that is a strong predictor of coronary artery disease in both pre- and postmenopausal women [58].

Through different mechanisms of activation, estrogen, caloric restriction and regular aerobic exercise enhance the synthesis and/or bioavailability of endothelial NO through a shared signaling cascade, the PI3K/Akt pathway [59–61]. Once formed, NO diffuses into the adjacent smooth muscle cells and causes vasodilation. Upon estrogen binding with its plasma-membrane receptor, estrogen rapidly activates endothelial NO synthase (eNOS) to produce NO via the PI3K/AKT pathway [61]. In addition to increasing production, estrogen also increases the bioavailablity of NO via anti-oxidant effects [61]. Physiologic shear stress, the frictional force of blood flow directly on the vessel wall, creates a mechanical signal that is sensed by endothelial cells and transduced to activate signaling cascades that activate eNOS to produce NO [60]. Dynamic exercise elicits marked increases in blood flow, resulting in large increases in shear stress and augmented production of NO. This effect is seen with both acute (e.g. a single bout of exercise) and chronic exercise training [60], the latter resulting in repeated exposure of the endothelium to increased shear stress and elevated NO production. This repeated exposure elicits vascular adaptations that are associated with attenuation of age-associated declines in endothelial function [62]. Life-long caloric restriction, defined as a reduction in energy intake without malnutrition, is also associated with the increased production and bioavailability of NO [63], and has been linked with attenuated age-associated declines in endothelial function [59].

## Vascucular Findings in EAA: Endothelial Function

Despite being aerobically trained and otherwise healthy, the beneficial effect of aerobic exercise training on resting endothelial function appears to be obviated in EAA. Specifically, despite similar resting brachial artery diameter, brachial artery flow mediated dilation (FMD), a shear stress NO-mediated event, is lower in EAA athletes compared with their regularly menstruating counterparts [64–69]. Importantly, FMD values in EAA are often very similar to that reported in older estrogen deficient exercise trained postmenopausal women [70], suggesting a profound effect of estrogen deficiency on endothelial function (see Fig. 18.1) and vascular adaptations to exercise training in both premenopausal and postmenopausal women.

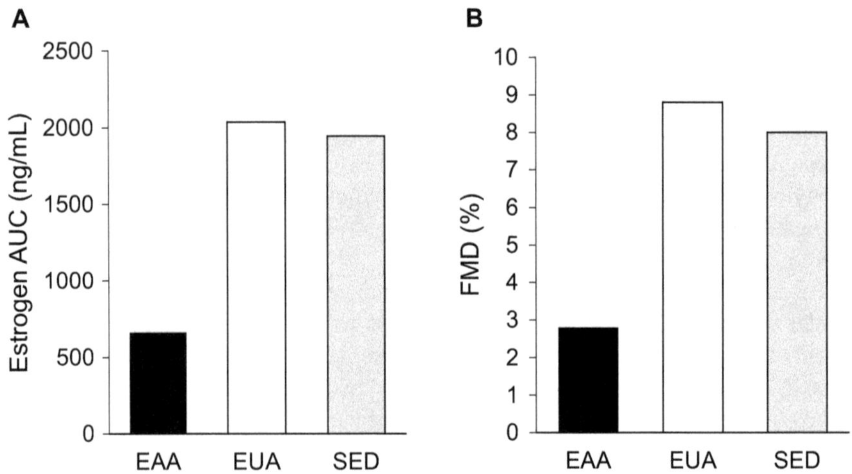


**Fig. 18.1** Estrogen exposure determined by daily urinary analysis of estrone glucuronide, the dominant metabolite of endogenous estrogen, across the menstrual cycle for eumenorrheic athletes (EUA) and sedentary women (SED), and across a 30-day monitoring period for EAA (A). FMD for EAA, EUA and SED (B). [71][67] Adapted from O'Donnell et al., 2007 and O'Donnell et al. 2014

Microvascular (i.e. small vessel) function is also reported to be impaired in EAA. Resting and peak-ischemic calf blood flow, estimated using venous occlusion strain-gauge plethysmography, is lower in physically active women with long-term EAA (average 271 days) versus physically active and sedentary ovulatory women [71]. Paradoxically, despite elevated vascular resistance in women with EAA, their blood pressure is significantly lower than that observed in their eumenorrheic counterparts [43, 71]. Of interest, several studies suggest that duration of estrogen deficiency may be important, with longer duration possibly resulting in both structural and functional vascular alterations [71]. This notion is supported in part by most [66, 67] but not all [65] studies reporting blunted endothelium-independent function in the brachial artery, assessed using exogenous sublingual NO administration (nitroglycerin). Blunted response to exogenous NO suggests vascular structure may be altered, and/or vascular smooth muscle cell responsiveness to NO may be impaired in EAA.

To investigate whether a single bout of exercise, a known NO stimulus, could restore macro-and micro-vascular endothelial function in EAA, O'Donnell et al. [67] assessed brachial artery FMD and calf blood flow before and one-hour after 45-min of dynamic exercise. Both FMD and calf blood flow were augmented in EAA after exercise, yet FMD remained lower while calf blood flow was 'restored' to levels seen in eumenorrheic ovulatory athletes, indicating possible differences in the vascular effects of estrogen deficiency in the: i) response to acute exercise in different vascular beds; and/or ii) vascular adaptations to chronic aerobic exercise training in the trained versus untrained limb (e.g. upper versus lower body).

In addition to increased vascular resistance, decreased vascular compliance and lower shear rate, other possible factors that may influence endothelial function in EAA include augmented muscle sympathetic nerve activity, and elevated circulating levels of oxidative stress and low-density lipoproteins, all of which have been reported in amenorrheic versus eumenorrheic athletes [65, 72–74]. Plasma low-density lipoprotein has been shown to be inversely associated with number of menstrual cycles per year and with brachial artery FMD [65]. Collectively, these findings are in keeping with the known vasomodulatory effects of the sympathetic nervous system [75], oxidative stress [76] and circulating lipids [77]. They are also in alignment with the known sympatho-inhibitory [78], anti-oxidant [79] and lipid metabolism [80] effects of estrogen.

## Restoration of Endothelial Function in EAA

Resumption of the normal menstrual cycle and therefore endogenous estrogen levels [66, 81], and administration of exogenous estrogen with oral contraceptive use [64] have both been demonstrated to improve FMD in EAA. Endothelium-independent function has also been reported to be restored in EAA after resumption of menses [66]. Importantly, these findings identify that impairment in endothelium-dependent and -independent function in EAA are reversible.

## Arterial Stiffness: Effects of Estrogen, Exercise and Caloric Restriction

Even in health, aging leads to progressive stiffening of the arterial tree in both men and women [82], contributing to the increased incidence of age-associated hypertension and atherosclerotic disease, and increased risk of CV events [57]. Stiffening of the arterial wall is caused by multiple mechanisms [83]. These age related changes lead to reduced ability of the large elastic arteries to expand and recoil in response to changes in pressure [57]. Changes in endothelium derived mediators, including decreased production of NO, also contribute to arterial stiffness [84].

Estrogen, exercise training and caloric restriction are independently associated with decreased arterial stiffness [59, 85, 86] predominantly through favorable effects on the endothelium. While exact mechanisms beyond the effects on endothelial function are unclear, estrogen, exercise training and caloric restriction are associated with decreased levels of oxidative stress [59, 61, 86]. Attenuation of the age-associated increase in oxidative stress is postulated to attenuate or prevent elastin degradation, protein oxidation (nitrotyrosine), and formation of advanced glycation end products [86]. The loss of estrogen due to menopause is associated with a steeper age-associated increase in arterial stiffness [82], with the first year after the final menstrual period identified as a critical window when marked increases in stiffness occur [87]. Notably, short-term (4 weeks) and long-term (~6 years) estrogen therapy lowers indices of arterial stiffness in postmenopausal women [88]. Aerobic exercise training has also been shown to attenuate the age-associated increase in stiffness in postmenopausal women [89].

## Vascular Findings in EAA: Arterial Stiffness

To date, only two studies have examined indices of arterial stiffness in EAA [69, 90]. In one study, central arterial stiffness assessed by pulsewave velocity was reported to be similar between EAA, eumenorrheic athletes and recreationally active controls [69]. In a separate study, indices of aortic stiffness, specifically augmentation index and augmentation pressure, were also demonstrated to be similar between EAA and eumenorrheic athletes [90]. In contrast, in another study using different indices, arterial stiffness, namely augmentation index adjusted for heart rate, was reported to be significantly lower in EAA (−11%) versus eumenorrheic athletes (−1%) [90]. Collectively, these findings suggest that despite estrogen deficiency, arterial structure is well preserved in EAA. This observation is in stark contrast to findings in estrogen deficient postmenopausal women [87].

## Intima Media Thickness: Effects of Estrogen, Exercise and Caloric Restriction

Increased intima-media thickness (IMT) is a recognised important surrogate marker for atherosclerosis [91]. Intima-media thickening in the femoral arteries is reported to occur earlier than in the carotid, and as such, is thought to better reflect the true extent of generalized atherosclerosis than that in the carotids [91]. Thus, IMT in the femoral arteries provides a useful tool for the detection of atherosclerosis in its early stages [91]. Mechanisms that increase IMT are shared factors that are responsible for the development and progression of atherosclerosis [91].

Through preservation of arterial structure and function, estrogen, exercise training and caloric restriction are independently associated with preserved or low (favourable) femoral and carotid IMT [92–96].

## Vascular Findings in EAA: Intima Media Thickness

In the only study to investigate IMT, femoral and brachial artery IMT in EAA was reported to not differ from that observed in eumenorrheic athletes [69]. In contrast, femoral, but not brachial, artery IMT was lower in EAA compared to eumenorrheic sedentary women [69]. However, in this study, 50% of EAA were taking oral contraceptives, making interpretation of the effects of estrogen deficiency on IMT difficult to determine. Thus, atherosclerotic risk in EAA remains unclear.

## Summary and Conclusion

Irregular menstrual cycles are associated with increased risk of premature development of atherosclerosis and CVD and events. Despite being exercise trained, otherwise healthy premenopausal women with EAA demonstrate abnormal vascular function, and possibly altered structure, including impaired endothelium-dependent and -independent function. Endothelial dysfunction is a recognized permissive factor for the development and progression of atherosclerosis. Loss of the cardioprotective effects of estrogen due to ovarian disruption is thought to play a key role. Notably, estrogen deficiency in women with EAA is similar to that observed in older estrogen deficient postmenopausal women. Importantly, endothelial dysfunction is reversible in EAA with resumption of normal menstrual cycles and normal estrogen exposure, and exogenous estrogen administration with hormonal contraceptives. Further, FHA in amenorrheic athletes is not due to exercise training per se, rather it is causally related to energy deficiency due to insufficient caloric intake to meet high energy expenditure demands. Thus, the observed vascular perturbations in these women are likely due to complex interactions between the vasomodulatory effects of estrogen, energy deficiency and exercise training. More studies are needed to determine the long-term cardiovascular consequences of estrogen deficiency in women with EAA.

## References

1. Bots SH, Peters SAE, Woodward M (2017) Sex differences in coronary heart disease and stroke mortality: a global assessment of the effect of ageing between 1980 and 2010. BMJ Glob Health 2(2):e000298
2. Leening MJ, Ferket BS, Steyerberg EW et al (2014) Sex differences in lifetime risk and first manifestation of cardiovascular disease: prospective population based cohort study. BMJ 349:g5992
3. Knowlton AA, Lee AR (2012) Estrogen and the cardiovascular system. Pharmacol Ther 135(1):54–70
4. Reckelhoff JF (2005) Sex steroids, cardiovascular disease, and hypertension: unanswered questions and some speculations. Hypertens 45(2):170–174
5. Figueroa-Vega N, Moreno-Frias C, Malacara JM (2015) Alterations in adhesion molecules, pro-inflammatory cytokines and cell-derived microparticles contribute to intima-media thickness and symptoms in postmenopausal women. PLoS ONE 10(5):e0120990
6. Hildreth KL, Kohrt WM, Moreau KL (2014) Oxidative stress contributes to large elastic arterial stiffening across the stages of the menopausal transition. Menopause 21(6):624–632
7. Moreau KL, Deane KD, Meditz AL, Kohrt WM (2013) Tumor necrosis factor-alpha inhibition improves endothelial function and decreases arterial stiffness in estrogen-deficient postmenopausal women. Atherosclerosis 230(2):390–396
8. Moreau KL, Stauffer BL, Kohrt WM, Seals DR (2013) Essential role of estrogen for improvements in vascular endothelial function with endurance exercise in postmenopausal women. J Clin Endocrinol Metab 98(11):4507–4515
9. Pulvirenti D, Signorelli S, Sciacchitano S et al (2007) Hyperhomocysteinemia, oxidative stress, endothelial dysfunction in postmenopausal women. Clin Ter 158(3):213–217
10. Wykretowicz J, Guzik P, Krauze T et al (2012) Fibrinogen and d-dimer in contrasting relation with measures of wave reflection and arterial stiffness. Scand J Clin Lab Invest 72(8):629–634
11. Pawlak K, Naumnik B, Brzosko S, Pawlak D, Mysliwiec M (2004) Oxidative stress—a link between endothelial injury, coagulation activation, and atherosclerosis in haemodialysis patients. Am J Nephrol 24(1):154–161
12. Kaplan JR (2008) Origins and health consequences of stress-induced ovarian dysfunction. Interdiscip Top Gerontol 36:162–185
13. Solomon CG, Hu FB, Dunaif A et al (2002) Menstrual cycle irregularity and risk for future cardiovascular disease. J Clin Endocrinol Metab 87(5):2013–2017
14. Cussons AJ, Stuckey BG, Watts GF (2006) Cardiovascular disease in the polycystic ovary syndrome: new insights and perspectives. Atherosclerosis 185(2):227–239
15. Liu JH (1990) Hypothalamic amenorrhea: clinical perspectives, pathophysiology, and management. Am J Obstet Gynecol 163(5 Pt 2):1732–1736
16. Gordon CM (2010) Clinical practice. Functional hypothalamic amenorrhea. N Engl J Med 363(4):365–371
17. Williams NI, Helmreich DL, Parfitt DB, Caston-Balderrama A, Cameron JL (2001) Evidence for a causal role of low energy availability in the induction of menstrual cycle disturbances during strenuous exercise training. J Clin Endocrinol Metab 86(11):5184–5193
18. Casiero D, Frishman WH (2006) Cardiovascular complications of eating disorders. Cardiol Rev 14(5):227–231
19. Brezinka V, Kittel F (1996) Psychosocial factors of coronary heart disease in women: a review. Soc Sci Med 42(10):1351–1365

20. Deanfield JE, Halcox JP, Rabelink TJ (2007) Endothelial function and dysfunction: testing and clinical relevance. Circulation 115(10):1285–1295
21. Kuehn BM (2019) Rising heart risks for young women linked to low Estrogen. Circulation 139(4):549–550
22. Gruber CJ, Tschugguel W, Schneeberger C, Huber JC (2002) Production and actions of estrogens. N Engl J Med 346(5):340–352
23. Zhu BT, Conney AH (1998) Functional role of estrogen metabolism in target cells: review and perspectives. Carcinogenesis 19(1):1–27
24. Kuhl H (2005) Pharmacology of estrogens and progestogens: influence of different routes of administration. Climacteric 8 Suppl 1:3–63
25. Labrie F, Luu-The V, Labrie C et al (2003) Endocrine and intracrine sources of androgens in women: inhibition of breast cancer and other roles of androgens and their precursor dehydroepiandrosterone. Endocr Rev 24(2):152–182
26. Giallauria F, Orio F, Palomba S, Lombardi G, Colao A, Vigorito C (2008) Cardiovascular risk in women with polycystic ovary syndrome. J Cardiovasc Med (Hagerstown) 9(10):987–992
27. Pauli SA, Berga SL (2010) Athletic amenorrhea: energy deficit or psychogenic challenge? Ann N Y Acad Sci 1205:33–38
28. Cline AD, Jansen GR, Melby CL (1998) Stress fractures in female army recruits: implications of bone density, calcium intake, and exercise. J Am Coll Nutr 17(2):128–135
29. De Souza MJ, Toombs RJ, Scheid JL, O'Donnell E, West SL, Williams NI (2010) High prevalence of subtle and severe menstrual disturbances in exercising women: confirmation using daily hormone measures. Hum Reprod 25(2):491–503
30. De Souza MJ, West SL, Jamal SA, Hawker GA, Gundberg CM, Williams NI (2008) The presence of both an energy deficiency and estrogen deficiency exacerbate alterations of bone metabolism in exercising women. Bone 43(1):140–148
31. Friedl KE, Nuovo JA, Patience TH, Dettori JR (1992) Factors associated with stress fracture in young army women: indications for further research. Mil Med 157(7):334–338
32. Shangold MM (1984) Menstrual disturbances in the athlete. Prim Care 11(1):109–114
33. Shangold MM, Levine HS (1982) The effect of marathon training upon menstrual function. Am J Obstet Gynecol 143(8):862–869
34. Otis CL, Drinkwater B, Johnson M, Loucks A, Wilmore J (1997) American college of sports medicine position stand. The Female Athlete Triad. Med Sci Sports Exerc 29(5):i–ix
35. De Souza MJ, Miller BE, Loucks AB et al (1998) High frequency of luteal phase deficiency and anovulation in recreational women runners: blunted elevation in follicle-stimulating hormone observed during luteal-follicular transition. J Clin Endocrinol Metab 83(12):4220–4232
36. Torstveit MK, Sundgot-Borgen J (2005) Participation in leanness sports but not training volume is associated with menstrual dysfunction: a national survey of 1276 elite athletes and controls. Br J Sports Med 39(3):141–147
37. Loucks AB, Verdun M, Heath EM (1998) Low energy availability, not stress of exercise, alters LH pulsatility in exercising women. J Appl Physiol 84(1):37–46
38. Warren MP, Voussoughian F, Geer EB, Hyle EP, Adberg CL, Ramos RH (1999) Functional hypothalamic amenorrhea: hypoleptinemia and disordered eating. J Clin Endocrinol Metab 84(3):873–877
39. De Souza MJ, Leidy HJ, O'Donnell E, Lasley B, Williams NI (2004) Fasting ghrelin levels in physically active women: relationship with menstrual disturbances and metabolic hormones. J Clin Endocrinol Metab 89(7):3536–3542
40. Laughlin GA, Yen SS (1996) Nutritional and endocrine-metabolic aberrations in amenorrheic athletes. J Clin Endocrinol Metab 81(12):4301–4309
41. Loucks AB, Mortola JF, Girton L, Yen SS (1989) Alterations in the hypothalamic-pituitary-ovarian and the hypothalamic-pituitary-adrenal axes in athletic women. J Clin Endocrinol Metab 68(2):402–411
42. O'Donnell E, De Souza MJ (2010) Increased serum adiponectin concentrations in amenorrheic physically active women are associated with impaired bone health but not with estrogen exposure. Bone 48(4):760–767

43. O'Donnell E, Harvey PJ, De Souza MJ (2009) Relationships between vascular resistance and energy deficiency, nutritional status and oxidative stress in oestrogen deficient physically active women. Clin Endocrinol (Oxf) 70(2):294–302
44. O'Donnell E, Goodman JM, Harvey PJ (2011) Cardiovascular consequences of ovarian disruption: a focus on functional hypothalamic amenorrhea in physically active women. J Clin Endocrinol Metab 96(12):3638–3648
45. Mountjoy M, Sundgot-Borgen JK, Burke LM et al (2018) IOC consensus statement on relative energy deficiency in sport (RED-S): 2018 update. Br J Sports Med 52(11):687–697
46. Goodman LR, Warren MP (2005) The female athlete and menstrual function. Curr Opin Obstet Gynecol 17(5):466–470
47. Cushman M, Shay CM, Howard VJ et al (2021) Ten-year differences in women's awareness related to coronary heart disease: results of the 2019 American heart association national survey: a special report from the American heart association. Circ 143(7):e239–e248
48. Norris CM, Yip CYY, Nerenberg KA et al (2020) State of the science in women's cardiovascular disease: a canadian perspective on the influence of sex and gender. J Am Heart Assoc 9(4):e015634
49. Hamelin BA, Methot J, Arsenault M et al (2003) Influence of the menstrual cycle on the timing of acute coronary events in premenopausal women. Am J Med 114(7):599–602
50. Mukamal KJ, Muller JE, Maclure M, Sherwood JB, Mittleman MA (2002) Variation in the risk of onset of acute myocardial infarction during the menstrual cycle. Am J Cardiol 90(1):49–51
51. Vaccarino V, Parsons L, Every NR, Barron HV, Krumholz HM (1999) Sex-based differences in early mortality after myocardial infarction. National registry of myocardial infarction 2 participants. N Engl J Med 341(4):217–225
52. La Vecchia C, Decarli A, Franceschi S, Gentile A, Negri E, Parazzini F (1987) Menstrual and reproductive factors and the risk of myocardial infarction in women under fifty-five years of age. Am J Obstet Gynecol 157(5):1108–1112
53. Bairey Merz CN, Johnson BD, Sharaf BL et al (2003) Hypoestrogenemia of hypothalamic origin and coronary artery disease in premenopausal women: a report from the NHLBI-sponsored WISE study. J Am Coll Cardiol 41(3):413–419
54. Merz CN, Johnson BD, Berga SL et al (2009) Total estrogen time and obstructive coronary disease in women: insights from the NHLBI-sponsored Women's Ischemia Syndrome Evaluation (WISE). J Womens Health (Larchmt) 18(9):1315–1322
55. Meyer C, McGrath BP, Teede HJ (2005) Overweight women with polycystic ovary syndrome have evidence of subclinical cardiovascular disease. J Clin Endocrinol Metab 90(10):5711–5716
56. Gimbrone MA Jr, Garcia-Cardena G (2016) Endothelial cell dysfunction and the pathobiology of atherosclerosis. Circ Res 118(4):620–636
57. Weber T, Chirinos JA (2018) Pulsatile arterial haemodynamics in heart failure. Eur Heart J 39(43):3847–3854
58. Zhou Y, Wang D, Yang X et al (2015) Effect of menopausal status on carotid intima-media thickness and presence of carotid plaque in Chinese women generation population. Sci Rep 5:8076
59. Donato AJ, Walker AE, Magerko KA et al (2013) Life-long caloric restriction reduces oxidative stress and preserves nitric oxide bioavailability and function in arteries of old mice. Aging Cell 12(5):772–783
60. Padilla J, Simmons GH, Bender SB, Arce-Esquivel AA, Whyte JJ, Laughlin MH (2011) Vascular effects of exercise: endothelial adaptations beyond active muscle beds. Physiology (Bethesda) 26(3):132–145
61. Mendelsohn ME, Karas RH (2005) Molecular and cellular basis of cardiovascular gender differences. Sci 308(5728):1583–1587
62. Pierce GL, Eskurza I, Walker AE, Fay TN, Seals DR (2011) Sex-specific effects of habitual aerobic exercise on brachial artery flow-mediated dilation in middle-aged and older adults. Clin Sci 120(1):13–23

63. Garcia-Prieto CF, Fernandez-Alfonso MS (2016) Caloric restriction as a strategy to improve vascular dysfunction in metabolic disorders. Nutr 8(6)
64. Rickenlund A, Eriksson MJ, Schenck-Gustafsson K, Hirschberg AL (2005) Oral contraceptives improve endothelial function in amenorrheic athletes. J Clin Endocrinol Metab 90(6):3162–3167
65. Rickenlund A, Eriksson MJ, Schenck-Gustafsson K, Hirschberg AL (2005) Amenorrhea in female athletes is associated with endothelial dysfunction and unfavorable lipid profile. J Clin Endocrinol Metab 90(3):1354–1359
66. Yoshida N, Ikeda H, Sugi K, Imaizumi T (2006) Impaired endothelium-dependent and -independent vasodilation in young female athletes with exercise-associated amenorrhea. Arterioscler Thromb Vasc Biol 26(1):231–232
67. O'Donnell E, Goodman JM, Mak S, Harvey PJ (2014) Impaired Vascular function in physically active premenopausal women with functional hypothalamic amenorrhea is associated with low shear stress and increased vascular tone. J Clin Endocrinol Metab 99(5):1798–1806
68. Zeni Hoch A, Dempsey RL, Carrera GF et al (2003) Is there an association between athletic amenorrhea and endothelial cell dysfunction? Med Sci Sports Exerc 35(3):377–383
69. Augustine JA, Lefferts WK, Dowthwaite JN, Brann LS, Brutsaert TD, Heffernan KS (2016) Subclinical atherosclerotic risk in endurance-trained premenopausal amenorrheic women. Atherosclerosis 244:157–164
70. Santos-Parker JR, Strahler TR, Vorwald VM, Pierce GL, Seals DR (2017) Habitual aerobic exercise does not protect against micro- or macrovascular endothelial dysfunction in healthy estrogen-deficient postmenopausal women. J Appl Physiol (1985) 122(1):11–19
71. O'Donnell E, Harvey PJ, Goodman JM, De Souza MJ (2007) Long-term estrogen deficiency lowers regional blood flow, resting systolic blood pressure, and heart rate in exercising premenopausal women. Am J Physiol Endocrinol Metab 292(5):E1401–1409
72. Ayres S, Baer J, Subbiah MT (1998) Exercised-induced increase in lipid peroxidation parameters in amenorrheic female athletes. Fertil Steril 69(1):73–77
73. O'Donnell E, Goodman JM, Mak S et al (2015) Discordant orthostatic reflex renin-angiotensin and sympathoneural responses in premenopausal exercising-hypoestrogenic women. Hypertens 65(5):1089–1095
74. Lamon-Fava S, Fisher EC, Nelson ME et al (1989) Effect of exercise and menstrual cycle status on plasma lipids, low density lipoprotein particle size, and apolipoproteins. J Clin Endocrinol Metab 68(1):17–21
75. Sheng Y, Zhu L (2018) The crosstalk between autonomic nervous system and blood vessels. Int J Physiol Pathophysiol Pharmacol 10(1):17–28
76. Sena CM, Leandro A, Azul L, Seica R, Perry G (2018) Vascular oxidative stress: impact and therapeutic approaches. Front Physiol 9:1668
77. Kawano H, Motoyama T, Hirai N, Kugiyama K, Yasue H, Ogawa H (2002) Endothelial dysfunction in hypercholesterolemia is improved by L-arginine administration: possible role of oxidative stress. Atherosclerosis 161(2):375–380
78. Saleh TM, Connell BJ (2007) Role of oestrogen in the central regulation of autonomic function. Clin Exp Pharmacol Physiol 34(9):827–832
79. Xiang D, Liu Y, Zhou S, Zhou E, Wang Y (2021) Protective effects of Estrogen on cardiovascular disease mediated by oxidative stress. Oxid Med Cell Longev 2021:5523516
80. Campos H, Walsh BW, Judge H, Sacks FM (1997) Effect of estrogen on very low density lipoprotein and low density lipoprotein subclass metabolism in postmenopausal women. J Clin Endocrinol Metab 82(12):3955–3963
81. Hoch AZ, Jurva JW, Staton MA et al (2007) Athletic amenorrhea and endothelial dysfunction. WMJ 106(6):301–306
82. Smulyan H, Asmar RG, Rudnicki A, London GM, Safar ME (2001) Comparative effects of aging in men and women on the properties of the arterial tree. J Am Coll Cardiol 37(5):1374–1380
83. Lacolley P, Regnault V, Laurent S (2020) Mechanisms of arterial stiffening: from Mechanotransduction to epigenetics. Arterioscler Thromb Vasc Biol 40(5):1055–1062

84. Wilkinson IB, McEniery CM (2004) Arterial stiffness, endothelial function and novel pharmacological approaches. Clin Exp Pharmacol Physiol 31(11):795–799
85. Miura T, Muraoka S, Ogiso T (1996) Inhibition of lipid peroxidation by estradiol and 2-hydroxyestradiol. Steroids 61(6):379–383
86. Santos-Parker JR, LaRocca TJ, Seals DR (2014) Aerobic exercise and other healthy lifestyle factors that influence vascular aging. Adv Physiol Educ 38(4):296–307
87. Samargandy S, Matthews KA, Brooks MM et al (2020) Arterial stiffness accelerates within 1 year of the final menstrual period: the swan heart study. Arterioscler Thromb Vasc Biol 40(4):1001–1008
88. Miura S, Tanaka E, Mori A et al (2003) Hormone replacement therapy improves arterial stiffness in normotensive postmenopausal women. Maturitas 45(4):293–298
89. Matsubara T, Miyaki A, Akazawa N et al (2014) Aerobic exercise training increases plasma Klotho levels and reduces arterial stiffness in postmenopausal women. Am J Physiol Heart Circ Physiol 306(3):H348–355
90. O'Donnell E, Goodman JM, Floras JS, Harvey PJ (2020) Indexes of aortic wave reflection are not augmented in estrogen-deficient physically active premenopausal women. Scand J Med Sci Sports 30(6):1054–1063
91. Stein JH, Korcarz CE, Hurst RT et al (2008) Use of carotid ultrasound to identify subclinical vascular disease and evaluate cardiovascular disease risk: a consensus statement from the American Society of Echocardiography Carotid Intima-Media Thickness Task Force. Endorsed by the Society for Vascular Medicine. J Am Soc Echocardiogr 21(2):93–111; quiz 189–190
92. Fontana L, Meyer TE, Klein S, Holloszy JO (2004) Long-term calorie restriction is highly effective in reducing the risk for atherosclerosis in humans. Proc Natl Acad Sci USA 101(17):6659–6663
93. Moreau KL, Donato AJ, Seals DR et al (2002) Arterial intima-media thickness: site-specific associations with HRT and habitual exercise. Am J Physiol Heart Circ Physiol 283(4):H1409–1417
94. Moreau KL, Silver AE, Dinenno FA, Seals DR (2006) Habitual aerobic exercise is associated with smaller femoral artery intima-media thickness with age in healthy men and women. Eur J Cardiovasc Prev Rehabil 13(5):805–811
95. Westendorp IC, in 't Veld BA, Bots ML et al (1999) Hormone replacement therapy and intima-media thickness of the common carotid artery: the Rotterdam study. Stroke 30(12):2562–2567
96. Wildman RP, Schott LL, Brockwell S, Kuller LH, Sutton-Tyrrell K (2004) A dietary and exercise intervention slows menopause-associated progression of subclinical atherosclerosis as measured by intima-media thickness of the carotid arteries. J Am Coll Cardiol 44(3):579–585

# Chapter 19
# Depression and Anxiety in Women with Coronary Artery Disease: Prevalence and Links to Adverse Cardiac Outcomes

**Karen Bouchard, Alexandra Chiarelli, Michael Dans, and Heather Tulloch**

**Abstract** Depression and anxiety are highly prevalent among patients with coronary artery disease. Symptoms can severely inhibit patients' recovery and potentially lead to adverse cardiac-related outcomes, including repeat hospitalizations, recurrent cardiac events, and cardiac-related mortality. Across existing meta-analyses and systematic reviews on the prognostic effects of depression and anxiety on cardiac outcomes, women comprise only a small subset of the population and sex and gender-based analyses are rarely conducted. This is problematic as rates of depression and anxiety post cardiac event are nearly two-times higher in women than men. This chapter will profile recent advances in research on sex differences in the prevalence and prognostic effects of depression and anxiety on adverse cardiac outcomes. Recommendations for improving the identification and appropriate management of women with comorbid coronary artery disease and depression and/or anxiety are proposed.

**Keywords** Depression · Anxiety · Mental health · Coronary artery disease · PTSD

## The Burden of Anxiety and Depression

The global estimates of depression and anxiety are 4.4% and 3.6%, respectively. This amounts to approximately 322 and 264 million people living with a depressive or anxiety disorder [1]. An additional 53 million and 76 million cases of depression and anxiety were reported in light of the Covid-19 pandemic [2]. As defined by the

K. Bouchard · A. Chiarelli · M. Dans · H. Tulloch (✉)
Division of Cardiac Prevention and Rehabilitation, University of Ottawa Heart Institute, 40 Ruskin Street, Ottawa, ON K1Y4W7, Canada
e-mail: hetulloch@ottawaheart.ca

K. Bouchard · H. Tulloch
University of Ottawa, Ottawa, Canada

© The Author(s), under exclusive license to Springer Nature Switzerland AG 2023 
L. Kirshenbaum and I. Rabinovich-Nikitin (eds.), *Biology of Women's Heart Health*, Advances in Biochemistry in Health and Disease 26,
https://doi.org/10.1007/978-3-031-39928-2_19

Diagnostic and Statistical Manual-5th edition (DSM-V), major depressive disorder includes persistent (at least two weeks) symptoms of sadness or a loss of interest or pleasure in activities, and at least 4 other symptoms: sleep or appetite disturbance, sense of worthlessness or excessive guilt, fatigue, impaired concentration, and thoughts of death or suicide. Anxiety disorder is a broad term that encapsulates several disorders, such as generalized anxiety disorder (GAD), panic disorder, and post-traumatic stress disorder (PTSD). GAD is defined by symptoms of excessive worry that is difficult to control and at least three other symptoms, such as muscle tension, fatigue, difficulty with concentration, restlessness, irritability, and sleep disturbance [3]. Individuals may also feel uneasy and nervous, with physical manifestations, such as increased heartrate, rapid breathing, and nausea. Panic disorder is characterized by recurrent, unexpected panic attacks during which an abrupt surge of intense fear occurs, during which the individual experiences physical sensations (e.g., heart palpitations, sweating, chest pain/discomfort, nausea, dizziness) and fears (e.g., fear of losing control or dying). PTSD involves an exposure to an actual or threatened death, serious injury or sexual violence. In response, the individual experiences intrusive symptoms (e.g., recurrent distressing memories, dreams, flashbacks), negative alterations in cognition or mood (e.g., fear, anger, shame, unable to remember part of the traumatic event), negative changes in arousal (e.g., hypervigilance, irritability, angry outburst), and persistently avoids stimuli associated with the event. To meet diagnostic criteria for a depressive or anxiety disorder, social or occupational/educational functioning must be impacted by these symptoms. Depression and anxiety possess substantial overlap in symptomology and are highly concomitant (50% of cases) but are recognized as distinct diagnostic entities that may be tied to specific chronic disease outcomes [4].

## Prevalence of Depression and Anxiety in Cardiac Disease

Cardiac disease can prompt a wide range of emotions. Patients are often fearful that they may experience another heart event, feelings that may become exacerbated when experiencing bodily pains or engaging in strenuous physical activity or emotionally stressful situations [5]. Adopting changes to behaviour and lifestyle factors may feel overwhelming and frustrating [6]. It is common for patients to feel a loss of identity [7] and it can be difficult to adjust to new roles within the family unit or place of employment [8]. Others are not able to return to work and must rely on social assistance for their income, resulting in financial strain. All of these experiences may lead to a depressed mood or anxiety. Further, some patients experience pre-existing depression and anxiety symptoms, placing them at higher risk of developing adverse cardiac outcomes [9] and recurrent mood or anxiety symptoms.

Approximately 40% of all patients with coronary artery disease (CAD) report symptoms of depression (9–50%) [10] or anxiety (14–55%) [11] during hospitalization or immediately post-event, with 15–20% meeting criteria for a major depressive or anxiety disorder [7]. At one-year post-cardiac event, 15% and 27% of patients suffer symptoms of depression and generalized anxiety, respectively [12], which is approximately twice as common as in primary care patients and three times more prevalent than in the general population. Prevalence rates of panic disorder are estimated to be 5–8% of patients with CAD [13] whereas rates of PTSD can be much higher, ranging from 6 to 24% [14, 15]. For some, symptoms are transient and resolve within the months following their event, but for others, symptoms can persist meeting diagnostic criteria for a mood or anxiety disorder. If left unaddressed, depression and anxiety can severely inhibit patients' recovery and potentially lead to adverse cardiac outcomes.

## Effects of Depression and Anxiety on Cardiac Outcomes

Depression has been identified as a risk factor for cardiac disease incidence and recurrence. To illustrate, a large, multi-centre, population-based cohort study of nearly 150,000 participants determined that depression increased the risk of incident myocardial infarction (MI) by 23% [16]. Depression is also a robust predictor of cardiac-related mortality in patients with established disease. A recent large prospective cohort study determined that depression is associated with nearly three times the risk of cardiac-related death within two years post-MI, even when controlling for age, smoking, diabetes, body-mass index, history of MI, and disease severity [17]. Furthermore, in comparison to patients who are not depressed, Meijer and colleagues noted that patients with depression post-MI are five times more likely to die of a cardiac-related illness [18]. A dose–response relationship appears to exist between depressive symptoms and adverse cardiac outcomes. Specifically, increases in scores from < 5 to > 5 on the Beck Depression Inventory (BDI) were associated with nearly twice the risk of cardiac-related death [19, 20]. This indicates that even mild symptoms of depression are related to worse outcomes for these patients. Depression also predicts recurrent cardiac events. For instance, in a meta-analysis of 4500 patients after percutaneous coronary intervention (PCI), depression was shown to double the risk of major adverse cardiovascular events (MACE; a composite outcome composed of cardiac mortality, non-fatal MI, non-fatal stroke, repeated coronary revascularization and cardiac readmission) [21]. Taken together, evidence from nearly four decades of research on depression and cardiac disease, has concluded that depression confers approximately a doubling to tripling in risk of cardiac events post-MI [10, 18]. This evidence has culminated into a scientific statement from the American Heart Association, which officially recognized depression as a risk factor for adverse cardiovascular clinical outcomes [10].

Although less studied than depression, the evidence supporting the connection between anxiety and post-event cardiac outcomes is accumulating. According to a meta-analytic review of 20 studies involving nearly 250,000 adults, anxiety is associated with 26% increased risk of incident heart disease and 48% increased risk of cardiac mortality in healthy populations, independent of demographic and biological risk factors and health behaviours [22]. There continues to be a dearth of research that examines the direct effects of anxiety on cardiac clinical endpoints in patients with established disease [23] and the data generated from this limited research are inconsistent, with some meta-analytic reviews noting a significant relationship between anxiety and adverse outcomes, while others noting a less-robust link. For example, a meta-analysis aggregating data from nearly 6000 patients with MI documented that post-event anxiety conferred a 71% increased risk of new cardiac events and a 23% increased risk of cardiac mortality [22]. A similar meta-analysis of 16 studies including 9373 patients with MI determined patients with anxiety had a 23% and 27% increased risk of short-term complications and long-term MACE, respectively, in comparison to those without anxiety [24]. Another meta-analysis of 44 studies involving 30,527 patients with CAD examined the association between anxiety and mortality [25]. In the unadjusted analyses, one standard deviation increase in anxiety resulted in a 20% increased risk of mortality. However, when adjusting for medical and psychological factors, all relationships between anxiety and clinical endpoints became insignificant. Further, a systematic review and meta-regression of anxiety and morbidity risk in patients with coronary heart disease noted a non-significant association between GAD and MACE ($p = 0.28$) [26]. The criteria for and measures used to determine levels of anxiety may explain the heterogeneity across the results reported in these meta-analyses. For example, Tully and colleagues [26] only included studies that used a structured diagnostic interview to estimate anxiety levels, while others were inclusive of research that used self-report instruments. In addition, depression may alter the association between anxiety and cardiovascular risk. For example, a 2020 meta-analysis of anxiety and clinical outcomes, including 17 studies that involved nearly 40,000 patients with acute coronary syndrome (ACS), determined that anxiety is independently associated with increased mortality risk (Relative Risk [RR] = 1.21; $p = 0.02$), but that depression attenuated this association to non-significant (RR = 0.88; $p = 0.37$) [27].

There continues to be minimal research linking other anxiety disorders to cardiac outcomes in patients with established disease. For instance, panic disorder has largely been studied with respect to incident MACE among healthy populations [28, 29]. A meta-analysis of twelve studies involving over one million adults determined that panic disorder is significantly associated with new-onset cardiac events (Hazard Ratio [HR] = 1.40) and MI (HR = 1.38) even after controlling for depression [29]. The link between panic disorder and recurrent events in patients is unclear as there is limited research. There is only one study investigating the effects of panic disorder on cardiac outcomes in patients with established CAD [30]. This longitudinal cohort study, which aims to recruit nearly 4000 participants, will be completed in 2023 so data is not yet available. There is comparatively a larger body of evidence to suggest that PTSD is linked to poor cardiac prognoses. For example, a meta-analyses of 6 studies with over 400,000 healthy adults reported a 55% increased risk of incident coronary heart disease associated with PTSD [31]. Similarly, a prospective study with 281 twin pairs demonstrated that, after a median follow-up period of 13 years, the incidence of coronary heart disease was double (OR = 2.2) among those with PTSD (22.6%) compared to those without (8.9%) [32]; this association remained significant after adjustment for lifestyle, other risk factors and depression. The risk is also present for those with established disease. To illustrate, a meta-analysis that aggregated data from 24 studies reported that clinically significant symptoms of PTSD are associated with a doubling of risk of mortality or an ACS recurrence [33]. Additional research is required to clarify the relationship between anxiety diagnoses above and beyond generalized anxiety and cardiac risk in healthy and clinical populations.

## Explanatory Mechanisms Linking Depression, Anxiety and Cardiac Outcomes

Plausible physiologic and behavioural pathways linking depression and anxiety to cardiac outcomes have been proposed and are discussed in detail elsewhere [34–35] but summarized below (Fig. 19.1). Briefly, depression and anxiety are associated with elevated inflammatory markers, endothelial dysfunction, increased platelet aggregation and activity, hypothalamic–pituitary–adrenal axis dysregulation, and disruptions to autonomic functioning, all of which are implicated in the incidence and progression of cardiac disease. Health behaviors, such as maintaining a healthy diet, engaging in physical activity, obtaining adequate sleep duration and quality, attending cardiac rehabilitation, adhering to prescribed medication regimes, smoking cessation, and social support utilization, are linked to reduced risk of future hospitalizations and MACE. Unfortunately, patients with depression and anxiety appear to be less likely to engage in health-promoting behaviours, which accumulates to increased risk of cardiovascular morbidity and mortality.

Taken together, current research has demonstrated that depression and anxiety confer risk to the patient with CAD and plausible mechanisms linking these psychological disorders to cardiac outcomes have been illuminated. It is noteworthy, however, that within studies profiled across existing meta-analyses and systematic reviews on the prognostic effects of depression and anxiety on cardiac outcomes, women comprised only a small subset of the population ($M = 26\%$). Table 19.1 provides a list of existing meta-analyses that examined the prognostic influence of depression and anxiety in patients with established CVD. The number of studies profiled within each review and the average percentage of women across all studies included in the review are reported. It should be noted that while rates of CAD in men are double the proportion of women in younger populations (<60 years) [38, 39] this gap diminishes with age, becoming on par in the > 80 age group [40]. In the 60–79 age group, the incidence rates of CVD are 77.2% for women and 78.2% for men. The proportion of women in extant research is not consistent with cardiac population norms.

Sex and gender-based analyses (i.e., testing hypotheses separately for males/men and females/women such as stratified analyses, gender-based interactions, etc.) were seldom integrated across the profiled studies within existing meta-analytic reviews. Indeed, Smaardijk and colleagues [41] determined that sex and gender-based analyses were conducted in only 24% of the 49 studies measuring the prognostic effects

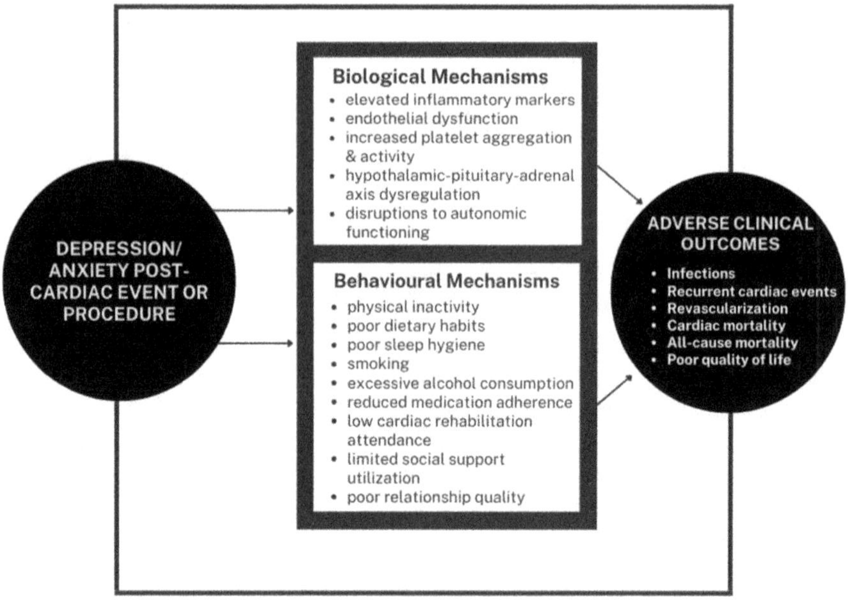


**Fig. 19.1** Mechanisms linking depression/anxiety to cardiac outcomes

of anxiety and depression on clinical outcomes in patients with heart disease. The underrepresentation of women and paucity of sex- and gender-based analyses across this research renders it difficult to draw definitive conclusions about the connections between depression and anxiety on cardiac prognosis in women. This gap has implications for the identification and appropriate management of women with comorbid CAD and depression and/or anxiety.

## Sex Differences in the Prevalence of Depression and Anxiety in Patients with Cardiac Disease

It is now acknowledged that sex- and gender-unique differences exist in the pathophysiology, symptoms, and outcomes of cardiac disease in women, including psychological variables, such as depression and anxiety. Table 19.2 displays a list of current meta-analyses examining sex and gender differences in depression and anxiety prevalence and link to CAD prognosis.

Several studies within these reviews have noted that women are disproportionally affected by depression and anxiety symptoms, post-CAD [42]. A meta-analysis of 16 prospective studies involving 10,175 patients (29% women) explored sex differences in depression and prognosis post-MI. The aggregated results indicated that 36% of women and 29% of men reported elevated levels of depression post-discharge [43].

**Table 19.1** Representation of women across studies profiled in current meta-analyses including patients with CAD

| First Author, Year | Construct studied | Cardiac diagnosis or procedure | Studies within review (N) | Mean % of women in profiled studies |
|---|---|---|---|---|
| Celano, 2015 | Anxiety | CAD | 44 | 24 |
| Edmonson, 2012 | Anxiety (PTSD) | ACS | 24 | 23 |
| Li, 2020 | Anxiety | ACS | 17 | 27 |
| Roest, 2010 | Anxiety | MI | 12 | 22 |
| Tully, 2014 | Anxiety | CHD | 40 | 26 |
| Wen, 2021 | Anxiety | MI | 16 | 25 |
| Barth, 2004 | Depression | CHD | 20 | 24 |
| Meijer, 2011 | Depression | MI | 29 | 25 |
| Meijer, 2013 | Depression | MI | 16 | 27 |
| Nicholson, 2006 | Depression | CHD | 34 | 22 |
| Song, 2020 | Depression | PCI | 9 | 27 |
| Van Melle, 2004 | Depression | MI | 22 | 27 |
| Zhang, 2019 | Depression | PCI | 8 | 28 |
| Takagi, 2017 | Depression & Anxiety | Cardiac Surgery | 16 | 28 |

Notes: *CAD* Coronary Artery Disease; *ACS* Acute Coronary Syndrome; *PTSD* Post Traumatic Stress Disorder; *MI* Myocardial Infarction; *CHD* Coronary Heart Disease; *PCI* Percutaneous Coronary Intervention

**Table 19.2** Meta-analyses of sex- and gender differences in depression and anxiety prevalence and CAD prognosis

| First Author, Year | Construct studied | Clinical population | Focus | Studies within review (N) |
|---|---|---|---|---|
| Shanmugasegaram, 2012 | Depression | CAD | Prevalence | 8 |
| Buckland, 2019 | Depression | CHD | Prevalence | 20 |
| Smaardijk, 2020 | Psychological factors (including, anger and hostility, anxiety, depression, social support, Type A behaviour pattern, Type D personality, and PTSD) | IHD | Prognostic Outcomes | 44 |
| Doyle, 2015 | Depression | MI | Prognostic Outcomes | 16 |

Notes: *CAD* Coronary Artery Disease; *CHD* Coronary Heart Disease; *PTSD* Post Traumatic Stress Disorder; *IHD* Ischemic Heart Disease; *MI* Myocardial Infarction

A systematic review of 20 longitudinal studies indicated that more women than men report depression during hospitalization (36% vs 23%) and this trend continued over two-years (23% vs 20%) [44]. Finally, another meta-analysis confirmed that the prevalence of depression is two-times greater in women than men with CAD, but that this doubling in prevalence is also found in the general population [45].

Similar to depression, anxiety levels are typically higher in women post-cardiac-event or procedure, but the literature base is markedly smaller and dated. A prospective, comparative study determined that women reported 25% higher levels of anxiety (*M* anxiety = 0.76 vs. 0.57 on the Brief Symptom Inventory) than men within the first 72 h of admission for MI, independent of other relevant sociodemographic or clinical variables. These higher rates of anxiety existed across a variety of geographic locations, including Australia, South Korea, Japan, England, and the United States. [46] Another study determined that anxiety levels are higher in women in comparison to men (*M* Men = 32.36 and Women = 49.00 as measured by the STAI) following MI, especially in the hours immediately preceding hospital discharge [47]. Bjerkset et al. [48] observed that women's risk for anxiety were highest in the first two years post-MI, but then symptoms reduced. Men's symptoms, however, increased two years post-MI. In addition, young women appear to be particularly susceptible to new-onset depression and anxiety, with one large prospective study (N = 306; 49% women) determining that young women are twice as likely as young men to develop mental stress post-MI [49].

It is still unclear why women with CAD are particularly susceptible to experiencing depression and anxiety, but several hypotheses have been proposed and tested. These vulnerabilities have been discussed in detail elsewhere [23, 42] but broadly include genetic, hormonal, demographic, psychological, social, and cultural factors. For instance, women may be more likely to inherit depression [50] although the genes responsible for this sex difference and its connection to CVD are still unclear. Pregnancy and the postpartum period may be responsible for the dysregulation of the hypothalamic–pituitary–adrenal axis and elevated inflammatory responses, which have been linked to post-partum depression and increased CVD risk [51, 52]. Women are less likely to be referred to cardiac rehabilitation than men and demonstrate lower rates of program completion [53], due to cited barriers such as childcare responsibilities and the underrepresentation of women-related factors in cardiac rehabilitation curricula (e.g., menopause, pregnancy) [54]. As cardiac rehabilitation has been shown to produce improvements in participants' mental health, including reducing symptoms of anxiety and depression and promoting quality of life [55, 56], women are missing this important intervention. Women-only cardiac rehabilitation programs may be a promising avenue to address cited barriers and adherence issues [57]. Lastly, women are also more likely to report more mental health symptoms than men in the general population, so it is possible that sex-differences may not be specific to those with a cardiac diagnosis.

## Sex Differences in the Prognostic Effects of Depression and Anxiety on Adverse Cardiac Outcomes

A systematic review and meta-analysis, pooling data from over 2 million healthy women and 3 million healthy men concluded that there are no significant group differences in the risk of psychological factors (including depression and anxiety) for IHD incidence between men and women [58]. Data on sex- and gender-differences in patients with established disease is comparatively less developed. Single studies report mixed findings as to whether women or men are disproportionally impacted by the effects of depression and anxiety on clinical outcomes. For instance, one study indicated that each standard deviation increase in psychological distress was associated with a 44% increased risk of CVD events in patients with CAD (cardiovascular mortality, MI, stroke, heart failure, unstable angina), but only among women [59]. A multi-centre prospective study involving 2411 patients (807 women) measuring the impact of depression on sex differences in the outcomes following MI, determined that depression was more prevalent in women than men (29% vs. 18%) [60] but these higher rates of depression in women contributed only slightly to their risk of rehospitalization (HR = 1.14) and symptoms of angina (HR = 1.22), but not mortality. In another study, 41% (N = 804) of patients with stable CAD had elevated scores on the Hospital Anxiety and Depression Scale-Anxiety (HADS-A; > 7), but the authors noted no sex differences in the prevalence of anxiety or anxiety-related risk for cardiac events [61].

As findings from primary research appear to be mixed, meta-analyses have been conducted to answer the question regarding the clinical risk of mental health symptoms between the sexes. Findings from two meta-analyses confirmed that there is limited data to support that the consequences of depression and anxiety are more severe for women than men; in fact, the risk associated with depression and anxiety may be more severe for men with CAD. To illustrate, a meta-analysis of data generated from 227,647 women and 321,894 men with ischemic heart disease determined that after adjustments (i.e., demographic and lifestyle factors, cardiac history and severity, psychological comorbidities, medication use, and cardiac treatment) the risk of MACE was stronger in men (HR = 1.37) than women (HR = 1.21) with psychological distress (i.e., anger, hostility, anxiety, depression, social support, Type A behaviour pattern, Type D personality, and PTSD) [41]. These results are consistent with another meta-analysis of 16 prospective studies where a stronger association was found between depression and poor cardiac prognosis in men (HR = 1.38) compared to women (HR = 1.22) post-ACS [43]. As such, it is evident that mental health symptoms place both sexes at higher risk for adverse outcomes, but it is unclear why men's cardiac health may be more adversely impacted by symptoms of depression and anxiety than women. Future research is required to investigate the potential mechanisms responsible for this potentially sex- and gender- dependent relationship as well as the factors that may protect anxious and/or depressed women from worse prognoses.

Studies using all women-samples may help to clarify the potency of depression and anxiety on cardiac outcomes for women with heart disease. Unfortunately, after a thorough review of the published literature (English-only; 2000–2022), only two studies were identified. Both studies were conducted prior to 2009 and both measured the prognostic effects of depression only. In one study [62] (N = 292), patients were assessed for depression within 6 months after admission for an acute coronary event and followed for 5 years. Outcome variables included recurrent events, including cardiovascular mortality, acute MI, and revascularization procedures (percutaneous transluminal coronary angioplasty and coronary artery bypass grafting). After adjusting for potential confounders, such as age and physical health parameters, the results indicated that patients with two or more depressive symptoms were nearly twice as likely to experience a recurrent cardiac event than those who reported only one or no depressive symptoms. Another study [63] (N = 67) examined the effects of depression while hospitalized for CABG on infections six-months later. One-month post-CABG, 41% of women were diagnosed with clinical depression via a diagnostic interview (50% of these women had a history of depression). More severe levels of depression were associated with more infections (e.g., pneumonia and upper-respiratory infections), and emergency room and primary care visits at six-months post-surgery. Taken together, these studies highlight that depression is a significant risk factor for poor cardiac prognoses in women, albeit possibly less so than men.

## Recommendations for Future Research and Clinical Practice

Accumulating research over the last two decades highlights the primacy of psychological factors in cardiac disease progression. Broadly, the research indicates that more women report symptoms of anxiety and depression post-event but the consequences of anxiety and depression, while clearly present for women, may be more severe for men, at least according to the currently published literature. There remains much uncertainty of the prognostic importance of depression on cardiac outcomes among women, and how it compares with that of men, due to the limited number of women involved in research and the paucity of studies using sex-and gender-based analyses. Research on sex-differences on the effects of anxiety on cardiac clinical endpoints is even less developed and is missing entirely in women-only studies. Based on the current state of the evidence, we outline priorities for future research and clinical practice.

First, the majority of studies did not measure "gender" as it is now defined, and the terms "sex" and "gender" were often used interchangeably (including "female" and "woman"). It is possible that gender-related factors (i.e., gender identity, gender roles, gender relations, and institutionalized gender) may help to explain differences in cardiac outcomes. A multicentre prospective follow-up study examined anxiety and depression as potential pathways between sex and gender and MACE among young patients with ACS (273 women; 636 men) [64]. The risk of events associated with sex were inconclusive. However, feminine characteristics were associated with higher rates of recurrent ACS in comparison to masculine characteristics. The relationship between femininity and recurrent ACS was partially explained by elevated levels of anxiety. This association was not found when examining sex as a binary construct (male/female), highlighting the potential importance of gender-related characteristics in determining cardiac outcomes. It is important to note that 75% of the studies measuring depression/anxiety and cardiac outcomes in women were conducted over 10 years ago, when robust definitions for sex and gender were only beginning to emerge. Some guidance is now available to assist researchers in the incorporation of gender into health research [65], as well as validated instruments used to measure gender (e.g., GENESIS-PRAXY; Stanford Gender-Related Variables for Health Research) and methods for incorporating gender-related variables into statistical analysis.

Second, there is a need for research to adopt an "intersectional" lens when examining cardiac health outcomes, which would involve considering patients' various identities (e.g., age, sex, gender, race, sexual orientation, social class, immigration status etc.) and their combination when determining risk [66]. For example, there is emerging evidence from large prospective studies (e.g., Virgo, GENESIS-PRAXY, YOUNG-MI) that younger women may be the most susceptible to the adverse cardiovascular effects of post-event depression [67]. Among a sample of men and women with obstructive CAD, an increased risk in death was observed in young women with depressive symptoms ($\leq$ 55 years) but not in men aged $\leq$ 55 or women aged > 55 years [68]. Other demographic and social determinants of health may combine to help explain worse outcomes among women [69]. For example, younger women with MI are often poor, of minority race, and with a history of trauma compared to men and older women [70]. Future research is tasked with untangling these complexities when examining the prognostic influence of psychological factors on cardiac outcomes among women.

Third, there continues to be wide heterogeneity in the clinical characteristics of the sample. It is plausible that the role of sex in depression predicting cardiac outcomes is more relevant in certain cardiac diagnostic subgroups, such as those with non-obstructive CAD. For example, Spontaneous Coronary Artery Dissection (SCAD) and Takotsubo Cardiomyopathy (TC) disproportionally afflict women, and accumulating research indicates that psychological triggers are prominent precursors and often remain prevalent post-event. To illustrate, a 2018 study [71] reported that 52% of patients with SCAD had a psychiatric history and 48% had an intense emotional stressor immediately preceding their SCAD [72]. Similarly, TC is often experienced by older women and triggered by severe emotional distress [73]. Not surprisingly, recent investigations report that depressed mood (20–41%), anxiety (14–17%), and traumatic stress symptoms (23–43%) are pervasive among patients with SCAD and TC [74, 75]. As women continue to be underrepresented in extant research on CAD, subgroup analyses on obstructive versus non-obstructive CAD amongst women are often not conducted. This research may be important as Smaardijk and colleagues [41] noted that the risk of MACE associated with psychological factors was more pronounced in women with non-obstructive CAD versus women with obstructive CAD.

Fourth, the Canadian Cardiovascular Society, American Heart Association, and European Society of Cardiology recommend that all patients with CAD be considered at risk for developing depression post-cardiac event or procedure. Therefore, patients should be screened using validated self-report questionnaires or diagnostic interviews by trained professionals. Similar recommendations are made for screening for anxiety, although anxiety has yet to be considered an established risk factor for poor cardiovascular prognoses, so definitive guidance is lacking. Screening for patients' mental health is recommended during hospitalization, at minimum, and it is suggested that screening continues at regular intervals throughout the recovery period [76]. There is evidence to suggest that measuring depression and anxiety during clinically stable periods may be more useful for predicting future cardiac events [25] Measuring symptoms over the longer term may also help to elucidate some sex- and gender differences.

Several validated screening tools are available to researchers and clinicians, many of which are available in a wide variety of languages and are easy and quick to use, with minimal patient and researcher or clinician burden (see Table 19.3 for examples). Scores serve as a proxy for a clinically significant level of impaired functioning in the individual but are not diagnostic on their own. Elevated scores signal the need for a clinical interview by a licensed professional to assess the presence and type of disorder. Although all patients ought to be screened for depression, there is particular urgency in younger women and those with a history of depression [77].Anxiety seems to be particularly pertinent in patients undergoing surgery, so screening for anxiety symptoms pre-and-post surgery would be helpful to identify those in need of additional support. There is also accumulating research to suggest that caregivers of patients with comorbid depression/anxiety and CVD ought to be screened for distress symptoms in conjunction with the patient, as their outcomes are highly interconnected [78–82]. It is worth noting that psychological problems are often underreported, so it is likely that accumulated data is skewed towards lower values. In other words, the problem of depression and anxiety in patients with CAD may be worse than it appears.

Fifth, the development and assessment of high-quality interventions targeting both sexes are required, but sex and gender-based analyses or women-specific interventions are lagging. Burgeoning evidence indicates that clinical and behavioural interventions may be beneficial for the treatment of depression and anxiety in patients with CAD [83–84] but women still represent a minority of the samples investigated (~25% in studies within the most recent meta-analysis [84]). Apart from one meta-analysis, sex- and gender-based analyses were not conducted, likely due to the small number of women included in the profiled observational studies. Linden and colleagues [83] aggregated data from 23 studies involving approximately 10,000 patients with CVD. Their results revealed that psychological treatments offered in addition to usual care for patients reduced mortality by 27% (up to 2 years) and cardiac recurrence (post-2 years) by 43%; the mortality benefits of psychological treatments were only seen in men. As there are recognized sex- and gender-differences in the

**Table 19.3** Examples of screening tools for depression and anxiety

| Tool | Author, Year | # of items | Time to complete (min) | Access |
|---|---|---|---|---|
| BDI | Beck, 1961 | 21 | < 10 | License and fee required for use |
| PHQ-9 | Spitzer, 1999 | 9 | < 5 | No license or fee required for use |
| HADS | Zigmond, 1983 | 14 | < 5 | License and fee required for use |
| GAD-7 | Spitzer, 2006 | 7 | < 2 | No license or fee required for use |
| DASS | Lovibond, 1995 | 42/21 | < 10 | No license or fee required for use |
| CAQ | Eifert, 2000 | 18 | < 2 | No license or fee required for use |
| STAI | Spielberger, 1970 | 40 | < 20 | License and fee required for use |
| BAI | Beck, 1988 | 21 | < 10 | License and fee required for use |
| BSI | Derogatis, 1993 | 53 | < 15 | License and fee required for use |

Notes: *BDI* Beck Depression Inventory; *PHQ-9* Patient Health Questionnaire; *HADS* Hospital Anxiety and Depression Scale; *GAD-7* Generalized Anxiety Disorder Assessment; *DASS* Depression Anxiety Stress Scales; *CAQ* Cardiac Anxiety Questionnaire; *STAI* State-Trait Anxiety Inventory; *BAI* Beck Anxiety Inventory; *BSI* Brief Symptom Inventory

pathophysiology, symptoms, and lived-experience of CAD, women-only interventions may be a promising approach [88]. There is limited evidence to date, however, as fewer than 16 interventions, using a wide range of intervention approaches, including psychotherapy, pharmacotherapy, cardiac rehabilitation, peer support, and stress-management, have been developed to target symptoms of anxiety and depression in women with CAD. Among these trials, there are mixed findings to their efficacy in reducing women's cardiac risk. For example, in the SWITCHD trial (Stockholm Women's Intervention Trial for Coronary Heart Disease), women were randomized to a group-based psychosocial intervention program or usual care [89]. Women participated in 20 sessions over the course of a year designed to improve knowledge about CVD risk factors, relaxation and self-monitoring skills, and clinical advice compliance (e.g., medication use). Participants were then assessed after several years ($M$ duration $= 7.1$ years) to investigate the program's ability to improve patients' survival. After controlling for clinical prognostic factors, the analysis indicated that the intervention yielded a threefold protective effect on women's survival. In another study, a 1-year cognitive-behavioural stress management program (20 2-h sessions) for women with ischaemic heart disease ($N = 159$) was found to improve stress behaviour and exhaustion [90] but did not improve biochemical indicators (hs-CRP, leptin, glucose and insulin) associated with cardiovascular risk [91] in comparison to usual care (i.e., general lifestyle advice and outpatient visits). Lastly, a 6-month telephone peer-support intervention for women with CHD (N = 108) was developed to reduce symptoms of depression and improve perceived social support [92]. At 6 months post-intervention, there was a general improvement in depression and perceived social support in both groups, but the intervention yielded no specific treatment effects. Although there are some promising findings, the intervention components and modalities that are most effective for supporting women's psychosocial health remain unclear. Therefore, providing concrete recommendations for interventions for women with CAD is still premature but would be a welcome addition to the literature following adequate investigation.

## Conclusion

Symptoms of depression and anxiety are highly prevalent amongst patients with established cardiac disease. Studies using sex- and gender-based analyses indicate that women with CAD experience higher levels of depression in comparison to men, but that the consequences of depression may be more severe for men. Research using all-women samples underscore the importance of identifying depression in women, post-cardiac event or procedure, as unmanaged depression may hinder women's recovery. If the results from women-only studies are taken in combination with the findings from studies using sex- and gender-based analyses, it can be concluded that depression is linked to adverse cardiac outcomes in both men and women. As the literature base is small and results are inconsistent, future studies examining the effects of anxiety on cardiac outcomes in both men and women are required. Screening for mental health symptoms is warranted and action must be taken to alleviate these symptoms when present. Additional interventions to reduce this distress are needed and preliminary research indicates that women-only programs may be successful but ought to be investigated further before investing in sex-or gender-specific programming.3 5

## References

1. World Health Organization (2017) Depression and other common mental disorders: global health estimates. Geneva: World Health Organization (https://apps.who.int/iris/handle/10665/254610)
2. Santomauro DF, Mantilla Herrera AM, Shadid J et al (2021) Global prevalence and burden of depressive and anxiety disorders in 204 countries and territories in 2020 due to the COVID-19 pandemic. The Lancet 398(10312):1700–1712. https://doi.org/10.1016/S0140-6736(21)02143-7
3. Battle DE (2013) Diagnostic and Statistical Manual of Mental Disorders (DSM). Codas 25(2):191–192
4. Hettema JM (2008) What is the genetic relationship between anxiety and depression? Am J Med Genet C Semin Med Genet 148C(2):140–146. https://doi.org/10.1002/ajmg.c.30171

5. Monane R, Sanchez GJ, Kronish IM, Edmondson D, Diaz KM (2018) Post-traumatic stress disorder symptoms and aversive cognitions regarding physical activity in patients evaluated for acute coronary syndrome. Eur J Prev Cardiol 25(4):402–403
6. Cohen SM, Kataoka-Yahiro M (2009) Themes in the literature related to cardiovascular disease risk reduction. J Cardiovasc Nurs 24(4):268–276. https://doi.org/10.1097/JCN.0b013e3181a6de90
7. Astin F, Horrocks J, Closs SJ (2014) Managing lifestyle change to reduce coronary risk: a synthesis of qualitative research on peoples' experiences. BMC Cardiovasc Disord 14:96. (In eng). https://doi.org/10.1186/1471-2261-14-96
8. Tulloch H, Bouchard K, Clyde MJ et al (2020) Learning a new way of living together: a qualitative study exploring the relationship changes and intervention needs of patients with cardiovascular disease and their partners. BMJ Open 10(5):e032948. https://doi.org/10.1136/bmjopen-2019-032948
9. Flaherty LB, Wood T, Cheng A, Khan AR (2017) Pre-existing psychological depression confers increased risk of adverse cardiovascular outcomes following cardiac surgery: a systematic review and meta-analysis. J Thorac Cardiovasc Surg 154(5):1578–1586 e1. https://doi.org/10.1016/j.jtcvs.2017.06.052
10. Lichtman JH, Froelicher ES, Blumenthal JA et al (2014) Depression as a risk factor for poor prognosis among patients with acute coronary syndrome: systematic review and recommendations: a scientific statement from the American Heart Association. Circ 129(12):1350–1369. https://doi.org/10.1161/CIR.0000000000000019
11. Chen YY, Xu P, Wang Y, Song TJ, Luo N, Zhao LJ (2019) Prevalence of and risk factors for anxiety after coronary heart disease: systematic review and meta-analysis. Medicine (Baltimore) 98(38):e16973. https://doi.org/10.1097/MD.0000000000016973
12. Murphy B, Le Grande M, Alvarenga M, Worcester M, Jackson A (2019) Anxiety and depression after a cardiac event: prevalence and predictors. Front Psychol 10:3010 (In eng). https://doi.org/10.3389/fpsyg.2019.03010
13. Celano CM, Suarez L, Mastromauro C, Januzzi JL, Huffman JC (2013) Feasibility and utility of screening for depression and anxiety disorders in patients with cardiovascular disease. Circ Cardiovasc Qual Outcomes 6(4):498–504. https://doi.org/10.1161/CIRCOUTCOMES.111.000049
14. Copeland LA, Sako EY, Zeber JE et al (2014) Mortality after cardiac or vascular operations by preexisting serious mental illness status in the Veterans Health Administration. Gen Hosp Psychiatry 36(5):502–508. https://doi.org/10.1016/j.genhosppsych.2014.04.003
15. Perkins-Porras L, Joekes K, Bhalla N, Sutherland C, Pollard M (2015) Reporting of posttraumatic stress disorder and cardiac misconceptions following cardiac rehabilitation. J Cardiopulm Rehabil Prev 35(4):238–245. https://doi.org/10.1097/HCR.0000000000000100
16. Rajan S, McKee M, Rangarajan S et al (2020) Association of symptoms of depression with cardiovascular disease and mortality in low-, middle-, and high-income countries. JAMA Psychiat 77(10):1052–1063. https://doi.org/10.1001/jamapsychiatry.2020.1351
17. Yusuf S, Joseph P, Rangarajan S et al (2020) Modifiable risk factors, cardiovascular disease, and mortality in 155 722 individuals from 21 high-income, middle-income, and low-income countries (PURE): a prospective cohort study. The Lancet 395(10226):795–808. https://doi.org/10.1016/S0140-6736(19)32008-2
18. Meijer A, Conradi HJ, Bos EH, Thombs BD, van Melle JP, de Jonge P (2011) Prognostic association of depression following myocardial infarction with mortality and cardiovascular events: a meta-analysis of 25 years of research. Gen Hosp Psychiatry 33(3):203–216 (In eng). https://doi.org/10.1016/j.genhosppsych.2011.02.007
19. Cohen BE, Edmondson D, Kronish IM (2015) State of the art review: depression, stress, anxiety, and cardiovascular disease. Am J Hypertens 28(11):1295–1302. https://doi.org/10.1093/ajh/hpv047
20. Fiedorowicz JG (2014) Depression and cardiovascular disease: an update on how course of illness may influence risk. Curr Psychiatry Rep 16(10):492. https://doi.org/10.1007/s11920-014-0492-6

21. Song X, Song J, Shao M et al (2020) Depression predicts the risk of adverse events after percutaneous coronary intervention: a meta-analysis. J Affect Disord 266:158–164. https://doi.org/10.1016/j.jad.2020.01.136
22. Roest AM, Martens EJ, Denollet J, de Jonge P (2010) Prognostic association of anxiety post myocardial infarction with mortality and new cardiac events: a meta-analysis. Psychosom Med 72(6):563–569. https://doi.org/10.1097/PSY.0b013e3181dbff97
23. O'Neil A, Russell JD, Murphy B (2021) How does mental health impact women's heart health? Heart Lung Circ 30(1):59–68. https://doi.org/10.1016/j.hlc.2020.05.111
24. Wen Y, Yang Y, Shen J, Luo S (2021) Anxiety and prognosis of patients with myocardial infarction: a meta-analysis. Clin Cardiol 44(6):761–770. https://doi.org/10.1002/clc.23605
25. Celano CM, Millstein RA, Bedoya CA, Healy BC, Roest AM, Huffman JC (2015) Association between anxiety and mortality in patients with coronary artery disease: a meta-analysis. Am Heart J 170(6):1105–1115. https://doi.org/10.1016/j.ahj.2015.09.013
26. Tully PJ, Cosh SM, Baumeister H (2014) The anxious heart in whose mind? A systematic review and meta-regression of factors associated with anxiety disorder diagnosis, treatment and morbidity risk in coronary heart disease. J Psychosom Res 77(6):439–448. https://doi.org/10.1016/j.jpsychores.2014.10.001
27. Li J, Ji F, Song J et al (2020) Anxiety and clinical outcomes of patients with acute coronary syndrome: a meta-analysis. BMJ Open 10(7):e034135. https://doi.org/10.1136/bmjopen-2019-034135
28. Edmondson D, Newman JD, Whang W, Davidson KW (2013) Emotional triggers in myocardial infarction: do they matter? Eur Heart J 34(4):300–306. https://doi.org/10.1093/eurheartj/ehs398
29. Tully PJ, Turnbull DA, Beltrame J et al (2015) Panic disorder and incident coronary heart disease: a systematic review and meta-regression in 1131612 persons and 58111 cardiac events. Psychol Med 45(14):2909–2920. https://doi.org/10.1017/S0033291715000963
30. Foldes-Busque G, Dionne CE, Turcotte S et al (2021) Epidemiology and prognostic implications of panic disorder and generalized anxiety disorder in patients with coronary artery disease: rationale and design for a longitudinal cohort study. BMC Cardiovasc Disord 21(1):26 (In eng). https://doi.org/10.1186/s12872-021-01848-3
31. Edmondson D, Kronish IM, Shaffer JA, Falzon L, Burg MM (2013) Posttraumatic stress disorder and risk for coronary heart disease: a meta-analytic review. Am Heart J 166(5):806–14 (In eng). https://doi.org/10.1016/j.ahj.2013.07.031
32. Vaccarino V, Goldberg J, Rooks C et al (2013) Post-traumatic stress disorder and incidence of coronary heart disease: a twin study. J Am Coll Cardiol 62(11):970–8 (In eng). https://doi.org/10.1016/j.jacc.2013.04.085
33. Edmondson D, Richardson S, Falzon L, Davidson KW, Mills MA, Neria Y (2012) Posttraumatic stress disorder prevalence and risk of recurrence in acute coronary syndrome patients: a meta-analytic review. PLoS One 7(6):e38915 (In eng). https://doi.org/10.1371/journal.pone.0038915
34. Celano CM, Daunis DJ, Lokko HN, Campbell KA, Huffman JC (2016) Anxiety disorders and cardiovascular disease. Curr Psychiatry Rep 18(11):101. https://doi.org/10.1007/s11920-016-0739-5
35. Celano CM, Huffman JC (2011) Depression and cardiac disease: a review. Cardiol Rev 19(3):130–142. https://doi.org/10.1097/CRD.0b013e31820e8106
36. Huffman JC, Celano CM, Beach SR, Motiwala SR, Januzzi JL (2013) Depression and cardiac disease: epidemiology, mechanisms, and diagnosis. Cardiovasc Psychiatry Neurol 2013:695925. https://doi.org/10.1155/2013/695925
37. Osborne MT, Shin LM, Mehta NN, Pitman RK, Fayad ZA, Tawakol A (2020) Disentangling the links between psychosocial stress and cardiovascular disease. Circ Cardiovasc Imaging 13(8):e010931. https://doi.org/10.1161/CIRCIMAGING.120.010931
38. Albrektsen G, Heuch I, Lochen ML et al (2016) Lifelong gender gap in risk of incident myocardial infarction: the Tromso study. JAMA Intern Med 176(11):1673–1679. https://doi.org/10.1001/jamainternmed.2016.5451
39. George J, Rapsomaniki E, Pujades-Rodriguez M et al (2015) How does cardiovascular disease first present in women and men? incidence of 12 cardiovascular diseases in a contemporary

cohort of 1,937,360 people. Circ 132(14):1320–1328. https://doi.org/10.1161/CIRCULATIONAHA.114.013797
40. Benjamin EJ, Virani SS, Callaway CW et al (2018) Heart disease and stroke statistics-2018 update: a report from the American heart association. Circulation 137(12):e67–e492. https://doi.org/10.1161/CIR.0000000000000558
41. Smaardijk VR, Maas A, Lodder P, Kop WJ, Mommersteeg PMC (2020) Sex and gender-stratified risks of psychological factors for adverse clinical outcomes in patients with ischemic heart disease: a systematic review and meta-analysis. Int J Cardiol 302:21–29. https://doi.org/10.1016/j.ijcard.2019.12.014
42. Liblik K, Mulvagh SL, Hindmarch CCT, Alavi N, Johri AM (2022) Depression and anxiety following acute myocardial infarction in women. Trends Cardiovasc Med 32(6):341–347. https://doi.org/10.1016/j.tcm.2021.07.005
43. Doyle F, McGee H, Conroy R et al (2015) Systematic review and individual patient data meta-analysis of sex differences in depression and prognosis in persons with myocardial infarction: a MINDMAPS study. Psychosom Med 77(4):419–428. https://doi.org/10.1097/PSY.0000000000000174
44. Buckland SA, Pozehl B, Yates B (2019) Depressive symptoms in women with coronary heart disease: a systematic review of the longitudinal literature. J Cardiovasc Nurs 34(1):52–59. https://doi.org/10.1097/JCN.0000000000000533
45. Shanmugasegaram S, Russell KL, Kovacs AH, Stewart DE, Grace SL (2012) Gender and sex differences in prevalence of major depression in coronary artery disease patients: a meta-analysis. Maturitas 73(4):305–311. https://doi.org/10.1016/j.maturitas.2012.09.005
46. Moser DK, Dracup K, McKinley S et al (2003) An international perspective on gender differences in anxiety early after acute myocardial infarction. Psychosom Med 65(4):511–516. https://doi.org/10.1097/01.psy.0000041543.74028.10
47. An K, De Jong MJ, Riegel BJ et al (2004) A cross-sectional examination of changes in anxiety early after acute myocardial infarction. Heart Lung 33(2):75–82. https://doi.org/10.1016/j.hrtlng.2003.12.007
48. Bjerkeset O, Nordahl HM, Mykletun A, Holmen J, Dahl AA (2005) Anxiety and depression following myocardial infarction: gender differences in a 5-year prospective study. J Psychosom Res 58(2):153–161. https://doi.org/10.1016/j.jpsychores.2004.07.011
49. Vaccarino V, Sullivan S, Hammadah M et al (2018) Mental stress-induced-myocardial ischemia in young patients with recent myocardial infarction: sex differences and mechanisms. Circ 137(8):794–805. https://doi.org/10.1161/CIRCULATIONAHA.117.030849
50. Kendler KS, Gardner CO, Neale MC, Prescott CA (2001) Genetic risk factors for major depression in men and women: similar or different heritabilities and same or partly distinct genes? Psychol Med 31(4):605–616. https://doi.org/10.1017/S0033291701003907
51. Li SH, Graham BM (2017) Why are women so vulnerable to anxiety, trauma-related and stress-related disorders? The potential role of sex hormones. The Lancet Psychiatry 4(1):73–82. https://doi.org/10.1016/s2215-0366(16)30358-3
52. Penninx BW (2017) Depression and cardiovascular disease: epidemiological evidence on their linking mechanisms. Neurosci Biobehav Rev 74(Pt B):277–286. https://doi.org/10.1016/j.neubiorev.2016.07.003
53. Colella TJ, Gravely S, Marzolini S et al (2015) Sex bias in referral of women to outpatient cardiac rehabilitation? A meta-analysis. Eur J Prev Cardiol 22(4):423–41 (In eng). https://doi.org/10.1177/2047487314520783
54. Parry M, Van Spall HGC, Mullen K-A et al (2022) The Canadian women's heart health alliance atlas on the epidemiology, diagnosis, and management of cardiovascular disease in women—Chapter 6: sex- and gender-specific diagnosis and treatment. CJC Open 4(7):589–608. https://doi.org/10.1016/j.cjco.2022.04.002
55. Rutledge T, Redwine LS, Linke SE, Mills PJ (2013) A meta-analysis of mental health treatments and cardiac rehabilitation for improving clinical outcomes and depression among patients with coronary heart disease. Psychosom Med 75(4):335–349 (In eng). https://doi.org/10.1097/PSY.0b013e318291d798

56. Shepherd CW, While AE (2012) Cardiac rehabilitation and quality of life: a systematic review. Int J Nurs Stud 49(6):755–771. https://doi.org/10.1016/j.ijnurstu.2011.11.019
57. Kuehn BM (2017) Women may benefit from cardiac rehabilitation programs tailored to their specific needs. Circ 135(6):612–613. https://doi.org/10.1161/CIRCULATIONAHA.116.027064
58. Smaardijk VR, Lodder P, Kop WJ, Gennep Bv, Maas AHEM, Mommersteeg PMC (2019) Sex and Gender stratified risks of psychological factors for incident ischemic heart disease: systematic review and meta analysis. J Am Heart Assoc 8(9):e010859. https://doi.org/10.1161/JAHA.118.010859
59. Pimple P, Lima BB, Hammadah M et al (2019) Psychological distress and subsequent cardiovascular events in individuals with coronary artery disease. J Am Heart Assoc 8(9):e011866 (In eng). https://doi.org/10.1161/jaha.118.011866
60. Parashar S, Rumsfeld JS, Reid KJ et al (2009) Impact of depression on sex differences in outcome after myocardial infarction. Circ Cardiovasc Qual Outcomes 2(1):33–40.https://doi.org/10.1161/CIRCOUTCOMES.108.818500
61. Frasure-Smith N, Lespérance F (2008) Depression and anxiety as predictors of 2-year cardiac events in patients with stable coronary artery disease. Arch Gen Psychiatry 65(1):62–71 (In eng). https://doi.org/10.1001/archgenpsychiatry.2007.4
62. Horsten M, Mittleman MA, Wamala SP, Schenck-Gustafsson K, Orth-Gomer K (2000) Depressive symptoms and lack of social integration in relation to prognosis of CHD in middle-aged women. The Stockholm Female Coronary Risk Study. Eur Heart J 21(13):1072–1080. https://doi.org/10.1053/euhj.1999.2012
63. Doering LV, Martínez-Maza O, Vredevoe DL, Cowan MJ (2008) Relation of depression, natural killer cell function, and infections after coronary artery bypass in women. Eur J Cardiovasc Nurs 7(1):52–58 (In eng). https://doi.org/10.1016/j.ejcnurse.2007.07.004
64. Pelletier R, Khan NA, Cox J et al (2016) Sex versus gender-related characteristics: which predicts outcome after acute coronary syndrome in the young? J Am Coll Cardiol 67(2):127–135 (In eng). https://doi.org/10.1016/j.jacc.2015.10.067
65. Gogovor A, Zomahoun HTV, Ekanmian G et al (2021) Sex and gender considerations in reporting guidelines for health research: a systematic review. Biol Sex Differ 12(1):62. https://doi.org/10.1186/s13293-021-00404-0
66. Allana S, Ski CF, Thompson DR, Clark AM (2021) Bringing intersectionality to cardiovascular health research in Canada. CJC Open 3(12 Suppl):S4–S8. https://doi.org/10.1016/j.cjco.2021.08.016
67. Andrikopoulos GK, Tzeis SE, Pipilis AG et al (2006) Younger age potentiates post myocardial infarction survival disadvantage of women. Int J Cardiol 108(3):320–325. https://doi.org/10.1016/j.ijcard.2005.05.016
68. Shah AJ, Ghasemzadeh N, Zaragoza-Macias E et al (2014) Sex and age differences in the association of depression with obstructive coronary artery disease and adverse cardiovascular events. J Am Heart Assoc 3(3):e000741. https://doi.org/10.1161/JAHA.113.000741
69. Koek HL, de Bruin A, Gast F et al (2006) Short- and long-term prognosis after acute myocardial infarction in men versus women. Am J Cardiol 98(8):993–999. https://doi.org/10.1016/j.amjcard.2006.05.016
70. Mallik S, Spertus JA, Reid KJ et al (2006) Depressive symptoms after acute myocardial infarction: evidence for highest rates in younger women. Arch Intern Med 166(8):876–883 (In eng). https://doi.org/10.1001/archinte.166.8.876
71. Adams H, Paratz E, Somaratne J et al (2018) Different patients, different outcomes: a case-control study of spontaneous coronary artery dissection versus acute coronary syndrome. J Interv Cardiol 31(1):41–47. https://doi.org/10.1111/joic.12447
72. Saw J, Humphries K, Aymong E et al (2017) Spontaneous coronary artery dissection: clinical outcomes and risk of recurrence. J Am Coll Cardiol 70(9):1148–1158 (In eng). https://doi.org/10.1016/j.jacc.2017.06.053

73. Salmoirago-Blotcher E, Rosman L, Wittstein IS et al (2016) Psychiatric history, post-discharge distress, and personality characteristics among incident female cases of takotsubo cardiomyopathy: a case-control study. Heart Lung 45(6):503–509. https://doi.org/10.1016/j.hrtlng.2016.07.008
74. Johnson A, Tweet MS, Best PJ, Gulati R, Hayes S (2019) Prevalance of post-traumatic stress disorder and psychological sequelae in patients who have experienced spontaneous coronary artery dissection. J Am Coll Cardiol 73(9):252
75. Liang JJ, Tweet MS, Hayes SE, Gulati R, Hayes SN (2014) Prevalence and predictors of depression and anxiety among survivors of myocardial infarction due to spontaneous coronary artery dissection. J Cardiopulm Rehabil Prev 34(2):138–142. https://doi.org/10.1097/HCR.0000000000000030
76. Post-Myocardial Infarction Depression Clinical Practice Guideline Panel (2009) AAFP guideline for the detection and management of post-myocardial infarction depression. Ann Fam Med 7(1):71–79
77. Jha MK, Qamar A, Vaduganathan M, Charney DS, Murrough JW (2019) Screening and management of depression in patients with cardiovascular disease: JACC state-of-the-art review. J Am Coll Cardiol 73(14):1827–1845. https://doi.org/10.1016/j.jacc.2019.01.041
78. Bouchard K, Gareau A, Gallant NL et al (2021) Dyadic effects of anxiety and depression on quality of life among couples facing cardiovascular disease. J Psychosom Res 149:110601. https://doi.org/10.1016/j.jpsychores.2021.110601
79. Bouchard K, Gareau A, McKee K, Lalande K, Greenman PS, Tulloch H (2021) Dyadic patterns of mental health and quality of life change in partners and patients during three months of cardiac rehabilitation. J Fam Psychol
80. Chung ML, Lennie TA, Mudd-Martin G, Dunbar SB, Pressler SJ, Moser DK (2016) Depressive symptoms in patients with heart failure negatively affect family caregiver outcomes and quality of life. Eur J Cardiovasc Nurs 15(1):30–38. https://doi.org/10.1177/1474515114535329
81. Dalteg T, Benzein E, Sandgren A, Malm D, Arestedt K (2016) Associations of emotional distress and perceived health in persons with atrial fibrillation and their partners using the actor-partner interdependence model. J Fam Nurs 22(3):368–391. https://doi.org/10.1177/1074840716656815
82. Thomson P, Howie K, Leslie SJ et al (2020) Evaluating emotional distress and health-related quality of life in patients with heart failure and their family caregivers: testing dyadic dynamics using the Actor-Partner Interdependence Model. PLoS ONE 15(1):e0227129. https://doi.org/10.1371/journal.pone.0227129
83. Linden W, Phillips MJ, Leclerc J (2007) Psychological treatment of cardiac patients: a meta-analysis. Eur Heart J 28(24):2972–2984 (In eng). https://doi.org/10.1093/eurheartj/ehm504
84. Magán I, Casado L, Jurado-Barba R et al (2021) Efficacy of psychological interventions on psychological outcomes in coronary artery disease: systematic review and meta-analysis. Psychol Med 51(11):1846–1860 (In eng). https://doi.org/10.1017/s0033291720000598
85. Rees K, Taylor RS, Singh S, Coats AJ, Ebrahim S (2004) Exercise based rehabilitation for heart failure. Cochrane Database Syst Rev (3):CD003331. https://doi.org/10.1002/14651858.CD003331.pub2
86. Richards SH, Anderson L, Jenkinson CE et al (2017) Psychological interventions for coronary heart disease. Cochrane Database Syst Rev 4:CD002902. https://doi.org/10.1002/14651858.CD002902.pub4
87. Whalley B, Rees K, Davies P et al (2011) Psychological interventions for coronary heart disease. Cochrane Database Syst Rev (8):CD002902. https://doi.org/10.1002/14651858.CD002902.pub3
88. Mosca L, Barrett-Connor E, Wenger NK (2011) Sex/gender differences in cardiovascular disease prevention: what a difference a decade makes. Circ 124(19):2145–2154. https://doi.org/10.1161/CIRCULATIONAHA.110.968792
89. Orth-Gomer K, Schneiderman N, Wang HX, Walldin C, Blom M, Jernberg T (2009) Stress reduction prolongs life in women with coronary disease: the Stockholm Women's Intervention Trial for Coronary Heart Disease (SWITCHD). Circ Cardiovasc Qual Outcomes 2(1):25–32. https://doi.org/10.1161/CIRCOUTCOMES.108.812859

90. Claesson M, Birgander LS, Lindahl B et al (2005) Women's hearts--stress management for women with ischemic heart disease: explanatory analyses of a randomized controlled trial. J Cardiopulm Rehabil 25(2):93–102 (In eng). https://doi.org/10.1097/00008483-200503000-00009
91. Claesson M, Birgander LS, Jansson JH et al (2006) Cognitive-behavioural stress management does not improve biological cardiovascular risk indicators in women with ischaemic heart disease: a randomized-controlled trial. J Intern Med 260(4):320–331. https://doi.org/10.1111/j.1365-2796.2006.01691.x
92. Kyaw Tha Tun E, Nagel J, Bosbach A et al (2021) Telephone-based peer support intervention to reduce depressive symptoms in women with coronary heart disease, a randomized controlled trial in Germany. WomenHealth 61(7):619–632. https://doi.org/10.1080/03630242.2021.1953208

# Chapter 20
# Cardiac Imaging in Women with Suspected Ischemic Heart Disease

**Sina O'Sullivan and Amy W. Pollak**

**Abstract** Cardiovascular imaging has undergone advances over the past decades and is increasingly used for the diagnosis of atherosclerotic disease in men and women. Recent guidelines suggest the use of stress imaging in symptomatic women at intermediate or high pretest probability of having ischemic heart disease, as well as the use of coronary computed tomography angiography to delineate the presence and location of coronary plaque. There are differences in pathophysiology of atherosclerotic disease in women, as women are more likely to have nonobstructive disease, often caused by abnormal coronary reactivity, microvascular dysfunction, as well as plaque erosion and distal microembolization. Myocardial perfusion imaging with positron emission tomography (PET) or magnetic resonance imaging (MRI) can aid in the diagnosis of microvascular dysfunction, making it even more valuable in women for an earlier and more accurate diagnosis leading to improved management and outcomes in women.

**Keywords** Imaging · Echocardiography · SPECT · PET · CT · MRI · Pregnancy

## Introduction

Ischemic heart disease remains the leading cause of death in women in the Western World [1]. Advances in cardiovascular imaging have led to improvement in diagnosis of atherosclerotic disease in women, and earlier and more accurate diagnosis of symptomatic women results in enhanced management with subsequent therapies.

Stress testing with EKG and the addition of imaging is recommended in symptomatic women at intermediate or high pretest probability of having coronary artery disease (CAD) based on the American College of Cardiology (ACC)/American Heart Association (AHA) practice guideline on exercise testing, which are generally defined as those with typical or atypical chest pain at $\geq$ 50 years of age and

S. O'Sullivan · A. W. Pollak (✉)
Department of Cardiovascular Medicine, Mayo Clinic Florida, Jacksonville, USA
e-mail: Pollak.Amy@mayo.edu

© The Author(s), under exclusive license to Springer Nature Switzerland AG 2023

L. Kirshenbaum and I. Rabinovich-Nikitin (eds.), *Biology of Women's Heart Health*, Advances in Biochemistry in Health and Disease 26,
https://doi.org/10.1007/978-3-031-39928-2_20

those < 50 years of age with typical angina [2]. In addition, women with symptoms and additional risk factors such as diabetes and metabolic syndrome are at increased CAD risk and should be considered for testing [3]. In general, women with low pretest probability should be assessed for non-ischemic etiology of their symptoms [4]. However, the commonly used pretest probability assessment determined by age, gender, chest pain characteristics and the presence of traditional risk factors can underestimate cardiovascular risk in women with atypical symptoms and non-traditional risk factors [5]. The 2021 ACC/AHA Chest Pain guidelines outlines the use of stress testing versus coronary computed tomography angiography (CTA) in patients who present with chest pain, acknowledging that for many women, symptoms of underlying CAD often include dyspnea, nausea, fatigue or discomfort in the neck, shoulder, arms or back. The recommendation has moved from classifying chest pain from typical angina versus atypical angina to cardiac chest pain versus possible cardiac chest pain or non-cardiac chest pain.

There are multiple challenges in non-invasive testing in women. Whereas men are more likely to have obstructive coronary artery disease, often left main or multivessel disease, women are more likely to have nonobstructive disease, often caused by abnormal coronary reactivity, microvascular dysfunction, as well as plaque erosion and distal microembolization [6]. These differences in pathophysiology should be reflected in the testing we choose for the diagnosis of cardiovascular disease in women with symptoms. Confounding the use of stress imaging is the higher false positive rates of stress testing in women compared to men. In addition, there is substantial underrepresentation of women in studies of non-invasive testing [7]. The advances in cardiovascular imaging and the development of novel imaging techniques have expanded their role in the detection of not only flow-limiting epicardial coronary atherosclerosis, but also microvascular disease and endothelial dysfunction.

## Stress Echocardiography

Stress echocardiography combines ultrasound images of the heart with exercise or pharmacologic stress testing to evaluate women with symptoms compatible with ischemic heart disease. It is recommended for identification of obstructive CAD and estimation of prognosis for symptomatic women at intermediate to high ischemic heart disease (IHD) risk and with any of the following: (1) resting ST-segment abnormalities, (2) functional disability, or (3) indeterminate or intermediate-risk stress ECG. (Class I; Level of Evidence B) (4).

Exercise echocardiography has improved diagnostic sensitivity and specificity for the detection of CAD in women compared to exercise ECG alone (79% and 83%, respectively) [8, 9]. If patients can exercise, it is preferred over pharmacologic stress testing, as exercise capacity has been shown to be a strong prognostic factor for cardiac events and mortality [10]. In patients that are unable to exercise, dobutamine stress testing has good sensitivity (80%) and specificity (84%) for the detection of IHD, with a slightly lower sensitivity in vasodilator stress testing with dipyridamole or adenosine [11]. Both abnormal exercise and dobutamine stress echocardiography has been shown to be a predictor of subsequent cardiac events and mortality [12, 13], whereas normal stress echocardiography has a good negative predictive value with an event rate of < 1% per year. High risk features of abnormal stress echocardiography are detailed in Table 20.1.

Stress echocardiography may also aid in the diagnosis of non-ischemic causes of cardiac symptoms. It has become widely established in the assessment of multiple clinical conditions including systolic or diastolic heart failure, non-ischemic cardiomyopathy, valvular heart disease, pulmonary hypertension, athletes' hearts and congenital heart disease [14]. The assessment of diastolic dysfunction using the ratio of early diastolic transmitral velocity to early diastolic tissue velocity (E/e') correlates with invasively measured LVDP during exercise and facilitates the recognition of likely abnormal filling pressures in response to exercise that may explain non-specific symptoms not caused by ischemic heart disease [15].

Novel techniques including the use of tissue doppler imaging and speckle tracking or strain imaging increase the sensitivity by early recognition of myocardial dysfunction compared to visual wall motion analysis alone [16].

**Table 20.1** High-risk features and diagnostic accuracy of imaging modalities in assessment of ischemic heart disease in women

| Imaging modality | Sensitivity (%) | Specificity (%) | High-risk features |
|---|---|---|---|
| Stress echocardiography | 79 | 83 | Baseline LVEF ≤ 40%, extensive resting wall motion abnormalities or extensive ischemia, right ventricular ischemia, increase in end-systolic size with stress, and decrease in LVEF with stress |
| Stress SPECT | 78–93 | 61–99 | Summed stress score > 8, ≥ 10% of abnormal myocardium at stress, left ventricular dilation, and peak stress or poststress LVEF ≤ 45 |
| Stress PET | 90 | 89 | Same features as stress SPECT and CFR < 2 |
| Stress MRI | 89 | 80 | Presence and extent of inducible wall motion abnormalities, or perfusion defects with stress imaging, low LVEF, presence and extent of infarct size |
| CT angiography | 90–99 | 79–91 | Coronary artery calcium ≥ 400, proximal left anterior descending artery stenosis ≥ 70%, 2- or 3-vessel CAD, left main stenosis ≥50%, and 3-vessel nonobstructive CAD |

*CAD* Coronary Artery Disease; *CFR* Coronary Flow Reserve; *CT* Computed Tomography; *LVEF* Left Ventricular Ejection Fraction; *MRI* Magnetic Resonance Imaging; *PET* Positron Emission Tomography, *SPECT* Single-Photon Emission Computerized Tomography

Given the absence of radiation exposure, stress echocardiography is the preferred imaging modality in younger women with symptoms of IHD, especially those of reproductive age [4].

## Myocardial Perfusion Imaging with SPECT and PET

Similar to stress echocardiography, myocardial perfusion imaging (MPI) with single photon emission computed tomography (SPECT) or positron emission tomography (PET is recommended for diagnosis of CAD and risk stratification in symptomatic women at intermediate to high IHD risk and with resting ST-segment abnormalities, functional disability, or indeterminate or intermediate-risk stress ECG, (Class I; Level of Evidence B) [4]. Stress myocardial perfusion imaging not only provides information about LVEF and wall motion abnormalities, but also about the extent and severity of myocardial perfusion at stress and rest.

### *Single Photon Emission Computed Tomography (SPECT)*

Stress myocardial perfusion imaging with single photon emission computed tomography (SPECT) is commonly performed in the evaluation of women with IHD symptoms. The sensitivity and specificity for the diagnosis of obstructive CAD in symptomatic women in exercise SPECT is reported as 81% and 78%, respectively [9]. Pharmacologic SPECT with a vasodilator (i.e., dipyridamole, adenosine, or regadenoson) improves sensitivity (91%) and specificity (86%) [4] and is used frequently in patients with inability to exercise or with a baseline left bundle branch block [17].

False positive findings, especially anterior and anterolateral defects due to breast attenuation, are common in women who are obese or have large breasts and reduce the diagnostic accuracy of myocardial perfusion SPECT. In addition, women have smaller heart volumes than men and a higher risk of false negative studies. Newer techniques including the use of ECG gating to allow functional assessment (LVEF and wall motion abnormalities), attenuation correction protocols, use of 2 position supine/prone or supine/upright imaging, and use of higher-energy radioisotope technetium ($^{99}$Tc instead of $^{201}$TI) have been shown to reduce artefacts and increase diagnostic accuracy [18]. Novel high-speed SPECT cameras provide increased sensitivity due to improved count detection and advanced reconstruction algorithms resulting in improved image quality and reduced radiation exposure [19].

There is robust data on the excellent prognostic accuracy of exercise and pharmacologic SPECT in both women and men. The risk of CAD events increases gradually based on size and severity of perfusion defects. A normal or low risk study (defined as < 5% of abnormal myocardium or summed stress score of < 4) is associated with a < 1% annual risk of CAD death or nonfatal myocardial infarction (MI), whereas a high-risk result (e.g., moderately to severely abnormal or multivessel perfusion pattern or summed stress score > 8) increases the annual CAD death or MI rate to 6% or greater [20]. Additional high-risk features include left ventricular dilation and left ventricular dysfunction at rest and stress [4] and should be noted on the final report. The finding of transient ischemic dilatation (TID) where the left ventricle dilates at stress (ratio of stress to rest greater than 1.18–1.35) varies in the literature, however this is a finding specific for obstructive left main disease or three vessel disease [21].

## *Positron Emission Tomography (PET)*

Stress myocardial perfusion positron emission tomography (PET) has several advantages compared to SPECT, including improved spatial and temporal resolution as well as the ability to quantify myocardial blood flow and coronary flow reserve, which leads to improved diagnostic and prognostic accuracy, especially in obese women.

The diagnostic sensitivity and specificity of PET is excellent at 92% and 85%, respectively [22], with an overall improvement in diagnostic accuracy of about 20% for stress PET MPI compared with SPECT MPI (88% versus 67%, respectively; P = 0.009) [23].

As in stress SPECT, an increase in the magnitude of stress and rest perfusion defect is associated with an increasing frequency of IHD events. Based on cumulative data of more than 7,000 patients, a normal scan indicates low risk (<1% annual cardiac event rate) while an abnormal scan is associated with worsening prognosis (> 4.2% annual event rate). There was also a graded increase in risk of CAD events with more extensive and severe perfusion defects (predicted CAD death increased by one-third for every 10% increase in percent of ischemic myocardium and by more than one-half for every 10% increase in percent of scarred myocardium) [18].

A distinct advantage of PET MPI is the ability to measure myocardial blood flow and calculate coronary flow reserve (CFR), which aids in the diagnosis of non-obstructive CAD and coronary microvascular dysfunction that is more prevalent in women (Figure 20.1) [24]. A CFR of less than 2 is associated with a higher rate of adverse events, with a hazard ratio ~ 5 for CAD events in a study with short term follow up of about 1 year [25]. In patients with ischemia on PET MPI, the addition of a reduced CFR doubled their CAD death rate [26].

PET MPI also improves the detection of multivessel CAD compared to SPECT by assessing left ventricular function at rest and during peak stress (as opposed to post-stress with gated SPECT). An abnormal LVEF reserve, defined as a decrease or diminished increase in the LVEF during peak stress even in the absence of apparent perfusion abnormalities is suggestive of multivessel or left main disease [27].

An additional advantage of PET MPI is the substantially lower radiation dose, with an effective radiation dose of 3.6 mSv using the perfusion tracers rubidium-82 and ammonia-13N versus 12.2 mSv for technetium-99 m SPECT [28].

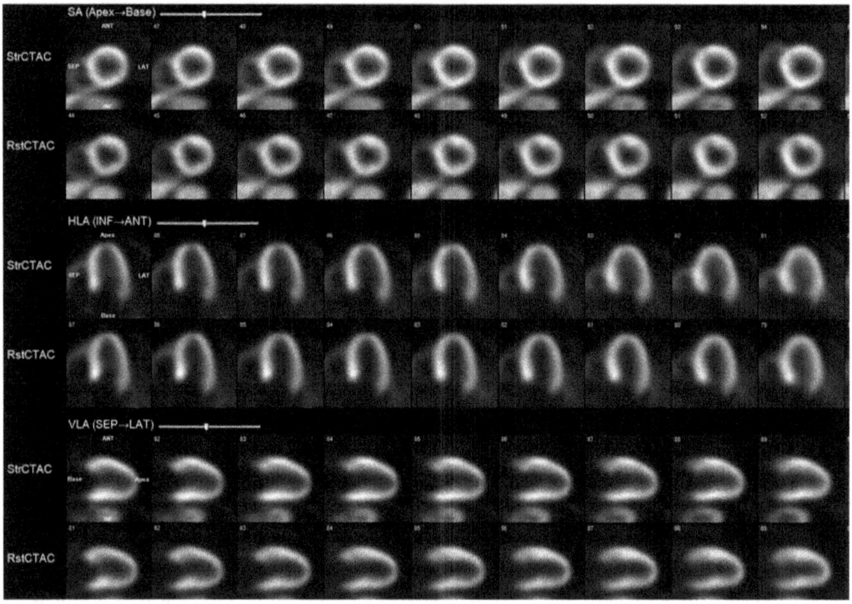


**Fig. 20.1** Vasodilator PET stress with global myocardial blood flow ratio of 2.3 which is mildly reduced from normal in a patient with a clinical history consistent with microvascular angina

## Computed Tomography (CT)

### *Coronary Artery Calcium Scoring*

Coronary CT has increased in use and significance over the past decade. Non-contrast coronary artery calcium (CAC) scoring quantifies coronary artery calcium which signifies atherosclerotic disease. It is often reported as the Agatston score which is calculated by multiplying the density of the plaque (in Hounsfield Units) by the area of the plaque. Thus, coronary artery calcium scoring is a marker for atherosclerotic burden, however it does not assess luminal obstruction. [3].

Based on the 2018 guidelines on the management of cholesterol, coronary artery calcium scoring can be used to evaluate cardiovascular risk for primary prevention and guide the use of statin therapy for individuals without diabetes, with LDL levels higher than 70 mg/dl but less than 190 mg/dl and atherosclerotic cardiovascular disease (ASCVD) risk score is between 7.5% and 19.9% [29]. Recommendations for use of statin therapy based on the CAC score are summarized in Table 20.2. Coronary artery calcium scores can also be added to the MESA (Multi-Ethnic Study of Atherosclerosis) risk calculator (https://www.mesa-nhlbi.org/MESACHDRisk/MesaRiskScore/RiskScore.aspx),which includes other high risk features such as family history of premature CAD, to better predict cardiovascular risk.

**Table 20.2** Recommendations on statin therapy based on coronary artery calcium scoring according to the 2018 Cholesterol guidelines {Grundy, 2019 #157}

| Coronary artery calcium score (Agatston units) | Addition | Recommendation on statin therapy |
|---|---|---|
| 0 | Without high-risk features | Not indicatedReassess in 5-10 years |
| 0 | With high-risk features | Reasonable to initiate statin |
| | Current tobacco use | |
| | Diabetes mellitus | |
| | Family history of premature ASCVD | |
| 1-99 | | Reasonable to initiate statin for patients > 55 years of age |
| ≥100 or ≥ 75th percentile | | Indicated (Unless deferred by patient-clinician risk discussion) |

When compared to invasive coronary angiograms, a normal calcium score (score of 0) has a high negative predictive value with an extremely low probability of stenosis (< 1%). In contrast, women with moderate (≥ 100) or higher (≥ 400) CAC scores had a greater prevalence of obstructive coronary disease [30]. Of note, women are more likely to have a relatively "low" coronary calcium score despite a high percentile of 75th percentile and the high percentile which will indicate an increased CV risk.

It is important to highlight that coronary calcium scoring are only appropriate as an initial test for asymptomatic individuals. If a woman is having possible cardiac chest pain, and CT is the chosen imaging strategy, coronary CT angiography is indicated.

## *Coronary CT Angiography*

CT angiography has undergone rapid improvements in technology, including dual-source CT, spectral imaging, gantry rotation time, and advanced reconstruction algorithms, which resulted in significantly improved diagnostic accuracy for the detection of both obstructive and nonobstructive coronary atherosclerotic burden [31]. Coronary CT (CCT) angiography can provide information about plaque location, plaque burden and luminal narrowing, as well as important insight about high-risk plaque features and plaque composition, such as non-calcified plaque, spotty calcification or low attenuation plaque (<30 Hounsfield units) and positive arterial remodeling [32] (Figure 20.2).

CCT may be used as the index procedure in symptomatic women at intermediate IHD risk with resting ST-segment abnormalities, functional disability, or indeterminate or intermediate-risk stress ECG, (Class IIb; Level of Evidence C) [4]. The

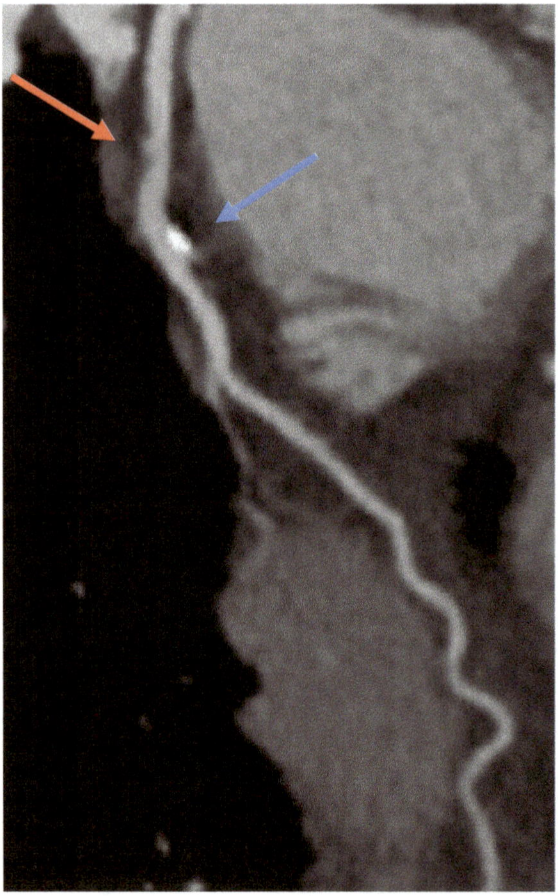

**Fig. 20.2** High-risk plaque in the proximal LAD. While the vessel is only 25-50% obstructed by the non-calcified plaque (red arrow), there is evidence of high-risk features such as low attenuation plaque with spotty calcification. In addition, there is a mixed plaque with calcification (blue arrow)

updated 2021 AHA/ACC Chest Pain guidelines recommend coronary CT for evaluation of intermediate to high-risk patients with chest pain with a possible cardiac cause, who are under the age of 65 years and with suspicion for less obstructive CAD. However, for intermediate-high risk individuals who are over age 65 or with obstructive CAD suspected, functional stress testing is preferred [33]. It may also be used for further investigation if prior functional stress testing was inconclusive.

Multiple clinical trials and meta-analyses have revealed a diagnostic sensitivity of 97% to 99% and a specificity of 88% to 91% for the detection of obstructive CAD compared with invasive angiography [4]. A recent secondary analysis of the Assessment by Coronary Computed Tomographic Angiography of Individuals Undergoing Invasive Coronary Angiography (ACCURACY) trial showed a diagnostic sensitivity and specificity to detect > 50% stenosis in women of 90% and 88%, respectively, which is comparable to men (Sensitivity 96%, specificity 78%) [34].

There is well established evidence about the prognostic accuracy of CCT on both IHD events and mortality. Results from the Coronary CT Angiography Evaluation for Clinical Outcomes International Multicenter (CONFIRM) registry demonstrated a proportional increase in adverse events based on the number of coronary arteries with obstructive disease for both men and women (adjusted hazard ratios of 1.6, 2.0, 2.9, and 3.7, respectively, for nonobstructive, 1-vessel, 2-vessel, and 3-vessel obstructive [> 50% stenosis] CAD compared with no CAD). Notably, the risk for adverse outcomes in patient with three-vessel or left main disease was higher in women compared to men (hazard ratio 4.21 vs 3.27). The absence of CAD was associated with a favorable outcome (annualized death rate: 0.28%) [35]. Non-obstructive CAD was found less in women than in men but was associated with an increased risk for mortality and major adverse cardiac events that was similar in both men and women (HR for women: 1.77, HR for men: 1.96) [36]. In addition, in patients with non-obstructive CAD, the presence of noncalcified and mixed coronary plaques were associated with worse clinical outcomes in women and men when compared to calcified plaque alone (6-year death rate of 1.4% for calcified plaque, 3.3% for mixed plaque, and 9.6% for noncalcified plaque) [37].

The ability of coronary CTA to assess hemodynamically significant lesions > 50% stenosis resulting in ischemia has traditionally been limited. Coronary computed tomography angiography-derived fractional flow reserve (FFR-CT) is a non-invasive physiological test that can assess flow limitation across coronary stenoses with high diagnostic accuracy and good correlation to invasive FFR [38]. In a study comparing FFR-CT with other stress testing a high sensitivity (90%) and moderate specificity (71%) for the detection of hemodynamically significant CAD was observed compared to invasive FFR testing [39]. The 2021 AHA/ACC guidelines recommend the addition of FFR-CT for intermediate-risk patients with acute chest pain and coronary artery stenosis of 40% to 90% in a proximal or middle segment on CCTA to guide decision-making regarding the use of coronary revascularization [33].

Current generation radiation dosimetry is low for CCTA, with effective doses for most patients in the 3 to 5 mSv range [33]. However, a disadvantage of cardiac CT is radiation exposure to breast tissue, especially in young women. A coronary CT angiogram is predicted to have a slight increase in lifetime relative risk for breast cancer of by 0.2–0.4% in women aged 25 and 55 years, respectively [40]. In younger women radiation exposure should be considered when the choice of testing is made.

The use of coronary CTA is limited in patients with underlying renal disease, allergy to iodinated contrast, respiratory instability or inability to follow breath-hold instructions as well as heart rate variability, arrhythmias or inability to tolerate beta blockers to achieve target heart rate [33].

## Cardiac Magnetic Resonance Imaging (MRI)

Cardiac magnetic resonance imaging (MRI) has developed significantly over the past decade. Advantages of MRI are high spatial and temporal resolution, lack of ionizing radiation, lack of attenuation artefacts, and lack of limitation by body habitus [32]. Cardiac MRI allows for the differentiation of a range of myocardial diseases causing chest pain in women, including myocarditis, stress-induced cardiomyopathy, sarcoidosis, and amyloidosis. In addition, in women at risk for or with suspected IHD, it has the ability to detect prior or recent myocardial infarctions with high sensitivity [31]. It may be reasonable to use stress cardiac MRI, especially stress perfusion cardiac MRI, as the index procedure for the diagnostic evaluation of symptomatic women at intermediate-high IHD risk and with (a) resting ST-segment abnormalities, (b) functional disability, or (c) indeterminate or intermediate risk (Class IIb; Level of Evidence B) [4].

Stress cardiac MRI can be performed with dobutamine, vasodilator or exercise and has been shown to have a high diagnostic accuracy for the detection of CAD in women [41]. A meta-analysis reported a diagnostic sensitivity of dobutamine stress MRI of 83% and specificity of 86%, with perfusion imaging 91% and 81%, respectively [42]. Vasodilator stress MRI is the preferred method due to shorter stress times and enhanced late gadolinium enhancement which facilitates detection of myocardial infarction [4]. A meta-analysis demonstrated a sensitivity of 89% and a specificity of 80% for vasodilator stress MRI, with adenosine having better sensitivity and specificity than dipyridamole (90% vs. 86% and 81% vs 77%, respectively) [43]. Specifically in women, sensitivity and specificity have been found to be 84% and 88%, respectively, based on a multicenter registry of 147 symptomatic women who underwent vasodilator stress MPI and late gadolinium enhancement [41].

The CE-MARC (Clinical Evaluation of Magnetic Resonance Imaging in Coronary Heart Disease) single center trial [44] as well as the MR-IMPACT II study (A Study of Gadodiamide Injection in Myocardial Perfusion Magnetic Resonance Imaging) [45] demonstrated superior sensitivity in detecting CAD in vasodilator stress MRI compared to SPECT imaging, which was especially significant in women (88.7% vs 50.9%, respectively).

A meta-analysis found that patients with a negative stress cardiac MRI had very low annual cardiovascular event rates (0.8% for combined annual events, 03% for cardiovascular death and 0.4% for MI). Patients with evidence of ischemia on MRI had a higher incidence of MI (OR 7.7) and cardiovascular death (OR 7.0) with annual event rates of 2.3% for cardiovascular death and 2.6% for MI. The presence of late gadolinium enhancement was also significantly associated with a worse prognosis [46]. Late gadolinium enhancement imaging is routinely included in the cardiac MRI examination and may help detect previously unrecognized myocardial infarction which will then lead to the use of anti-ischemic and risk factor modifying therapies [32].

An advantage of stress MRI is the ability to assess for perfusion defects due to obstructive CAD, but also due to impaired coronary vasoreactivity and endothelial dysfunction. In a study of 113 symptomatic women without obstructive CAD on invasive coronary angiography, 57% demonstrated subendocardial hypoperfusion abnormalities on adenosine stress MRI with normal resting perfusion images consistent with coronary microvascular disease, making stress MRI a useful an important diagnostic tool in these women [47]. In a subgroup of patients of the WISE (Women's Ischemia Syndrome Evaluation) study consisting of 118 women, patients with abnormal myocardial perfusion reserve determined by advanced semiquantitative assessment on stress MRI were more likely to demonstrate abnormal invasive coronary reactivity consistent with endothelial dysfunction [48]. The presence of perfusion defects in the absence of obstructive CAD were associated with an increased risk of cardiac events. One study showed that women experienced a 56-times higher annual rate of major adverse cardiac events (MACE) compared to those with normal stress MRI (0.3% vs 15.1%, P < 0.001) [49]. Similarly, another sub analysis of the WISE study demonstrated a 2.3-fold higher 3-year risk of MACE in women with MRI-detected ischemia with no obstructive coronary artery disease (INOCA) compared to those with normal MRIs (43% with INOCA and 13% with normal MRI) [50].

The 2021 AHA/ACC Chest Pain Guidelines indicate that for approximately 6% to 15% of troponin-positive acute coronary syndrome (ACS) occurring in the absence of obstructive CAD and importantly cardiac MRI can differentiate infarct-related scar (MINOCA) from alternative causes such as myocarditis and nonischemic cardiomyopathy, when performed within 2 weeks of ACS [33].

## Cardiac Imaging in Pregnancy

Cardiac imaging in pregnancy is less common but poses a unique challenge. There are two important concerns in pregnancy, especially for the fetus: radiation exposure and contrast exposure. Careful consideration of differential diagnoses in pregnant women with chest pain should be made and imaging modalities should be chosen based on these with the goal to minimize exposure to contrast and radiation. Radiation exposure beyond 50 mGy can double the relative risk of childhood cancer from 0.1to 0.2% [51]. If a threshold of 150 mGy is reached, significant damage of the fetus can occur, and assessment of intervention including pregnancy termination should be made. Notably, however, fetal exposure is well below 50 mGy for all cardiac imaging modalities. Contrast agents may cross the placenta and enter fetal circulation, however fetal teratogenic or mutagenic effects have not been reported. Intravenous iodinated contrast agents are classified as US Food and Drug Administration Category B, gadolinium-based contrast agent as US Food and Drug Administration Category class C [52].

## Conclusion

Ischemic heart disease remains a major threat to women and early diagnosis and treatment is important to decrease cardiovascular events and mortality. Contemporary imaging techniques have high sensitivity and specificity for the diagnosis and prognostic evaluation of IHD in women, including obstructive and non-obstructive CAD as well as more female predominant microvascular and endothelial dysfunction.

Advances in cardiac imaging will further establish its role in the non-invasive risk assessment of symptomatic women with suspected IHD to help guide management and decrease cardiovascular mortality.

## References

1. Wenger NK, Lloyd-Jones DM, Elkind MSV, Fonarow GC, Warner JJ, Alger HM et al (2022) Call to action for cardiovascular disease in women: epidemiology, awareness, access, and delivery of equitable health care: a presidential advisory from the American heart association. Circ 145(23):e1059–e1071
2. Gibbons RJ, Balady GJ, Bricker JT, Chaitman BR, Fletcher GF, Froelicher VF et al (2002) ACC/AHA 2002 guideline update for exercise testing: summary article. Circ 106(14):1883–1892
3. Mieres JH, Shaw LJ, Arai A, Budoff MJ, Flamm SD, Hundley WG et al (2005) Role of noninvasive testing in the clinical evaluation of women with suspected coronary artery disease: consensus statement from the cardiac imaging committee, council on clinical cardiology, and the cardiovascular imaging and intervention committee, council on cardiovascular radiology and intervention American Heart Association. Circ 111(5):682–696
4. Mieres JH, Gulati M, Bairey Merz N, Berman DS, Gerber TC, Hayes SN et al (2014) Role of noninvasive testing in the clinical evaluation of women with suspected ischemic heart disease: a consensus statement from the American Heart Association. Circ 130(4):350–379
5. Agarwala A, Michos ED, Samad Z, Ballantyne CM, Virani SS (2020) The Use of sex-specific factors in the assessment of women's cardiovascular risk. Circ 141(7):592–599
6. Bairey Merz CN, Shaw LJ, Reis SE, Bittner V, Kelsey SF, Olson M et al (2006) Insights from the NHLBI-sponsored women's ischemia syndrome evaluation (WISE) study: Part II: gender differences in presentation, diagnosis, and outcome with regard to gender-based pathophysiology of atherosclerosis and macrovascular and microvascular coronary disease. J Am College Cardiol 47(3, Supplement):S21–S9
7. Charney P (1999) Coronary artery disease in women: what all physicians need to know. American College of Physicians, Philadelphia
8. Marwick TH, Anderson T, Williams MJ, Haluska B, Melin JA, Pashkow F et al (1995) Exercise echocardiography is an accurate and cost-efficient technique for detection of coronary artery disease in women. J Am Coll Cardiol 26(2):335–341
9. Sanders GD, Patel MR, Chatterjee R, Ross AK, Bastian LA, Coeytaux RR, Heidenfelder BL, Musty MD, Dolor RJ (2013) Noninvasive Technologies for the diagnosis of coronary artery disease in women: future research needs: identification of future research needs from comparative effectiveness review no. 58. Rockville (MD): Agency for Healthcare Research and Quality (US). Report No.: 13-EHC072-EF
10. Roger VL, Jacobsen SJ, Pellikka PA, Miller TD, Bailey KR, Gersh BJ (1998) Prognostic value of treadmill exercise testing: a population-based study in Olmsted County Minnesota. Circ 98(25):2836–2841
11. Kim C, Kwok YS, Heagerty P, Redberg R (2001) Pharmacologic stress testing for coronary disease diagnosis: a meta-analysis. Am Heart J 142(6):934–944
12. Chuah SC, Pellikka PA, Roger VL, McCully RB, Seward JB (1998) Role of dobutamine stress echocardiography in predicting outcome in 860 patients with known or suspected coronary artery disease. Circ 97(15):1474–1480
13. Arruda-Olson AM, Juracan EM, Mahoney DW, McCully RB, Roger VL, Pellikka PA (2002) Prognostic value of exercise echocardiography in 5798 patients: is there a gender difference? J Am Coll Cardiol 39(4):625–631
14. Lancellotti P, Pellikka PA, Budts W, Chaudhry FA, Donal E, Dulgheru R et al (2017) The clinical use of stress echocardiography in non-ischaemic heart disease: recommendations

from the European Association of Cardiovascular Imaging and the American Society of Echocardiography. J Am Soc Echocardiogr 30(2):101–138
15. Burgess MI, Jenkins C, Sharman JE, Marwick TH (2006) Diastolic stress echocardiography: hemodynamic validation and clinical significance of estimation of ventricular filling pressure with exercise. J Am Coll Cardiol 47(9):1891–1900
16. Uusitalo V, Luotolahti M, Pietilä M, Wendelin-Saarenhovi M, Hartiala J, Saraste M et al (2016) Two-dimensional speckle-tracking during dobutamine stress echocardiography in the detection of myocardial ischemia in patients with suspected coronary artery disease. J Am Soc Echocardiogr 29(5):470–9.e3
17. Amanullah AM, Berman DS, Hachamovitch R, Kiat H, Kang X, Friedman JD (1997) Identification of severe or extensive coronary artery disease in women by adenosine technetium-99m sestamibi SPECT. Am J Cardiol 80(2):132–137
18. Taqueti VR, Dorbala S, Wolinsky D, Abbott B, Heller GV, Bateman TM et al (2017) Myocardial perfusion imaging in women for the evaluation of stable ischemic heart disease-state-of-the-evidence and clinical recommendations. J Nucl Cardiol 24(4):1402–1426
19. Perrin M, Djaballah W, Moulin F, Claudin M, Veran N, Imbert L et al (2015) Stress-first protocol for myocardial perfusion SPECT imaging with semiconductor cameras: high diagnostic performances with significant reduction in patient radiation doses. Eur J Nucl Med Mol Imaging 42(7):1004–1011
20. Shaw LJ, Iskandrian AE (2004) Prognostic value of gated myocardial perfusion SPECT. J Nucl Cardiol 11(2):171–185
21. Alama M, Labos C, Emery H, Iwanochko RM, Freeman M, Husain M et al (2018) Diagnostic and prognostic significance of transient ischemic dilation (TID) in myocardial perfusion imaging: A systematic review and meta-analysis. J Nucl Cardiol 25(3):724–737
22. Nandalur KR, Dwamena BA, Choudhri AF, Nandalur SR, Reddy P, Carlos RC (2008) Diagnostic performance of positron emission tomography in the detection of coronary artery disease: a meta-analysis. Acad Radiol 15(4):444–451
23. Bateman TM, Heller GV, McGhie AI, Friedman JD, Case JA, Bryngelson JR et al (2006) Diagnostic accuracy of rest/stress ECG-gated Rb-82 myocardial perfusion PET: comparison with ECG-gated Tc-99m sestamibi SPECT. J Nucl Cardiol 13(1):24–33
24. Gould KL, Johnson NP, Bateman TM, Beanlands RS, Bengel FM, Bober R et al (2013) Anatomic versus physiologic assessment of coronary artery disease. Role of coronary flow reserve, fractional flow reserve, and positron emission tomography imaging in revascularization decision-making. J Am Coll Cardiol 62(18):1639–1653
25. Murthy VL, Lee BC, Sitek A, Naya M, Moody J, Polavarapu V et al (2014) Comparison and prognostic validation of multiple methods of quantification of myocardial blood flow with 82Rb PET. J Nucl Med 55(12):1952–1958
26. Murthy VL, Naya M, Foster CR, Hainer J, Gaber M, Di Carli G et al (2011) Improved cardiac risk assessment with noninvasive measures of coronary flow reserve. Circ 124(20):2215–2224
27. Dorbala S, Vangala D, Sampson U, Limaye A, Kwong R, Di Carli MF (2007) Value of vasodilator left ventricular ejection fraction reserve in evaluating the magnitude of myocardium at risk and the extent of angiographic coronary artery disease: a 82Rb PET/CT study. J Nucl Med 48(3):349–358
28. Desiderio MC, Lundbye JB, Baker WL, Farrell MB, Jerome SD, Heller GV (2018) Current Status of patient radiation exposure of cardiac positron emission tomography and single-photon emission computed tomographic myocardial perfusion imaging. Circ: Cardiovasc Imaging 11(12):e007565
29. Grundy SM, Stone NJ, Bailey AL, Beam C, Birtcher KK, Blumenthal RS et al (2019) 2018 AHA/ACC/AACVPR/AAPA/ABC/ACPM/ADA/AGS/APhA/ASPC/NLA/PCNA guideline on the management of blood cholesterol. J Am Coll Cardiol 73(24):e285–e350
30. Haberl R, Becker A, Leber A, Knez A, Becker C, Lang C et al (2001) Correlation of coronary calcification and angiographically documented stenoses in patients with suspected coronary artery disease: results of 1764 patients. J Am Coll Cardiol 37(2):451–457

31. Aggarwal NR, Bond RM, Mieres JH (2018) The role of imaging in women with ischemic heart disease. Clin Cardiol 41(2):194–202
32. Baldassarre LA, Raman SV, Min JK, Mieres JH, Gulati M, Wenger NK et al (2016) Noninvasive imaging to evaluate women with stable ischemic heart disease. JACC Cardiovasc Imaging 9(4):421–435
33. Gulati M, Levy PD, Mukherjee D, Amsterdam E, Bhatt DL, Birtcher KK et al (2021) 2021 AHA/ACC/ASE/CHEST/SAEM/SCCT/SCMR guideline for the evaluation and diagnosis of chest pain. J Am Coll Cardiol 78(22):e187–e285
34. Tsang JC, Min JK, Lin FY, Shaw LJ, Budoff MJ (2012) Sex comparison of diagnostic accuracy of 64-multidetector row coronary computed tomographic angiography: results from the multicenter ACCURACY trial. J Cardiovasc Comput Tomogr 6(4):246–251
35. Min JK, Dunning A, Lin FY, Achenbach S, Al-Mallah M, Budoff MJ et al (2011) Age- and sex-related differences in all-cause mortality risk based on coronary computed tomography angiography findings results from the International Multicenter CONFIRM (Coronary CT Angiography Evaluation for Clinical Outcomes: an International Multicenter Registry) of 23,854 patients without known coronary artery disease. J Am Coll Cardiol 58(8):849–860
36. Leipsic J, Taylor CM, Gransar H, Shaw LJ, Ahmadi A, Thompson A et al (2014) Sex-based prognostic implications of nonobstructive coronary artery disease: results from the international multicenter confirm study. Radiol 273(2):393–400
37. Ahmadi N, Nabavi V, Hajsadeghi F, Flores F, French WJ, Mao SS et al (2011) Mortality incidence of patients with non-obstructive coronary artery disease diagnosed by computed tomography angiography. Am J Cardiol 107(1):10–16
38. Nørgaard BL, Hjort J, Gaur S, Hansson N, Bøtker HE, Leipsic J et al (2017) Clinical use of coronary CTA-derived FFR for decision-making in stable CAD. JACC Cardiovasc Imaging 10(5):541–550
39. Danad I, Szymonifka J, Twisk JWR, Norgaard BL, Zarins CK, Knaapen P et al (2017) Diagnostic performance of cardiac imaging methods to diagnose ischaemia-causing coronary artery disease when directly compared with fractional flow reserve as a reference standard: a meta-analysis. Eur Heart J 38(13):991–998
40. Hurwitz LM, Reiman RE, Yoshizumi TT, Goodman PC, Toncheva G, Nguyen G et al (2007) Radiation dose from contemporary cardiothoracic multidetector CT protocols with an anthropomorphic female phantom: implications for cancer induction. Radiol 245(3):742–750
41. Klem I, Greulich S, Heitner JF, Kim H, Vogelsberg H, Kispert E-M et al (2008) Value of cardiovascular magnetic resonance stress perfusion testing for the detection of coronary artery disease in women. JACC: Cardiovascular Imaging 1(4):436–445
42. Nandalur KR, Dwamena BA, Choudhri AF, Nandalur MR, Carlos RC (2007) Diagnostic performance of stress cardiac magnetic resonance imaging in the detection of coronary artery disease: a meta-analysis. J Am Coll Cardiol 50(14):1343–1353
43. Hamon M, Fau G, Née G, Ehtisham J, Morello R, Hamon M (2010) Meta-analysis of the diagnostic performance of stress perfusion cardiovascular magnetic resonance for detection of coronary artery disease. J Cardiovasc Magn Reson 12(1):29
44. Greenwood JP, Maredia N, Younger JF, Brown JM, Nixon J, Everett CC et al (2012) Cardiovascular magnetic resonance and single-photon emission computed tomography for diagnosis of coronary heart disease (CE-MARC): a prospective trial. The Lancet 379(9814):453–460
45. Schwitter J, Wacker CM, Wilke N, Al-Saadi N, Sauer E, Huettle K et al (2012) Superior diagnostic performance of perfusion-cardiovascular magnetic resonance versus SPECT to detect coronary artery disease: The secondary endpoints of the multicenter multivendor MR-IMPACT II (Magnetic Resonance Imaging for Myocardial Perfusion Assessment in Coronary Artery Disease Trial). J Cardiovasc Magn Reson 14(1):61
46. Lipinski MJ, McVey CM, Berger JS, Kramer CM, Salerno M (2013) Prognostic value of stress cardiac magnetic resonance imaging in patients with known or suspected coronary artery disease: a systematic review and meta-analysis. J Am Coll Cardiol 62(9):826–838
47. Jalnapurkar S, Zarrini P, Mehta PK, Thomson LEJ, Agarwal M, Samuels BA et al (2017) Role of stress cardiac magnetic resonance imaging in women with suspected ischemia but no obstructive coronary artery disease. J Radiol Nurs 36(3):180–183

48. Thomson LE, Wei J, Agarwal M, Haft-Baradaran A, Shufelt C, Mehta PK, Gill EB, Johnson BD, Kenkre T, Handberg EM, Li D, Sharif B, Berman DS, Petersen JW, Pepine CJ, Bairey Merz CN (2015) Cardiac magnetic resonance myocardial perfusion reserve index is reduced in women with coronary microvascular dysfunction. A National Heart, Lung, and Blood Institute-sponsored study from the Women's Ischemia Syndrome Evaluation. Circ Cardiovasc Imaging 8(4):e002481. https://doi.org/10.1161/CIRCIMAGING.114.002481
49. Coelho-Filho OR, Seabra LF, Mongeon FP, Abdullah SM, Francis SA, Blankstein R et al (2011) Stress myocardial perfusion imaging by CMR provides strong prognostic value to cardiac events regardless of patient's sex. JACC Cardiovasc Imag 4(8):850–861
50. Johnson BD, Shaw LJ, Buchthal SD, Bairey Merz CN, Kim HW, Scott KN et al (2004) Prognosis in women with myocardial ischemia in the absence of obstructive coronary disease: results from the National Institutes of Health-National Heart, Lung, and Blood Institute-Sponsored Women's Ischemia Syndrome Evaluation (WISE). Circ 109(24):2993–2999
51. Acr practice guideline for imaging pregnant or potentially pregnant adolescents and women with ionizing radiation.
52. Litmanovich DE, Tack D, Lee KS, Shahrzad M, Bankier AA (2014) Cardiothoracic imaging in the pregnant patient. J Thorac Imaging 29(1):38–49

# Chapter 21
# Heart Ischemia/Reperfusion Injury—Is the Female Equally Protected Compared to Male?

**Delphine Baetz and Marie Vedere**

**Abstract** Despite the decrease in cardiovascular mortality over the past 3 decades, it still remains the leading cause of death in women. Young women have a lower risk of cardiovascular disease (CVD), but this trend is reversed after menopause. There are many reasons for this difference between men and women, including traditional risk factors such as diabetes, smoking, dyslipidemia, or aging for which women are clearly more impacted than men. Additionally, there are female-specific risk factors, called non-traditional risk factors, that are associated with increased risk of cardiovascular disease in women. These so-called non-traditional risk factors concern women with pre-eclampsia, recurrent pre-eclampsia, gestational diabetes and premature delivery. In addition, there is also an increased risk for women who use contraceptives, who have suffered recurrent miscarriages, premature ovarian failure and early menopause. There are also psychological, social and cultural aspects related to sex. Indeed, lower level of education is more frequently observed in women. Further, numerous pre-clinical animal studies have highlighted some of the cellular mechanisms involved in the differences in the cardiovascular risk in females. These studies have shown a link between high estradiol levels, calcium handling and cardioprotection in young females. In addition, it seems that the mitochondria, which are essential to cardiac function by providing ATP for contraction and play a central role in the management of oxygen and calcium, are also, differentially regulated between males and females. Therefore, it is important to better understand the origin of these differences between men and women in order to improve the diagnosis, prevention and management of CVD in the future.

**Keywords** Ischemia reperfusion · Hypoxia · Coronary heart disease · Acute coronary syndromes · STEMI

---

D. Baetz (✉) · M. Vedere
CarMeN Laboratory, INSERM, Université Claude Bernard Lyon 1, 69100 Villeurbanne, France
e-mail: delphine.baetz@univ-lyon1.fr

© The Author(s), under exclusive license to Springer Nature Switzerland AG 2023

L. Kirshenbaum and I. Rabinovich-Nikitin (eds.), *Biology of Women's Heart Health*, Advances in Biochemistry in Health and Disease 26,
https://doi.org/10.1007/978-3-031-39928-2_21

# Introduction

For many years, cardiovascular diseases (CVD) have been, in the common imagination, a male pathology related to their "bad habits" such as smoking, drinking, eating fat-enriched foods... Indeed, the differences in hormones and genes between males and females underlie sex-related differences in the regulation of physiological and pathological cardiovascular function.

In 2019, cardiovascular diseases were the leading cause of death worldwide, with 18.6 million people which represent 32% of all death [1]. Of these numbers, World Health Organization (WHO) estimated that 7.4 million people died from coronary heart disease (CHD) alone (https://www.who.int/news-room/fact-sheets/detail/cardiovascular-diseases-(cvds)), although it seems that over the last decades, deaths related to CVDs significantly declined [2]. Moreover, CVDs are responsible for the majority of deaths in females [3], which also have worse outcomes following acute coronary heart disease, compared with males [4]. In general, among CVDs, CHD and stroke are the main killer of women worldwide [5] (https://vizhub.healthdata.org/gbd-compare/). Nevertheless, there are some variations between males and females' CVDs. In fact, in pre-menopausal women, CHD and cardiac remodeling following myocardial infarction is at lower incidence when compared to age matched males. On the contrary, in post-menopausal females or in pre-menopausal females who underwent oophorectomy, the incidence of CVDs is increased when compare to males [3, 4].

Sex differences in the prognosis, treatment, related disabilities and death related to Acute Coronary Syndromes (ACS) were first described in the Framingham Study in 1979 [6]. Nevertheless, the underlying mechanisms of these differences are miscellaneous and not all known yet. This lack of knowledge is partly linked to the underrepresentation of women in clinical trials [7] and the use of male animals in most basic research studies [8].

This chapter will discuss these differences and their origins.

### (1) Epidemiology of ischemia reperfusion injury in humans

In the last 4 decades, we observed a clear decline in mortality from CHD. This decline is attributed to a significant progress in prevention and treatment, a decline in cigarette smoking, a better control of hypertension, and more importantly, the development and timely use of thrombolysis and stents in acute coronary syndrome to limit or prevent cardiac infarction.

Among CVDs, CHD results mostly from ACS. ACS is divided into three categories: ST-segment myocardial infarction (STEMI), non-ST-segment elevation myocardial infarction (NSTEMI) and unstable angina. Recent analysis of 78,254 patients from the Get With The Guidelines–Coronary Artery Disease registry pointed out that fewer women suffered from STEMI compare to men (28.2% vs. 35.1%) [9]. Moreover, women were less likely than men to have NSTEMI or unstable angina (36.6% vs. 47.6%) [10]. However, when comparing young STEMI patients under the age of 45, 22% of women have spontaneous aortic dissection compared to 3% of men which is also in accordance with greater risk for cardiac re-infarction (15% vs. 1%) [11].

Over the past 3 decades, mortality from heart disease has declined, but cardiovascular disease remains the leading cause of death among women. Moreover, recent reports show that there is even a trend of increasing CVD related mortality in past several years [12] and especially among young women [2, 13].

A recent study that examining the average rates of recurrent MI, recurrent CHD events, heart failure hospitalization, and mortality at the first year after MI between the years 2008 and 2017 found a dramatic drop in recurrent events for both men and women. However, mortality, recurrent MI and CHD events are still superior in men compared to women, with greater decrease in recurrence in women, compared with men [14]. On the contrary, survival rates of women with STEMI and NSTEMI are worse than that of men in a 10 years follow-up post infarction. The excess mortality rates for women are still present after adjusting for age and comorbidities, but it decreases after adjusting for the use of evidence-based treatments [15]. In parallel, a very recent publication that compiled the results of 15 different studies showed that short-term mortality continues to be higher in female with STEMI compared with male [16].

This chapter is dealing with female's disadvantages when facing CVD, compared with males. Premenopausal women have significantly lower CVD risk than men of the same age, but women lose their benefit at the time of menopause. In fact, women are at higher risk for CVD 7 to 10 years later than men. These sex differences have been largely attributed to the protective effects of sex steroid hormones during reproductive years. The mechanisms involved in the cardioprotective effects of estrogen are varied; they include antioxidant properties, increased angiogenesis and vasodilatation, reduced fibroblast proliferation and antiapoptotic properties [17].

This book chapter will cover specific areas where women are disadvantaged compared to men, and include traditional risk factors for women, sex specific treatment and diagnosis, and lastly, female's personal knowledge and behavior, that can contribute in a negative way to cardiovascular disease development and management [18] (Fig. 21.1).

## (2) **Traditional risk factors for women**

While in younger ages, women are at lower risk for CVD, in older ages women are at higher risk for CVD than men. A significant part for CVD risk could be prevented by maintaining a healthy lifestyle and controlling glucose, cholesterol and blood pressure at normal levels for both men and women. Most of the CHD risk factors have been descried in both sexes, but some of them are linked with different risk for men or women [18]. Sex differences have been pointed out in diabetes, dyslipidemia and obesity [19]. For example, the prevalence of diabetes is higher in women [20]. However, on the contrary, women tend to have better controlled blood pressure and lower cholesterol [21, 22]. Indeed, total cholesterol had more pronounced adverse effect on CHD in men than women [23]. But a number of known cardiovascular risk factors were shown to have a higher relative effect in women, including diabetes [24], atrial fibrillation [25] smoking [26], and low socioeconomic status [27].

(a) **Diabetes**

Cardiovascular complications are one of the most common causes of morbidity and mortality among diabetic patients [28]. Surprisingly, during the last 30 years, a reduction in CVD mortality was observed only in men and not in women. In the 1970s, women with diabetes had better survival rate than men, that was largely suppressed between 1988 and 2000 [29]. The lack of improvement among women is concerning but not well documented.

While type 2 diabetes is more prevalent in men compared to women (9.6 vs. 7.9%) [30], cardiovascular risk is higher in diabetic women than men, suggesting worse CV consequences and mortality in diabetic women, independently of age [31, 32]. For example, a meta-analysis from 64 cohorts that included 858,507 individuals, 28,203 of them with CHD demonstrated that diabetic women have 40% increased risk for CHD compared to diabetic men [24]. Moreover, obesity, which is a related risk factor to diabetes, is more prominent in women [33]. Interestingly, it seems that women with diabetes lose their sex-related protection in the development of CVD [34, 35]. An analysis of pooled data from the Framingham Heart Study and the Framingham Offspring Study with follow-up of 20 years showed that the adjusted hazard ratios (HR) with 95% confidence intervals for death from CHD were 2.1 (95% CI, 1.3–3.3) in men with diabetes only, and 4.2 (95% CI, 3.2–5.6) in men with CHD only compared with men without diabetes or CHD. This is reversed in women with hazard ratio for CHD death being 3.8 (95% CI, 2.2–6.6) in women with diabetes, and 1.9 (95% CI, 1.1–3.4) in women with CHD [36].

Different studies in humans and animal models suggest that diabetic female have endothelium impairment [37]. While men have a two-fold increased risk for CHD, diabetic women have four-fold increased risk for CHD [38]. This difference can be partly explained by the endothelial dysfunction observed in women with diabetes that is more pronounced than in diabetic men [39]. Moreover, evidence suggests that impaired glucose tolerance combined with vascular dysfunction in pre-diabetic individuals present a greater risk in women than in men [40].

Notably, ex disparities in diabetes pharmacotherapy also exists, but they cannot explain in full the excess risk in women. Additional studies are needed to better elucidate the different mechanisms responsible for this disparity in diabetes-related risk of CHD [24]. It is reasonable to speculate that treatment recommendations for women with diabetes should be different from those for men.

(b) Smoking

Smoking represents a major risk factor for myocardial infarction for women both before and after menopause [41]. When comparing the risk factor for myocardial infarction between men and women, the relative risk is 1.9 versus 3.3, respectively [42]. Among adults with known CVD, nearly one third are users of tobacco. Furthermore, women are even more at risk when using oral contraceptives with tobacco products. In fact, smoking is a risk factor for venous thrombosis with a bigger relative effect among young women using oral contraceptives [43].

To date, we do not have enough information on the cardiovascular impact of electronic cigarettes usage, and the sex dependent risk, although one study revealed greater usage of electronic cigarettes in women with prevalent cardiovascular disease when compared with men [44]. Nevertheless, it is also important to evaluate the influence of dual use (cigarettes and e-cigarettes) regardless of sex [45].

(c) Hypertension

In average, women develop hypertension a decade later than men do, and the effect of hypertension becomes more prevalent in elderly women than in elderly men. Although there are no sex differences in the clinical manifestation of hypertension [46], hypertension in older women is often less controlled. In this regard, only 23% of women vs. 38% of men over 80 years have a blood pressure inferior of 140/90 mm Hg [47].

(d) Dyslipidemia

Atherosclerosis related CVD are strongly linked to dyslipidemia, and represent the highest population-adjusted risk of 47.1% among women after menopause, compared with all other known risk factors for CVD [19]. This risk is typically not observed prior to menopause, even if cholesterol levels are high.

A study called INTERHEART aimed to measure Apolipoprotein B (ApoB)/ Apolipoprotein A1 (ApoA1) and TC/HDL-C levels in men compared to women. The study found a strong association between higher blood lipid levels to acute myocardial infarction in women, compared with men [48].Therefore, in the last decades, statins are used extensively for lowering serum cholesterol and associated risk for CVD. However, a recent study that assessed the association between absolute reductions in LDL-C levels with statin therapy and all-cause mortality suggested that association between absolute reductions in cholesterol levels and individual clinical outcomes is not well established [49].

Importantly, most of the clinical studies that evaluated the effects of lipid-lowering medications on CHD were done mainly in men, and not in women. Therefore, extrapolated recommendations were made for clinical use for women. Recently, a new meta-analysis study pointed out the observed effects of statins treatment in women, showing that statins therapy reduced CHD in primary prevention in men but not in women [50, 51]. More specifically, for women without CVD, statins use does not have any impact on all-cause mortality or CHD mortality, however, for women with known cardiovascular disease, statin treatment is effective in reducing CHD events and CHD mortality [52, 53].

(e) Aging

Age is linked to a deterioration in heart function which is linked to an increased risk for developing CVD [33]. As already mentioned, sex is another risk factor in aging. In the AHA 2019 Heart Disease Statistical Update, the incidence of CVD was reported to be greater in females than males at ages 60–79 ( 78.2% versus 77.2%) and 91.8% versus 89.3% for ages over 80 [54]. The risks associated with cardiovascular disease increase with age regardless, of sex, and this is related to a significant decrease in testosterone and estrogen levels [33]. Within the older population, cardiovascular disease is still the leading cause of death, and, with increasing life expectancy, the incidence of cardiovascular disease is likely to remain very high among older people, representing an increasing high sociological and financial cost to our society [55].

(3) **Non-traditional Risk Factors for Women**

In 2018 and 2019 the American Heart Association/American College of Cardiology Multi-Society cholesterol guideline and American College of Cardiology/American Heart Association guideline first introduced the concept of risk-enhancing factors that are specific to women [56, 57]. These guidelines included premature menopause, preeclampsia, gestational diabetes and polycystic ovary syndrome as CVD risk-enhancing factors [58].

In order to quantify the associations that may exist between reproductive factors in women at childbearing age and their cardiovascular risk, a meta-analysis published in 2020 compiled the results of 32 review papers. According to this study, the risk for ischemic heart disease among women with pre-eclampsia, recurrent pre-eclampsia, gestational diabetes, and preterm birth increased by 2-folds. The risk for women who use combined oral contraceptives (estrogen and progesterone), experienced recurrent miscarriage, premature ovarian insufficiency, and early menopause have increased by 1.5–1.9 folds. Miscarriage, polycystic ovary syndrome and menopausal symptoms also represent a risk factor, but the associations with cardiovascular risk is less prominent [59]. According to these findings, from menarche to menopause, some hormonal factors linked to reproduction are associated with increased risk of CVD, therefore, it is of importance to consider these hormonal factors in order to propose new guidelines to women that will to improve their care and better predict their risks, especially in younger women.

(a) Early Menopause and Premature Menarche

In 2016, a systematic review revealed that women who experience premature or early-onset menopause had a higher risk of CHD, CVD mortality, and overall mortality [60]. More recently, in 2019 and 2020 two other meta-analysis studies linked the type and age of menopause onset to the risk of CVD. The first study, from the InterLACE consortium, compared natural menopause and surgical menopause, and showed that surgical menopause was associated with over 20% increased risk of CVD [61]. The second pooled individual-level data from 15 observational studies performed across five countries (Australia, Scandinavia, the USA, Japan, and the UK) between 1946 and 2013, revealed an increased risk of non-fatal CVD before the age of 60 years for women who experienced premature and early menopause. Interestingly, this risk was no more apparent after the age of 70 years [62].

These results led to the idea that hormone replacement therapy (HRT) could provide a solution for post-menopausal women to reduce risk for CVD. A study conducted by The Women's Health Initiative that included 16,608 postmenopausal women who were randomized to HRT versus placebo was ended prematurely due to a strong signal of increased CVD risk in women on the combined estrogen-progestin pill [63]. On the contrary, another study pointed out to a lower risk for CHD among women who experienced surgical menopause before 50 years old and were treated with HRT, compared to women that were not treated with HRT [61]. Altogether, most studies failed to show any cardiovascular benefit for the use of HRT and several studies even reported increased risk. To date, HRT is not recommended for prevention of CVD [64] and women with premature menopause need specific monitoring that takes into account the increased risk factor.

Recently, the link between CVD and the age of menarche was examined. A meta-analysis study revealed an increased risk of death from all causes for women who had menarche before the age of 12 years. Moreover, early menarche was associated with increased risk of death from ischemic heart disease, among nonsmoking women [65]. Notably, early menarche was also linked to development of hypertension, diabetes and hypercholesterolemia which all increase CVD risk [66].

(b) Pre-eclampsia

Pre-eclampsia, a hypertensive disorder of pregnancy, affects around 10% of all pregnancies. Women who experience pre-eclampsia have a twofold increased risk for CVD later in life, compared with women with normotensive pregnancies. Moreover, these women display an increased risk of chronic hypertension, dyslipidemia, diabetes and atherosclerosis [67]. Interestingly, when the pre-eclampsia is recurrent, the risk is even higher (RR 2.40, 95% CI 2.15 to 2.68) when compared with women with subsequent uncomplicated pregnancy [68]. NT-proBNP levels are increased in patients with preeclampsia, which is also an indication for increased risk for CVD [69]. To date, there is no consensus on the guidelines addressing postpartum cardiovascular risk assessment for women with pre-eclampsia [67, 70, 71].

(c) Gestational diabetes

Gestational diabetes (GD) affects 7–10% of all pregnancies worldwide. GD is correlated with an increased risk for developing type 2 diabetes (T2DM) which is a major risk factor of CVD, as discussed previously [72]. 19% of women who experienced GD will develop T2DM compared to 4.8% in women who did not experience GD [73].

According to a Canadian meta-analysis, women with GD had an increased cumulative incidence of hospitalization for CVD 25 years after delivery that is associated with higher risk of myocardial infarction (HR 2.14, 95% CI 1.15–2.47), coronary angioplasty (HR 2.23, 95% CI 1.87–2.65), and coronary artery bypass graft (HR 3.16, 95% CI 2.24–4.47) [74]. In 2019, another meta-analysis that pooled data from 5,390,591 women with GD showed an elevated risk of cardiovascular disease in the first decade after pregnancy that was not affected by the incidence of T2DM, that is more frequent after GD [75].

The reason for increased risk for CVD among women with a history of GD, may be explained by alterations to endothelial function. In fact, glucose intolerance has been related to endothelial changes during GD. These changes can last years after pregnancy as shown by a significant reduction in coronary blood flow which is correlated with a coronary microvascular dysfunction [76].

Furthermore, poor blood sugar control during pregnancy, even without GD diagnosis, was also found to be associated with a future risk for CVD. Every 1 mmol/L increment in the glucose tolerance test during pregnancy, was associated with a 13% higher risk of CVD, after adjustment for age, ethnicity, income, rurality, and women with gestational diabetes [77].

Therefore, it is of importance to consider GD as a potential CVD risk factor, and to implement a regular follow up to women with GD following pregnancy.

(d) Polycystic ovary syndrome

5–20% of women suffer from polycystic ovary syndrome (PCOS) that is characterized by excessive androgen hormone production by the ovaries and is associated with insulin resistance and compensatory hyperinsulinaemia [78].

A recent study showed an association between subclinical CVD markers, such as endothelial dysfunction and coronary artery calcium stores and PCOS. Nevertheless, because PCOS is currently associated with metabolic disorders, it is difficult to distinguish the relative importance of metabolic risk factors from PCOS when considering CVD risk factors [79]. There is a lack of population-based long-term studies examining cardiometabolic morbidity and mortality in PCOS with a need for further research to better understand the long-term cardiometabolic impacts on CVD among women with PCOS [80].

(e) Oral contraceptives (COCs)

The use of COCs is associated with 2- to 4-fold increased risk for arterial and venous thromboembolic events. The risk for venous thromboembolism in women using oral contraceptives aged < 30 years is estimated to be 3.7/10,000 cases annually, compared to 1.2/10,000 in women who do not use COC. Moreover, the risk for venous thromboembolism is rising with age [81]. The dose of COCs is in relation to the risk of AMI. A meta-analysis that compiled data from 1298 publications showed that the risk of myocardial infarction or ischemic stroke was 1.6-fold higher in women using > 50 μg estrogen COCs [82].

(4) **Clinical presentation, diagnosis and management**

Clinical presentation, diagnosis and therapeutic strategies used for women suffering from acute coronary syndrome may explain partially sex-related differences observed in ischemia/reperfusion injury. There is a large amount of data describing the differences involved with ACS between women and men and their outcomes.

The management of ACS is based on a rapid diagnosis. Different studies evaluated women specific symptoms of AMI, in order to promote knowledge and better diagnosis of AMI in women. One meta-analysis of fifteen prospective studies with a total sample size of 10,730 pointed out that during a suspected ACS, women present with dyspnea, arm pain, nausea and vomiting, fatigue, heart palpitations and pain at the shoulder. Moreover, the average age of AMI in women is four years older, compared to men [83]. Another study indicated that chest pain is the predominant symptom for both sexes, however, women present a greater number of additional non-chest pain, and both women and their healthcare providers were less likely to attribute these prodromal symptoms to heart disease, in comparison with men [84]. Prodromal cardiac symptoms are warning signals that precede cardiac disease. Previous studies have shown sex differences in prodromal symptoms as well as established risk factors for MI. Recently, 213 middle-aged men and women, all diagnosed with myocardial infarction were enrolled in a study that mapped possible sex differences and established a list of risk factors that precede myocardial infarction. In this regard, the dominant prodromal symptom for both males and females was chest pain at the time of onset of MI, however, hyperlipidemia status was more frequent in females. Females also reported to have higher stress load the year preceding myocardial infarction with serious life events, followed by reports on sadness/depression. These findings may contribute to a better understanding of different diagnostic approach between sexes, as well as a sex-oriented cardiovascular preventive approach [85]. Insights from the PROMISE trial revealed that although women had more frequently normal noninvasive testing compared with men, those with abnormal results were less likely to be referred for catheterization or to receive statins therapy. These results strongly support the need for a sex-specific algorithm to guide noninvasive testing and chest pain management [86].

For women experiencing STEMI, the door-to-device time and first medical contact-to-device time are longer and are associated with increased mortality rates [87]. Indeed, different studies indicated that women traditionally delay seeking medical help, compared to men. This may be attributed to misinterpretation of symptoms, difficulties to access care, embarrassment or even a low awareness of personal risk [88]. Recently, a study performed on 6330 patients with STEMI (21% women) compared the differences in delays for emergency call services and door-to-device time and reported that female patients undergone excess delays in prehospital system, and hospital delays, even after adjustment for confounders that resulted in an adjusted excess delay of 10 min, compared with men [89].

Notably, the anatomic coronary structure of women is different from the one in men, with smaller blood vessels in women, compared to men [72]. It is correlated with a more frequent coronary plaque erosion with distal embolization in women while men are more often suffering from plaque rupture with thrombosis [90]. Diffuse atherosclerotic is typical in CAD in women and often associated with microvascular and endothelial dysfunction, while epicardial coronary stenosis is more frequent in men [91]. Notably, women diagnosed with STEMI continue to experience suboptimal treatment delays related to worse adverse outcomes. It could be in part related to a misperception of CVD risk in females by the medical staff, because women have commonly been considered more protected, or, it could be related to presentation with atypical symptoms [87]. Indeed, coronary interventions are less frequent in women [92], and have a lower rate of -prescribed guideline-directed medical therapy and a greater mortality in hospital [13]. Recently, it has been shown that the treatments offered to women are less aggressive, and a lower rate of interventional procedures are performed in women, when compared to men [93]. In addition, a recent study showed that after multivariable adjustments, evidence-based acute treatments for ACS, including early dual antiplatelet therapy, heparin, and reperfusion therapy, are less frequent in women than in men. Furthermore, women that are suitable for secondary prevention are less likely to receive dual antiplatelet therapy, angiotensin-converting enzyme inhibitors/angiotensin receptor blockers, or statins at discharge [94]. Moreover, a study performed on 7878 patients hospitalized following MI or angina, demonstrated that women less frequently receive immediate invasive management [95]. It appears that sex disparities in timeliness to reperfusion in young patients with STEMI are also persisted [96].

In addition to delayed diagnosis and lack of optimal treatment, women with ACS that undergone successful percutaneous coronary intervention still have a 20% higher adjusted mortality risk in the short term. 1 year after ACS, those over age 45 have a 74% survival rate vs. 81% in men [97]. Furthermore, residual angina is more frequent after myocardial revascularization in women [98], and young women in particular, have worse short-term and long-term outcomes than men [4]. Indeed, spontaneous coronary artery dissection is an increasingly recognized cause of acute coronary syndromes (ACS) afflicting predominantly younger to middle-aged women. Spontaneous coronary artery dissection (SCAD) is responsible for a particular and atypical form of myocardial infarction, affecting mostly women in apparent good health (80 to 90% of patients are women) [99]. Recent studies show that this poorly understood disease is not as rare as previously thought: it is in fact responsible for almost a third of acute myocardial infarctions occurring in women under 60 years old. The number of these heart attacks is clearly underestimated because the patients do not present the classic risk factors, such as high cholesterol and overweight, which may lead to preventive treatment. The first genetic risk factor for spontaneous coronary artery dissection was identified in the largest study conducted to date for this condition, but more studies are needed to better understand the mechanisms involved in this atypical form of ACS [100].

Recently, the Lancet women and cardiovascular disease Commission summarized the main differences in prevention, diagnosis, treatment and access to care for women when compared to men. They concluded that cardiovascular disease in women remains understudied, under-recognized, underdiagnosed, and undertreated [101].

(5) **Personal knowledge and behavior disadvantaged**

It is evident that sex, greatly influences physiological functions and pathophysiology. However, psychological, social and cultural aspects linked to sex can also play a role in pathophysiology. Indeed, these factors can exacerbate or inhibit biological processes, and cardiovascular risks are not spared by this [102]. In most of the world, women live longer than men. However, this advantage tends to disappear. The reason for that may rely on the fact that in many countries women have less access to education, which gives them less access to health, and consequently, may lead to opposite tendency in survival rates [103].

(a) Low social status

Several studies have already shown that there is a relationship between a low socio-educational level and an increased risk for CVD. This may be related to other factors, such as biological, behavioral (lack of sleep, obesity, etc.) or even psychological (anxiety, depression, etc.), which are known to be more frequent in women with low socio-educational level [104]. Moreover, adverse childhood experiences is also associated with higher CVD risk in adults [105]. In a systematic review that collected data from 116 cohorts, a strong link between risk of CHD was associated with lower educational achievement in women, compared with men [27]. In order to explain this link, two recent studies demonstrated that low socio-economic status is a source of chronic stress that promotes inflammation and atherogenesis [106, 107].

(b) Obesity, overweight, and physical inactivity

In the last 3 decades, the prevalence of overweight (BMI > 25 $kg/m^2$) and obesity has increased significantly in the general population. In 2015, 1.9 billion and 609 million adults were estimated to be overweight or obese, respectively. This corresponds to almost 39% of the world's population. Global prevalence for obesity have increased from 8.9 to 14.8% in women, and from 5 to 10.1% in men between the years 1980 and 2015 [108–110]. These sex-related differences in obesity prevalence are not only related to nutrition, but also to lifestyle discrepancies between men and women and are important to be considered [111, 112].

Generally, overweight and obesity are strongly correlated with CVD mortality, especially in women [113]. The Framingham Heart Study found that obesity increased the relative risk of CAD by 64% in women, as opposed to 46% in men [114]. It is well known that adiposity is a source of sterile inflammation that can be strongly associated with CVD [115, 116]. Moreover, the hormonal changes occurring after menopause, especially estrogen decrease, result in an increased risk of metabolic syndrome and cardiovascular complications in obese postmenopausal women [117].

To counteract the effect of obesity, exercise training is beneficial. Indeed, exercise training is a good physiological stimulus, which was shown to reduce primary and secondary cardiovascular events. The WHO organization reported recently that inactivity or insufficient physical activity is correlated with an increased risk of CVD and shortens lifespan by three to five years. In addition, data from 2016 reported that 31.7% of adult females and 23.4% of adult males are not active enough. This can certainly be attributable to historical habits where males have been more likely involved in specific sports/physical activities that are different from those practiced by females [118]. It is therefore important to reconsider the recommendations for physical activity, especially in women to promote better prevention.

(c) Psychological health

There is an increasing evidence that psychological health can represent a risk factor in the development of CVD [119].

A meta-analyze study from the last decade showed a significant increased risk for coronary heart disease among adults experiencing stress at work and in personal life. The proposed mechanisms are inflammation, metabolic changes and hemostatic disturbances [120]. Different studies have defined anxiety as a risk factor for hypertension [121]and metabolic syndrome [122]. Moreover, a recent study looking at 2,017,276 participants from 46 cohorts reveal a close association between anxiety and increased risk of CHD mortality of 41% [123]. Depressive disorders, that affect more women than men, are also associated with increased risk for developing and dying from CVD [124], especially myocardial infarction which stands at 30% [125]. About 20–25% of women go through depression during their life, therefore, it begs the question- how is it linked to CVDs? it seems that biological risk factors, including sympathetic nervous system, hyperactivity and impairment in hypothalamic–pituitary–adrenal function are involved [126]. In this regard, an observational study which included 560 women, showed that women with depression had a 2.53-fold risk for MI during the 16 years follow-up [127].

On the contrary, positive psychological health has its own independent associations with lower risk of CVD [128]. The American Heart Association proposed a definition of cardiovascular health that comprises of 7 components: 4 health behaviors (physical activity, healthy diet, no smoking, and normal BMI) and 3 health factors (normal blood pressure, total cholesterol, and plasma glucose). It is well recognized that optimism is associated with healthier behaviors, such as more physical activity, no smoking, healthy diet score, better sleep quality, and higher composite cardiovascular health scores [119]. In women in particular, it has been shown that positive state of mind is associated with lower risk of CVD [129] and slower progression of atherosclerosis [130] (Fig. 21.1).

(6) **Mechanisms of CVD in women**

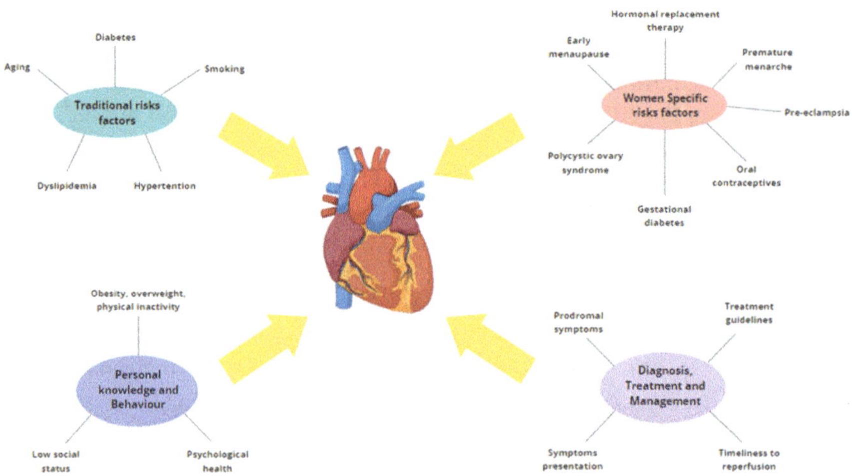


**Fig. 21.1 Why CVDs are responsible for the majority of deaths in the female population?** Specific factors in which women are considered disadvantaged compared to men in terms of cardiovascular health, including traditional risk factors, women-specifics risk factors, diagnosis, treatment and management, and personal knowledge and behavior

Clinical studies show that many biological or non-biological determinants affect cardiovascular health differently in women and men. Many in vivo animal studies try to decipher the cellular mechanisms involved in the different responses to the risk factors of cardiovascular disease in females. However, a limiting factor in these studies is that they have long been performed in male mice only. Nevertheless, many studies deliberately trying to search for sex-related differences in the cardiovascular system discovered genes that exhibited sexual dimorphism [8].

Many in vivo animals based studies are contradictory when comparing females and males. For example, a study focusing on IHD in women demonstrated that female rat hearts were more resistant to oxygen deprivation than hearts of male rats [131]. The same protection was shown in female dogs, rats, mice and rabbits hearts subjected to ischemia–reperfusion [132]. There are two others examples where females have better resistance to heart ischemia. The first one shows in a model of rat Langendorff perfused heart that ischemia causes smaller infarct size in females compared to males [134]. The second study performed in mice describe a better functional recovery of the heart [133]. However, it is important to note that there are also many experimental models that did not show any differences between males and females; this was the case of an in vivo model of coronary artery ligation in adult dogs subjected to 1 h of ischemia and 4 h of reperfusion [134]. Two others studies reported no differences in the infarct size between males and females in rat models in vivo [135] or ex vivo [136]. It is important to note that all these studies were performed with young animals.

(a) Estrogen

Epidemiological data suggest that premenopausal women are protected from CHD compared with age-matched men, but this protection is lost after menopause. This led to the generally accepted conclusion that the sex hormones, and especially estradiol (E2) can protect against CHD in women. Most studies in humans found a link between high estradiol levels and cardioprotection in young women. 17β-Estradiol ($E_2$) is the principal estrogenic hormone that exerts its cardioprotective effect via different receptors, including estrogen receptors α (ERα) and β (ERβ) and the G protein-coupled estrogen receptor 1 (GPER). However, the literature reports some conflicting results, which may be due to different experimental models considered, acute versus chronical effects, species, age, etc. Moreover, because of the dramatic changes in cardiac risk factors between premenopausal and postmenopausal women, estrogen has been extensively studied for its potential cardioprotective effect. In the early 90th an evolutionary study pointed out to the beneficial role of estrogen use in reducing the incidence of CHD and the associated mortality from CVD [137]. However; at the same time, the Framinghan heart study showed a 50% increased risk of cardiovascular morbidity from estrogen consumption [138]. This controversy has not been fully settled to that day, in part due to the potentially severe side effects and contradicting benefits [17].

Ovaries are the primary site of E2 production in premenopausal women, and E2 is produced in small amounts by the testes in men [139]. It was shown that E2 can be synthesized in extragonadal tissues, through the conversion of testosterone by cytochrome p450 aromatase in both sexes [140]. Among these E2-producing tissues, heart is expressing the aromatase and E2 can also be produced locally in cardiac cells. This suggests that local cardiac E2 synthesis could play a role in the E2-mediated effects on CHD [141]. Erα, ERβ and GPER are all expressed in the heart [142]. ERα and ERβ are also detected in vascular smooth muscle cells (VSMC) [133]. The multiplicity of receptors and sites of action of E2 makes it difficult for understanding the mechanisms involved in cardioprotection and therefore, this field of research is not fully elucidated yet [143, 144].

In order to better understand the role of estradiol, scientists mostly used ovariectomized females or transgenic models that do not express the E2 receptors. E2 acts through interaction with the different receptors (ERα, ERβ or GPER) that induce different responses, which can be genomic and/or non-genomic. The physiological responses may be different because E2 levels varies between sexes and in accordance with age in females. Notably, it has been shown that E2 levels in women after menopause may be similar to those in men. In order to understand the importance of E2 levels, Quesada et al. demonstrated in rat Langendorff perfused hearts, that infarct size was reduced in hearts from intact females compared to ovariectomized ones, where E2 production was inhibited [11]. Interestingly, the signaling pathway of the protective effect of estrogen may be due to suppressed calcium-calmodulin-dependent protein kinase II (CaMKII) activity. Indeed, CaMKII was up-regulated in the hearts of ovariectomized rats and was related to bigger infarct size [145]. Furthermore, some studies have shown a reciprocal link between E2 and calcium signaling. ERα and GPER are regulated by $Ca^{2+}$ at the receptor level and downstream signaling via a feedforward loop [146]. E2 also plays a role in the wound healing response following myocardial infarction, but the results of different studies are contradictory. Genomic responses via ERα and ERβ were shown to be mostly in favor of decreased infarct size, while the GPER response is mostly non-genomic and is associated with smaller infarct size. Surprisingly, however treatment with E2 to ovariectomized females also leads to conflicting results, showing higher mortality and infarct size in some studies, or improved heart remodeling and increased survival in others, as described in the review by Matarrese [147]. Another contradictory is seen with the type of receptors implicated, such as shown in a recent work demonstrating that activation of ERα reduces mortality with increasing infarction extension, while stimulation of ERβ accounts for reduced infarction area and increased cardiac hypertrophy and mortality [148, 149].

(b) Calcium handling

Calcium homeostasis is highly important in maintaining the principal activity of the heart, pumping blood. Its regulation is very complex and is the result of many signaling pathways with many factors involved. In this regard, E2 regulates numerous calcium-dependent activities in cardiac tissues. For example, it has been shown that ovariectomy decreased the abundance of beta1-Adrenergic receptor and L-type $Ca^{2+}$ channel protein without affecting the other factor involved in calcium regulation, such as sarcoplasmic reticular ATPase, phospholamban, and ryanodine receptor.

However, hormone-dependent modifications in protein contents were not associated with changes in contractile function [150]. Nevertheless, during stress conditions, such as during ischemia–reperfusion, the extent of injury was greater in males and associated with increased sarcoplasmic reticulum $Ca^{2+}$ [151]. Therefore, it was postulated that lower expression of β1-AR in intact females would provide a mechanism for the reduced calcium overload and reduced ischemia–reperfusion injury observed, but it still cannot explain the differences in extracellular calcium between males and females [151]. Another mechanism than can account for the differences observed in $Ca^{2+}$ handling may be mediated in part by nitric oxide synthase (NOS) [152]. Estrogen upregulated NOS in female hearts and caused S-nitrosylation of the L-type-calcium channel which resulted in decreased calcium entry via the L-type calcium channel and therefore, less calcium loading during ischemia [153].

(c) Genetic and proteomic regulation of cardiac metabolism

Different studies related to modification in a number of genes implicated in metabolism could account for the different capacity of protection from ischemia–reperfusion. Some studies showed that beta estrogen receptor play a role in this protection and that its absence is correlated with altered expression of certain metabolic genes that may be important in ischemic injury as the relative use of carbohydrate versus fatty acid in the production of ATP. Indeed, female hearts have increased ratio of carbohydrate to fatty acid metabolism, compared with males [149, 154]. In addition, cytoskeletal, contractile, and mitochondrial [155]. Vijay et al. [156] proteins have been shown to differentiate between the sexes [157].

(d) Role of mitochondria

A large body of evidence indicates that mitochondria are essential for the heart function by providing ATP for the contraction and are central in the oxygen and calcium handling. Mitochondrial dysfunction plays a major role in the development of ischemia–reperfusion injury [158]. Nevertheless, the role of mitochondria in the differences observed in tolerance to ischemia–reperfusion between male and females is still unclear [159]. There is no difference in mitochondrial function and metabolism between male and female rats until the end of weaning period (day 30); however, during the period of sexual maturation the cardiac tolerance decreases in males, but remains unchanged in females. This suggests that the adult female heart is more resistant to oxygen deprivation than male heart [132], which may correlate with differences in mitochondrial content or function. Indeed, it has been shown that cardiomyocytes from female rat hearts exhibit lower mitochondrial content but better function [160] coupled with improved lipid utilization, higher oxidative capacity, lower ROS production, and better calcium retention capacity [161] in comparison with males [162, 163]. Moreover, mitochondria play a role in cell death. In a Langendorff-perfused rat heart model, caspase activation was higher in males compared to females, suggesting a mechanism for increased cardiac cell death in males [164].

The inversion in the risk to CHD after menopause can be explained by reduction in heart mitochondrial mass and functionality, which decreases with age and correlates with increased $H_2O_2$ generation. In particular, aged females exhibited significant loss of mitochondrial function and increased relative $H_2O_2$ production compared with their male counterparts. This results demonstrated sex dimorphism in the age-associated defects on cardiac mitochondrial function [165].

## Conclusion

In order to provide better care to women, it is important to consider all risk factors that are specific to women. With all the recent data showing the differences between females and males in the management and consequences of cardiovascular disease, it is reasonable to think that in the future, strategies will be put in place that take these intersex differences into account in order to better support females. For this purpose, it is important to focus on sex-specific differences to deliver optimal preventive medical care. Promisingly, a new approach was described in 2020 by Agarwala et al. in order to stratified risk factors in women [57]. This is supported also by The Framingham Risk Score (FRS), which was first developed in 1987 based on a sample population from northeastern USA. The risk factors that are not specific to women which are usually considered include age, gender, smoking status, total cholesterol, high-density lipoprotein (HDL), blood pressure and diabetes mellitus [166]. However, recent analysis showed that using the FRS alone classifies up to 90% of asymptomatic women as low risk, underestimating the true incidence of CVD in women [167]. This is why; Ridker et al. develop a new tool of risk stratification that takes into account risks factors specific to women. These new stratification named Reynolds Risk Score (RRS) allows to reclassify 40–50% of women at intermediate risk into higher- or lower-risk categories [168].

We are hoping that research can continue to bring us new knowledge that will allow us to better consider the differences between male and females in order to improve the diagnosis, prevention and management of CHD.

## References

1. Roth GA et al (2020) Global burden of cardiovascular diseases and risk factors, 1990–2019: update from the GBD 2019 study. J Am Coll Cardiol 76:2982–3021
2. Mensah GA et al (2017) Decline in cardiovascular mortality: possible causes and implications. Circ Res 120:366–380
3. Mozaffarian D et al (2015) Heart disease and stroke statistics–2015 update: a report from the American Heart Association. Circ 131:e29-322
4. Pagidipati NJ, Peterson ED (2016) Acute coronary syndromes in women and men. Nat Rev Cardiol 13:471–480
5. Woodward M (2019) Cardiovascular disease and the female disadvantage. Int J Environ Res Public Health 16(7):1165
6. Kannel WB, Sorlie P, Mcnamara PM (1979) Prognosis after initial myocardial infarction: the framingham study. Am J Cardiol 44:53–59

7. Kim ESH, Carrigan TP, Menon V (2008) Enrollment of women in national heart, lung, and blood institute-funded cardiovascular randomized controlled trials fails to meet current federal mandates for inclusion. J Am Coll Cardiol 52:672–673
8. Leinwand LA (2003) Sex is a potent modifier of the cardiovascular system. J Clin Invest 112:302–307
9. Jneid H et al (2008) Sex differences in medical care and early death after acute myocardial infarction. Circ 118:2803–2810
10. Hochman JS et al (1999) Sex, clinical presentation, and outcome in patients with acute coronary syndromes. N Engl J Med 341:226–232
11. Quesada O, Henry TD (2020) STEMI in young women: don't miss spontaneous coronary artery dissection! Catheter Cardiovasc Interv 96:1231–1232
12. Khandelwal A et al (2021) Managing ischemic heart disease in women: role of a women's heart center. Curr Atheroscler Rep 23:56
13. Virani SS, Alonso A, Aparicio HJ, Benjamin EJ, Bittencourt MS, Callaway CW, Carson AP, Chamberlain AM, Cheng S, Delling FN, Elkind MSV, Evenson KR, Ferguson JF, Gupta DK, Khan SS, Kissela BM, Knutson KL, Lee CD, Lewis TT, Liu J, Loop MS, Lutsey PL, Ma J, Mackey J, Martin SS, Matchar DB, Mussolino ME, Navaneethan SD, Perak AM, Roth GA, Samad Z, Satou GM, Schroeder EB, Shah SH, Shay CM, Stokes A, VanWagner LB, Wang NY, Tsao CW (2021) American Heart Association Council on Epidemiology and Prevention Statistics Committee and Stroke Statistics Subcommittee. Heart disease and stroke statistics-2021 update: a report from the American Heart Association. Circ 143(8):e254–e743
14. Peters SAE et al (2021) Trends in recurrent coronary heart disease after myocardial infarction among US women and men between 2008 and 2017. Circ 143:650–660
15. Alabas OA et al (2017) Sex differences in treatments, relative survival, and excess mortality following acute myocardial infarction: national cohort study using the SWEDEHEART registry. J Am Heart Assoc 6:e007123
16. Xi Z et al (2022) Contemporary sex differences in mortality among patients with ST-segment elevation myocardial infarction: a systematic review and meta-analysis. BMJ Open 12:e053379
17. Iorga A et al (2017) The protective role of estrogen and estrogen receptors in cardiovascular disease and the controversial use of estrogen therapy. Biol Sex Differ 8:33
18. Woodward M (2019) Cardiovascular disease and the female disadvantage. Int J Environ Res Public Health 16:1165
19. Yusuf S et al (2004) Effect of potentially modifiable risk factors associated with myocardial infarction in 52 countries (the INTERHEART study): case-control study. Lancet Lond Engl 364:937–952
20. NCD Risk Factor Collaboration (NCD-RisC) (2016). Trends in adult body-mass index in 200 countries from 1975 to 2014: a pooled analysis of 1698 population-based measurement studies with 19·2 million participants. Lancet 387(10026):1377–1396
21. Kondo T, Nakano Y, Adachi S, Murohara T (2019) Effects of tobacco smoking on cardiovascular disease. Circ J Off J Jpn Circ Soc 83:1980–1985
22. Gillis EE, Sullivan JC (2016) Sex differences in hypertension: recent advances. Hypertens Dallas Tex 1979(68):1322–1327
23. Peters SAE, Singhateh Y, Mackay D, Huxley RR, Woodward M (2016) Total cholesterol as a risk factor for coronary heart disease and stroke in women compared with men: a systematic review and meta-analysis. Atherosclerosis 248:123–131
24. Peters SAE, Huxley RR, Woodward M (2014) Diabetes as risk factor for incident coronary heart disease in women compared with men: a systematic review and meta-analysis of 64 cohorts including 858,507 individuals and 28,203 coronary events. Diabetologia 57:1542–1551
25. Emdin CA, Wong CX, Hsiao AJ, Altman DG, Peters SA, Woodward M, Odutayo AA (2016) Atrial fibrillation as risk factor for cardiovascular disease and death in women compared with men: systematic review and meta-analysis of cohort studies. BMJ 532:h7013

26. Huxley RR, Woodward M (2011) Cigarette smoking as a risk factor for coronary heart disease in women compared with men: a systematic review and meta-analysis of prospective cohort studies. The Lancet 378:1297–1305
27. Backholer K et al (2017) Sex differences in the relationship between socioeconomic status and cardiovascular disease: a systematic review and meta-analysis. J Epidemiol Community Health 71:550–557
28. American Diabetes Association (2015) 8. Cardiovascular Disease and Risk Management. Diabetes Care 38:S49–S57
29. Gregg EW, Gu Q, Cheng YJ, Venkat Narayan KM, Cowie CC (2007) Mortality trends in men and women with diabetes, 1971 to 2000. Ann Intern Med 147:149
30. Huebschmann AG et al (2019) Sex differences in the burden of type 2 diabetes and cardiovascular risk across the life course. Diabetologia 62:1761–1772
31. Toedebusch R, Belenchia A, Pulakat L (2018) Diabetic Cardiomyopathy: impact of biological sex on disease development and molecular signatures. Front Physiol 9:453
32. Fourny N et al (2021) Sex differences of the diabetic heart. Front Physiol 12:661297
33. Rodgers JL et al (2019) Cardiovascular risks associated with gender and aging. J Cardiovasc Dev Dis 6:19
34. Murphy E (2011) Estrogen signaling and cardiovascular disease. Circ Res 109:687–696
35. Kautzky-Willer A, Harreiter J, Pacini G (2016) Sex and gender differences in risk, pathophysiology and complications of type 2 diabetes mellitus. Endocr Rev 37:278–316
36. Natarajan S, Liao Y, Cao G, Lipsitz SR, McGee DL (2003) Sex differences in risk for coronary heart disease mortality associated with diabetes and established coronary heart disease. Arch Intern Med 163:1735
37. Ranucci M et al (2019) Gender-based differences in platelet function and platelet reactivity to P2Y12 inhibitors. PLoS ONE 14:e0225771
38. Haffner SM, Miettinen H, Stern MP (1997) Relatively more atherogenic coronary heart disease risk factors in prediabetic women than in prediabetic men. Diabetologia 40:711–717
39. Steinberg HO et al (2000) Type II diabetes abrogates sex differences in endothelial function in premenopausal women. Circ 101:2040–2046
40. Donahue RP et al (2007) Sex differences in endothelial function markers before conversion to pre-diabetes: does the clock start ticking earlier among women? Diabetes Care 30:354–359
41. Thomas D (2017) Risque cardiovasculaire du tabagisme selon le genre. Presse Med 46:681–687
42. Njølstad I, Arnesen E, Lund-Larsen PG (1996) Smoking, serum lipids, blood pressure, and sex differences in myocardial infarction: a 12-year follow-up of the finnmark study. Circ 93:450–456
43. Pomp ER, Rosendaal FR, Doggen CJM (2008) Smoking increases the risk of venous thrombosis and acts synergistically with oral contraceptive use. Am J Hematol 83:97–102
44. Reynolds LM et al (2021) Tobacco use prevalence and transitions from 2013 to 2018 among adults with a history of cardiovascular disease. J Am Heart Assoc 10:e021118
45. Allagbé I et al (2021) Tobacco-related cardiovascular risk in women: New issues and therapeutic perspectives. Arch Cardiovasc Dis 114:694–706
46. Gender in cardiovascular diseases: impact on clinical manifestations, management, and outcomes. Eur Heart J 37:24–34 (2016)
47. Lloyd-Jones DM, Evans JC, Levy D (2005) Hypertension in adults across the age spectrum: current outcomes and control in the community. JAMA 294:466
48. McQueen MJ et al (2008) Lipids, lipoproteins, and apolipoproteins as risk markers of myocardial infarction in 52 countries (the INTERHEART study): a case-control study. The Lancet 372:224–233
49. Byrne P et al (2022) Evaluating the association between low-density lipoprotein cholesterol reduction and relative and absolute effects of statin treatment: a systematic review and meta-analysis. JAMA Intern Med 182:474
50. Petretta M, Costanzo P, Perrone-Filardi P, Chiariello M (2010) Impact of gender in primary prevention of coronary heart disease with statin therapy: a meta-analysis. Int J Cardiol 138:25–31

51. Brugts JJ et al (2009) The benefits of statins in people without established cardiovascular disease but with cardiovascular risk factors: meta-analysis of randomised controlled trials. BMJ 338:b2376–b2376
52. Walsh JME (2004) Drug treatment of hyperlipidemia in women. JAMA 291:2243
53. Faubion SS, Kapoor E, Moyer AM, Hodis HN, Miller VM (2019) Statin therapy: does sex matter? Menopause 26:1425–1435
54. Benjamin EJ, Muntner P, Alonso A, Bittencourt MS, Callaway CW, Carson AP, Chamberlain AM, Chang AR, Cheng S, Das SR, Delling FN, Djousse L, Elkind MSV, Ferguson JF, Fornage M, Jordan LC, Khan SS, Kissela BM, Knutson KL, Kwan TW, Lackland DT, Lewis TT, Lichtman JH, Longenecker CT, Loop MS, Lutsey PL, Martin SS, Matsushita K, Moran AE, Mussolino ME, O'Flaherty M, Pandey A, Perak AM, Rosamond WD, Roth GA, Sampson UKA, Satou GM, Schroeder EB, Shah SH, Spartano NL, Stokes A, Tirschwell DL, Tsao CW, Turakhia MP, VanWagner LB, Wilkins JT, Wong SS, Virani SS (2019) American Heart Association Council on Epidemiology and Prevention Statistics Committee and Stroke Statistics Subcommittee. Heart disease and stroke statistics-2019 update: a report from the American Heart Association. Circ 139(10):e56–e528
55. Yazdanyar A, Newman AB (2009) The burden of cardiovascular disease in the elderly: morbidity, mortality, and costs. Clin Geriatr Med 25:563–577
56. Arnett DK, Blumenthal RS, Albert MA, Buroker AB, Goldberger ZD, Hahn EJ, Himmelfarb CD, Khera A, Lloyd-Jones D, McEvoy JW, Michos ED, Miedema MD, Muñoz D, Smith SC Jr, Virani SS, Williams KA Sr, Yeboah J, Ziaeian B (2019) 2019 ACC/AHA guideline on the primary prevention of cardiovascular disease: a report of the American College of Cardiology/American Heart Association Task Force on Clinical Practice Guidelines. Circ 140(11):e596–e646
57. Agarwala A, Michos ED, Samad Z, Ballantyne CM, Virani SS (2020) The use of sex-specific factors in the assessment of women's cardiovascular risk. Circ 141:592–599
58. Young L, Cho L (2019) Unique cardiovascular risk factors in women. Heart 105:1656–1660
59. Okoth K, Chandan JS, Marshall T, Thangaratinam S, Thomas GN, Nirantharakumar K, Adderley NJ (2020) Association between the reproductive health of young women and cardiovascular disease in later life: umbrella review. BMJ 371:m3502
60. Muka T et al (2016) Association of age at onset of menopause and time since onset of menopause with cardiovascular outcomes, intermediate vascular traits, and all-cause mortality: a systematic review and meta-analysis. JAMA Cardiol 1:767
61. Zhu D et al (2020) Type of menopause, age of menopause and variations in the risk of incident cardiovascular disease: pooled analysis of individual data from 10 international studies. Hum Reprod 35:1933–1943
62. Zhu D et al (2019) Age at natural menopause and risk of incident cardiovascular disease: a pooled analysis of individual patient data. Lancet Public Health 4:e553–e564
63. Rossouw JE, Anderson GL, Prentice RL, LaCroix AZ, Kooperberg C, Stefanick ML, Jackson RD, Beresford SA, Howard BV, Johnson KC, Kotchen JM, Ockene J (2002) Writing group for the women's health initiative investigators. Risks and benefits of estrogen plus progestin in healthy postmenopausal women: principal results from the women's health Initiative randomized controlled trial. JAMA 288(3):321–333
64. Prabakaran S, Schwartz A, Lundberg G (2021) Cardiovascular risk in menopausal women and our evolving understanding of menopausal hormone therapy: risks, benefits, and current guidelines for use. Ther Adv Endocrinol Metab 12:204201882110139
65. Charalampopoulos D, McLoughlin A, Elks CE, Ong KK (2014) Age at menarche and risks of all-cause and cardiovascular death: a systematic review and meta-analysis. Am J Epidemiol 180:29–40
66. Canoy D et al (2015) Age at menarche and risks of coronary heart and other vascular diseases in a large UK cohort. Circ 131:237–244
67. Benschop L, Duvekot JJ Roeters van Lennep JE (2019) Future risk of cardiovascular disease risk factors and events in women after a hypertensive disorder of pregnancy. Heart 105:1273–1278

68. Brouwers L et al (2018) Recurrence of pre-eclampsia and the risk of future hypertension and cardiovascular disease: a systematic review and meta-analysis. BJOG Int J Obstet Gynaecol 125:1642–1654
69. Nguyen TX, Nguyen VT, Nguyen-Phan HN, Hoang Bui B (2022) Serum levels of NT-Pro BNP in patients with preeclampsia. Integr Blood Press Control 15:43–51
70. Webster K, Fishburn S, Maresh M, Findlay SC, Chappell LC (2019) Guideline committee. Diagnosis and management of hypertension in pregnancy: summary of updated NICE guidance. BMJ 366:l5119
71. Murphy M, Smith G (2016) Pre-eclampsia and cardiovascular disease risk assessment in women. Am J Perinatol 33:723–731
72. Garcia M, Mulvagh SL, Bairey Merz CN, Buring JE, Manson JE (2016) Cardiovascular disease in women: clinical perspectives. Circ Res 118:1273–1293
73. Tobias DK et al (2017) Association of history of gestational diabetes with long-term cardiovascular disease risk in a large prospective cohort of US women. JAMA Intern Med 177:1735
74. McKenzie-Sampson S, Paradis G, Healy-Profitós J, St-Pierre F, Auger N (2018) Gestational diabetes and risk of cardiovascular disease up to 25 years after pregnancy: a retrospective cohort study. Acta Diabetol 55:315–322
75. Kramer CK, Campbell S, Retnakaran R (2019) Gestational diabetes and the risk of cardiovascular disease in women: a systematic review and meta-analysis. Diabetologia 62:905–914
76. Caliskan M et al (2015) Previous gestational diabetes history is associated with impaired coronary flow reserve. Ann Med 47:615–623
77. Retnakaran R, Shah BR (2019) Glucose screening in pregnancy and future risk of cardiovascular disease in women: a retrospective, population-based cohort study. Lancet Diabetes Endocrinol 7:378–384
78. Azziz R et al (2016) Polycystic ovary syndrome. Nat Rev Dis Primer 2:16057
79. Osibogun O, Ogunmoroti O, Michos ED (2020) Polycystic ovary syndrome and cardiometabolic risk: opportunities for cardiovascular disease prevention. Trends Cardiovasc Med 30:399–404
80. Kakoly N, Moran L, Teede H, Joham A (2019) Cardiometabolic risks in PCOS: a review of the current state of knowledge. Expert Rev Endocrinol Metab 14:23–33
81. Morimont L, Haguet H, Dogné J-M, Gaspard U, Douxfils J (2021) Combined oral contraceptives and venous thromboembolism: review and perspective to mitigate the risk. Front Endocrinol 12:769187
82. Roach RE, Helmerhorst FM, Lijfering WM, Stijnen T, Algra A, Dekkers OM (2015) Combined oral contraceptives: the risk of myocardial infarction and ischemic stroke. Cochrane Database Syst Rev 2015(8):CD011054
83. Cardeillac M, Lefebvre F, Baicry F, Le Borgne P, Gil-Jardiné C, Cipolat L, Peschanski N, Abensur Vuillaume L (2022) Symptoms of infarction in women: is there a real difference compared to men? A systematic review of the literature with meta- analysis. J Clin Med 11(5):1319
84. Lichtman JH et al (2018) Sex differences in the presentation and perception of symptoms among young patients with myocardial infarction: evidence from the VIRGO study (variation in recovery: role of gender on outcomes of young AMI patients). Circ 137:781–790
85. Nyström A, Strömberg S, Jansson K, Faresjö ÅO, Faresjö T (2022) Cardiovascular risks before myocardial infarction differences between men and women. BMC Cardiovasc Disord 22:110
86. Pagidipati NJ et al (2019) Sex differences in management and outcomes of patients with stable symptoms suggestive of coronary artery disease: insights from the PROMISE trial. Am Heart J 208:28–36
87. Kaul P et al (2011) Temporal trends in patient and treatment delay among men and women presenting with ST-elevation myocardial infarction. Am Heart J 161:91–97

88. Pacheco C et al (2021) Impact of STEMI diagnosis and catheterization laboratory activation systems on sex- and age-based differences in treatment delay. CJC Open 3:723–732
89. Stehli J et al (2021) Sex differences in prehospital delays in patients with st-segment–elevation myocardial infarction undergoing percutaneous coronary intervention. J Am Heart Assoc 10:e019938
90. Kim HO et al (2020) Relative risk of plaque erosion among different age and sex groups in patients with acute coronary syndrome. J Thromb Thrombolysis 49:352–359
91. Vancheri F, Longo G, Vancheri S, Henein M (2020) Coronary Microvascular Dysfunction. J Clin Med 9:2880
92. Ostadal B, Ostadal P (2014) Sex-based differences in cardiac ischaemic injury and protection: therapeutic implications. Br J Pharmacol 171:541–554
93. Lucà F et al (2022) Update on management of cardiovascular diseases in women. J Clin Med 11:1176
94. Hao Y et al (2019) Sex differences in in-hospital management and outcomes of patients with acute coronary syndrome: findings from the CCC project. Circ 139:1776–1785
95. Jackson AM et al (2020) Healthcare disparities for women hospitalized with myocardial infarction and angina. Eur Heart J - Qual Care Clin Outcomes 6:156–165
96. Gupta A et al (2018) Sex differences in timeliness of reperfusion in young patients with ST-segment–elevation myocardial infarction by initial electrocardiographic characteristics. J Am Heart Assoc 7:e007021
97. Mehilli J, Presbitero P (2020) Coronary artery disease and acute coronary syndrome in women. Heart 106:487–492
98. Vogel B et al (2019) Residual angina in female patients after coronary revascularization. Int J Cardiol 286:208–213
99. Lebrun S, Bond RM (2018) Spontaneous coronary artery dissection (SCAD): the underdiagnosed cardiac condition that plagues women. Trends Cardiovasc Med 28:340–345
100. Adlam D et al (2019) Association of the PHACTR1/EDN1 genetic locus with spontaneous coronary artery dissection. J Am Coll Cardiol 73:58–66
101. Vogel B et al (2021) The lancet women and cardiovascular disease commission: reducing the global burden by 2030. The Lancet 397:2385–2438
102. Connelly PJ et al (2021) The importance of gender to understand sex differences in cardiovascular disease. Can J Cardiol 37:699–710
103. O'Neil A, Russell JD, Thompson K, Martinson ML, Peters SAE (2020) The impact of socioeconomic position (SEP) on women's health over the lifetime. Maturitas 140:1–7
104. Schultz WM et al (2018) Socioeconomic status and cardiovascular outcomes: challenges and interventions. Circ 137:2166–2178
105. Mannoh I, Hussien M, Commodore-Mensah Y, Michos ED (2021) Impact of social determinants of health on cardiovascular disease prevention. Curr Opin Cardiol 36:572–579
106. Miller GE, Chen E, Shimbo D (2019) Mechanistic understanding of socioeconomic disparities in cardiovascular disease. J Am Coll Cardiol 73:3256–3258
107. Appleton AA, Holdsworth E, Ryan M, Tracy M (2017) Measuring childhood adversity in life course cardiovascular research: a systematic review. Psychosom Med 79:434–440
108. Chooi YC, Ding C, Magkos F (2019) The epidemiology of obesity. Metabolism 92:6–10
109. Liu B, Du Y, Wu Y, Snetselaar LG, Wallace RB, Bao W (2021) Trends in obesity and adiposity measures by race or ethnicity among adults in the United States 2011–18: population based study. BMJ 372:n365
110. GBD 2015 Obesity Collaborators; Afshin A, Forouzanfar MH, Reitsma MB, Sur P, Estep K, Lee A, Marczak L, Mokdad AH, Moradi-Lakeh M, Naghavi M, Salama JS, Vos T, Abate KH, Abbafati C, Ahmed MB, Al-Aly Z, Alkerwi A, Al-Raddadi R, Amare AT, Amberbir A, Amegah AK, Amini E, Amrock SM, Anjana RM, Ärnlöv J, Asayesh H, Banerjee A, Barac A, Baye E, Bennett DA, Beyene AS, Biadgilign S, Biryukov S, Bjertness E, Boneya DJ, Campos-Nonato I, Carrero JJ, Cecilio P, Cercy K, Ciobanu LG, Cornaby L, Damtew SA, Dandona L, Dandona R, Dharmaratne SD, Duncan BB, Eshrati B, Esteghamati A, Feigin VL, Fernandes JC, Fürst T, Gebrehiwot TT, Gold A, Gona PN, Goto A, Habtewold TD, Hadush

KT, Hafezi-Nejad N, Hay SI, Horino M, Islami F, Kamal R, Kasaeian A, Katikireddi SV, Kengne AP, Kesavachandran CN, Khader YS, Khang YH, Khubchandani J, Kim D, Kim YJ, Kinfu Y, Kosen S, Ku T, Defo BK, Kumar GA, Larson HJ, Leinsalu M, Liang X, Lim SS, Liu P, Lopez AD, Lozano R, Majeed A, Malekzadeh R, Malta DC, Mazidi M, McAlinden C, McGarvey ST, Mengistu DT, Mensah GA, Mensink GBM, Mezgebe HB, Mirrakhimov EM, Mueller UO, Noubiap JJ, Obermeyer CM, Ogbo FA, Owolabi MO, Patton GC, Pourmalek F, Qorbani M, Rafay A, Rai RK, Ranabhat CL, Reinig N, Safiri S, Salomon JA, Sanabria JR, Santos IS, Sartorius B, Sawhney M, Schmidhuber J, Schutte AE, Schmidt MI, Sepanlou SG, Shamsizadeh M, Sheikhbahaei S, Shin MJ, Shiri R, Shiue I, Roba HS, Silva DAS, Silverberg JI, Singh JA, Stranges S, Swaminathan S, Tabarés-Seisdedos R, Tadese F, Tedla BA, Tegegne BS, Terkawi AS, Thakur JS, Tonelli M, Topor-Madry R, Tyrovolas S, Ukwaja KN, Uthman OA, Vaezghasemi M, Vasankari T, Vlassov VV, Vollset SE, Weiderpass E, Werdecker A, Wesana J, Westerman R, Yano Y, Yonemoto N, Yonga G, Zaidi Z, Zenebe ZM, Zipkin B, Murray CJL (2017) Health effects of overweight and obesity in 195 countries over 25 years. N Engl J Med 377(1):13–27

111. Jones-Smith JC, Gordon-Larsen P, Siddiqi A, Popkin BM (2011) Cross-national comparisons of time trends in overweight inequality by socioeconomic status among women using repeated cross-sectional surveys from 37 developing countries, 1989–2007. Am J Epidemiol 173:667–675
112. Roskam A-JR et al (2010) Comparative appraisal of educational inequalities in overweight and obesity among adults in 19 European countries. Int J Epidemiol 39:392–404
113. Dudina A et al (2011) Relationships between body mass index, cardiovascular mortality, and risk factors: a report from the SCORE investigators. Eur J Cardiovasc Prev Rehabil 18:731–742
114. Wilson PWF, D'Agostino RB, Sullivan L, Parise H, Kannel WB (2002) Overweight and obesity as determinants of cardiovascular risk: the Framingham experience. Arch Intern Med 162:1867
115. Manrique-Acevedo C, Chinnakotla B, Padilla J, Martinez-Lemus LA, Gozal D (2020) Obesity and cardiovascular disease in women. Int J Obes 44:1210–1226
116. Stapel B, Jelinic M, Drummond GR, Hartung D, Kahl KG (2022) Adipose tissue compartments, inflammation, and cardiovascular risk in the context of depression. Front Psychiatry 13:831358
117. Maas AHEM, Appelman YEA (2010) Gender differences in coronary heart disease. Neth Heart J 18:598–603
118. Vaccarezza M et al (2020) Sex/gender-specific imbalance in CVD: could physical activity help to improve clinical outcome targeting CVD molecular mechanisms in women? Int J Mol Sci 21:1477
119. Levine GN, Cohen BE, Commodore-Mensah Y, Fleury J, Huffman JC, Khalid U, Labarthe DR, Lavretsky H, Michos ED, Spatz ES, Kubzansky LD (2021) Psychological health, well-Being, and the mind-heart-body connection: a scientific statement from the American Heart Association. Circ 143(10):e763–e783
120. Steptoe A, Kivimäki M (2012) Stress and cardiovascular disease. Nat Rev Cardiol 9:360–370
121. Pan Y, Cai W, Cheng Q, Dong W, An T, Yan J (2015) Association between anxiety and hypertension: a systematic review and meta-analysis of epidemiological studies. Neuropsychiatr Dis Treat 11:1121–1130
122. Tang F, Wang G, Lian Y (2017) Association between anxiety and metabolic syndrome: A systematic review and meta-analysis of epidemiological studies. Psychoneuroendocrinology 77:112–121
123. Emdin CA et al (2016) Meta-Analysis of Anxiety as a Risk Factor for Cardiovascular Disease. Am J Cardiol 118:511–519
124. Hasin DS et al (2018) Epidemiology of Adult *DSM-5* Major Depressive Disorder and Its Specifiers in the United States. JAMA Psychiat 75:336
125. Gan Y et al (2014) Depression and the risk of coronary heart disease: a meta-analysis of prospective cohort studies. BMC Psychiatry 14:371

126. Bucciarelli V et al (2020) Depression and cardiovascular disease: The deep blue sea of women's heart. Trends Cardiovasc Med 30:170–176
127. Gafarov VV, Panov DO, Gromova EA, Gagulin IV, Gafarova AV (2013) The influence of depression on risk development of acute cardiovascular diseases in the female population aged 25–64 in Russia. Int J Circumpolar Health 72:21223
128. Kubzansky LD et al (2018) Reprint of: positive psychological well-being and cardiovascular disease. J Am Coll Cardiol 72:3012–3026
129. Tindle HA et al (2009) Optimism, cynical hostility, and incident coronary heart disease and mortality in the women's health initiative. Circ 120:656–662
130. Matthews KA, Räikkönen K, Sutton-Tyrrell K, Kuller LH (2004) Optimistic attitudes protect against progression of carotid atherosclerosis in healthy middle-aged women. Psychosom Med 66:640–644
131. Ostádal B, Procházka J, Pelouch V, Urbanová D, Widimský J (1984) Comparison of cardiopulmonary responses of male and female rats to intermittent high altitude hypoxia. Physiol Bohemoslov 33:129–138
132. Ostadal B, Netuka I, Maly J, Besik J, Ostadalova I (2009) Gender differences in cardiac ischemic injury and protection—experimental aspects. Exp Biol Med 234:1011–1019
133. Wang M, Crisostomo P, Wairiuko GM, Meldrum DR (2006) Estrogen receptor-alpha mediates acute myocardial protection in females. Am J Physiol Heart Circ Physiol 290:H2204-2209
134. Przyklenk K, Ovize M, Bauer B, Kloner RA (1995) Gender does not influence acute myocardial infarction in adult dogs. Am Heart J 129:1108–1113
135. Li N, Kloner N (1995) Is there a gender difference in infarct size and arrhythmias following experimental coronary occlusion and reperfusion? J Thromb Thrombolysis 2:221–225
136. Cross HR, Lu L, Steenbergen C, Philipson KD, Murphy E (1998) Overexpression of the cardiac Na+/Ca2+ exchanger increases susceptibility to ischemia/reperfusion injury in male, but not female, transgenic mice. Circ Res 83:1215–1223
137. Stampfer MJ et al (1991) Postmenopausal estrogen therapy and cardiovascular disease: ten-year follow-up from the nurses' health study. N Engl J Med 325:756–762
138. Wilson PWF, Garrison RJ, Castelli WP (1985) Postmenopausal estrogen use, cigarette smoking, and cardiovascular morbidity in women over 50: the Framingham study. N Engl J Med 313:1038–1043
139. Grohé C et al (1997) Cardiac myocytes and fibroblasts contain functional estrogen receptors. FEBS Lett 416:107–112
140. Morselli E et al (2017) The effects of oestrogens and their receptors on cardiometabolic health. Nat Rev Endocrinol 13:352–364
141. Bell JR et al (2011) Aromatase deficiency confers paradoxical postischemic cardioprotection. Endocrinology 152:4937–4947
142. Pugach EK, Blenck CL, Dragavon JM, Langer SJ, Leinwand LA (2016) Estrogen receptor profiling and activity in cardiac myocytes. Mol Cell Endocrinol 431:62–70
143. Mahmoodzadeh S, Dworatzek E (2019) The Role of 17β-estradiol and estrogen receptors in regulation of $Ca^{2+}$ channels and mitochondrial function in cardiomyocytes. Front Endocrinol 10:310
144. Kam KWL, Qi JS, Chen M, Wong TM (2004) Estrogen reduces cardiac injury and expression of $\beta_1$-adrenoceptor upon ischemic insult in the rat heart. J Pharmacol Exp Ther 309:8–15
145. Ma Y, Cheng WT, Wu S, Wong TM (2009) Oestrogen confers cardioprotection by suppressing $Ca^{2+}$/calmodulin-dependent protein kinase II. Br J Pharmacol 157:705–715
146. Tran Q-K (2020) Reciprocality between estrogen biology and calcium signaling in the cardiovascular system. Front Endocrinol 11:568203
147. Matarrese P et al (2021) Role of β-adrenergic receptors and estrogen in cardiac repair after myocardial infarction: an overview. Int J Mol Sci 22:8957
148. Babiker FA et al (2007) Oestrogen modulates cardiac ischaemic remodelling through oestrogen receptor-specific mechanisms. Acta Physiol Oxf Engl 189:23–31
149. Gabel SA et al (2005) Estrogen receptor beta mediates gender differences in ischemia/reperfusion injury. J Mol Cell Cardiol 38:289–297

150. Chu SH, Goldspink P, Kowalski J, Beck J, Schwertz DW (2006) Effect of estrogen on calcium-handling proteins, beta-adrenergic receptors, and function in rat heart. Life Sci 79:1257–1267
151. Chen J et al (2003) Gender differences in sarcoplasmic reticulum calcium loading after isoproterenol. Am J Physiol-Heart Circ Physiol 285:H2657–H2662
152. Cross HR, Murphy E, Steenbergen C (2002) Ca(2+) loading and adrenergic stimulation reveal male/female differences in susceptibility to ischemia-reperfusion injury. Am J Physiol Heart Circ Physiol 283:H481–489
153. Sun J et al (2006) Hypercontractile female hearts exhibit increased S-nitrosylation of the L-type $Ca^{2+}$ channel alpha1 subunit and reduced ischemia/reperfusion injury. Circ Res 98:403–411
154. Tepavcevic S et al (2011) Interaction between insulin and estradiol in regulation of cardiac glucose and free fatty acid transporters. Horm Metab Res 43:524–530
155. Chen J-Q, Cammarata PR, Baines CP, Yager JD (2009) Regulation of mitochondrial respiratory chain biogenesis by estrogens/estrogen receptors and physiological, pathological and pharmacological implications. Biochim Biophys Acta BBA–Mol Cell Res 1793:1540–1570
156. Vijay V et al (2015) Sexual dimorphism in the expression of mitochondria-related genes in rat heart at different ages. PLoS ONE 10:e0117047
157. Barta BA et al (2021) Sex-related differences of early cardiac functional and proteomic alterations in a rat model of myocardial ischemia. J Transl Med 19:507
158. Bonora M et al (2019) Targeting mitochondria for cardiovascular disorders: therapeutic potential and obstacles. Nat Rev Cardiol 16:33–55
159. Ostadal B et al (2019) Developmental and sex differences in cardiac tolerance to ischemia-reperfusion injury: the role of mitochondria (1). Can J Physiol Pharmacol 97:808–814
160. Colom B, Oliver J, Roca P, Garciapalmer F (2007) Caloric restriction and gender modulate cardiac muscle mitochondrial $H_2O_2$ production and oxidative damage. Cardiovasc Res 74:456–465
161. Pavón N et al (2012) Sexual hormones: effects on cardiac and mitochondrial activity after ischemia–reperfusion in adult rats. Gender difference. J Steroid Biochem Mol Biol 132:135–146
162. Murphy E, Steenbergen C (2007) Gender-based differences in mechanisms of protection in myocardial ischemia-reperfusion injury. Cardiovasc Res 75:478–486
163. Ventura-Clapier R et al (2017) Mitochondria: a central target for sex differences in pathologies. Clin Sci 131:803–822
164. Morkuniene R, Arandarcikaite O, Ivanoviene L, Borutaite V (2010) Estradiol-induced protection against ischemia-induced heart mitochondrial damage and caspase activation is mediated by protein kinase G. Biochim Biophys Acta BBA - Bioenerg 1797:1012–1017
165. Colom B, Oliver J, Garcia-Palmer FJ (2015) Sexual dimorphism in the alterations of cardiac muscle mitochondrial bioenergetics associated to the ageing process. J Gerontol A Biol Sci Med Sci 70:1360–1369
166. D'Agostino RB et al (2008) General cardiovascular risk profile for use in primary care: the Framingham heart study. Circ 117:743–753
167. Michos ED et al (2006) Framingham risk equation underestimates subclinical atherosclerosis risk in asymptomatic women. Atherosclerosis 184:201–206
168. Ridker PM, Buring JE, Rifai N, Cook NR (2007) Development and validation of improved algorithms for the assessment of global cardiovascular risk in women: the Reynolds risk score. JAMA 297:611

# Chapter 22
# Engaging Women in Decisions About Their Heart Health

**Krystina B. Lewis, Faria Ahmed, Sandra Lauck, Sandra Carroll, and Dawn Stacey**

**Abstract** Women living with cardiovascular disease face many health decisions throughout their journey. Most women want more information and greater involvement in decision-making about their health in partnership with their clinicians. Yet, facing these decisions can lead to a sense of personal uncertainty about the best course of action. The individualized and intentional communication offered by a shared decision-making approach is particularly important for women with cardiovascular disease, given the limited availability of scientific data about women regarding the risks and benefits for screening and treatment options, making the elicitation and incorporation of personal preferences and values in their decision-making critical. Further, sex and gender considerations are important in shared decision-making particularly as they are associated with various decision-making styles, communication styles, and values and preferences which can influence an individual's preferred option. In this chapter, we begin by providing a definition of shared decision-making and discuss the evidence related to women's involvement in health decisions, particularly for cardiovascular conditions. We present the evidence supporting interventions to facilitate shared decision-making in clinical practice such as patient decision

---

K. B. Lewis (✉) · F. Ahmed · D. Stacey
School of Nursing, University of Ottawa, Ottawa, Canada
e-mail: Krystina.Lewis@uottawa.ca

K. B. Lewis
University of Ottawa Heart Institute, Ottawa, Canada

K. B. Lewis · D. Stacey
Ottawa Hospital Research Institute, Ottawa, Canada

S. Lauck
St. Paul's Hospital and Heart & Stroke Professorship, University of British Columbia, Vancouver, Canada

St. Paul's Hospital, Vancouver, Canada

University of British Columbia, Vancouver, Canada

S. Carroll
School of Nursing, McMaster University, Hamilton, Canada

© The Author(s), under exclusive license to Springer Nature Switzerland AG 2023 
L. Kirshenbaum and I. Rabinovich-Nikitin (eds.), *Biology of Women's Heart Health*, Advances in Biochemistry in Health and Disease 26,
https://doi.org/10.1007/978-3-031-39928-2_22

aids, decision coaching and question prompt lists. Finally, we present considerations for shared decision-making implementation in clinical practice. Throughout, we highlight opportunities for meaningful patient engagement amidst the challenges of shared decision-making to achieve true patient-centered care for women living with cardiovascular disease.

**Keywords** Shared decision making · Follow-up · Healthcare systems · Policies · Patients

## Introduction

Women living with cardiovascular disease (CVD) face standard and unique health decisions throughout their journey. Making these decisions can lead to a sense of personal uncertainty about the best course of action. The best option is not always clear, either because the evidence is insufficient, information is not readily available or unclear, or there is a need to trade off benefits and harms across options, including the option of maintaining status quo. When women are engaged in decisions about their health, they have better outcomes and experiences [1, 2].

The majority of women want more information and greater involvement in decision-making about their health in partnership with their clinicians [3]. Women with CVD are encouraged to take charge of their health by enhancing their knowledge about their condition, and advocate for their own health [4]. Healthcare providers have been urged to enhance their understanding of women's experience with CVD and create a culture that supports open communication and respect for patients' perspectives [4]. These invitations aligns with key shared decision-making principles. Shared decision-making (SDM) is a collaborative process between patients and healthcare providers whereby they work together to achieve a health decision that is based on the best evidence and the patient's values and preferences [5, 6]. Active patient participation in health decisions is essential for patient-centered care; to ensure that care is attentive to the needs, values and preferences of patients [7].

Increasingly, a SDM approach, which integrates patients' values and preferences, is recommended in clinical practice guidelines and expert consensus documents for patients with cardiovascular conditions. Canadian, American and European guidelines have endorsed SDM to guide decisions across cardiac care; in the management of patients with heart failure [8] atrial fibrillation [9] ventricular arrhythmias [10] implantable cardioverter-defibrillators [11] and valvular heart disease [12]. The Centres for Medicare and Medicaid in the United States have mandated use of patient decision aids to facilitate a SDM approach before percutaneous left atrial appendage closure and implantable cardioverter-defibrillator for primary prevention as a condition of reimbursement [13]. These endorsements are driven by the supporting evidence that SDM aligns with the quadruple aim [14, 15] (better outcomes, better experiences for both patients and providers, and optimized costs) and that it fosters value-based healthcare [16–18].

The overall aim of this chapter is to discuss the use of SDM and effective interventions to engage women with CVD in decisions about their health. We begin by providing a definition of SDM and discuss the evidence related to women's involvement in health decisions, particularly for cardiovascular conditions. We present the evidence supporting interventions to facilitate shared decision-making in clinical practice such as patient decision aids, decision coaching and question prompt lists. Finally, we present considerations for SDM implementation in clinical practice. Throughout, we highlight opportunities for meaningful patient engagement amidst the challenges of SDM to achieve true patient-centered care for women living with CVD.

## Shared Decision Making: Defined

In 1982, the President's Commission for the Study of Ethical Problems in Medicine and Biomedical and Behavioral Research coined the term "shared decision-making" to revitalize the approach to obtaining informed consent [19]. The report stated: "*The ethical foundation of informed consent can be traced to the promotion of two values: personal well-being and self-determination. To ensure that these values are respected and enhanced, the Commission finds that patients… must have all relevant information regarding their condition and alternative treatments, including possible benefits, risks, costs, other consequences, and significant uncertainties.... Ethically valid consent is a process of shared decision making based upon mutual respect and participation, not a ritual to be equated with reciting the contents of a form.*" (p. 2) Forty years later, informed consent remains largely viewed as a legal process that acquires patients' permission to pursue one specific treatment, often without consideration of all possible options and their attributes. Experts agree that SDM is a more comprehensive approach because it invites patients to engage in a conversation about all the feasible treatment options and outcomes based on the best available evidence, to share and incorporate their preferences, enabling a jointly made decision [20–22]. Patients also agree that current informed consent approaches achieve suboptimal comprehension and do not guide decision-making [23].

Shared decision making (SDM) recognizes the expertise of both patients and their healthcare providers to make health-related decisions together. It differs from patient education which is primarily aimed at unilaterally "giving" information to people to help them understand their conditions and management in general terms with little or any particular attention to decision-making needs [24]. Healthcare providers share information about the condition, the treatment options, the risks and benefits of each, and the potential outcomes. Patients contribute their expertise related to their lived experiences, their priorities and goals, attitudes and tolerance towards risk, and the value they place on those various outcomes. A decision is then made jointly, incorporating what is known about the patient's priorities all while engaging them to their desired level of involvement [25]. When choices are informed by the best available evidence and reflect what matters most to the person, a quality decision can be made. A SDM approach is often initiated with an invitation from healthcare providers, asking patients to share their goals, values, and preferences. Derived from a comprehensive review of 161 studies [6], nine SDM-related behaviors were identified (Box 22.1). If goals, values, and preferences are not elicited from the patients, clinicians run the risk of incorrectly - and inappropriately - assuming what patients prefer and value [26, 27].

**Box 22.1: Essential Shared Decision-Making Behaviours for Healthcare Providers**

- Define and explain the healthcare problem
- Present the options
- Discuss the pros and cons (benefits, risks, costs)
- Elicit values and preferences for possible outcomes
- Discuss patient ability and self-efficacy to follow agreed upon plan
- Present what is known and make recommendations, informed by patient's goals, values and preferences
- Check and clarify the patient's understanding
- Make or explicitly defer a decision, and
- Arrange follow-up

*Based on Makoul and Clayman [6].

## Shared Decision-Making and Women's Heart Health

The individualized and intentional communication offered by a SDM approach is particularly important for women with CVD, given the relative paucity of scientific data about women's heart health [28]. When the level of research evidence is low, as is the case for women with CVD, SDM is arguably even more critical to elicit and incorporate personal preferences and values for the options and potential outcomes [29]. Further, sex and gender considerations are important in SDM, particularly as they are associated with various decision-making styles, communication styles, and values and preferences which can influence an individual's preferred option [30–34]. The lived experience with certain treatments can also be different for women. For example, with implantable cardioverter-defibrillators, being female is a risk factor for adjustment difficulties, reasons that have been attributed to body image concerns, social support and roles, caregiver role(s) and activity limitations [35]. Females with ICDs experience higher levels of anxiety and depressive symptoms, worse functional outcomes, and more ICD-related concerns. In one study, females were considerably more indecisive when asked to engage in deactivation discussions and express preferences about ICD therapy [36].

Despite recent gains in the awareness of sex and gender differences in the diagnosis, treatment, management, and rehabilitation for women living with CVD, women's specific decision-making needs and experiences in these contexts have received little attention [4, 37, 38]. In fact, much of the SDM literature for and about women is focused on sexual and reproductive health decisions; also known as the *bikini boundaries* [39]. In a review of 590 Choosing Wisely recommendations encouraging conversations between women and their healthcare providers about (in)effective, high (and low) value practices pertinent to women's health, none were related to heart health [40]. A recent content analysis of Canadian health policies about conditions with known gendered inequities identified 30 policies, 10 of which were focused on cardiac rehabilitation. Of these, only one addressed women's cardiac health [41].

First steps in tailoring SDM approaches for women with CVD's specific decision-making needs include acknowledging the sex differences in pathophysiology, clinical manifestations, efficacy of diagnostic tests, availability and efficacy of treatments for females with CVD, including the known gendered disparities in care, health behaviours, lived experience(s), and clinical outcomes [42, 43]. For example, several risk factors for CVD have a stronger contribution to its development in women than in men (e.g., hypertension, diabetes, stress, depression, and family history of CVD) which may, for example, influence an individual's preference for early vs. delayed treatment [38]. Other conditions and risk factors are unique to women, such as conditions occurring during pregnancy and related to hormonal status. Social gender roles should also be considered including caregiving, transportation, and/or work-related responsibilities as these can often shape values and preferences for or against certain options. Hence, we need to adopt standardized use of sex and gender terms to clearly articulate what drives the issue of interest. In SDM research specifically, misappropriate use of the terms, sex and gender, limits our understanding of the impacts of these variables on decision-making outcomes [44].

**Women's Decisional Needs**. When women are faced with difficult health decisions, decisional needs often arise. This can be because the decision has more than one reasonable option, uncertain outcomes, or known outcomes that people value differently [45, 46]. A decisional need is defined as a deficit that can adversely affect the quality of a decision [47]. Examples of modifiable decisional needs include inadequate knowledge about the condition, problem, options, or outcomes, unrealistic expectations, unclear personal values, inadequate support from others, inadequate resources, and unreceptive decision stage due to denial or emotions [47]. Complex decision characteristics, and personal and clinical needs such as a patient or healthcare provider's gender can influence decisional needs and manifestations of decisional conflict. The Ottawa Decision Support Framework delineates these needs, and asserts that when they go unresolved the quality of a decision and its implementation can be adversely affected [48].

In a systematic review of 45 decisional needs studies [47], only two included studies focused on the needs of patients facing cardiovascular decisions: (1) implantable cardioverter-defibrillator implantation where 8 participants (40%) were female[49], and (2) left ventricular assist device placement (sex/gender was not reported) [50]. In these two studies, numerous decisional needs were reported. Participants had unrealistic expectations as they did not believe that outcome probabilities applied to them, and desired knowing about the experiences of others. They experienced difficult decision roles as they valued outcomes differently than their family members; a need often reported in the context of decisions regarding life prolongation. Participants also experienced negative emotions given their progressive condition of heart failure, feeling distress, fear, anxiety, worry, uncertainty, frustration and out of control [49, 50]. Unrealistic expectations were also noted in an interview-based study with older women about CVD screening [51]. The women believed screening did not apply to them as they felt healthy and were symptom-free. Others feared finding out about a potentially incurable condition, which would impact their lifestyle and quality of life with little perceived control over the disease outcome [51].

When decisional needs are assessed and identified, healthcare providers can tailor their decision support to their patients' needs. When not addressed, decisional conflict is likely to persist and can result in consequences such as vacillation between options, delays in decision-making, blaming others for bad outcomes, and regret [52, 53]. Decision regret, defined as distress or remorse after making a decision [54], has been identified in patients with implantable cardioverter-defibrillators [55] and left ventricular assist devices [56]. In one study, female LVAD recipients had significantly higher decision regret than males [56]. Evidence-based decision support interventions (e.g., patient decision aids, decision coaching and/or question prompts) aim to address decisional needs and improve decision-making outcomes [48, 57].

## Interventions to Facilitate Shared Decision-Making

Evidence-based interventions to facilitate SDM in clinical practice include patient decision aids, decision coaching, and question prompt lists. Counselling can also be used by physicians and other practitioners to help patients understand options, benefits, harms, probabilities, and scientific uncertainties; clarify the personal value of the ratio of benefit to harm, and participate in decision making according to their preferred level [58].

**Patient Decision Aids**. Patient decision aids are evidence-based interventions that aim to support patients' participation in decision making [59]. According to international standards, a patient decision aid must: (1) make explicit the decision to be made, (2) provide information on the condition, (3) list the options, (4) provide the benefits, (5) provide the harms, and (6) help patients clarify their values in association with the options [60]. Patient decision aids can take various formats (e.g., booklets, videos, web-based) and are designed to be used either ahead or during the consultation. Either way they are intended to supplement, not replace, the clinical encounter.

A 2017 systematic review evaluating the effectiveness of patient decision aids for adults facing treatment or screening decisions identified 105 randomized controlled trials involving 31,043 people [59]. Sex and/or gender was reported in 102 trials (30, 642 participants) with 54.7% women [61]. As compared to usual care, patient decision aids were associated with higher decision quality. Patients have significant better knowledge, have more realistic expectations of the benefits and harms, and make choices that are more aligned with their values and preferences [59]. Patient decision aids were also associated with improved decision-making processes. People exposed to the decision aids experienced less decisional conflict, felt more informed, and had clearer values. There was more active participation in the decision-making process; patient-clinician communication improved, with no difference in the length of the consultation. There was no difference compared to usual care on anxiety, depression, quality of life, or satisfaction. Importantly, patient decision aids are not associated with worsened health outcomes or harms. Organizational and healthcare system level outcomes warrant further investigation.

Numerous patient decision aids exist to support CVD prevention and treatment decisions. A systematic review of patient decision aids for stroke prevention in atrial fibrillation identified 10 studies with female representation ranging from 1 to 49.4%) [62]. Pooled results demonstrated that patients exposed to a patient decision aid experienced reduced decisional conflict [mean difference (MD) = −0.10; 95% confidence interval (CI): −0.18 to −0.02; P = 0.01], including on both the Decisional Conflict Scale informed and values clarity subscales. Significant differences in knowledge were reported in 4 of the 5 studies which measured it. Participants exposed to the patient decision aids were more likely to choose oral anticoagulant [risk ratio: 1.03; 95% CI: 1.01–1.05; P = 0.004]. Not included in this review, is a recent randomized controlled trial of 922 patients using an encounter conversation tool (compared with no tool) which increased patient involvement in decision making and clinicians' satisfaction, with no significant effect on treatment decisions or encounter duration. [63] Participants in both groups reported better communication quality, knowledge, and lower decisional conflict. For primary CVD prevention, Bonner et al. [64] identified 25 publicly accessible online decision aids, the majority of which focused on a single medication for CVD prevention. These patient decision aids were evaluated using the Patient Education Materials Assessment Tool (PEMAT-P) ratings and found to be very understandable, yet their readability levels were higher than the general population's health literacy level. This could potentially make it difficult for people to participate in decision making based on their content. Single studies have demonstrated the effectiveness, feasibility, and acceptability of patient decision aids for implantable cardioverter-defibrillator implantation [65, 66] and replacement decisions [67]. Individuals exposed to a patient decision aid prior to specialist consultation experienced less decisional conflict, improved knowledge scores and better values-choice concordance compared to usual care. At the time of ICD replacement, individuals exposed to the patient decision aid were also more knowledgeable [67].

Online repositories house freely accessible patient decision aids for cardiovascular decisions (Table 22.1). There is also the Ottawa Personal Decision Guide or the Ottawa Personal Decision Guide for Two, which are generic decision support tools that can be used for any health or social decisions [68]. It has also been culturally adapted with Indigenous women [69].

**Table 22.1** Online repositories of freely accessible patient decision aids for cardiovascular decisions

| Sources | Country | Website | Topics of patient decision aids |
|---|---|---|---|
| The Ottawa Hospital Research Institute (Inventory of decision aids) | Canada | https://decisionaid.ohri.ca/AZlist.html | Angiogram<br>Aspirin<br>Antihypertension medication<br>Cardiac ablation<br>Cardiac catheterization<br>Cardiac implantable electronic devices<br>Statins |
| Laval University (Decision Boxes) | Canada | https://www.boitedecision.ulaval.ca/en/ | Aspirin<br>Statins |
| American College of Cardiology, CardioSmart | United States | https://www.cardiosmart.org/SDM/Decision-Aids/Find-Decision-Aids | Aortic stenosis<br>Atrial fibrillation<br>Cardiac resynchronization therapy<br>Implantable cardioverter-defibrillator<br>Left ventricular assist device |
| Shared Cardiology | United States | https://sharedcardiology.wordpress.com/tools/ | Atrial fibrillation<br>Aortic stenosis<br>Primary prevention<br>Stable angina |
| Colorado program for patient centered decisions | United States | https://patientdecisionaid.org/ | Aortic stenosis<br>Atrial fibrillation<br>Heart Failure<br>Medications<br>Implantable cardioverter-defibrillator<br>Left ventricular assist device |

**Decision Coaching**. Decision coaching is a non-directive approach to help patients to prepare for making health decisions [70]. Any member of the healthcare team (e.g., nurses, social workers, pharmacists, genetic counsellors) trained in decision coaching, or those using a protocol, can assume the role of decision coach either face to face or on the telephone [71]. Decision coaches: (a) assess patients' decision-making needs; (b) provide information on options, benefits and harms (verbally or using evidence-based resource such as a patient decision aid); (c) assess patients' understanding; (d) clarify patients' values on features and outcomes of options; and (e) screen to determine patients' needs relevant to implementing the chosen option (e.g., motivation, self-confidence, barriers, commitment). Decision coaching aims to: develop the patient's skills in thinking about the options, prepare for discussing the decision in a consultation with his or her healthcare provider, and implement the chosen option [72, 73]. The non-directive approach allows decision coaches to be "mediators" of information, listeners of patients' preferences, and messengers of expressed preferences to the health care provider, often physicians, ultimately responsible for prescribing or implementing treatment with the chosen option [74]. Systematic reviews on shared decision-making interventions have shown that decision coaching is important to empower patients and enhance their autonomy [72, 75]. A decision coaching protocol with video is available online [76]. When paired with a patient decision aid, decision coaching improved patients' understanding of, and participation in their care, enhanced informed decisions and reduced costs [70, 72]. A recently published systematic review of 28 randomized controlled trials synthesizing the effectiveness of decision coaching interventions yielded low certainty evidence on knowledge [73]. It was not possible to establish strong conclusions for other outcomes such as preparation for decision making, decision self-confidence, feeling informed, clear values, or feeling supported. When combined with evidence-based information, participants' knowledge may improve. There were no adverse effects (e.g., decision regret, anxiety). None of the included trials were about cardiovascular decisions.

**Question Prompts**. Question prompts are delivered to patients before consultations and are designed to help patients identify their informational needs and prepare questions to ask in the consultation. A systematic review of 33 studies, two of which conducted in the cardiovascular setting, showed a 27% improvement in patients asking questions and 9% improvement in overall satisfaction, without any difference in knowledge, anxiety, or consultation length [77]. An example of a question prompt is the AskShareKnow intervention which is composed of three generic questions specific to facilitating shared decision making that patients can ask their healthcare provider (Box 22.2). When simulated patients asked these questions in the consultation, clinicians provided them with more information and were more likely to involve them in decision making [78]. When actual patients observed a 4-min video-clip about the 3-questions in the waiting room of family physicians, 87% of patients making a decision asked one or more question in the consultation [79].

**Box 22.2: The AskShareKnow Intervention**

What are my options?
What are the possible benefits and harms of those options?
How likely are each of those benefits and harms to happen to me?

## Implementation of Shared Decision-Making in Clinical Practice

Widespread SDM implementation has not been without challenges [80]. The implementation of health interventions is more successful when interventions are designed to overcome identified barriers [81]. Hence, to increase the use of SDM interventions in clinical practice, it can be useful to: (1) identify barriers interfering with their use, (2) determine effective approaches for increasing their uptake in clinical practice, and (3) measure their impact on patient, provider, and system-level outcomes.

**Identifying Barriers to Shared Decision-Making for Women**. Barriers and facilitators of implementing shared decision-making in various settings and populations have been systematically investigated from the perspectives of patients and providers and at the level of organizations and systems [74, 82–86]. Examples of barriers at the patient, healthcare provider, organizational and system levels are listed in Table 22.2. Interestingly, for many of these perceived barriers, there is evidence to counter them [87]. Women face additional barriers, over and above the ones identified. In certain cultures, familial paternalism challenges women's ability to be engaged in their health decisions [88]. This is especially true for elderly women and those with low health literacy. For example, in a Tawainese study, an 89-year-old illiterate woman with non-ST elevated myocardial infarction deferred decisions and receipt of education to her eldest son, despite preferring non-invasive therapies. The deferral of decisions to her son was a strategy to honor the head of the household and wish to maintain harmony in the family over the need to exercise her individual choices and agency. Many women have expressed challenges accessing acute cardiovascular care, concerns substantiated by research. When women seek care for cardiovascular concerns, they are often misunderstood, their symptoms misinterpreted, resulting in misdiagnoses, and consequently do not receive the needed treatment [4].

The implementation of SDM and supportive interventions are more likely to be successful when interventions are co-produced with end-users, entire teams are trained in SDM, integrated deliberately into existing interprofessional teams' workflows, and outcomes are measured to demonstrate impact [89]. In some specialized cardiovascular centres, Heart Teams structures offer an existing platform to integrate and enhance interprofessional SDM in clinical practice [90]. For women with CVD specifically, tailoring SDM interventions to their decisional needs is warranted.

**Effective Approaches to Increase Shared Decision-Making Uptake**. Interventions that aim to increase SDM use in clinical practice target patients, healthcare providers, or both, with best results when integrated within supportive organizational and healthcare systems [75, 89]. The most recent Cochrane review on interventions for increasing the use of SDM in clinical practice identified 87 studies, 14 of which were about cardiovascular diseases [75]. Forty-four studies evaluated SDM interventions targeting patients, which included patient decision aids, question prompt lists and training for patients, among others. Fifteen studies evaluated interventions targeting healthcare providers and the most common were education and training to enhance SDM knowledge and skills [75]. A previous update of this review found

**Table 22.2** Barriers to shared decision-making

| Level | Barriers | Interventions/strategies to overcome barriers |
|---|---|---|
| Patient [74] | • Lack of knowledge (e.g., health condition, options, outcomes and values and preferences)<br>• Power imbalance in the patient-physician relationship<br>• Lack of confidence in knowledge<br>• Lack of self-efficacy in SDM skills<br>• Limited health literacy | • Patient decision aids<br>• Decision coaching<br>• Question Prompts<br>• Involve nurses to help them understand their options, discuss values/preferences, and advocate for them within the interprofessional healthcare team |
| Healthcare Providers [82] | • Lack of SDM knowledge<br>• Believe it is inappropriate due to patient characteristics or the clinical situation<br>• Indifference/resistance<br>• Perception they already do it<br>• Perception that patients do not want it<br>• Concerns about workflow disruptions<br>• Lack of time | • Dispel myths<br>• Training for use with various populations<br>• Training for use across various situations<br>• Team-based training<br>• Endorsement of SDM interventions by specialty societies |
| Organization [86] | • Lack of leadership support for SDM<br>• Culture does not value patients' voices<br>• Disruption to workflow<br>• Lack of resources<br>• Infrastructure challenges<br>• Unaware of added value<br>• Lack of incentives | • Supportive leadership<br>• Alignment of SDM with organizational priorities (i.e., quality and safety)<br>• Integration into existing workflow<br>• Investment in resources<br>• Measure to demonstrate benefits |
| System [86] | • Culture of healthcare delivery not acknowledging value of patients<br>• Lack of policies and guidelines<br>• Lack of incentives or enforcement of incentives<br>• Inadequate healthcare provider education and licensing | • Integration of SDM into policies and guidelines<br>• Implementation of incentives for correct/quality SDM use<br>• Provide education for pre and post licensure healthcare providers |

that there was greater uptake of SDM when interventions targeted both patients and healthcare professionals or targeted the patient alone [91].

A systematic review of 44 studies identified 36 unique training programs for healthcare providers [92]. Teaching methods in these studies included case-based learning, small group sessions, role play, printed materials, audit and feedback and online tutorials. An international group of experts reached consensus that SDM training programs must focus on, at a minimum, building relational competencies and risk communication competencies [93]. When interprofessional SDM approach is the goal, training in teams and socially supportive environments is recommended [94]. SDM training content aligns well with the call by the Canadian Women Heart Health Alliance for mandatory skill and capacity building training for healthcare providers in active listening, understanding, and empathy to improve cardiovascular outcomes for women [4].

Currently, there is a systematic review and meta-analysis underway to investigate the effectiveness of SDM interventions for cardiovascular decisions (PROSPERO registration CRD42021290164). Its results will contribute to our understanding of these interventions specifically for cardiovascular health decisions. Depending on the results of this review, there may still be a pressing need for SDM evaluation and implementation studies for cardiovascular decisions [95]. Future studies should ensure adequate representation of women with other underrepresented groups in research such as people with low literacy and numeracy, and people from diverse racial and ethnic groups.

**Measuring for Impact**. To monitor whether SDM is occurring and reinforce its use, numerous patient-reported, provider-reported, and/or observer-based instruments exist [96, 97]. Frequently measured outcomes in SDM research involve the extent, quality, and outcomes of the SDM that occurred in preparation for, during, and/or after the encounter. Examples include knowledge, risk perception, satisfaction with the decision-making process, decisional conflict, and decision regret. Patient-reported outcomes (PROs) measure various aspects of patients' health and healthcare, from their perspective, free of interpretation from others [98]. These are increasingly being used to enhance and personalize SDM approaches [99, 100].

The SURE test is a simple patient-reported tool that can be used in clinical practice to monitor patients' decisional needs (Table 22.3) [101]. The four-item SURE test is used to screen patients for decisional conflict and is based on the original version of the Decisional Conflict Scale [102]. Each 'yes' response is worth one point; a total score of less than four indicates the presence of decisional conflict.

## Conclusion and Future Directions

SDM has the potential to benefit women with CVD because it tailors care to the individuals' clinical and decisional needs, life circumstances, and personal preferences; a process that is particularly appropriate in the setting of limited evidence. It encourages an exchange of expertise between women and their healthcare providers,

**Table 22.3** Screening for decisional conflict—The SURE test

| | SURE test item |
|---|---|
| Sure of myself | Do you feel SURE about the best choice for you? |
| Understand information | Do you know the benefits and risks of each option? |
| Risk/benefit ratio | Are you clear about which benefits and risks matter most to you? |
| Encouragement | Do you have enough support and advice to make a choice? |

Scoring per item: Yes = 1; No = 0; Total score < 4 indicates decisional conflict

meaningful participation, and active listening to better understand and achieve quality decisions that are informed and based on what matters most to these women. Yet, currently, most SDM interventions do not encompass the unique decisional and clinical needs of women with CVD. As the state of the evidence evolves on the prevention, screening, treatment, rehabilitation, and end of life decisions relevant to women with CVD, opportunities should be sought and encouraged for collaboration to develop decision support tools and strategies, with women having CVD, that rely on that same evidence to increase the quality of women's decision-making and decision-making processes.

## References


1. Shay LA, Lafata JE (2015) Where is the evidence? A systematic review of shared decision making and patient outcomes. Med Decis Making 35(1):114–131
2. Hauser K, Koerfer A, Kuhr K, Albus C, Herzig S, Matthes J (2015) Outcome-relevant effects of shared decision making. Dtsch Arztebl Int 112(40):665–671
3. Chewning B, Bylund CL, Shah B, Arora NK, Gueguen JA, Makoul G (2012) Patient preferences for shared decisions: a systematic review. Patient Educ Couns 86(1):9–18
4. Colella TJF, Hardy M, Hart D, Price JAD, Sarfi H, Mullen KA et al (2021) The Canadian women's heart health alliance atlas on the epidemiology, diagnosis, and management of cardiovascular disease in women-chapter 3: Patient Perspectives. CJC Open. 3(3):229–235
5. Charles C, Gafni A, Whelan T (1997) Shared decision-making in the medical encounter: what does it mean? (or it takes at least two to tango). Soc Sci Med 44(5):681–692
6. Makoul G, Clayman ML (2006) An integrative model of shared decision making in medical encounters. Patient Educ Couns 60(3):301–312
7. Barry MJ, Edgman-Levitan S (2012) Shared decision making–pinnacle of patient-centered care. N Engl J Med 366(9):780–781
8. Yancy CW, Jessup M, Bozkurt B, Butler J, Casey DE Jr, Drazner MH et al (2013) 2013 ACCF/AHA guideline for the management of heart failure: executive summary: a report of the American College of Cardiology Foundation/American Heart Association Task Force on practice guidelines. Circ 128(16):1810–1852
9. Andrade JG, Aguilar M, Atzema C, Bell A, Cairns JA, Cheung CC et al (2020) The 2020 canadian cardiovascular society/canadian heart rhythm society comprehensive guidelines for the management of atrial fibrillation. Can J Cardiol 36(12):1847–1948
10. Al-Khatib SM, Stevenson WG, Ackerman MJ, Bryant WJ, Callans DJ, Curtis AB, Deal BJ, Dickfeld T, Field ME, Fonarow GC, Gillis AM, Granger CB, Hammill SC, Hlatky MA, Joglar JA, Kay GN, Matlock DD, Myerburg RJ, Page RL (2018). 2017 AHA/ACC/HRS guideline for management of patients with ventricular arrhythmias and the prevention of sudden cardiac

death: executive summary: a report of the American College of Cardiology/American Heart Association Task Force on Clinical Practice Guidelines and the Heart Rhythm Society. Heart Rhythm 15(10):e190–e252

11. Bennett M, Parkash R, Nery P, Senechal M, Mondesert B, Birnie D et al (2017) Canadian cardiovascular society/canadian heart rhythm society 2016 implantable cardioverter-defibrillator guidelines. The Can J Cardiol. 33:174–188
12. Asgar AW, Ouzounian M, Adams C, Afilalo J, Fremes S, Lauck S et al (2019) Canadian cardiovascular society position statement for transcatheter aortic valve implantation. Can J Cardiol 35(11):1437–1448
13. Jensen T, Chin J, Ashby L, Long K, Schafer J, Hakim R (2016) Decision memo for percutaneous left atrial appendage (LAA) closure therapy (CAG-00445N)
14. Bodenheimer T, Sinsky C (2014) From triple to quadruple aim: care of the patient requires care of the provider. Ann Fam Med 12(6):573–576
15. Sikka R, Morath JM, Leape L (2015) The quadruple aim: care, health, cost and meaning in work. BMJ Qual Saf 24(10):608–610
16. Porter ME, Teisberg EO (2006) Redefining health care: creating value-based competition on results. Harvard Business School Press, Boston
17. Veroff D, Marr A, Wennberg DE (2013) Enhanced support for shared decision making reduced costs of care for patients with preference-sensitive conditions. Health Aff (Millwood). 32(2):285–293
18. Oshima Lee E, Emanuel EJ (2013) Shared decision making to improve care and reduce costs. N Engl J Med 368(1):6–8
19. President's commission for the study of ethical problems in medicine and biomedical and behavioral research. Making health care decisions: a report on the ethical and legal implications of informed consent in the patient-practitioner relationship (1982)
20. Kunneman M, Montori VM (2017) When patient-centred care is worth doing well: informed consent or shared decision-making. BMJ Qual Saf 26(7):522–524
21. Spatz ES, Krumholz HM, Moulton BW (2016) The new era of informed consent: getting to a reasonable-patient standard through shared decision making. JAMA 315(19):2063–2064
22. King JS, Moulton BW (2006) Rethinking informed consent: the case for shared medical decision-making. Am J Law Med 32(4):429–501
23. Manta CJ, Ortiz J, Moulton BW, Sonnad SS (2021) From the patient perspective, consent forms fall short of providing information to guide decision making. J Patient Saf 17(3):e149–e154
24. Holmes-Rovner M (2007) International patient decision aid standards (IPDAS): beyond decision aids to usual design of patient education materials. Health Expect 10(2):103–107
25. Kon AA (2010) The shared decision-making continuum. JAMA 304(8):903–904
26. Mulley AG, Trimble C, Elwyn G (2012) Stop the silent misdiagnosis: patients' preferences matter. BMJ 345:e6572
27. Lee CN, Hultman CS, Sepucha K (2010) What are patients' goals and concerns about breast reconstruction after mastectomy? Ann Plast Surg 64(5):567–569
28. Heart and Stroke Foundation of Canada. Ms. Understood. Heart & Stroke 2018 heart report. 2018. Available from: https://www.heartandstroke.ca/-/media/pdf-files/canada/2018-heart-month/hs_2018-heart-report_en.ashx. Retrieved July 13, 2022
29. Friesen-Storms JH, Bours GJ, van der Weijden T, Beurskens AJ (2015) Shared decision making in chronic care in the context of evidence based practice in nursing. Int J Nurs Stud 52(1):393–402
30. Kim DJ, Choo EK, Ranney ML (2014) Impact of gender on patient preferences for technology-based behavioral interventions. West J Emerg Med 15(5):593–599
31. Wessels H, de Graeff A, Wynia K, de Heus M, Kruitwagen CL, Woltjer GT et al (2010) Gender-related needs and preferences in cancer care indicate the need for an individualized approach to cancer patients. Oncologist 15(6):648–655
32. Delaney R, Strough J, Parker AM, de Bruin WB (2015) Variations in decision-making profiles by age and gender: a cluster-analytic approach. Pers Individ Dif 85:19–24

33. Saeed F, Hoerger M, Norton SA, Guancial E, Epstein RM, Duberstein PR (2018) Preference for palliative care in cancer patients: are men and women alike? J Pain Symptom Manage 56(1):1–6 e1
34. Bowling A, Culliford L, Smith D, Rowe G, Reeves BC (2008) What do patients really want? Patients' preferences for treatment for angina. Health Expect 11(2):137–147
35. Walker RL, Campbell KA, Sears SF, Glenn BA, Sotile R, Curtis AB et al (2004) Women and the implantable cardioverter defibrillator: a lifespan perspective on key psychosocial issues. Clin Cardiol 27(10):543–546
36. Thylén I, Moser DK, Chung ML, Miller J, Fluur C, Stromberg A (2013) Are ICD recipients able to foresee if they want to withdraw therapy or deactivate defibrillator shocks? IJC Heart & Vessels. 1:22–31
37. Angus JE, King-Shier KM, Spaling MA, Duncan AS, Jaglal SB, Stone JA et al (2015) A secondary meta-synthesis of qualitative studies of gender and access to cardiac rehabilitation. J Adv Nurs 71(8):1758–1773
38. Mulvagh SL, Mullen KA, Nerenberg KA, Kirkham AA, Green CR, Dhukai AR et al (2022) The canadian women's heart health alliance atlas on the epidemiology, diagnosis, and management of cardiovascular disease in women - Chapter 4: sex- and gender-unique disparities: CVD across the lifespan of a woman. CJC Open 4(2):115–132
39. Gulati M (2017) Improving the cardiovascular health of women in the nation: moving beyond the bikini boundaries. Circ 135(6):495–498
40. Murphy J, Tanner T, Komorowski J (2019) Shared decision-making with choosing wisely. Nurs Womens Health 23(3):253–264
41. Gagliardi AR, Dunn S, Foster AM, Grace SL, Khanlou N, Stewart DE et al (2020) Is patient-centred care for women a priority for policy-makers? content analysis of government policies. Health Res Policy Syst 18(1):23
42. Moore JE, Mompe A, Moy E (2018) Disparities by sex tracked in the 2015 national healthcare quality and disparities report: trends across national quality strategy priorities, health conditions, and access measures. Womens Health Issues 28(1):97–103
43. Canadian Institute of Health Information (2003) Women's health surveillance report. Ottawa
44. Adisso EL, Zomahoun HTV, Gogovor A, Legare F (2020) Sex and gender considerations in implementation interventions to promote shared decision making: a secondary analysis of a Cochrane systematic review. PLoS ONE 15(10):e0240371
45. O'Connor AM, Drake ER, Wells GA, Tugwell P, Laupacis A, Elmslie T (2003) A survey of the decision-making needs of Canadians faced with complex health decisions. Health Expect 6(2):97–109
46. Bunn H, Lange I, Urrutia M, Campos MS, Campos S, Jaimovich S et al (2006) Health preferences and decision-making needs of disadvantaged women. J Adv Nurs 56(3):247–260
47. Hoefel L, O'Connor AM, Lewis KB, Boland L, Sikora L, Hu J et al (2020) 20th anniversary update of the ottawa decision support framework Part 1: a systematic review of the decisional needs of people making health or social decisions. Med Decis Making 40(5):555–581
48. O'Connor AM, Tugwell P, Wells GA, Elmslie T, Jolly E, Hollingworth G et al (1998) A decision aid for women considering hormone therapy after menopause: decision support framework and evaluation. Patient Educ Couns 33(3):267–279
49. Matlock DD, Nowels CT, Masoudi FA, Sauer WH, Bekelman DB, Main DS et al (2011) Patient and cardiologist perceptions on decision making for implantable cardioverter-defibrillators: a qualitative study. Pacing Clin Electrophysiol 34(12):1634–1644
50. Wilhelms LA, Blumenthal-Barby JS, Kostick KM, Estep JD, Bruce CR (2017) Patients' perspectives on transplantation while undergoing left ventricular assist device support. ASAIO J 63(6):740–744
51. Dahl M, Lindholt J, Sogaard R, Frost L, Andersen LS, Lorentzen V (2018) An interview-based study of nonattendance at screening for cardiovascular diseases and diabetes in older women: Nonattendees' perspectives. J Clin Nurs 27(5–6):939–948
52. Gattellari M, Ward JE (2005) A community-based randomised controlled trial of three different educational resources for men about prostate cancer screening. Patient Educ Couns 57(2):168–182

53. Sun Q (2004) Predicting downstream effects of high decisional conflict: meta-analysis of the Decisional Conflict Scale: University of Ottawa
54. Brehaut JC, O'Connor AM, Wood TJ, Hack TF, Siminoff L, Gordon E et al (2003) Validation of a decision regret scale. Med Decis Making 23(4):281–292
55. Green AR, Jenkins AMY, Masoudi FA, Magid DJ, Kutner JS, Leff B et al (2016) Decision-making experiences of patients with implantable cardioverter defibrillators. Pacing Clin Electrophysiol 39(10):1061–1069
56. Stahl EP, Dickert NW, Cole RT, Laskar SR, Morris AA, Smith AL et al (2019) Decisional regret in left ventricular assist device patient-caregiver dyads. Heart Lung 48(5):400–404
57. Stacey D, Legare F, Boland L, Lewis KB, Loiselle MC, Hoefel L et al (2020) 20th anniversary Ottawa decision support framework: part 3 overview of systematic reviews and updated framework. Med Decis Making 40(3):379–398
58. O'Connor AM, Legare F, Stacey D (2003) Risk communication in practice: the contribution of decision aids. BMJ 327(7417):736–740
59. Stacey D, Legare F, Lewis K, Barry MJ, Bennett CL, Eden KB et al (2017) Decision aids for people facing health treatment or screening decisions. Cochrane Database Syst Rev 4:CD001431
60. Joseph-Williams N, Newcombe R, Politi M, Durand MA, Sivell S, Stacey D et al (2013) Toward minimum standards for certifying patient decision aids: a modified Delphi consensus process. Med Decis Making 34(6):699–710
61. Stacey D, Legare F, Lewis KB (2017) Patient decision aids to engage adults in treatment or screening decisions. JAMA 318(7):657–658
62. Song D, Zhou J, Fan T, Chang J, Qiu Y, Zhuang Z et al (2021) Decision aids for shared decision-making and appropriate anticoagulation therapy in patients with atrial fibrillation: a systematic review and meta-analysis. Eur J Cardiovasc Nurs 21(2):97–106
63. Kunneman M, Branda ME, Hargraves IG, Sivly AL, Lee AT, Gorr H et al (2020) Assessment of shared decision-making for stroke prevention in patients with atrial fibrillation: a randomized clinical trial. JAMA Intern Med 180(9):1215–1224
64. Bonner C, Patel P, Fajardo MA, Zhuang R, Trevena L (2019) Online decision aids for primary cardiovascular disease prevention: systematic search, evaluation of quality and suitability for low health literacy patients. BMJ Open 9(3):e025173
65. Carroll SL, Stacey D, McGillion M, Healey JS, Foster G, Hutchings S et al (2017) Evaluating the feasibility of conducting a trial using a patient decision aid in implantable cardioverter defibrillator candidates: a randomized controlled feasibility trial. Pilot Feasibility Stud. 3:49
66. Wallace BC, Jones J, Masoudi FA, Nowels CT, Varosy P, Thomson R et al (2021) Development and piloting of four decision aids for implantable cardioverter-defibrillators in different media formats. Pacing Clin Electrophysiol 44(11):1842–1852
67. Lewis KB, Birnie D, Carroll SL, Brousseau-Whaley C, Clark L, Green M et al (2021) Decision support for implantable cardioverter-defibrillator replacement: a pilot feasibility randomized controlled trial. J Cardiovasc Nurs 36(2):143–150
68. Ottawa Personal Decision Guide (2022) Available from: https://decisionaid.ohri.ca/decguide.html. Retrieved July 13
69. Jull J, Giles A, Minwaashin Lodge TAWsSC, Boyer Y, Stacey D (2015) Cultural adaptation of a shared decision making tool with Aboriginal women: a qualitative study. BMC Med Inform Decis Mak 15:1
70. Rahn AC, Jull J, Boland L, Finderup J, Loiselle MC, Smith M et al (2021) Guidance and/or decision coaching with patient decision aids: scoping reviews to inform the international patient decision aid standards (IPDAS). Med Decis Making 41(7):938–953
71. Legare F, Stacey D, Gagnon S, Dunn S, Pluye P, Frosch D et al (2011) Validating a conceptual model for an inter-professional approach to shared decision making: a mixed methods study. J Eval Clin Pract 17(4):554–564
72. Stacey D, Kryworuchko J, Bennett C, Murray MA, Mullan S, Legare F (2012) Decision coaching to prepare patients for making health decisions: a systematic review of decision coaching in trials of patient decision AIDS. Med Decis Making 32(3):E22-33

73. Jull J, Kopke S, Smith M, Carley M, Finderup J, Rahn AC et al (2021) Decision coaching for people making healthcare decisions. Cochrane Database Syst Rev 11:CD013385
74. Joseph-Williams N, Elwyn G, Edwards A (2014) Knowledge is not power for patients: a systematic review and thematic synthesis of patient-reported barriers and facilitators to shared decision making. Patient Educ Couns 94(3):291–309
75. Legare F, Adekpedjou R, Stacey D, Turcotte S, Kryworuchko J, Graham ID et al (2018) Interventions for increasing the use of shared decision making by healthcare professionals. Cochrane Database Syst Rev 7:CD006732
76. Decision coaching (2022) Available from: https://decisionaid.ohri.ca/coaching.html. Retrieved July 13
77. Kinnersley P, Edwards A, Hood K, Ryan R, Prout H, Cadbury N et al (2008) Interventions before consultations to help patients address their information needs by encouraging question asking: systematic review. BMJ 337:a485–a494
78. Shepherd HL, Barratt A, Trevena LJ, McGeechan K, Carey K, Epstein RM et al (2011) Three questions that patients can ask to improve the quality of information physicians give about treatment options: a cross-over trial. Patient Educ Couns 84(3):379–385
79. Shepherd HL, Barratt A, Jones A, Bateson D, Carey K, Trevena LJ et al (2016) Can consumers learn to ask three questions to improve shared decision making? A feasibility study of the ASK (AskShareKnow) Patient-Clinician Communication Model((R)) intervention in a primary health-care setting. Health Expect 19(5):1160–1168
80. Bravo P, Harter M, McCaffery K, Giguere A, Hahlweg P, Elwyn G (2022) Editorial: 20 years after the start of international Shared Decision-Making activities: Is it time to celebrate? Probably. Z Evid Fortbild Qual Gesundhwes 171:1–4
81. Baker R, Camosso-Stefinovic J, Gillies C, Shaw EJ, Cheater F, Flottorp S et al (2015) Tailored interventions to address determinants of practice. Cochrane Database Syst Rev (4):CD005470
82. Legare F, Ratte S, Gravel K, Graham ID (2008) Barriers and facilitators to implementing shared decision-making in clinical practice: update of a systematic review of health professionals' perceptions. Patient Educ Couns 73(3):526–535
83. Alsulamy N, Lee A, Thokala P, Alessa T (2020) What influences the implementation of shared decision making: an umbrella review. Patient Educ Couns 11:S0738-3991(20)30436-5.
84. Pel-Littel RE, Snaterse M, Teppich NM, Buurman BM, van Etten-Jamaludin FS, van Weert JCM et al (2021) Barriers and facilitators for shared decision making in older patients with multiple chronic conditions: a systematic review. BMC Geriatr 21(1):112
85. Waddell A, Lennox A, Spassova G, Bragge P (2021) Barriers and facilitators to shared decision-making in hospitals from policy to practice: a systematic review. Implement Sci 16(1):74
86. Scholl I, LaRussa A, Hahlweg P, Kobrin S, Elwyn G (2018) Organizational- and system-level characteristics that influence implementation of shared decision-making and strategies to address them - a scoping review. Implement Sci 13(1):40
87. Legare F, Thompson-Leduc P (2014) Twelve myths about shared decision making. Patient Educ Couns 96(3):281–286
88. Chao HY, Chen HM, Lin EC (2022) Ethical challenges of nonreading older adult women's autonomy in receiving percutaneous coronary intervention under familial paternalism in Taiwan. J Transcult Nurs 33(1):110–117
89. Joseph-Williams N, Abhyankar P, Boland L, Bravo P, Brenner AT, Brodney S et al (2021) What works in implementing patient decision aids in routine clinical settings? a rapid realist review and update from the international patient decision aid standards collaboration. Med Decis Making 41(7):907–937
90. Mesana T (2019) Heart teams for treatment of cardiovascular disease: a guide for advancing patient-centered cardiac care. Springer, Ottawa
91. Legare F, Stacey D, Turcotte S, Cossi MJ, Kryworuchko J, Graham ID et al (2014) Interventions for improving the adoption of shared decision making by healthcare professionals. Cochrane Database Syst Rev 9:CD006732

92. Legare F, Politi M, Drolet R, Desroches S, Stacey D, Bekker H et al (2012) Training health professionals in shared decision making: An international environmental scan. Patient Educ Couns 88(2):159–169
93. Legare F, Moumjid-Ferdjaoui N, Drolet R, Stacey D, Harter M, Bastian H et al (2013) Core competencies for shared decision making training programs: Insights from an international, interdisciplinary working group. J Contin Educ Health Prof 33(4):267–273
94. Boland L, Lawson ML, Graham ID, Legare F, Dorrance K, Shephard A et al (2019) Post-training shared decision making barriers and facilitators for Pediatric healthcare providers: a mixed-methods study. Acad Pediatr 19(1):118–129
95. Chung MK, Fagerlin A, Wang PJ, Ajayi TB, Allen LA, Baykaner T et al (2021) Shared decision making in cardiac electrophysiology procedures and arrhythmia management. Circ Arrhythm Electrophysiol 14(12):e007958
96. Scholl I, Koelewijn-van Loon M, Sepucha K, Elwyn G, Legare F, Harter M et al (2011) Measurement of shared decision making - a review of instruments. Z Evid Fortbild Qual Gesundhwes 105(4):313–324
97. Gartner FR, Bomhof-Roordink H, Smith IP, Scholl I, Stiggelbout AM, Pieterse AH (2018) The quality of instruments to assess the process of shared decision making: a systematic review. PLoS ONE 13(2):e0191747
98. Valderas JM, Kotzeva A, Espallargues M, Guyatt G, Ferrans CE, Halyard MY et al (2008) The impact of measuring patient-reported outcomes in clinical practice: a systematic review of the literature. Qual Life Res 17(2):179–193
99. Damman OC, Jani A, de Jong BA, Becker A, Metz MJ, de Bruijne MC et al (2020) The use of PROMs and shared decision-making in medical encounters with patients: an opportunity to deliver value-based health care to patients. J Eval Clin Pract 26(2):524–540
100. Lin E, Uhler LM, Finley EP, Jayakumar P, Rathouz PJ, Bozic KJ et al (2022) Incorporating patient-reported outcomes into shared decision-making in the management of patients with osteoarthritis of the knee: a hybrid effectiveness-implementation study protocol. BMJ Open 12(2):e055933
101. Legare F, Kearing S, Clay K, Gagnon S, D'Amour D, Rousseau M et al (2010) Are you SURE? Assessing patient decisional conflict with a 4-item screening test. Can Fam Physician 56(8):e308–e314
102. O'Connor AM (1995) Validation of a decisional conflict scale. Med Decis Making 15(1):25–30

# Chapter 23
# Sex Differences in the Function of Cardiac Sodium-Calcium Exchanger in Physiological and Pathophysiological Settings: Implications for Cardiac Arrhythmias

**Norbert Nagy and István Baczkó**

**Abstract** In the past decades, it has become increasingly clear that women and men significantly differ in the epidemiology, pathophysiology and outcome of cardiovascular diseases that also include certain cardiac electrophysiological aspects leading to different disease phenotypes and disparate outcomes of pharmacological interventions. These dissimilarities stem from numerous differences in ion channel expression, kinetics and regulation. One of the first observations of sex-related differences was the longer QT-interval in women measured on the ECG. Sex hormones can influence the electrophysiological parameters on the genomic level altering gene expression. In this regard, the reduced expression of various ion channels carrying repolarizing currents, including $I_{to}$, $I_{K1}$, $I_{Kr}$, $I_{Ks}$ and $I_{K,ATP}$, have been described in women. Sex hormones can also change ion channel functions by non-genomic effects, including the modulation of specific signalling pathways (such as eNOS). Furthermore, direct effects of sex hormones on ion channels were also described. For example, 17β-oestradiol directly reduced $I_{Kr}$, while testosterone increased $I_{Kr}$ and progesterone enhanced $I_{Ks}$. In addition to repolarizing ion currents, sex hormones can influence a large number of transmembrane ion channels and exchangers in various ways, therefore, in this chapter, the sex-related differences regarding an important component of intracellular $Ca^{2+}$ handling, the $Na^+/Ca^{2+}$ exchanger are discussed.

**Keywords** Arrhythmia · eNOS · Estrogen · Ion channels · Calcium

---

N. Nagy · I. Baczkó (✉)
Department of Pharmacology and Pharmacotherapy, Albert Szent-Györgyi Medical School, University of Szeged, Dóm Tér 12, P.O. Box 427, Szeged 6720, Hungary
e-mail: baczko.istvan@med.u-szeged.hu

N. Nagy
ELKH-SZTE Research Group of Cardiovascular Pharmacology, Szeged, Hungary

© The Author(s), under exclusive license to Springer Nature Switzerland AG 2023 
L. Kirshenbaum and I. Rabinovich-Nikitin (eds.), *Biology of Women's Heart Health*, Advances in Biochemistry in Health and Disease 26,
https://doi.org/10.1007/978-3-031-39928-2_23

## Overview of Cardiac $Ca^{2+}$ Handling and Its Gender Differences

Intracellular $Ca^{2+}$ has a crucial role in the excitation–contraction coupling. A complex interplay of intracellular $Ca^{2+}$ fluxes under strict control provides the integrity of the intracellular $Ca^{2+}$ homeostasis. The actual membrane potential intimately influences $Ca^{2+}$ handling and vice versa, the intracellular $Ca^{2+}$ has important roles in the function and kinetics of several ion channels [1].

During depolarization, the opening of the L-type $Ca^{2+}$ channels provides large $Ca^{2+}$ influx that triggers $Ca^{2+}$-induced $Ca^{2+}$ release by opening the ryanodine receptors (RyR). The RyRs and L-type $Ca^{2+}$ channels locate in close proximity by the abundant expression in the extensive T-tubule network forming nanodomains, called dyads. The released $Ca^{2+}$ ($Ca^{2+}$-transient) interacts with the contractile proteins and initiates several $Ca^{2+}$-dependent signalling pathways (such as calmodulin-Kinase II signaling) [2]. During relaxation, the intracellular $Ca^{2+}$ is sequestered to the sarcoplasmic reticulum by the ATP-dependent sarcoplasmic reticulum $Ca^{2+}$ ATP-ase (SERCA), and extruded to the extracellular space by the $Na^+/Ca^{2+}$ exchanger. In a small extent, the ATP-dependent $Ca^{2+}$ pump also contributes to the relaxation (Fig. 23.1) [3].

Several studies investigated the possible gender related differences in $Ca^{2+}$ handling, however, data are often controversial. A study comparing expression levels demonstrated that female ventricular myocytes have markedly higher levels of RyR compared to male animals. Similarly, RyR mRNA was increased in female animals [4]. Ovariectomy caused hyperactivity of the ryanodine receptor, and this increased flux could be reversed by replacement of estrogen and inhibition of protein-kinase A (PKA). This result suggests that estrogen has a role in controlling the $Ca^{2+}$ flux through the modulation of the ryanodine receptor [5]. Experiments carried out on streptozotocin-induced diabetic rats revealed that expression levels of RyR2 and FKBP12.6 was higher in control females than in control males. In contrast, in diabetes, RyR2 phosphorylation and FKBP12.6 unbinding was lower in females [6].

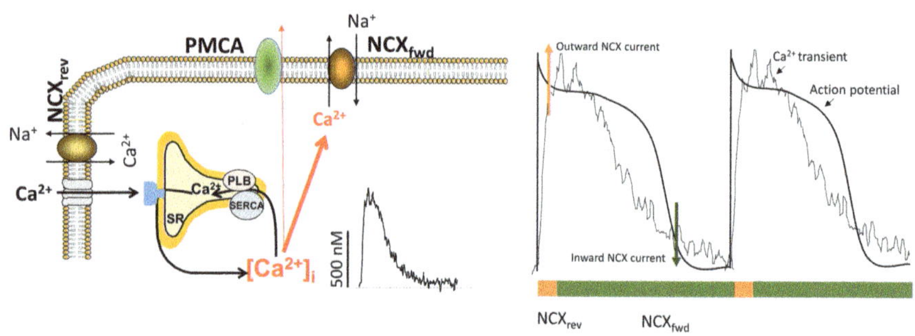


**Fig. 23.1** Schematic illustration of ventricular intracellular $Ca^{2+}$ handling and the suggested operation of NCX during action potential. See text for detailed description

In 10-week ovariectomized rats the maximum $Ca^{2+}$ uptake activity of sarcoplasmic reticulum $Ca^{2+}$ ATPase (SERCA) was reduced together with SERCA protein downregulation and reduction of SERCA mRNA levels. Since supplementation of estrogen and progesterone effectively antagonized the effects of ovariectomy it was concluded that female sex hormones have an important role in SERCA-mediated $Ca^{2+}$ uptake [7].

It was demonstrated that disruption of the FKBP12.6 gene in mice led to $Ca^{2+}$ handling mismanagement in both sexes, however, cardiac hypertrophy was observed only in male animals. When female animals were treated with tamoxifen, an estrogen receptor antagonist, similar cardiac hypertrophy could be observed as in the case of male mice. Therefore, it seems possible that estrogen could be protective against hypertrophic response [8].

In contrast, in human atrial tissue it was found that L-type $Ca^{2+}$ current, RyR, calsequestrin and phospholamban did not show gender differences on the expression level [9].

## The Role of the $Na^+/Ca^{2+}$ Exchanger in Ventricular Myocytes

The mammalian $Na^+/Ca^{2+}$ exchanger (NCX) consists of 10 transmembrane segments. In the myocardium, the NCX1 isoform is a critical modulator of cardiomyocyte $Ca^{2+}$ cycling [10]. A large loop between the 5th and 6th segments has regulatory functions [11–13] providing allosteric regulations by cytoplasmic $Na^+$ and $Ca^{2+}$ ions. It has been found that high intracellular $Na^+$ inactivates NCX [14], however, its physiological significance is questionable since relatively high levels of intracellular $Na^+$ (>20 mM) are required.

The NCX transports three $Na^+$ together with one $Ca^{2+}$ ion where the $Na^+$ concentration gradient provides the driving force for the exchange. Depending on the intracellular and extracellular $Na^+$ and $Ca^{2+}$ concentrations, as well as the actual membrane potential, the NCX can work in two operational modes even during the same action potential. When intracellular $Na^+$ is high, the intracellular $Ca^{2+}$ is low and the membrane potential is depolarized, the reverse mode is favoured where $Ca^{2+}$ influx takes place. In contrast, the high intracellular $Ca^{2+}$ and the hyperpolarized membrane potential facilitate forward mode and NCX extrudes the intracellular $Ca^{2+}$ (Fig. 23.1).

The NCX is abundantly expressed in the sarcolemma, however, it is suggested that the expression level is higher in the t-tubules, having important consequences in $Ca^{2+}$ handling [15]. 15% of the NCX may be located in close proximity to the ryanodine receptors, therefore, can sense microdomain $Ca^{2+}$ levels [16]. While the role of the forward mode in the relaxation is clear, the possible role of the reverse mode in the $Ca^{2+}$ induced $Ca^{2+}$ release is controversial. There are studies demonstrating that $Na^+$

influx facilitates the reverse mode of NCX and this $Ca^{2+}$ influx is able to contribute to the $Ca^{2+}$-induced $Ca^{2+}$ release mechanism [17–19].

Since NCX generates net current, it is feasible that it contributes to the action potential waveform, although the available data are controversial. There are results showing that forward NCX-mediated inward current is a crucial component of the action potential [20]. In contrast, experiments with the novel selective NCX inhibitor ORM-10962 indicate that action potential duration remained unchanged following NCX inhibition [21].

Altered NCX function is described in various pathological conditions. In heart failure, the NCX is upregulated and becomes a better competitor for the SERCA in $Ca^{2+}$ removal [22]. Therefore, it can contribute to the reduced intracellular $Ca^{2+}$ content and the generated inward current can be an important source of arrhythmias [22–25]. In the setting of myocardial ischaemia–reperfusion, the reverse mode of NCX can contribute to the $Ca^{2+}$-load during ischemia, and the rapid onset of the forward NCX might generate large inward current evoking arrhythmic triggers [26–28].

## Sex Differences in the NCX-Expression and Genomic Regulation

Chen et al. investigated the sex-related regional expressional differences of NCX in adult rabbits. Furthermore, both the reverse and forward modes of the exchanger were compared [29].

The study found that during the intersex comparison, that in the case of female rabbits, the outward NCX current was larger in the base region, but was smaller in the apex region compared to males. The inward current was identical between genders in both regions. When NCX was compared within the same sex, they observed that in the case of female rabbits, both the outward and inward NCX was larger in the base region. In male animals, the outward NCX was larger in the apex, but the inward NCX had higher amplitude in the base.

Western blot analysis revealed that NCX1 expression was higher in females obtained from the base region of the heart compared to males, as well as was higher compared to apex from both genders (Fig. 23.2). This enhancement of NCX1 could be the consequence of estrogen-induced genomic mechanism.

The authors also found that NCX and Cav1.2α were upregulated in the base of female hearts that could contribute to the sex-differences in the manifestation of LQT2 syndrome. The higher $Ca^{2+}$ influx in the base region due to larger $I_{CaL}$ is suggested to be compensated for higher NCX to maintain stable $Ca^{2+}$ balance [29]. However, the higher $I_{CaL}$ prolongs the action potential and causes sarcoplasmic reticulum $Ca^{2+}$-overload, and the spontaneous releases elicits early afterdepolarizations via NCX activity. The authors concluded that the apex-base heterogeneity of

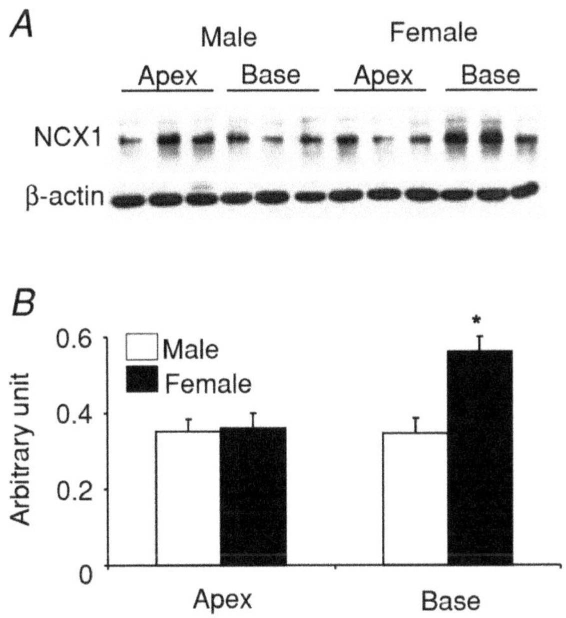


**Fig. 23.2** Distribution of NCX1 protein in male and female rabbits obtained from the apex and base region. Asterisk denotes that NCX1 protein was more abundant in female base compared to male base, and the apex of both sexes (with permission, [29])

NCX expression could be an important arrhythmogenic factor in the development of various arrhythmias (Fig. 23.3) [29].

Golden et al. have demonstrated that testosterone regulates the expression of the major proteins involved in $Ca^{2+}$-handling such as NCX. It was found that ventricular myocytes isolated from two-day old rats after 24 h of testosterone treatment had maximal increase in NCX expression. Therefore, the male sex hormone may play a significant role in the gender-related differences of cardiac performance [30].

Furthermore, it has been found that estrogen upregulates the $I_{CaL}$ and NCX in female rabbits, therefore increases the risk of LQT2-type arrhythmias [31]. The same group also investigated whether these results could be confirmed in the human heart. It was found that Cav1.2 and NCX1 protein levels were higher in women than in men or in postmenopausal women in the apex. $I_{NCX}$ and $I_{CaL}$ were measured from female and male cardiomyocyte derived human induced pluripotent stem cells, where both $I_{CaL}$ and $I_{NCX}$ amplitude were higher in the case of women-derived cells. It was concluded that estrogen upregulated $I_{CaL}$ and $I_{NCX}$ in female human ventricular myocytes. These sex-related differences could be attributable, at least in part, to the increased sex-related differences in $Ca^{2+}$-handling, and arrhythmia propensity [31].

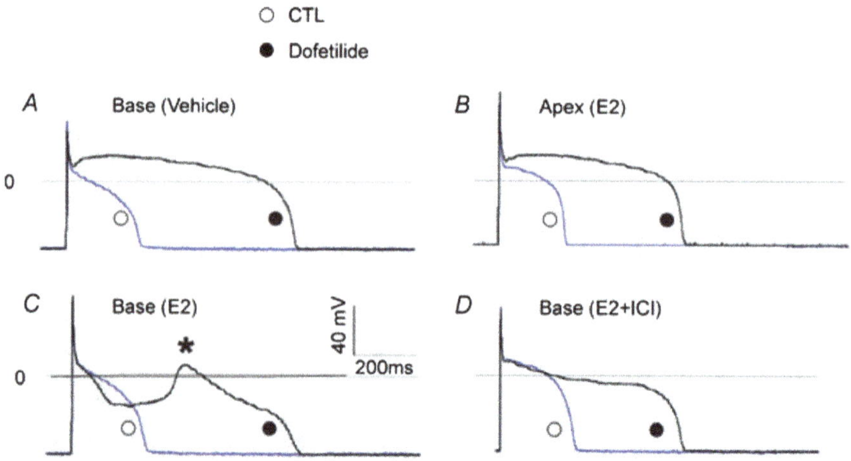


**Fig. 23.3** Genomic effect of estrogen on early afterdepolarization (EAD) development. Panel A and B show action potentials from the base (A) and apex (B) regions where the cells were incubated in vehicle and estrogen, respectively. In both cases the $I_{Kr}$ inhibitor dofetilide largely prolonged action potentials without eliciting early afterdepolarizations. Panel C: when female base myocytes were incubated in estrogen, the dofetilide induced early afterdepolarizations. Panel D: when female base myocytes were treated with estrogen and its antagonists (ICI), the application of dofetilide prolonged the action potential without evoking early afterdepolarizations (with permission, [29])

## Sex-Related Role of NCX in $Ca^{2+}$ Handling Balance

In NCX-overexpressed transgenic and wild-type mice, Sugishita et al. investigated the effect of metabolic inhibition on $[Ca^{2+}]_i$ and $[Na^+]_i$. It was found that metabolic inhibition induced higher $[Ca^{2+}]_i$ rise in male transgenic (Tg) animals compered to wild-type, however, in contrast, in female Tg mice, the $[Ca^{2+}]_i$ increase was not significant. The increase of $[Na^+]_i$ was also larger in male animals than in females. The non-selective NCX inhibitor KB-R7943 abolished the effect of NCX overexpression, however, failed to annul all gender differences. In contrast, estrogen significantly decreased the $[Ca^{2+}]_i$ and $[Na^+]_i$ rise in male mice and attenuated gender differences, indicating that estrogen is able to protect cardiac myocytes against $[Na^+]_i$ and $[Ca^{2+}]_i$ elevation during metabolic inhibition [32].

The $Ca^{2+}$-handling of ovariectomized rats was investigated by Kravtsov et al. [5]. The ovariectomy did not influence the expression level of the NCX, however, increased $Ca^{2+}$ flux was found via NCX and ryanodine receptor together with enhanced expression of protein-kinase A. These changes suggest that ovariectomy increases contractility, and left ventricular developed pressure [5].

Comparison of the expression levels of different $Ca^{2+}$-handling proteins in healthy male and female rats revealed that female ventricular myocytes have markedly higher level of CaV1.2, RyR and NCX proteins compared to male animals. Similarly, RyR and NCX mRNA were increased in female animals. Contractile properties were compared by using right ventricular papillary muscles which demonstrated faster maximal rate of force development in female rats [4].

In healthy male and female rats, the key $Ca^{2+}$-handling proteins were examined. It was found that NCX, RyR and L-type $Ca^{2+}$ channel mRNA content was higher in female rats [33].

In line with the previous results, it was found that the base of female rabbit hearts exhibited larger $I_{CaL}$ than female apex or males. Estrogen also upregulated $I_{CaL}$ in cultured female myocytes. Mathematical modeling indicated that increased $I_{CaL}$ level increased action potential duration (APD) and promoted arrhythmias. Experimental and modeling data indicates that estrogen upregulates $I_{CaL}$ that promotes APD lengthening and EAD formation [34].

NCX has a crucial role in beat-to-beat $Ca^{2+}$- handling balance, therefore, it intimately influences the contractile force. The effect of male sex hormones on $Ca^{2+}$-balance was investigated on orchidectomized male rats, where it was found that the hypogonadal condition caused 50% decrease in the contraction force which could be partially restored by testosterone supplementation. The orchidectomized rats also exerted lower expression levels of NCX with prolonged relaxation of contraction [35].

## Sex-Related Differences of NCX in Myocardial Ischemia

Myocardial ischemia often develops following occlusion of a coronary artery establishing serious imbalance between blood supply and demand. The deficit in blood flow initiates several alterations in the kinetics of ion channels, intracellular pH, intracellular $Na^+$ and $Ca^{2+}$ levels, extracellular $K^+$ level, changes in the secondary messenger system, release of free radicals that altogether largely increase arrhythmia propensity in the heart [36]. Cross et al. investigated the sex-related effects of NCX overexpression during ischemia–reperfusion in transgenic mice. It was found that transgenic male mice exerted lower cardiac performance than male wild type mice, however, there was no difference among female transgenic versus wild type animals. When bilateral ovariectomized and sham-operated female mice were subjected to ischemia, the cardiac performance was lower in the case of ovariectomized mice indicating the role of sex-related hormones [37]. In another study, arrhythmia incidence between left anterior descending artery (LAD) ligation and sham-operated rats from both sexes was also compared. It was found that male gender was a strong predictor of increased arrhythmia vulnerability [38].

## Sex-Related Changes of NCX in Heart Failure

Heart failure is a complex clinical syndrome leading to impairment of cardiac performance, pump failure and increased susceptibility to serious cardiac arrhythmias. Various structural (hypertrophy, fibrosis), metabolic as well as electrical alterations (including changes in ion channel protein expression and regulation) can be observed in heart failure, collectively termed 'remodeling' [39]. These changes together lead to heterogeneous repolarization and impaired impulse conduction. Repolarizing currents that normally form a strong safety margin by redundant activation ("repolarization reserve") [40] are seriously compromised and attenuated and the resultant impaired repolarization could serve as a substrate for arrhythmias.

Sex differences also exist in the case of heart failure. Heart failure with reduced ejection fraction is more frequent in males together with ischemic etiology, however in the case of females, heart failure with preserved ejection fraction coupled with hypertension or diastolic dysfunction is more often observed. Transgenic overexpressing TNF1.6 mice exhibit heart failure and increased mortality [41]. It was found that female transgenic mice have slower decay of the $Ca^{2+}$-transients, however, the transient amplitude, contraction and response to isoproterenol were identical to wild-type mice. In the case of male mice, the transient decline, amplitude, as well as the contraction and isoproterenol response was significantly reduced compared to wild-type animals.

A ventricular tachypacing-induced heart failure swine model was used to investigate the sex differences in NCX function in heart failure. The control (non-failing) ventricular myocytes exerted identical NCX current and beta-adrenergic responsiveness. However, NCX was upregulated in HF and this remodeling was more pronounced in males than in females, however, the beta-adrenergic responsiveness was smaller in male animals (Fig. 23.4) [42].

## Conclusion

The expression and function of NCX, a crucial component of cardiac intracellular $Ca^{2+}$ handling seems to be significantly influenced by gender. It is feasible that in some pathophysiological settings in females, the upregulated NCX function shows heterogeneous distribution via estrogen-mediated genomic mechanisms, and can be an important contributor to increased risk of delayed afterdepolarization development and therefore, increased arrhythmia propensity.

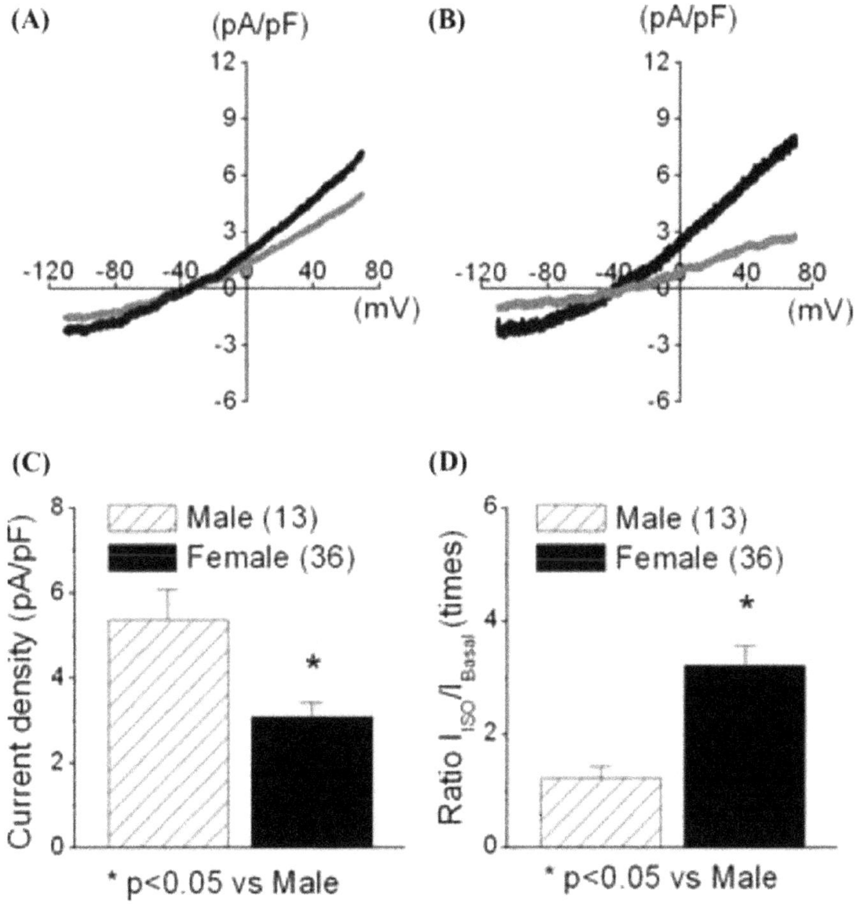


**Fig. 23.4** Gender differences in the NCX current and beta-adrenergic responsiveness in pig heart failure myocytes. Panel A shows NCX currents from male myocytes, panel B from female myocytes. Gray curve illustrates control, black curve illustrates NCX current after isoproterenol treatment. Panel C demonstrates the NCX current density, panel D demonstrates the ratio of outward isoproterenol-induced and basal NCX current. Results indicate that male myocytes have larger NCX current but reduced isoproterenol response in heart failure ([42] with permission)

**Acknowledgements** This work was supported by grants from the Hungarian National Research, Development and Innovation Office (NKFIH FK-142949, K-128851).

## References


1. Eisner D, Bode E, Venetucci L, Trafford A (2013) Calcium flux balance in the heart. J Mol Cell Cardiol 58:110–117
2. Bers DM, Grandi E (2009) Calcium/calmodulin-dependent kinase II regulation of cardiac ion channels. J Cardiovasc Pharmacol 54:180–187
3. Despa S, Bers DM (2013) Na(+) transport in the normal and failing heart - remember the balance. J Mol Cell Cardiol 61:2–10
4. Chu SH, Sutherland K, Beck J, Kowalski J, Goldspink P, Schwertz D (2005) Sex differences in expression of calcium-handling proteins and beta-adrenergic receptors in rat heart ventricle. Life Sci 76:2735–2749
5. Kravtsov GM, Kam KW, Liu J, Wu S, Wong TM (2007) Altered $Ca^{(2+)}$ handling by ryanodine receptor and $Na^{(+)}$-$Ca^{(2+)}$ exchange in the heart from ovariectomized rats: role of protein kinase A. Am J Physiol Cell Physiol 292:C1625–C1635
6. Yaras N, Tuncay E, Purali N, Sahinoglu B, Vassort G, Turan B (2007) Sex-related effects on diabetes-induced alterations in calcium release in the rat heart. Am J Physiol Heart Circ Physiol 293:H3584–H3592
7. Bupha-Intr T, Wattanapermpool J (2006) Regulatory role of ovarian sex hormones in calcium uptake activity of cardiac sarcoplasmic reticulum. Am J Physiol Heart Circ Physiol 291:H1101–H1108
8. Xin HB, Senbonmatsu T, Cheng DS, Wang YX, Copello JA, Ji GJ et al (2002) Oestrogen protects FKBP12.6 null mice from cardiac hypertrophy. Nature 416:334–338
9. Lai LP, Su MJ, Lin JL, Lin FY, Tsai CH, Chen YS et al (1999) Down-regulation of L-type calcium channel and sarcoplasmic reticular $Ca^{(2+)}$-ATPase mRNA in human atrial fibrillation without significant change in the mRNA of ryanodine receptor, calsequestrin and phospholamban: an insight into the mechanism of atrial electrical remodeling. J Am Coll Cardiol 33:1231–1237
10. Dong H, Dunn J, Lytton J (2002). Stoichiometry of the Cardiac $Na^+/Ca^{2+}$ exchanger NCX1.1 measured in transfected HEK cells. Biophys J 82(4):1943–1952
11. Liao J, Li H, Zeng W, Sauer DB, Belmares R, Jiang Y (2012) Structural insight into the ion-exchange mechanism of the sodium/calcium exchanger. Sci 335:686–690
12. Ren X, Philipson KD (2013) The topology of the cardiac $Na^+/Ca^{2+}$ exchanger, NCX1. J Mol Cell Cardiol 57:68–71
13. Philipson KD, Nicoll DA, Ottolia M, Quednau BD, Reuter H, John S et al (2002) The $Na^+/Ca^{2+}$ exchange molecule: an overview. Ann N Y Acad Sci 976:1–10
14. Hilgemann DW, Matsuoka S, Nagel GA, Collins A (1992) Steady-state and dynamic properties of cardiac sodium-calcium exchange. Sodium-dependent inactivation. J Gener Physiol 100:905–932
15. Despa S, Brette F, Orchard CH, Bers DM (2003) Na/Ca exchange and Na/K-ATPase function are equally concentrated in transverse tubules of rat ventricular myocytes. Biophys J 85:3388–3396
16. Acsai K, Antoons G, Livshitz L, Rudy Y, Sipido KR (2011) Microdomain $[Ca^{(2)(+)}]$ near ryanodine receptors as reported by L-type $Ca^{(2)(+)}$ and $Na^+/Ca^{(2)(+)}$ exchange currents. J Physiol 589:2569–2583
17. Larbig R, Torres N, Bridge JH, Goldhaber JI, Philipson KD (2010) Activation of reverse $Na^+$-$Ca^{2+}$ exchange by the Na+ current augments the cardiac $Ca^{2+}$ transient: evidence from NCX knockout mice. J Physiol 588:3267–3276
18. Neco P, Rose B, Huynh N, Zhang R, Bridge JH, Philipson KD et al (2010) Sodium-calcium exchange is essential for effective triggering of calcium release in mouse heart. Biophys J 99:755–764
19. Torres NS, Larbig R, Rock A, Goldhaber JI, Bridge JH (2010) Na+ currents are required for efficient excitation-contraction coupling in rabbit ventricular myocytes: a possible contribution of neuronal Na+ channels. J Physiol 588:4249–4260

20. Armoundas AA, Hobai IA, Tomaselli GF, Winslow RL, O'Rourke B (2003) Role of sodium-calcium exchanger in modulating the action potential of ventricular myocytes from normal and failing hearts. Circ Res 93:46–53
21. Kohajda Z, Farkas-Morvay N, Jost N, Nagy N, Geramipour A, Horvath A et al (2016) The effect of a novel highly selective inhibitor of the sodium/calcium exchanger (NCX) on cardiac arrhythmias in in vitro and in vivo experiments. PLoS ONE 11:e0166041
22. Pogwizd SM, Qi M, Yuan W, Samarel AM, Bers DM (1999) Upregulation of $Na^{(+)}/Ca^{(2+)}$ exchanger expression and function in an arrhythmogenic rabbit model of heart failure. Circ Res 85:1009–1019
23. Studer R, Reinecke H, Bilger J, Eschenhagen T, Bohm M, Hasenfuss G et al (1994) Gene expression of the cardiac $Na^{(+)}$-$Ca^{2+}$ exchanger in end-stage human heart failure. Circ Res 75:443–453
24. Dipla K, Mattiello JA, Margulies KB, Jeevanandam V, Houser SR (1999) The sarcoplasmic reticulum and the $Na^{+}/Ca^{2+}$ exchanger both contribute to the $Ca^{2+}$ transient of failing human ventricular myocytes. Circ Res 84:435–444
25. Hobai IA, O'Rourke B (2000) Enhanced $Ca^{(2+)}$-activated $Na^{(+)}$-$Ca^{(2+)}$ exchange activity in canine pacing-induced heart failure. Circ Res 87:690–698
26. Imahashi K, Kusuoka H, Hashimoto K, Yoshioka J, Yamaguchi H, Nishimura T (1999) Intracellular sodium accumulation during ischemia as the substrate for reperfusion injury. Circ Res 84:1401–1406
27. Takahashi K, Takahashi T, Suzuki T, Onishi M, Tanaka Y, Hamano-Takahashi A et al (2003) Protective effects of SEA0400, a novel and selective inhibitor of the $Na^{+}/Ca^{2+}$ exchanger, on myocardial ischemia-reperfusion injuries. Eur J Pharmacol 458:155–162
28. Kormos A, Nagy N, Acsai K, Váczi K, Ágoston S, Pollesello P et al (2014) Efficacy of selective NCX inhibition by ORM-10103 during simulated ischemia/reperfusion. Eur J Pharmacol 740:539–551
29. Chen G, Yang X, Alber S, Shusterman V, Salama G (2011) Regional genomic regulation of cardiac sodium-calcium exchanger by oestrogen. J Physiol 589:1061–1080
30. Golden KL, Marsh JD, Jiang Y (2004) Testosterone regulates mRNA levels of calcium regulatory proteins in cardiac myocytes. Hormone and metabolic research = Hormon- und Stoffwechselforschung = Hormones et metabolism 36:197–202
31. Papp R, Bett GCL, Lis A, Rasmusson RL, Baczko I, Varro A et al (2017) Genomic upregulation of cardiac Cav1.2alpha and NCX1 by estrogen in women. Biol Sex differ 8:26
32. Sugishita K, Su Z, Li F, Philipson KD, Barry WH (2001) Gender influences $[Ca^{(2+)}](i)$ during metabolic inhibition in myocytes overexpressing the $Na^{(+)}$-$Ca^{(2+)}$ exchanger. Circ 104:2101–2106
33. Tappia PS, Dent MR, Aroutiounova N, Babick AP, Weiler H (2007) Gender differences in the modulation of cardiac gene expression by dietary conjugated linoleic acid isomers. Can J Physiol Pharmacol 85:465–475
34. Kalik ZM, Mike JL, Slipski C, Wright M, Jalics JZ, Womble MD (2017) Sex and regional differences in rabbit right ventricular L-type calcium current levels and mathematical modelling of arrhythmia vulnerability. Exp Physiol 102:804–817
35. Witayavanitkul N, Woranush W, Bupha-Intr T, Wattanapermpool J (2013) Testosterone regulates cardiac contractile activation by modulating SERCA but not NCX activity. Am J Physiol Heart Circ Physiol 304:H465–H472
36. Carmeliet E (1999) Cardiac ionic currents and acute ischemia: from channels to arrhythmias. Physiol Rev 79:917–1017
37. Cross HR, Lu L, Steenbergen C, Philipson KD, Murphy E (1998) Overexpression of the cardiac $Na^{+}/Ca^{2+}$ exchanger increases susceptibility to ischemia/reperfusion injury in male, but not female, transgenic mice. Circ Res 83:1215–1223
38. Okninska M, Paterek A, Bierla J, Czarnowska E, Maczewski M, Mackiewicz U (2021) Effect of age and sex on the incidence of ventricular arrhythmia in a rat model of acute ischemia. Biomed Pharmacother = Biomedecine & pharmacotherapie 142:111983

39. Husti Z, Varró A, Baczkó I (2021). Arrhythmogenic remodeling in the failing heart. Cell 10(11):3203
40. Varro A, Baczko I (2011) Cardiac ventricular repolarization reserve: a principle for understanding drug-related proarrhythmic risk. Br J Pharmacol 164:14–36
41. Janczewski AM, Kadokami T, Lemster B, Frye CS, McTiernan CF, Feldman AM (2003) Morphological and functional changes in cardiac myocytes isolated from mice overexpressing TNF-alpha. Am J Physiol Heart Circ Physiol 284:H960–H969
42. Wei SK, McCurley JM, Hanlon SU, Haigney MC (2007) Gender differences in Na/Ca exchanger current and beta-adrenergic responsiveness in heart failure in pig myocytes. Ann N Y Acad Sci 1099:183–189

# Chapter 24
# Role of Estrogen in Attenuating Apoptosis and Cardiac Dysfunction in Female Heart Failure

**Sukhwinder K. Bhullar, Karina Oliveira Mota, Carla Maria Lins de Vasconcelos, and Naranjan S. Dhalla**

**Abstract** Cardiovascular diseases including heart failure are the major cause of morbidity and mortality in both males and females of all ages and ethnicities. Premenopausal females are relatively protected from cardiovascular diseases compared to age-matched males; however, the risk of these health hazards increases in females substantially after menopause and becomes equal to that in males. Given the accelerated heart failure risk associated with the onset of menopause, sex hormones such as estrogen have been postulated to explain cardioprotection in premenopausal females. By acting on specific receptors through genomic and non-genomic signaling mechanisms, estrogen is considered to protect against cardiac dysfunction and associated apoptotic cell death in females under pathological conditions including myocardial infarction, pressure overload and volume overload. Accordingly, this chapter is intended to highlight some observations which support the viewpoint regarding the protective role of estrogen in the development of cardiac dysfunction and prevention or slowing of apoptosis in female failing hearts. Since the clinical impact of estrogen therapy in heart failure is of a significant concern, some experimental evidence for the estrogen treatment on apoptosis and cardiac dysfunction in female heart failure due to volume overload with or without ovariectomy has been presented. It is concluded that estrogen therapy offers partial cardioprotective effects against heart failure and cardiac remodeling due to volume overload in females. Accordingly, it is proposed to develop a combination therapy with other ovarian hormones to delay the progression of heart failure in postmenopausal females.

---

S. K. Bhullar · N. S. Dhalla (✉)
Department of Physiology and Pathophysiology, Max Rady College of Medicine, Institute of Cardiovascular Sciences, St. Boniface Hospital Albrechtsen Research Centre, University of Manitoba, Winnipeg, MB R2H 2A6, Canada
e-mail: nsdhalla@sbrc.ca

K. O. Mota · C. M. L. de Vasconcelos
Department of Physiology, Heart Biophysics Laboratory, Center for Biological and Health Sciences, Federal University of Sergipe, Sergipe, Brazil

© The Author(s), under exclusive license to Springer Nature Switzerland AG 2023 
L. Kirshenbaum and I. Rabinovich-Nikitin (eds.), *Biology of Women's Heart Health*, Advances in Biochemistry in Health and Disease 26,
https://doi.org/10.1007/978-3-031-39928-2_24

**Keywords** Female heart failure · Estrogen therapy · Ovariectomy · Menopause · Cardiac dysfunction · Cardiomyocyte apoptosis

## Introduction

It is well known that cardiac dysfunction occurs in response to several molecular, cellular, and interstitial alterations that manifest cardiac remodeling under both ischemic and non-ischemic conditions [1–3]. Furthermore, cardiac dysfunction leads to heart failure due to a variety of mechanisms including cell death [4–6], changes in energy metabolism [7], oxidative stress [8], inflammation and fibrosis [9], defects in contractile proteins [10], calcium transport abnormalities [11], alterations in ventricular structure [12, 13] and prolonged neurohormonal activation [14, 15]. It is noteworthy that heart failure is a growing public health hazard [16] and the worldwide prevalence of death due to this cause of illness is estimated to be 64.34 million people annually. In fact, heart failure poses a considerable burden after 60 years of age as the global expenditure for managing patients suffering from this disease is estimated to reach approximately $400 billion in 2030. The overall lifetime risk of heart failure was considered to be 21% for men and 20% for women at the age of 40 years, 33% for men, and 29% for women at the age of 55 years. However, over the past several years, the prevalence of morbidity and mortality in patients with heart failure has been constantly higher in females but its impact in different age groups is significantly varied [16, 17].

It is now evident that the prevalence of cardiovascular disease risk in young females is low and the symptoms for heart failure usually occur more than a decade later than in males [18–21]. However, females after natural or surgical menopause or in mid-life during the menopausal transition experience adverse changes in cardiovascular and cardiometabolic health simultaneously [22–29]; the prevalence of premature (3.7%) and early menopause (12.2%) in women is considerable. Estimated data has shown that 1.2 billion women worldwide will be perimenopausal or postmenopausal by 2030 with 47 million women becoming menopausal each year and at risk of heart disease [30, 31]. In fact, monitoring and management of women's health during midlife is considered to be a key in preventing the increased incidence and severity of cardiac disease [24, 32–34]. The mechanisms responsible for cardioprotection observed in young females and its loss after menopause and subsequent cessation of ovarian function are not clear; however, it has been suggested that estrogen may be responsible in this regard [27, 35–44]. Additionally, compared to males, there is also a lower incidence of cardiomyocyte apoptosis in different cardiac diseases in female failing hearts, which is considered to contribute to slowing the development of cardiac dysfunction and delaying the progress of heart failure [45–51]. It is believed that the occurrence of lower rates of apoptosis in premenopausal females could either be owing to the presence of apoptosis-related genes on the Y chromosomes or attributed to the estrogen [52, 53].

This chapter is intended to summarize the knowledge concerning differences between males and females for the occurrence of cardiomyocytes apoptosis, changes in the pro-and anti-apoptotic proteins as well as the development of cardiac dysfunction in various types of cardiovascular diseases, including heart failure. Since menopause contributes significantly to enhancing the cardiovascular risk in females, and the beneficial effects of hormone therapy in patients are controversial, we have attempted to provide some information regarding the role of estrogen in cardioprotection against the volume overload induced heart failure in animals with or without ovariectomy [34, 54–56]. In view of the observations that cardiomyocyte apoptosis may be a potential therapeutic target for delaying the development of cardiac dysfunction in female heart failure [57–59], some evidence has been presented to show association between alterations in the degree of apoptosis and cardiac dysfunction in female failing hearts upon the induction of volume overload.

## Sex Differences in Heart Failure

Since male and female cardiovascular systems differ significantly in many characteristics under pathological conditions, there are dramatic sex differences in the development and incidence of heart disease [60–70]. Almost all facets of heart failure from epidemiology to pathophysiology as well as responses to therapy and ultimate outcomes in females are affected differently compared to males [16, 54, 55, 71–77]. Figure 24.1 shows considerable differences of pathological events and general characteristics in female failing hearts in comparison to those in male hearts under various stressful situations. Although some risk factors such as menopause, pregnancy and hypertensive disorders are the major determinants of differences in female and male failing hearts, several other risk factors including obesity [78, 79], smoking [80], socioeconomic status [81, 82], aging [22, 83], cancer therapy [84, 85] as well as various cardiomyopathies [86], including Takotsubo syndrome [87, 88] and peripartum [89] are common for developing heart failure in both sexes.

It should also be mentioned that there is a prevalence of preserved ejection fraction (HFpEF) and occurrence of concentric ventricular remodeling in female failing hearts. Primarily, females are older than men when they have their first episode of myocardial infarction [90–92]. Similarly, heart failure of ischemic and nonischemic origin is generally considered to be a male disease, and in response to prolonged coronary ischemia, female hearts are protected in comparison to males [92, 93]. Women predominantly present cardiac dysfunction and heart failure in association with hypertension and diabetes [94–96]. Compared to males, cardiac function [57, 97], left ventricular size, and ejection fraction are preserved [98, 99] in females when challenged with myocardial infarction [90, 100], dilated cardiomyopathy [101–103], atherosclerosis [104], myocardial ischemia and ischemia/reperfusion injury [105–107], pressure overload [50, 108, 109] and volume overload [57, 110, 111]. Since, the hormonal profile varies between males and females, protective effects of estrogen in the failing heart at multiple levels, as well as on cardiac dysfunction are more

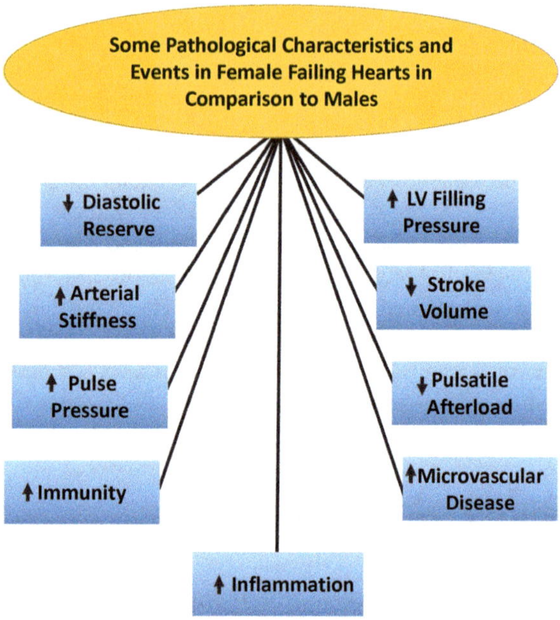


**Fig. 24.1** Some pathological characteristics and events of female hearts

evident in females than males of comparable age [112–114]. Furthermore, the degree of apoptotic myocyte death is markedly lower in female failing hearts than in males [48–51]. These observations suggest that the development of heart failure in females occurs differently than in males and estrogen plays a critical role in determining the pathophysiology of cardiac disease in females.

## Effects of Estrogen and Cardiac Dysfunction in Female Heart Failure

Estrogen is known to affect female cardiovascular function during the development of different diseases in several manners [36, 42, 115]. By acting on specific G-protein coupled estrogen receptors (α and β) as well as through genomic and non-genomic signaling mechanisms, this hormone protects females against cardiac dysfunction and progression of heart failure due to various pathological stimuli. These receptors are widely expressed in different species ranging from mice to humans. The binding of estrogen (particularly, 17β-estradiol) to the estrogen receptors has been shown to promote homo/hetero dimers that translocate to the nucleus and bind directly to gene estrogen response elements for the regulation of gene transcription [116–119]. At the plasma membrane level or in the cytosol, estrogen receptors interact with various kinases and scaffolding molecules to modulate multiple signal pathways for

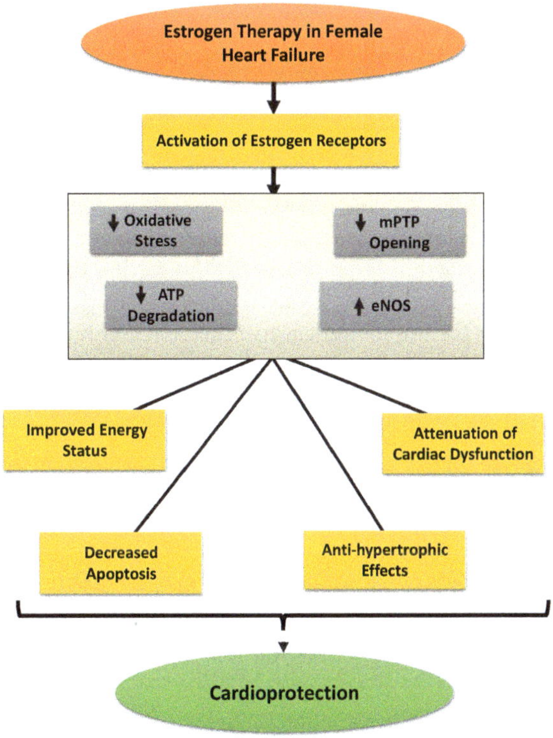


**Fig. 24.2** Mechanism of the protective effects of estrogen in the development of heart failure. ATP, adenosine triphosphate; mPTP, mitochondrial permeability transition pores; eNOS, extracellular nitric oxide synthase

exerting protective effects against the development of heart failure (Fig. 24.2) [117, 120–122].

It is noteworthy that estrogen exerts protective effects on the myocardium in females against various pathological conditions such as myocardial infarction as well as ischemia/reperfusion injury and promotes delay in heart failure in addition to improving survival and a better prognosis after myocardial infarction [123, 124]. Since female myocardium shows greater resistance to ischemia/reperfusion injury in different animal models, enhanced recovery of contractile function and lesser arrhythmias during reperfusion have been observed in young adult female rats than the age-matched males [107, 123, 125–127]. Also, estrogen has been shown to exert protection against ischemia/reperfusion injury via activation of the mitochondrial and sarcolemma ATP-sensitive $K^+$ channels [128, 129]. Estrogen treatment reduces oxidative stress [130, 131] and mitochondrial permeability transition pore opening (mPTP) [132–134] in addition to increasing the endothelial nitric oxide synthase (eNOS) activity [135]. Inhibition of nuclear factor-κB (NF-κB) as well as reduction of infarcted area and neutrophil accumulation were also reported by estrogen therapy [136]. Additionally, estrogen has been shown to modulate manganese superoxide dismutase phosphorylation through mitochondrial p38β at Threonine 79 and Serine 106 levels [123]. Furthermore, estrogen-induced anti-hypertrophic effects were demonstrated [137–139] in slowing down cardiac dysfunction, development of cardiac hypertrophy and progression of heart failure [140]. By suppressing $Ca^{2+}$/calmodulin-dependent protein kinase II [137, 141], estrogen was observed to inhibit cardiac hypertrophy. Moreover, cardiomyocyte apoptosis was attenuated by estrogen and the activation of estrogen receptors [52, 142–144] through signaling system such as PI3K/Akt pathway and estrogen-mitogen-activated protein kinase (MAPK) and protein kinase G interaction [145, 146]. Since the protection of female from heart disease is lost after menopause or during menopause transition [37, 38], estrogen has been claimed to be protective against cardiac dysfunction [39, 147]. These observations are consistent with the view that estrogen plays a role in cardioprotection in females for improving cardiac function and reducing the incidence of heart failure in females.

## Mechanisms of Modification of Cardiomyocyte Apoptosis in Female Failing Hearts by Estrogen Treatment

Although the pathogenesis of heart failure remains unclear, cardiomyocyte apoptosis is considered to be among the critical factors involved and implicated in cardiac dysfunction [46, 47]. Since the transition from compensated to decompensated hypertrophy is associated with cardiac remodeling upon restructuring of the ventricular wall and chamber dilation of the postinfarcted heart [90, 148], prevention of cell death can be seen to affect the ischemic damage to maintain ventricular structure and function [12, 13, 149]. It has also been observed that under pathological conditions, there is a lower incidence of cardiomyocyte apoptosis in the female's failing heart, attenuation of cardiac dysfunction, and delay in the progression of heart failure compared to males [45, 49, 50, 66–68, 150–153]. The possibility of reduced myocyte death in women due to stressful stimuli may either be due to inactivation of pro-apoptotic signal pathway or activation of anti-apoptotic pathway by the endogenous sex hormones such as estrogen. The role of insulin-like growth factor-1 (IGF-1) has also been suggested in improving cell survival under these conditions [154]. It needs to be emphasized that among all the programmed cell death mechanisms, which eliminate abnormal cells and maintain cellular homeostasis during the development of different diseases [155–161], apoptosis is believed to play the most significant role in cardiovascular disease [162–165]. Various stimuli such as deprivation of growth factors [166], hypoxia [167], oxidative stress [168], intracellular $Ca^{2+}$-overload [11], inflammatory cytokines [169, 170] and nitric oxide [171], DNA injury [172] heat shock [173], radiation [174], viral infection [175] and aging [176] are involved in the utilization of complex molecular machinery for the initiation, transduction, and execution of cell death as well as the occurrence of apoptosis [177–183]. Progressive apoptosis in the myocardium, limited capacity of the myocardium for self-renewal and inability of the heart to sustain efficient contractile function leading to ischemic damage as well as adverse cardiac remodeling and cardiac dysfunction in failing hearts have been well documented [46, 47, 184–189].

A variety of pathological conditions including hypertension [190], myocardial infarction [191–193], ischemia/reperfusion [194, 195] and dilated cardiomyopathy or end-stage heart failure [46] have been shown to initiate abnormalities in various signaling pathways for the occurrence of cardiomyocyte apoptosis and cardiac dysfunction in failing hearts [196–207]. It has been indicated that the inhibition of apoptotic cardiomyocyte death upon modulation of signaling pathways by different interventions improves cardiac function in diseased hearts [208–224]. Indeed, the role of estrogen in inhibiting apoptosis and cardiac dysfunction in female heart failure has been indicated under several pathological conditions [50, 51, 103, 202, 225–228]. In this regard, estrogen therapy was shown to attenuate cardiac remodeling, lower content of pro-apoptotic proteins, such as caspases 3 and 9, and suppress cardiomyocyte apoptosis in female heart failure after pressure overload [50, 227]. Estrogen was also demonstrated to markedly reduce the severe pulmonary hypertension in rats by reducing endothelial cell apoptosis due to pressure overload [229]. Moreover, estrogen demonstrated a reduced degree of apoptosis [202] and prevention of cardiac dysfunction as well as adverse ventricular remodeling in female rats subjected to volume overload [51]. This hormone has been shown to attenuate cardiomyocyte apoptosis and the development of congestive heart failure upon improving cardiac contractility [140]. Furthermore, it was found to inhibit apoptosis signal-regulating kinase 1/c-Jun N-terminal kinase/p38 as well as MAPK through antioxidative effects [203]; long-term 17β-estradiol treatment was also shown to improve congestive heart failure by these mechanisms [103]. By activating thioredoxin, inhibiting NADPH oxidase activity, and reducing oxidative stress in the heart, estrogen was observed to prevent progressive cardiac enlargement in a genetic mouse model of congestive heart failure [151]. Administration of estrogen attenuated the generation of reactive oxygen species and reduced apoptosis by suppressing ischemia/reoxygenation-induced oxidative stress in the myocardium [140, 230, 231].

Under the ischemic or hemodynamic overload conditions, administration of 17β-estradiol was observed to reduce cardiac apoptosis and related adverse ventricular remodeling in post-menopausal females because antiapoptotic actions of estrogen promote the prevention of oxidative stress that produce unstable, reactive oxygen species to trigger apoptosis [231]. Differential effects of 17β-estradiol have shown to suppress oxidative stress to some extent via modulation of p38 MAPK isoforms in rat cardiomyocytes [232]. Furthermore, 17β-estradiol reduced pro-apoptotic p38α and its downstream target p53, which are the primary regulator of cell death while promoting the activity of pro-survival protein p38β through PI3K/Akt signaling [145, 231]. Estrogen was also found to reduce myocyte apoptosis by activating pro-survival PI3K/Akt signaling in cultured neonatal cardiomyocytes, suggesting that estrogen may improve survival partially in female heart failure [145]. The interaction of estrogen receptor α with miR-22, and Sp-1 was reported to enhance cystathionine-lyase expression in the myocardium of female rats, increase antioxidative defense and promote cell survival, which may prevent apoptosis in cardiomyocytes [131]. Increased cardiomyocyte hypertrophy [226] may also prevent the deterioration of cardiac function upon treatment with 17-β estradiol in female rats subsequent to myocardial infarction [233]. Furthermore, 17-β estradiol attenuated the hypertrophic response and cardiomyocyte apoptosis due to pressure overload [50, 139, 226, 227] and volume overload in females [51, 57, 228]. Thus, estrogen therapy in females may prevent cardiomyocyte apoptosis and improve cardiac function by an interplay of several signal transduction mechanisms, which may serve to delay the progression of heart failure.

Given the protection of the heart from the oxidative stress and ischemia–reperfusion insults in females, there is a possibility that estrogen may produce beneficial effects at the level of individual cardiomyocytes in several experimental models of heart failure [50, 51, 57, 145, 225, 234, 235]. Over the past three decades, the incidence of heart failure has been higher in different age groups of female patients and thus there is a great interest in instituting appropriate hormone therapy for the control of risk of this heart disease [236–238]. It is pointed out that several factors such as type of heart failure, menopause timings, age of patients, as well as route and dose of hormone administration are considered carefully before initiating the therapy in females in order to avoid cardiovascular complications. Particularly, contraceptives and postmenopausal hormone therapy are believed to promote the increase in cardiovascular complications in females [239–244]. It has been indicated that oral estrogen may increase circulating angiotensin II, which may cause adverse cardiovascular effects [242]. Furthermore, it has been observed that initiating therapy with estrogen to younger women in their menopause transition period exhibits a lower event rate of coronary disease events than in women of the age 60 and over [243]. It has also been demonstrated that the loss of estrogen may contribute to left ventricular dilation in females receiving postmenopausal hormone therapy [244, 245]. On the basis of some observational research, it has been argued that there a potential survival advantage in female heart failure with postmenopausal hormone treatment [246]. Furthermore, hormone replacement therapy has been shown to reduce myocardial hypertrophy in females. On the other hand, estrogen treatment has been reported to increase the risk of congestive heart failure and it has been indicated that hormone therapy may be initiated for the relief of menopausal symptoms but should not dispensed for the purpose of heart disease protection [247, 248]. Since there is a controversy regarding the estrogen therapy in postmenopausal females, it is suggested that a combination therapy with other ovarian hormones or some existing cardiovascular drugs may prove beneficial for suppressing the incidence of heart failure.

## Experimental Evidence for the Effects of Ovariectomy and Estrogen Therapy on Cardiac Dysfunction and Apoptosis in Female Heart Failure Due to Volume Overload

In order to gain some information regarding the role of estrogen in protecting the development of heart failure in females, we have examined the effects of estrogen therapy on cardiac dysfunction, cardiac remodeling, degree of apoptosis and plasma catecholamines in failing hearts due to volume overload. Heart failure for these experiments was induced by arteriovenous (AV) shunt in female rats with intact ovaries as well as in ovariectomized animals for a period of 16 weeks [51, 57] and the results are shown in Figs. 24.3 and 24.4 as well as Tables 24.1 and 24.2. It can be seen from Fig. 24.3 that the volume overload induced cardiac hypertrophy (as represented by heart weight/body weight ratio) and fractional shortening were attenuated whereas cardiac output was increased without any significant changes in heart rate by ovariectomy. These alterations in ovariectomized animals were fully or partially reversed by estrogen treatment. It can also be observed that heart weight to body weight ratio, cardiac output and heart rate in sham control animals were not affected by ovariectomy or estrogen therapy except that fractional shortening in ovariectomized animals was depressed significantly, which effect was not reversed by estrogen treatment. Ovariectomy has also been reported to decrease +dP/dt, −dP/dt and left ventricular pressure as well as increase left ventricular end diastolic pressure in volume overload female hearts; these effects were also fully or partially reversed by treatment with estrogen [57].

Figure 24.4 shows the effects of ovariectomy and estrogen therapy on some parameters of cardiac remodeling in volume overload female rats with intact ovaries and ovariectomized animals. Both systolic and diastolic ventricular internal diameters were increased in volume overload ovariectomized animals and estrogen therapy suppressed the diastolic, but not the systolic, internal diameter significantly. Furthermore, ovariectomy significantly depressed the systolic, but not the diastolic, thickness of posterior wall significantly; this effect of ovariectomy on systolic posterior wall thickness was partially reversed by estrogen treatment. Since the status of cardiac function and cardiac remodeling is determined by changes in the sympathetic activity in heart failure [57, 249, 250], we have also determined the plasma levels of catecholamines. The data shown in Table 24.1 indicate that ovariectomy depressed the plasma levels of both epinephrine and norepinephrine but increase the plasma level of dopamine in the volume overload animals. These changes in both epinephrine and dopamine levels, unlike that for norepinephrine levels, were significantly reversed by estrogen treatment.

In view of the important role of cardiomyocyte apoptosis in determining the status of cardiac dysfunction in diseased hearts, we determined changes in the degree of apoptosis and associated caspases enzymes as well as both pro-apoptotic and anti-apoptotic protein contents due to ovariectomy and estrogen therapy in the volume overload hearts. The results in Table 24.2 show the degree of apoptosis as well

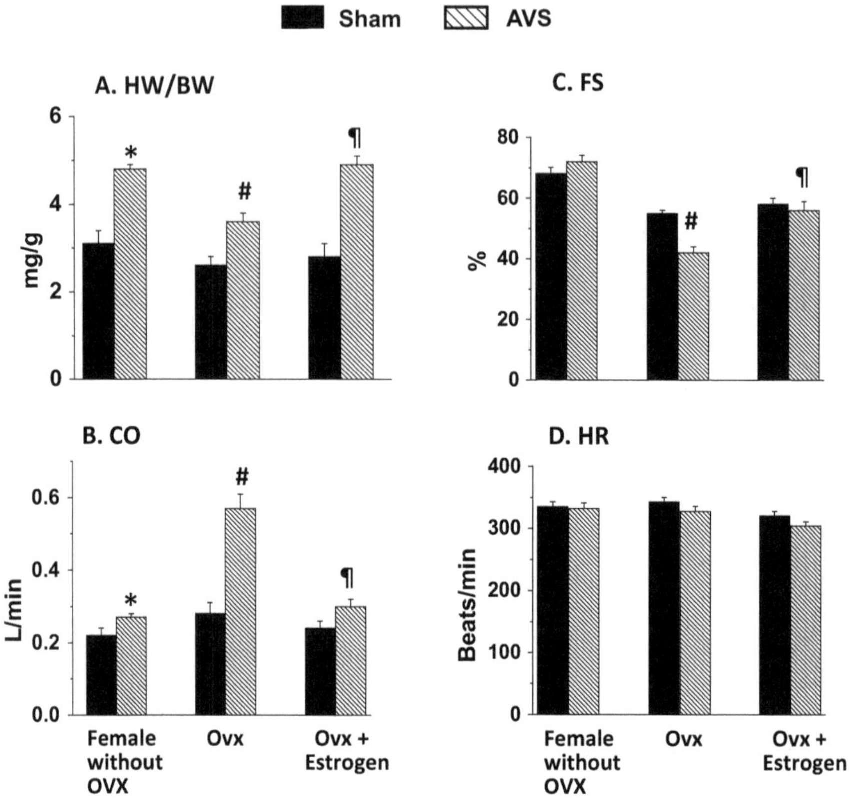


**Fig. 24.3** Effects of OVX in control female rats as well as estrogen therapy on cardiac dysfunction in OVX animals 16 weeks post AV shunt. Values are mean ± S.E. of 12 animals in each group. AV, arteriovenous shunt; OVX, ovariectomized; BW, body weight; HW, heart weight; CO, cardiac output; FS, fractional shortening; HR, heart rate. $*P < 0.05$ versus corresponding values for female sham without OVX; $\#P < 0.05$ versus corresponding value for AV female without OVX; $P < 0.05$ versus corresponding value for AV OVX animals without estrogen treatment. Data are based on the results from our paper—Dent et al. [57]

as protein content for both caspase 3 and 9 were increased by ovariectomy and these effects were fully reversed by estrogen treatment. Furthermore, protein content for pro-apoptotic proteins, such as Bax and phosphorylated Bad, were increased whereas anti-apoptotic protein, phosphorylated Bcl-2, content were decreased due to ovariectomy in volume overload hearts. These effects on pro- and anti-apoptotic proteins were fully or partially reversible by estrogen treatment. It should also be pointed out that the degree of apoptosis as well as caspase 3 and 9 protein content in volume overload females were less than those in the age-matched males [51]. Furthermore, mRNA levels for both caspases 3 and 9 in volume overload females were also less than those in the age-matched males [51]. Although these studies support the role of estrogen in cardioprotection during heart failure due to volume

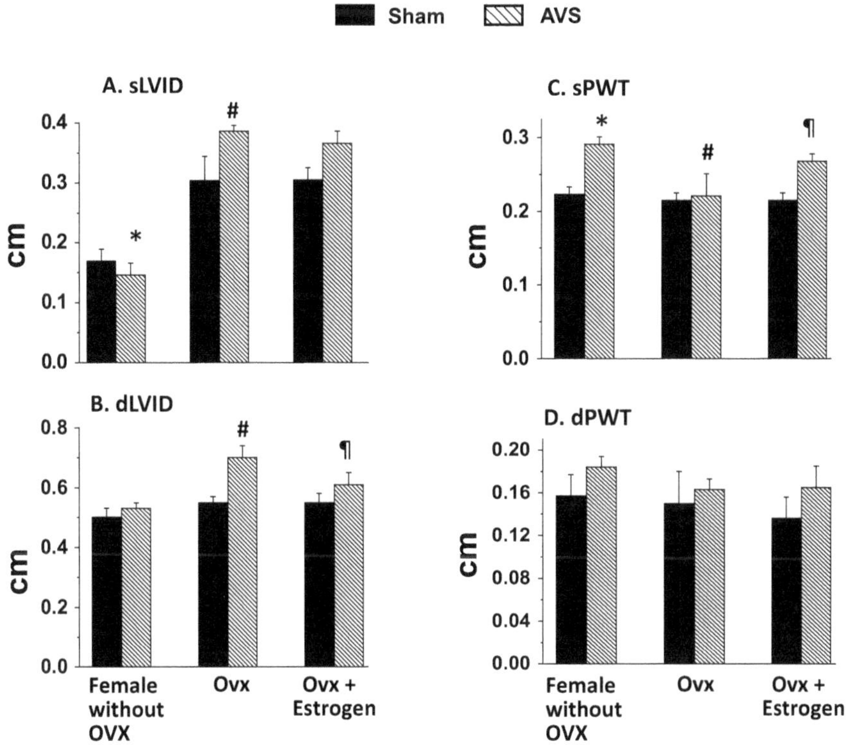


**Fig. 24.4** Effects of OVX in control female rats as well as estrogen therapy on cardiac remodeling in OVX animals 16 weeks post AV shunt. Values are mean ± S.E. of 12 animals in each group. AV, arteriovenous shunt; OVX, ovariectomized; sLVID, systolic left ventricular internal diameter; dLVID, diastolic left ventricular internal diameter; sPWT, systolic posterior wall thickness; dPWT, diastolic posterior wall thickness. *$P < 0.05$ versus corresponding value for female sham without OVX; #P < 0.05 versus corresponding value for AV female without OVX; $P < 0.05$ versus corresponding value for AV OVX animals without estrogen treatment. Data are based on the results from our paper—Dent et al. [57]

**Table 24.1** Catecholamine plasma levels in female without OVX as well as in OVX animals treated with or without estrogen post AV shunt at 16 weeks

| | Without OVX | | OVX | | OVX + Estrogen | |
|---|---|---|---|---|---|---|
| | Sham | AV | Sham | AV | Sham | AV |
| Epinephrine | 745 ± 64 | 687 ± 32 | 556 ± 30 | 247 ± 25# | 375 ± 27# | 322 ± 15 |
| Norepinephrine | 679 ± 9.2 | 457 ± 38* | 394 ± 25 | 236 ± 18# | 476 ± 25 | 297 ± 26 |
| Dopamine | 101 ± 5 | 19 ± 6 | 101 ± 6 | 137 ± 7# | 173 ± 10 | 100 ± 6 |

AV, arteriovenous shunt; OVX, ovariectomized. Data are mean ± S.E. of 5 experiments performed with different preparations. Data are based on the results from our paper—Dent et al. [57]. * $P < 0.05$ versus corresponding value for sham female without OVX; #$P < 0.05$ versus corresponding value for AV female without OVX; $P < 0.05$ versus corresponding value for OVX without estrogen treatment

**Table 24.2** Changes in the degree of apoptosis (% of control), caspases 3 and 9 as well as pro-and anti-apoptotic proteins content and (Densitometric Units) in female without OVX, OVX animals treated with or without estrogen post AV shunt at 16 weeks

| | Without OVX | | OVX | | OVX + Estrogen | |
|---|---|---|---|---|---|---|
| | Sham | AV | Sham | AV | Sham | AV |
| Degree of Apoptosis | 100 ± 0.02 | 72 ± 7* | 100 ± 0.02 | 215.3 ± 24# | 100 ± 0.07 | 49.9 ± 8 |
| Caspase3(35 kDa) | 2.16 ± 0.12 | 1.54 ± 0.08* | 1.45 ± 0.07 | 1.87 ± 0.05# | 1.94 ± 0.04 | 1.56 ± 0.05 |
| Caspase9(51 kDa) | 2.74 ± 0.1 | 2.0 ± 0.11* | 2.83 ± 0.2 | 3.03 ± 0.4# | 3 ± 0.4 | 1.56 ± 0.15 |
| BAX (20 kDa) | 5.5 ± 0.22 | 5.36 ± 0.22 | 5.08 ± 0.35 | 7.45 ± 0.42# | 6.46 ± 0.4 | 6 ± 0.4 |
| Bad (23 kDa) | 4 ± 0.2 | 4.5 ± 0.4 | 4.6 ± 0.3 | 4.4 ± 0.22 | 4.2 ± 0.4 | 2.9 ± 0.11 |
| P-Bad(23 kDa) | 4 ± 0.2 | 4.2 ± 0.22 | 4.4 ± 0.3 | 5.1 ± 0.22# | 2.89 ± 0.1 | 2.97 ± 0.11 |
| Bcl-2 (28 kDa) | 1.95 ± 0.09 | 1.85 ± 0.1 | 2.18 ± 0.2 | 1.78 ± 0.2 | 1.065 ± 0.11 | 0.9 ± 0.15 |
| P-Bcl-2(28 kDa) | 3 ± 0.18 | 4.6 ± 0.2* | 5.1 ± 0.3 | 3.2 ± 0.4# | 3.75 ± 0.3 | 4.8 ± 0.4 |

AV, arteriovenous shunt; OVX, ovariectomized. Data are mean ± S.E. of 5 experiments performed with different preparations. Data are based on the results from our paper – Dent MR, Tappia PS and Dhalla NS. Apoptosis 15: 499–510, 2010 [51]. * $P < 0.05$ versus corresponding value for sham female without OVX; #$P < 0.05$ versus corresponding value for AV female without OVX; $P < 0.05$ versus corresponding value for AV OVX without estrogen treatment

overload in females, it is noteworthy that alterations in hemodynamic parameters, cardiac remodeling and changes in the degree of apoptosis as well as content of pro-apoptotic and anti-apoptotic proteins due to ovariectomy in volume overload animals were either fully or partially reversible by estrogen. However, the role of various other hormones in offering cardioprotection due to heart failure in females cannot be ruled out at present.

## Conclusion

From the foregoing discussion, it is evident that premenopausal women are less prone to heart disease including the development of heart failure in comparison to men. Such a difference is reflected by several characteristics such as decreased diastole reserve, stroke volume and pulsatile afterload as well as increased arterial stiffness, pulse pressure and left ventricular filling pressure in addition to increased immunity, inflammation and microvascular disease in females under various pathological conditions. Furthermore, these differences between males and females for the development of cardiac remodeling, loss of cardiomyocyte due to apoptosis and heart failure as a consequence of various pathological stimuli including myocardial infarction, pressure overload, and volume overload have been attributed to differences in sex hormones such as estrogen. This viewpoint is supported by the fact that the plasma levels of estrogen in young female are markedly higher than those in males. Furthermore, the incidence of heart failure in females becomes equal or greater than that of males after the initiation of menopause, which is associated with development of ovarian dysfunction and depression in the level of plasma estrogen.

The role of estrogen as a cardioprotective hormone is also attested in several experimental studies with female animals which show greater resistance for the development of heart failure due to risk factors such as diabetes, hypertension and aging in comparison to males. Female animal hearts were also observed to exhibit less cardiac defects such as cardiac dysfunction, metabolic changes, cellular damage upon subjecting to myocardial ischemia, ischemia–reperfusion injury and coronary occlusion (myocardial infarction). The development of heart failure in female animals due to the induction of volume overload was found to be considerably reduced in comparison to males because several parameters for cardiac dysfunction as well as signal transduction systems for apoptosis were suppressed in comparison to males. It is noteworthy that ovariectomy was demonstrated to promote the development of volume overload induced characteristics of heart failure in female animals whereas these alterations in different hemodynamic parameters as well as in both pro-apoptotic and anti-apoptotic signal transduction pathways due to volume overload were fully or partially reversed by estrogen therapy in ovariectomized animals. The increase in plasma levels of epinephrine and dopamine, unlike that for norepinephrine, in volume overload ovariectomized animals were also partially reversed by estrogen therapy. These observations indicate that ovariectomy may affect different hormones other than sex hormones in the body.

The mechanisms for the beneficial effects of estrogen therapy are considered to be mediated through the activation of estrogen receptors and subsequent depressions in oxidation stress, ATP degradation and mPTP opening in the myocardium. It is pointed out that the treatment of estrogen in postmenopausal state is considered to offer cardioprotection by improving energy status and anti-hypertrophic effects as well as by decreasing the degree of apoptosis and attenuation of cardiac dysfunction in failing hearts. Although such beneficial effects of estrogen were seen in both clinical and experimental observations with different types of heart failure, some of the clinical studies have indicated contradictory results and have cautioned against the use of estrogen therapy in delaying the progression of heart failure as well as associated adverse cardiovascular abnormalities in postmenopausal females. Nonetheless, in view of the partial beneficial effects of estrogen treatment, the loss of some other ovarian hormones may also be involved in promoting the development of heart failure under postmenopausal conditions. Thus, it would be prudent to use a combination therapy with different ovarian hormones for the treatment of heart failure after menopause in females.

**Acknowledgements** Infrastructural support for this project was provided by St. Boniface Hospital Albrechtsen Research Centre, Winnipeg, Canada.

## References

1. Pfeffer MA, Braunwald E (1990) Ventricular remodeling after myocardial infarction. Experimental observations and clinical implications. Circ 81:1161–1172
2. Azevedo PS, Polegato BF, Minicucci MF et al (2016) Cardiac remodeling: concepts, clinical impact, pathophysiological mechanisms and pharmacologic treatment. Arq Bras Cardiol 106:62–69
3. Maytin M, Colucci WS (2002) Molecular and cellular mechanisms of myocardial remodeling. J Nucl Cardiol 9:319–327
4. Moe GW, Marín-García J (2016) Role of cell death in the progression of heart failure. Heart Fail Rev 21:157–167
5. Zhou L, Sun J, Gu L et al (2021) Programmed cell death: complex regulatory networks in cardiovascular disease. Front Cell Dev Biol 9:794879
6. Del Re DP, Amgalan D, Linkermann A et al (2019) Fundamental mechanisms of regulated cell death and implications for heart disease. Physiol Rev 99:1765–1817
7. Azevedo PS, Minicucci MF, Santos PP et al (2013) Energy metabolism in cardiac remodeling and heart failure. Cardiol Rev 21:135–140
8. Münzel T, Gori T, Keaney JF Jr et al (2015) Pathophysiological role of oxidative stress in systolic and diastolic heart failure and its therapeutic implications. Eur Heart J 36:2555–2564
9. Leask A (2015) Getting to the heart of the matter: new insights into cardiac fibrosis. Circ Res 116:1269–1276
10. Feridooni HA, Dibb KM, Howlett SE (2015) How cardiomyocyte excitation, calcium release and contraction become altered with age. J Mol Cell Cardiol 83:62–72
11. Luo M, Anderson ME (2013) Mechanisms of altered $Ca^{2+}$ handling in heart failure. Circ Res 113:690–708
12. Buckberg GD, Hoffman JIE, Coghlan HC, Nanda NC (2015) Ventricular structure–function relations in health and disease: Part I. The normal heart. Eur J Cardio-Thoracic Surg 47:587–601

13. Buckberg GD, Hoffman JIE, Coghlan HC, Nanda NC (2015) Ventricular structure–function relations in health and disease: Part II clinical considerations. Eur J Cardio-Thoracic Surg 47:778–787
14. Sayer G, Bhat G (2014) The renin-angiotensin-aldosterone system and heart failure. Cardiol Clin 32:21–32
15. Florea VG, Cohn JN (2014) The autonomic nervous system and heart failure. Circ Res 114:1815–1826
16. Tsao CW, Aday AW, Almarzooq ZI et al (2022) Heart disease and stroke statistics—2022 update: a report from the American Heart Association. Circ 145:e153–e639
17. Lippi G, Sanchis-Gomar F (2020) Global epidemiology and future trends of heart failure. AME Med J 5:1–5
18. Maxwell SRJ (1998) Women and heart disease. Basic Res Cardiol 93:s079–s084
19. Lloyd-Jones DM, Larson MG, Leip EP et al (2002) Lifetime risk for developing congestive heart failure: the Framingham Heart Study. Circ 106:3068–3072
20. Okoth K, Chandan JS, Marshall T et al (2020) Association between the reproductive health of young women and cardiovascular disease in later life: umbrella review. BMJ 2020:371. https://doi.org/10.1136/bmj.m3502
21. Muka T, Oliver-Williams C, Kunutsor S et al (2016) Association of age at onset of menopause and time since onset of menopause with cardiovascular outcomes, intermediate vascular traits, and all-cause mortality: a systematic review and meta-analysis. JAMA Cardiol 1:767–776
22. van der Schouw YT, van der Graaf Y, Steyerberg EW et al (1996) Age at menopause as a risk factor for cardiovascular mortality. Lancet 347:714–718
23. Kannel WB, Hjortland MC, McNAMARA PM, Gordon T (1976) Menopause and risk of cardiovascular disease: the Framingham study. Ann Int Med 85:447–452
24. El Khoudary SR, Aggarwal B, Beckie TM et al (2020) Menopause transition and cardiovascular disease risk: implications for timing of early prevention: a scientific statement from the American Heart Association. Circ 142:e506–e532
25. Yeh JS, Cheng H-M, Hsu P-F et al (2013) Hysterectomy in young women associates with higher risk of stroke: a nationwide cohort study. Int J Cardiol 168:2616–2621
26. Honigberg MC, Zekavat SM, Aragam K et al (2019) Association of premature natural and surgical menopause with incident cardiovascular disease. JAMA 322:2411–2421
27. El Khoudary SR, Thurston RC (2018) Cardiovascular implications of the menopause transition: endogenous sex hormones and vasomotor symptoms. Obstet Gynecol Clin 45:641–661
28. Zhu D, Chung H-F, Dobson AJ et al (2020) Type of menopause, age of menopause and variations in the risk of incident cardiovascular disease: pooled analysis of individual data from 10 international studies. Hum Reprod 35:1933–1943
29. Clayton GL, Soares AG, Kilpi F et al (2022) Cardiovascular health in the menopause transition: a longitudinal study of up to 3892 women with up to four repeated measures of risk factors. BMC Med 20:1–13
30. Golezar S, Ramezani Tehrani F, Khazaei S et al (2019) The global prevalence of primary ovarian insufficiency and early menopause: a meta-analysis. Climacteric 22:403–411
31. Schneider HPG, Birkhäuser M (2017) Quality of life in climacteric women. Climacteric 20:187–194
32. Nappi RE, Simoncini T (2021) Menopause transition: a golden age to prevent cardiovascular disease. Lancet Diabetes Endocrinol 9:135–137
33. Genazzani AR, Simoncini T (2013) Benefits of menopausal hormone therapy—timing is key. Nat Rev Endocrinol 9:5–6
34. Liu L, Klein L, Eaton C et al (2020) Menopausal hormone therapy and risks of first hospitalized heart failure and its subtypes during the intervention and extended postintervention follow-up of the women's health initiative randomized trials. J Card Fail 26:2–12
35. Bachmann G (2001) Physiologic aspects of natural and surgical menopause. J Reprod Med 46:307–315

36. Willemars M, Nabben M, Verdonschot JAJ, Hoes MF (2022) Evaluation of the interaction of sex hormones and cardiovascular function and health. Curr Heart Fail Rep 1–13
37. Dam V, Van Der Schouw YT, Onland-Moret NC et al (2019) Association of menopausal characteristics and risk of coronary heart disease: a pan-European case–cohort analysis. Int J Epidemiol 48:1275–1285
38. Zhu D, Chung H-F, Dobson AJ et al (2019) Age at natural menopause and risk of incident cardiovascular disease: a pooled analysis of individual patient data. Lancet Public Heal 4:e553–e564
39. Maslov PZ, Kim JK, Argulian E et al (2019) Is cardiac diastolic dysfunction a part of postmenopausal syndrome? JACC Heart Fail 7:192–203
40. Mendelsohn ME, Karas RH (1999) The protective effects of estrogen on the cardiovascular system. N Engl J Med 340:1801–1811
41. Roger VL (2021) Epidemiology of heart failure: a contemporary perspective. Circ Res 128:1421–1434
42. Blenck CL, Harvey PA, Reckelhoff JF, Leinwand LA (2016) The importance of biological sex and estrogen in rodent models of cardiovascular health and disease. Circ Res 118:1294–1312
43. Tepper PG, Randolph JF Jr, McConnell DS et al (2012) Trajectory clustering of estradiol and follicle-stimulating hormone during the menopausal transition among women in the Study of Women's Health across the Nation (SWAN). J Clin Endocrinol Metab 97:2872–2880
44. Dos Santos RL, da Silva FB, Ribeiro RF, Stefanon I (2014) Sex hormones in the cardiovascular system. Horm Mol Biol Clin Investig 18:89–103
45. Biondi-Zoccai GGL, Abate A, Bussani R et al (2005) Reduced post-infarction myocardial apoptosis in women: a clue to their different clinical course? Heart 91:99–101
46. Narula J, Haider N, Virmani R et al (1996) Apoptosis in myocytes in end-stage heart failure. N Engl J Med 335:1182–1189
47. Olivetti G, Abbi R, Quaini F et al (1997) Apoptosis in the failing human heart. N Engl J Med 336:1131–1141
48. Guerra S, Leri A, Wang X et al (1999) Myocyte death in the failing human heart is gender dependent. Circ Res 85:856–866
49. Schipke J, Grimm C, Arnstein G et al (2016) Cardiomyocyte loss is not required for the progression of left ventricular hypertrophy induced by pressure overload in female mice. J Anat 229:75–81
50. Fliegner D, Schubert C, Penkalla A et al (2010) Female sex and estrogen receptor-β attenuate cardiac remodeling and apoptosis in pressure overload. Am J Physiol Integr Comp Physiol 298:R1597–R1606
51. Dent MR, Tappia PS, Dhalla NS (2010) Gender differences in apoptotic signaling in heart failure due to volume overload. Apoptosis 15:499–510; Erratum in (2011) Apoptosis 16:757–758
52. Luo T, Kim JK (2016) The role of estrogen and estrogen receptors on cardiomyocytes: an overview. Can J Cardiol 32:1017–1025
53. Winham SJ, de Andrade M, Miller VM (2015) Genetics of cardiovascular disease: importance of sex and ethnicity. Atherosclerosis 241:219–228
54. Kessler EL, Rivaud MR, Vos MA, van Veen TAB (2019) Sex-specific influence on cardiac structural remodeling and therapy in cardiovascular disease. Biol Sex Differ 10:1–11
55. Tamargo J, Rosano G, Walther T et al (2017) Gender differences in the effects of cardiovascular drugs. Eur Heart J Cardiovasc Pharmacother 3:163–182
56. Liu L, Miura K, Kadota A et al (2019) The impact of sex on risk of cardiovascular disease and all-cause mortality in adults with or without diabetes mellitus: a comparison between the US and Japan. J Diabetes Complications 33:417–423
57. Dent MR, Tappia PS, Dhalla NS (2010) Gender differences in cardiac dysfunction and remodeling due to volume overload. J Card Fail 16:439–449; Erratum in (2011) J Card Fail 17:179.
58. Chiong M, Wang ZV, Pedrozo Z et al (2011) Cardiomyocyte death: mechanisms and translational implications. Cell Death Dis 2:e244

59. He X, Du T, Long T et al (2022) Signaling cascades in the failing heart and emerging therapeutic strategies. Signal Transduct Target Ther 7:1–36
60. Leinwand LA (2003) Sex is a potent modifier of the cardiovascular system. J Clin Invest 112:302–307
61. Luczak ED, Leinwand LA (2009) Sex-based cardiac physiology. Annu Rev Physiol 71:1–18
62. Barrett-Connor E (2013) Gender differences and disparities in all-cause and coronary heart disease mortality: epidemiological aspects. Best Pract Res Clin Endocrinol Metab 27:481–500
63. Vallabhajosyula S, Verghese D, Desai VK et al (2022) Sex differences in acute cardiovascular care: a review and needs assessment. Cardiovasc Res 118:667–685
64. Lam CSP, Arnott C, Beale AL et al (2019) Sex differences in heart failure. Eur Heart J 40:3859–3868
65. Regitz-Zagrosek V (2020) Sex and gender differences in heart failure. Int J Heart Fail 2:157–181
66. Carroll JD, Carroll EP, Feldman T et al (1992) Sex-associated differences in left ventricular function in aortic stenosis of the elderly. Circ 86:1099–1107
67. Crabbe DL, Dipla K, Ambati S et al (2003) Gender differences in post-infarction hypertrophy in end-stage failing hearts. J Am Coll Cardiol 41:300–306
68. Regitz-Zagrosek V, Oertelt-Prigione S, Seeland U, Hetzer R (2010) Sex and gender differences in myocardial hypertrophy and heart failure. Circ J 74:1265–1273
69. Mendelsohn ME, Karas RH (2005) Molecular and cellular basis of cardiovascular gender differences. Science 308:1583–1587
70. Khamis RY, Ammari T, Mikhail GW (2016) Gender differences in coronary heart disease. Heart 102:1142–1149
71. Pandey A, Omar W, Ayers C et al (2018) Sex and race differences in lifetime risk of heart failure with preserved ejection fraction and heart failure with reduced ejection fraction. Circ 137:1814–1823
72. Belcher P, Boerboom LE, Olinger GN (1985) Standardization of end-systolic pressure-volume relation in the dog. Am J Physiol Circ Physiol 249:H547–H553
73. Mitchell GF, Parise H, Benjamin EJ et al (2004) Changes in arterial stiffness and wave reflection with advancing age in healthy men and women: the Framingham Heart Study. Hypertension 43:1239–1245
74. Melchiorre K, Sharma R, Thilaganathan B (2014) Cardiovascular implications in preeclampsia: an overview. Circ 130:703–714
75. Regitz-Zagrosek V, Roos-Hesselink JW, Bauersachs J et al (2018) 2018 ESC guidelines for the management of cardiovascular diseases during pregnancy: the task force for the management of cardiovascular diseases during pregnancy of the European Society of Cardiology (ESC). Eur Heart J 39:3165–3241
76. Santema BT, Ouwerkerk W, Tromp J et al (2019) Identifying optimal doses of heart failure medications in men compared with women: a prospective, observational, cohort study. Lancet 394:1254–1263
77. He J, Ogden LG, Bazzano LA et al (2001) Risk factors for congestive heart failure in US men and women: NHANES I epidemiologic follow-up study. Arch Intern Med 161:996–1002
78. Kitzman DW, Lam CSP (2017) Obese heart failure with preserved ejection fraction phenotype: from pariah to central player. Circ 136:20–23
79. Ambikairajah A, Walsh E, Tabatabaei-Jafari H, Cherbuin N (2019) Fat mass changes during menopause: a metaanalysis. Am J Obstet Gynecol 221:393–409
80. Kondo T, Nakano Y, Adachi S, Murohara T (2019) Effects of tobacco smoking on cardiovascular disease. Circ J 83(10):1980–1985
81. Organization WH (2008) Tobacco free initiative-TFI
82. Hawkins NM, Jhund PS, McMurray JJV, Capewell S (2012) Heart failure and socioeconomic status: accumulating evidence of inequality. Eur J Heart Fail 14:138–146
83. Dewan P, Rørth R, Jhund PS et al (2019) Income inequality and outcomes in heart failure: a global between-country analysis. JACC Heart Fail 7:336–346

84. Reffelmann T, Sensebat O, Birnbaum Y et al (2004) A novel minimal-invasive model of chronic myocardial infarction in swine. Coron Artery Dis 15:7–12
85. Darby SC, Ewertz M, McGale P et al (2013) Risk of ischemic heart disease in women after radiotherapy for breast cancer. N Engl J Med 368:987–998
86. Mehta LS, Watson KE, Barac A et al (2018) Cardiovascular disease and breast cancer: where these entities intersect: a scientific statement from the American Heart Association. Circ 137:e30–e66
87. Pelliccia F, Limongelli G, Autore C et al (2019) Sex-related differences in cardiomyopathies. Int J Cardiol 286:239–243
88. Gori T, Anadol R (2018) Tako-Tsubo syndrome, spontaneous coronary dissection and microvascular disease: sex-differences. Clin Hemorheol Microcirc 70:375–379
89. Deshmukh A, Kumar G, Pant S et al (2012) Prevalence of takotsubo cardiomyopathy in the United States. Am Heart J 164:66–71
90. Sliwa K, Fett J, Elkayam U (2006) Peripartum cardiomyopathy. Lancet 368:687–693
91. Cavasin MA, Tao Z, Menon S, Yang X-P (2004) Gender differences in cardiac function during early remodeling after acute myocardial infarction in mice. Life Sci 75:2181–2192
92. Shaw LJ, Bairey Merz CN, Pepine CJ et al (2006) Insights from the NHLBI-sponsored women's ischemia syndrome evaluation (WISE) study: Part I: gender differences in traditional and novel risk factors, symptom evaluation, and gender-optimized diagnostic strategies. J Am Coll Cardiol 47:S4–S20
93. Dunlay SM, Roger VL (2012) Gender differences in the pathophysiology, clinical presentation, and outcomes of ischemic heart failure. Curr Heart Fail Rep 9:267–276
94. Pfeffer MA, Swedberg K, Granger CB et al (2003) Effects of candesartan on mortality and morbidity in patients with chronic heart failure: the CHARM-Overall programme. Lancet 362:759–766
95. Krumholz HM, Larson M, Levy D (1993) Sex differences in cardiac adaptation to isolated systolic hypertension. Am J Cardiol 72:310–313
96. Chandramouli C, Teng TK, Tay WT et al (2019) Impact of diabetes and sex in heart failure with reduced ejection fraction patients from the ASIAN-HF registry. Eur J Heart Fail 21:297–307
97. Gregg EW, Gu Q, Cheng YJ et al (2007) Mortality trends in men and women with diabetes, 1971–2000. Ann Intern Med 147:149–155
98. Kolar F, Ostadal B (2013) Sex differences in cardiovascular function. Acta Physiol 207:584–587
99. Tromp J, Westenbrink BD, Ouwerkerk W et al (2018) Identifying pathophysiological mechanisms in heart failure with reduced versus preserved ejection fraction. J Am Coll Cardiol 72:1081–1090
100. Eaton CB, Pettinger M, Rossouw J et al (2016) Risk factors for incident hospitalized heart failure with preserved versus reduced ejection fraction in a multiracial cohort of postmenopausal women. Circ Heart Fail 9:e002883
101. Blomkalns AL, Chen AY, Hochman JS et al (2005) Gender disparities in the diagnosis and treatment of non–ST-segment elevation acute coronary syndromes: large-scale observations from the CRUSADE (Can rapid risk stratification of unstable angina patients suppress adverse outcomes with early implementation). J Am Coll Cardiol 45:832–837
102. Chen L, Jiang C (2019) Bioinformatics analysis of sex differences in arrhythmogenic right ventricular cardiomyopathy. Mol Med Rep 19:2238–2244
103. Ware JS, Li J, Mazaika E et al (2016) Shared genetic predisposition in peripartum and dilated cardiomyopathies. N Engl J Med 374:233–241
104. Fairweather D, Cooper LT Jr, Blauwet LA (2013) Sex and gender differences in myocarditis and dilated cardiomyopathy. Curr Probl Cardiol 38:7–46
105. Schulman-Marcus J, Hartaigh O, Gransar BH et al (2016) Sex-specific associations between coronary artery plaque extent and risk of major adverse cardiovascular events: the CONFIRM long-term registry. JACC Cardiovasc Imaging 9:364–372
106. Ostadal B, Netuka I, Maly J et al (2009) Gender differences in cardiac ischemic injury and protection—Experimental aspects. Exp Biol Med 234:1011–1019

107. Wang M, Baker L, Tsai BM et al (2005) Sex differences in the myocardial inflammatory response to ischemia-reperfusion injury. Am J Physiol Metab 288:E321–E326
108. Smetana M, Besik J, Netuka I et al (2017) Sensitivity to perioperative ischemia/reperfusion injury in male and female donor myocardium. Physiol Res 66:949–957
109. Villari B, Campbell SE, Schneider J et al (1995) Sex-dependent differences in left ventricular function and structure in chronic pressure overload. Eur Heart J 16:1410–1419
110. Prévilon M, Pezet M, Vinet L et al (2014) Gender-specific potential inhibitory role of $Ca^{2+}$/calmodulin dependent protein kinase phosphatase (CaMKP) in pressure-overloaded mouse heart. PLoS ONE 9:e90822
111. Gardner JD, Brower GL, Janicki JS (2002) Gender differences in cardiac remodeling secondary to chronic volume overload. J Card Fail 8:101–107
112. Beaumont C, Walsh-Wilkinson É, Drolet M-C et al (2017) Female rats with severe left ventricle volume overload exhibit more cardiac hypertrophy but fewer myocardial transcriptional changes than males. Sci Rep 7:1–12
113. Von Eiff AW, Gogolin E, Jacobs U, Neus H (1986) Ambulatory blood pressure in children followed for 3 years: influence of sex and family history of hypertension. Clin Exp Hypertens Part A Theory Pract 8:577–581
114. Sayed Y, Taxel P (2003) The use of estrogen therapy in men. Curr Opin Pharmacol 3:650–654
115. Nabulsi AA, Folsom AR, White A et al (1993) Association of hormone-replacement therapy with various cardiovascular risk factors in postmenopausal women. N Engl J Med 328:1069–1075
116. Wren BG, Wren BG (1992) The effect of oestrogen on the female cardiovascular system. Med J Aust 157:204–208
117. Grohé C, Kahlert S, Löbbert K et al (1997) Cardiac myocytes and fibroblasts contain functional estrogen receptors. FEBS Lett 416:107–112
118. Fuentes N, Silveyra P (2019) Estrogen receptor signaling mechanisms. Adv Protein Chem Struct Biol 116:135–170
119. Wang H, Sun X, Chou J et al (2017) Cardiomyocyte-specific deletion of the G protein-coupled estrogen receptor (GPER) leads to left ventricular dysfunction and adverse remodeling: a sex-specific gene profiling analysis. Biochim Biophys Acta-Molecular Basis Dis 1863:1870–1882
120. Wang H, Sun X, Lin MS et al (2018) G protein-coupled estrogen receptor (GPER) deficiency induces cardiac remodeling through oxidative stress. Transl Res 199:39–51
121. Murphy E (2011) Estrogen signaling and cardiovascular disease. Circ Res 109:687–696
122. Menazza S, Murphy E (2016) The expanding complexity of estrogen receptor signaling in the cardiovascular system. Circ Res 118:994–1007
123. Ueda K, Karas RH (2013) Emerging evidence of the importance of rapid, non-nuclear estrogen receptor signaling in the cardiovascular system. Steroids 78:589–596
124. Luo T, Liu H, Kim JK (2016) Estrogen protects the female heart from ischemia/reperfusion injury through manganese superoxide dismutase phosphorylation by mitochondrial p38β at threonine 79 and serine 106. PLoS ONE 11:e0167761
125. Vaccarino V, Krumholz HM, Berkman LF, Horwitz RI (1995) Sex differences in mortality after myocardial infarction: is there evidence for an increased risk for women? Circ 91:1861–1871
126. Ross JL, Howlett SE (2012) Age and ovariectomy abolish beneficial effects of female sex on rat ventricular myocytes exposed to simulated ischemia and reperfusion. PLoS ONE 7:e38425
127. Furukawa T, Kurokawa J (2008) Non-genomic regulation of cardiac ion channels by sex hormones. Cardiovasc Haematol Disord Targets 8:245–251
128. Kolodgie FD, Farb A, Litovsky SH et al (1997) Myocardial protection of contractile function after global ischemia by physiologic estrogen replacement in the ovariectomized rat. J Mol Cell Cardiol 29:2403–2414
129. Lauro F-V, Francisco D-C, Elodia G-C et al (2015) Evaluation of activity of an estrogen-derivative as cardioprotector drug using an ischemia-reperfusion injury model. Int J Clin Exp Med 8:12041
130. Zhai P, Eurell TE, Cotthaus R et al (2000) Effect of estrogen on global myocardial ischemia-reperfusion injury in female rats. Am J Physiol Circ Physiol 279:H2766–H2775

131. Wang F, He Q, Sun Y et al (2010) Female adult mouse cardiomyocytes are protected against oxidative stress. Hypertension 55:1172–1178
132. Wang L, Tang Z-P, Zhao W et al (2015) MiR-22/Sp-1 links estrogens with the up-regulation of cystathionine $\gamma$-lyase in myocardium, which contributes to estrogenic cardioprotection against oxidative stress. Endocrinology 156:2124–2137
133. Patterson SD, Spahr CS, Daugas E et al (2000) Mass spectrometric identification of proteins released from mitochondria undergoing permeability transition. Cell Death Differ 7:137–144
134. Rattanasopa C, Phungphong S, Wattanapermpool J, Bupha-Intr T (2015) Significant role of estrogen in maintaining cardiac mitochondrial functions. J Steroid Biochem Mol Biol 147:1–9
135. Sbert-Roig M, Bauzá-Thorbrügge M, Galmés-Pascual BM et al (2016) GPER mediates the effects of 17β-estradiol in cardiac mitochondrial biogenesis and function. Mol Cell Endocrinol 420:116–124
136. Nuedling S, Kahlert S, Loebbert K et al (1999) 17β-Estradiol stimulates expression of endothelial and inducible NO synthase in rat myocardium in-vitro and in-vivo. Cardiovasc Res 43:666–674
137. Pelzer T, Neumann M, de Jager T et al (2001) Estrogen effects in the myocardium: inhibition of NF-κB DNA binding by estrogen receptor-α and-β. Biochem Biophys Res Commun 286:1153–1157
138. Pedram A, Razandi M, Aitkenhead M, Levin ER (2005) Estrogen inhibits cardiomyocyte hypertrophy in vitro: antagonism of calcineurin-related hypertrophy through induction of MCIP1. J Biol Chem 280:26339–26348
139. Bell JR, Porrello ER, Huggins CE et al (2008) The intrinsic resistance of female hearts to an ischemic insult is abrogated in primary cardiac hypertrophy. Am J Physiol Circ Physiol 294:H1514–H1522
140. Van Eickels M, Grohé C, Cleutjens JPM et al (2001) 17β-Estradiol attenuates the development of pressure-overload hypertrophy. Circ 104:1419–1423
141. Kim JK, Pedram A, Razandi M, Levin ER (2006) Estrogen prevents cardiomyocyte apoptosis through inhibition of reactive oxygen species and differential regulation of p38 kinase isoforms. J Biol Chem 281:6760–6767
142. Jiao L, Machuki JO, Wu Q et al (2020) Estrogen and calcium handling proteins: new discoveries and mechanisms in cardiovascular diseases. Am J Physiol Circ Physiol 318:H820–H829
143. Shen T, Ding L, Ruan Y et al (2014) SIRT1 functions as an important regulator of estrogen-mediated cardiomyocyte protection in angiotensin II-induced heart hypertrophy. Oxid Med Cell Longev 2014:713894. https://doi.org/10.1155/2014/713894
144. Pelzer T, Schumann M, Neumann M et al (2000) 17β-estradiol prevents programmed cell death in cardiac myocytes. Biochem Biophys Res Commun 268:192–200
145. Kong D, Zheng T, Zhang M et al (2013) Static mechanical stress induces apoptosis in rat endplate chondrocytes through MAPK and mitochondria-dependent caspase activation signaling pathways. PLoS ONE 8:e69403
146. Patten RD, Pourati I, Aronovitz MJ et al (2004) 17β-Estradiol reduces cardiomyocyte apoptosis in vivo and in vitro via activation of phospho-inositide-3 kinase/Akt signaling. Circ Res 95:692–699
147. Morkuniene R, Arandarcikaite O, Ivanoviene L, Borutaite V (2010) Estradiol-induced protection against ischemia-induced heart mitochondrial damage and caspase activation is mediated by protein kinase G. Biochim Biophys Acta -Bioenergetics 1797:1012–1017
148. Hayward CS, Kelly RP, Collins P (2000) The roles of gender, the menopause and hormone replacement on cardiovascular function. Cardiovasc Res 46:28–49
149. Petrov G, Regitz-Zagrosek V, Lehmkuhl E et al (2010) Regression of myocardial hypertrophy after aortic valve replacement: faster in women? Circ 122:S23–S28
150. Goussev A, Sharov VG, Shimoyama H et al (1998) Effects of ACE inhibition on cardiomyocyte apoptosis in dogs with heart failure. Am J Physiol Circ Physiol 275:H626–H631
151. Cesselli D, Jakoniuk I, Barlucchi L et al (2001) Oxidative stress–mediated cardiac cell death is a major determinant of ventricular dysfunction and failure in dog dilated cardiomyopathy. Circ Res 89:279–286

152. Xie D, Liao Y, Wu B et al (2018) Cardiac nestin+ cells derived from early stage of dilated cardiomyopathy enhanced the survival of the doxorubicin-injured cardiac muscle hl-1 cells. Int Heart J 59:180–189
153. Cleland JGF, Swedberg K, Follath F et al (2003) The EuroHeart failure survey programme—A survey on the quality of care among patients with heart failure in Europe: Part 1: patient characteristics and diagnosis. Eur Heart J 24:442–463
154. Olivetti G, Giordano G, Corradi D et al (1995) Gender differences and aging: effects on the human heart. J Am Coll Cardiol 26:1068–1079
155. Li Q, Li B, Wang X et al (1997) Overexpression of insulin-like growth factor-1 in mice protects from myocyte death after infarction, attenuating ventricular dilation, wall stress, and cardiac hypertrophy. J Clin Invest 100:1991–1999
156. Boya P, Reggiori F, Codogno P (2013) Emerging regulation and functions of autophagy. Nat Cell Biol 15:713–720
157. Zhu H, Sun A (2018) Programmed necrosis in heart disease: molecular mechanisms and clinical implications. J Mol Cell Cardiol 116:125–134
158. Jiang X, Stockwell BR, Conrad M (2021) Ferroptosis: mechanisms, biology and role in disease. Nat Rev Mol Cell Biol 22:266–282
159. Wang Y, Kanneganti T-D (2021) From pyroptosis, apoptosis and necroptosis to PANoptosis: a mechanistic compendium of programmed cell death pathways. Comput Struct Biotechnol J 19:4641–4657
160. Bedoui S, Herold MJ, Strasser A (2020) Emerging connectivity of programmed cell death pathways and its physiological implications. Nat Rev Mol Cell Biol 21:678–695
161. Rowe VL, Stevens SL, Reddick TT et al (2000) Vascular smooth muscle cell apoptosis in aneurysmal, occlusive, and normal human aortas. J Vasc Surg 31:567–576
162. Ucker DS (2016) Exploiting death: apoptotic immunity in microbial pathogenesis. Cell Death Differ 23:990–996
163. Kerr JFR, Wyllie AH, Currie AR (1972) Apoptosis: a basic biological phenomenon with wideranging implications in tissue kinetics. Br J Cancer 26:239–257
164. Knaapen MWM, Davies MJ, De Bie M et al (2001) Apoptotic versus autophagic cell death in heart failure. Cardiovasc Res 51:304–312
165. Kim N-H, Kang PM (2010) Apoptosis in cardiovascular diseases: mechanism and clinical implications. Korean Circ J 40:299–305
166. Bennett MR (2002) Apoptosis in the cardiovascular system. Heart 87:480–487
167. Araki S, Shimada Y, Kaji K, Hayashi H (1990) Apoptosis of vascular endothelial cells by fibroblast growth factor deprivation. Biochem Biophys Res Commun 168:1194–1200
168. Sendoel A, Hengartner MO (2014) Apoptotic cell death under hypoxia. Physiology 29:168–176
169. Kannan K, Jain SK (2000) Oxidative stress and apoptosis. Pathophysiology 7:153–163
170. Lotem J, Sachs L (1999) Cytokines as suppressors of apoptosis. Apoptosis 4:187–196
171. Uberti F, Caimmi PP, Molinari C et al (2011) Levosimendan modulates programmed forms of cell death through KATP channels and nitric oxide. J Cardiovasc Pharmacol 57:246–258
172. Arstall MA, Sawyer DB, Fukazawa R, Kelly RA (1999) Cytokine-mediated apoptosis in cardiac myocytes: the role of inducible nitric oxide synthase induction and peroxynitrite generation. Circ Res 85:829–840
173. Norbury CJ, Zhivotovsky B (2004) DNA damage-induced apoptosis. Oncogene 23:2797–2808
174. Song AS, Najjar AM, Diller KR (2014) Thermally induced apoptosis, necrosis, and heat shock protein expression in three-dimensional culture. J Biomech Eng 136:71006
175. Watters D (1999) Molecular mechanisms of ionizing radiation-induced apoptosis. Immunol Cell Biol 77:263–271
176. Hardwick JM (2001) Apoptosis in viral pathogenesis. Cell Death Differ 8:109–110
177. Tower J (2015) Programmed cell death in aging. Ageing Res Rev 23:90–100
178. Elmore S (2007) Apoptosis: a review of programmed cell death. Toxicol Pathol 35:495–516
179. King KL, Cidlowski JA (1998) Cell cycle regulation and apoptosis. Annu Rev Physiol 60:601

180. Ohsawa S, Vaughen J, Igaki T (2018) Cell extrusion: a stress-responsive force for good or evil in epithelial homeostasis. Dev Cell 44:284–296
181. Danial NN, Korsmeyer SJ (2004) Cell death: critical control points. Cell 116:205–219
182. Zou H, Li Y, Liu X, Wang X (1999) An APAF-1·cytochrome c multimeric complex is a functional apoptosome that activates procaspase-9. J Biol Chem 274:11549–11556
183. Thomas MP, Liu X, Whangbo J et al (2015) Apoptosis triggers specific, rapid, and global mRNA decay with 3′ uridylated intermediates degraded by DIS3L2. Cell Rep 11:1079–1089
184. Wernig F, Xu Q (2002) Mechanical stress-induced apoptosis in the cardiovascular system. Prog Biophys Mol Biol 78:105–137
185. Abbate A, Narula J (2012) Role of apoptosis in adverse ventricular remodeling. Heart Fail Clin 8:79–86
186. Narula J, Arbustini E, Chandrashekhar Y, Schwaiger M (2001) Apoptosis and the systolic dysfunction in congestive heart failure: story of apoptosis interruptus and zombie myocytes. Cardiol Clin 19:113–126
187. Narula J, Kharbanda S, Khaw B-A (1997) Apoptosis and the heart. Chest 112:1358–1362
188. Kostin S, Pool L, Elsässer A et al (2003) Myocytes die by multiple mechanisms in failing human hearts. Circ Res 92:715–724
189. Orogo AM, Gustafsson ÅB (2013) Cell death in the myocardium: my heart won't go on. IUBMB Life 65:651–656
190. Abbate A, Biondi-Zoccai GGL, Bussani R et al (2003) Increased myocardial apoptosis in patients with unfavorable left ventricular remodeling and early symptomatic post-infarction heart failure. J Am Coll Cardiol 41:753–760
191. Jurasz P, Courtman D, Babaie S, Stewart DJ (2010) Role of apoptosis in pulmonary hypertension: from experimental models to clinical trials. Pharmacol Ther 126:1–8
192. Hofstra L, Liem H, Dumont EA et al (2000) Visualisation of cell death in vivo in patients with acute myocardial infarction. Lancet 356:209–212
193. Saraste A, Pulkki K, Kallajoki M et al (1997) Apoptosis in human acute myocardial infarction. Circ 95:320–323
194. Cheng W, Kajstura J, Nitahara JA et al (1996) Programmed myocyte cell death affects the viable myocardium after infarction in rats. Exp Cell Res 226:316–327
195. Hochhauser E, Kivity S, Offen D et al (2003) Bax ablation protects against myocardial ischemia-reperfusion injury in transgenic mice. Am J Physiol Circ Physiol 284:H2351–H2359
196. Freude B, Masters TN, Robicsek F et al (2000) Apoptosis is initiated by myocardial ischemia and executed during reperfusion. J Mol Cell Cardiol 32:197–208
197. Zhou H, Yang H-X, Yuan Y et al (2013) Paeoniflorin attenuates pressure overload-induced cardiac remodeling via inhibition of TGFβ/Smads and NF-κB pathways. J Mol Histol 44:357–367
198. Satoh M, Matter CM, Ogita H et al (2007) Inhibition of apoptosis-regulated signaling kinase-1 and prevention of congestive heart failure by estrogen. Circ 115:3197–3204
199. Poller W, Dimmeler S, Heymans S et al (2018) Non-coding RNAs in cardiovascular diseases: diagnostic and therapeutic perspectives. Eur Heart J 39:2704–2716
200. Sallam T, Sandhu J, Tontonoz P (2018) Long noncoding RNA discovery in cardiovascular disease: decoding form to function. Circ Res 122:155–166
201. Das S, Babick AP, Xu Y et al (2010) TNF-α-mediated signal transduction pathway is a major determinant of apoptosis in dilated cardiomyopathy. J Cell Mol Med 14:1988–1997
202. Xu H, Li J, Zhao Y, Liu D (2017) TNFα-induced downregulation of microRNA-186 contributes to apoptosis in rat primary cardiomyocytes. Immunobiology 222:778–784
203. Dent MR, Das S, Dhalla NS (2007) Alterations in both death and survival signals for apoptosis in heart failure due to volume overload. J Mol Cell Cardiol 43:726–732
204. Cook SA, Sugden PH, Clerk A (1999) Activation of c-Jun N-terminal kinases and p38-mitogen-activated protein kinases in human heart failure secondary to ischaemic heart disease. J Mol Cell Cardiol 31:1429–1434
205. Das DK (2003) Protein kinase C isozymes signaling in the heart. J Mol Cell Cardiol 8:887–889

206. Wang S, Zhang F, Zhao G et al (2017) Mitochondrial PKC-ε deficiency promotes I/R-mediated myocardial injury via GSK 3β-dependent mitochondrial permeability transition pore opening. J Cell Mol Med 21:2009–2021
207. Bueno OF, Molkentin JD (2002) Involvement of extracellular signal-regulated kinases 1/2 in cardiac hypertrophy and cell death. Circ Res 91:776–781
208. Xu L, He D, Wu Y et al (2022) Tanshinone IIA inhibits cardiomyocyte apoptosis and rescues cardiac function during doxorubicin-induced cardiotoxicity by activating the DAXX/MEK/ERK1/2 pathway. Phytomedicine 107:154471
209. Zhang X, Javan H, Li L et al (2013) A modified murine model for the study of reverse cardiac remodelling. Exp Clin Cardiol 18:e115
210. Xu L-N, Wang S-H, Su X-L et al (2021) Targeting glycogen synthase kinase 3 beta regulates cd47 expression after myocardial infarction in rats via the NF-κB signaling pathway. Front Pharmacol 12:662726
211. Nie S, Cui X, Guo J et al (2021) Long non-coding RNA AK006774 inhibits cardiac ischemia-reperfusion injury via sponging miR-448. Bioengineered 12:4972–4982
212. Wu C, Zhou X-X, Li J-Z et al (2021) Pretreatment of cardiac progenitor cells with bradykinin attenuates H2O2-induced cell apoptosis and improves cardiac function in rats by regulating autophagy. Stem Cell Res Ther 12:1–15
213. Lu C, Liu L, Chen S et al (2021) Azathioprine pretreatment ameliorates myocardial ischaemia reperfusion injury in diabetic rats by reducing oxidative stress, apoptosis, and inflammation. Clin Exp Pharmacol Physiol 48:1621–1632
214. Cheng X-J, Li L, Xin B-Q (2021) MiR-124 regulates the inflammation and apoptosis in myocardial infarction rats by targeting STAT3. Cardiovasc Toxicol 21:710–720
215. Guo X, Yin H, Li L et al (2017) Cardioprotective role of tumor necrosis factor receptor-associated factor 2 by suppressing apoptosis and necroptosis. Circ 136:729–742
216. Jiang W, Song J, Zhang S et al (2021) CTRP13 protects H9c2 cells against hypoxia/reoxygenation (H/R)-induced injury via regulating the AMPK/Nrf2/ARE signaling pathway. Cell Transplant 30:09636897211033275
217. Groeneweg JA, van der Heijden JF, Dooijes D et al (2014) Arrhythmogenic cardiomyopathy: diagnosis, genetic background, and risk management. Netherlands Heart J 22:316–325
218. Wu Q, Yao Q, Hu T et al (2022) Tax1 banding protein 1 exacerbates heart failure in mice by activating ITCH-P73-BNIP3-mediated cardiomyocyte apoptosis. Acta Pharmacol Sin 43:2562–2572
219. Katare PB, Nizami HL, Paramesha B et al (2020) Activation of toll like receptor 4 (TLR4) promotes cardiomyocyte apoptosis through SIRT2 dependent p53 deacetylation. Sci Rep 10:1–15
220. Shen Z, Shen A, Chen X et al (2020) Huoxin pill attenuates myocardial infarction-induced apoptosis and fibrosis via suppression of p53 and TGF-β1/Smad2/3 pathways. Biomed Pharmacother 130:110618
221. Hua F, Li JY, Zhang M et al (2022) Kaempferol-3-O-rutinoside exerts cardioprotective effects through NF-κ B/NLRP3/Caspase-1 pathway in ventricular remodeling after acute myocardial infarction. J Food Biochem e14305
222. Malhotra A, Kang BPS, Hashmi S, Meggs LG (2005) PKCε inhibits the hyperglycemia-induced apoptosis signal in adult rat ventricular myocytes. Mol Cell Biochem 268:169–173
223. Mohamed BA, Schnelle M, Khadjeh S et al (2016) Molecular and structural transition mechanisms in long-term volume overload. Eur J Heart Fail 18:362–371
224. He Z, Zeng X, Zhou D et al (2021) LncRNA Chaer prevents cardiomyocyte apoptosis from acute myocardial infarction through AMPK activation. Front Pharmacol 12:649398. https://doi.org/10.3389/fphar.2021.649398
225. Xiong X, Liu J, He Q et al (2021) Long non-coding RNA NORAD aggravates acute myocardial infarction by promoting fibrosis and apoptosis via miR-577/COBLL1 axis. Environ Toxicol 36:2256–2265
226. Arslan F, Smeets MB, Riem Vis PW et al (2011) Lack of fibronectin-EDA promotes survival and prevents adverse remodeling and heart function deterioration after myocardial infarction. Circ Res 108:582–592

227. Patten RD, Pourati I, Aronovitz MJ et al (2008) 17 Beta-estradiol differentially affects left ventricular and cardiomyocyte hypertrophy following myocardial infarction and pressure overload. J Card Fail 14:245–253
228. Kararigas G, Fliegner D, Gustafsson J-Å, Regitz-Zagrosek V (2011) Role of the estrogen/estrogen-receptor-beta axis in the genomic response to pressure overload-induced hypertrophy. Physiol Genomics 43:438–446
229. Voloshenyuk TG, Gardner JD (2010) Estrogen improves TIMP-MMP balance and collagen distribution in volume-overloaded hearts of ovariectomized females. Am J Physiol Integr Comp Physiol 299:R683–R693
230. Chaudhary KR, Deng Y, Yang A et al (2021) Penetrance of severe pulmonary arterial hypertension in response to vascular endothelial growth factor receptor 2 blockade in a genetically prone rat model is reduced by female sex. J Am Heart Assoc 10:e019488
231. Lagranha CJ, Deschamps A, Aponte A et al (2010) Sex differences in the phosphorylation of mitochondrial proteins result in reduced production of reactive oxygen species and cardioprotection in females. Circ Res 106:1681–1691
232. Liu H, Pedram A, Kim JK (2011) Oestrogen prevents cardiomyocyte apoptosis by suppressing p38α-mediated activation of p53 and by down-regulating p53 inhibition on p38β. Cardiovasc Res 89:119–128
233. Nuedling S, Kahlert S, Loebbert K et al (1999) Differential effects of 17β-estradiol on mitogen-activated protein kinase pathways in rat cardiomyocytes. FEBS Lett 454:271–276
234. Cavasin MA, Sankey SS, Yu A-L et al (2003) Estrogen and testosterone have opposing effects on chronic cardiac remodeling and function in mice with myocardial infarction. Am J Physiol Circ Physiol 284:H1560–H1569
235. Kanda N, Watanabe S (2003) 17β-estradiol inhibits oxidative stress-induced apoptosis in keratinocytes by promoting Bcl-2 expression. J Invest Dermatol 121:1500–1509
236. Dent MR, Dhalla NS, Tappia PS (2004) Phospholipase C gene expression, protein content, and activities in cardiac hypertrophy and heart failure due to volume overload. Am J Physiol Circ Physiol 287:H719–H727
237. Turino Miranda K, Kalenga CZ, Saad N et al (2022) Gender-affirming estrogen therapy route of administration and cardiovascular risk: a systematic review and narrative synthesis. Am J Physiol Circ Physiol 323:H861–H868
238. Kok HS, van Asselt KM, van der Schouw YT et al (2006) Heart disease risk determines menopausal age rather than the reverse. J Am Coll Cardiol 47:1976–1983
239. Oliver-Williams C, Glisic M, Shahzad S et al (2019) The route of administration, timing, duration and dose of postmenopausal hormone therapy and cardiovascular outcomes in women: a systematic review. Hum Reprod Update 25:257–271
240. Zhu D, Chung H-F, Pandeya N et al (2019) Premenopausal cardiovascular disease and age at natural menopause: a pooled analysis of over 170,000 women. Eur J Epidemiol 34:235–246
241. Canonico M, Carcaillon L, Plu-Bureau G et al (2016) Postmenopausal hormone therapy and risk of stroke: impact of the route of estrogen administration and type of progestogen. Stroke 47:1734–1741
242. Perez MV, Wang PJ, Larson JC et al (2012) Effects of postmenopausal hormone therapy on incident atrial fibrillation: the women's health initiative randomized controlled trials. Circ Arrhythmia Electrophysiol 5:1108–1116
243. Ahmed SB, Kang AK, Burns KD et al (2004) Effects of oral contraceptive use on the renal and systemic vascular response to angiotensin II infusion. J Am Soc Nephrol 15:780–786
244. Rossouw JE, Prentice RL, Manson JE et al (2007) Postmenopausal hormone therapy and risk of cardiovascular disease by age and years since menopause. JAMA 297:1465–1477
245. Alecrin IN, Aldrighi JM, Caldas MA et al (2004) Acute and chronic effects of oestradiol on left ventricular diastolic function in hypertensive postmenopausal women with left ventricular diastolic dysfunction. Heart 90:777–781
246. Pines A, Fisman EZ, Drory Y et al (1992) Menopause-induced changes in Doppler-derived parameters of aortic flow in healthy women. Am J Cardiol 69:1104–1106

247. Liu L, Yin X, Chen M et al (2018) Geographic variation in heart failure mortality and its association with hypertension, diabetes, and behavioral-related risk factors in 1723 counties of the United States. Front Public Health 6:132. https://doi.org/10.3389/fpubh.2018.00132
248. Wassertheil-Smoller S, Hendrix S, Limacher M et al (2003) Effect of estrogen plus progestin on stroke in postmenopausal women: the women's health initiative: a randomized trial. JAMA 289:2673–2684
249. Vilatoba M, Eckstein C, Bilbao G et al (2005) 17β-estradiol differentially activates mitogen-activated protein-kinases and improves survival following reperfusion injury of reduced-size liver in mice. In: Transplantation proceedings. Elsevier, pp 399–403
250. Wang XI, Ren B, Liu S et al (2003) Characterization of cardiac hypertrophy and heart failure due to volume overload in the rat. J Appl Physiol 94:752–763

# Chapter 25
# Prevention of β-Adrenoceptor-Mediated Alterations in Female Heart Failure by Estrogen

**Paramjit S. Tappia, Adriana Adameova, Vijayan Elimban, and Naranjan S. Dhalla**

**Abstract** Recent interest in Women Heart Health has led to increased attention to the influence and importance of sex in the incidence of cardiac dysfunction under a wide variety of stressful situations. It is now well known that pre-menopausal females are protected from several types of pathological stimuli leading to heart dysfunction. While the morbidity and mortality rates due to heart failure in pre-menopausal women are much less than their male counterparts, the mechanisms responsible are far from clear. Since the β-adrenergic system, which regulates heart function, is an integral pathway in the transition from cardiac hypertrophy to heart failure, this article is intended to present the evidence on sex differences in the β-adrenoceptor (β-AR) signaling mechanisms in a rat model of heart failure due to volume overload for 16 weeks. Both mRNA levels and protein content for $\beta_1$-AR and $\beta_2$-AR in female volume-overloaded hearts were elevated whereas these values were depressed in male failing hearts. Unlike female hearts, mRNA levels for adenylyl cyclase as well as β-arrestin 2 were depressed in male failing hearts whereas unlike male hearts, protein content for adenylyl cyclase and mRNA levels for GRK2 were increased in volume-overloaded female hearts. Ovariectomy for 16 weeks increased protein contents for $\beta_1$-AR, $\beta_2$-AR and adenylyl cyclase in control hearts whereas values for these components were depressed in the ovariectomized volume-overloaded hearts. These alterations due to ovariectomy were reversed by 17β-estradiol. Although ovariectomy depressed epinephrine-stimulated adenylyl cyclase activity in both control and

P. S. Tappia (✉)
Asper Clinical Research Institute, CR3129-369 Tache Avenue, Winnipeg, MB R2H 2A6, Canada
e-mail: ptappia@sbrc.ca

A. Adameova
Department of Pharmacology and Toxicology, Faculty of Pharmacy, Comenius University, Bratislava, Slovakia

V. Elimban · N. S. Dhalla
Institute of Cardiovascular Sciences and Department of Physiology & Pathophysiology, Max Rady College of Medicine, University of Manitoba, Winnipeg, MB, Canada

© The Author(s), under exclusive license to Springer Nature Switzerland AG 2023 
L. Kirshenbaum and I. Rabinovich-Nikitin (eds.), *Biology of Women's Heart Health*, Advances in Biochemistry in Health and Disease 26,
https://doi.org/10.1007/978-3-031-39928-2_25

volume-overloaded hearts, this change in control animals was not affected by 17β-estradiol. These observations support the view that estrogen plays an important role in augmenting β-AR-mediated signal transduction in female heart failure.

**Keywords** Sex differences · Volume overload · Heart failure · Ovariectomy · β-adrenoceptors · Adenylyl cyclase · Signaling transduction

## Introduction

It is now established that heart failure (HF) with preserved ejection fraction (HFpEF) represents a large proportion of total worldwide HF [1, 2]. In fact, women have a higher incidence of HFpEF, while men present with HF with reduced ejection fraction (HFrEF). Such differences in the incidence of HF have been attributed mainly to the female sex hormone, estrogen [3, 4]. It is also evident that the incidence of cardiovascular disease events including coronary artery disease, hypertension and HF in young women is substantially lower than in men, but increases disproportionately in women after menopause [5, 6]. Furthermore, hypertension, diabetes and cigarette smoking are reported to be the highest risk factors for the occurrence of HF in women [7]. It should be mentioned that low income as well as low level of education have been linked to increased incidence of hospitalizations due to HF in post-menopausal women [8]. Women have also been reported to exhibit worse symptoms and poor well-being as compared to men [9].

The resistance to cardiovascular disease events in females have frequently been attributed to the female sex hormones. Particularly, the role for estrogen and estrogen receptors (ER) in the development of HF has been suggested in view of the observation that hormone replacement therapy (HRT) in women attenuates cardiac hypertrophy [10]. While cellular remodeling associated with left ventricle (LV) dysfunction is also attenuated in women, outcome is improved due to greater preservation of ventricular function, as well as reduced cellular hypertrophy and dilatation. However; the exact cellular signaling mechanisms responsible for the cardioprotection observed in young females is far from clear. It should be noted that estrogen appears to be protective in women since the increased risk of HF is associated with a reduction in estrogen levels during and after menopause. On the other hand, clinical trials have demonstrated that estrogen replacement therapy in post-menopausal women results in adverse cardiac events and increases the risk of breast cancer [11]. Thus, additional experiments and investigation to fully understand the beneficial/protective actions of estrogen are required for improved therapies and prevention strategies in women. It is also noteworthy that over the last few years the importance of sex and susceptibility to HF is increasingly being recognized with respect to clinical presentation and outcomes [12]. Unfortunately, women were underrepresented in many clinical and experimental studies in the field of cardiovascular disease. However, the lack of good trial evidence concerning sex-specific outcomes has led to assumptions about heart disease treatment in women, which in turn has resulted in inadequate diagnoses and suboptimal management, and impacting outcomes [13–18].

While a sustained adrenergic drive seen in different cardiac pathologies is considered to result in several cellular and molecular defects including a downregulation of the β-adrenoceptor (β-AR) associated with cardiac dysfunction (Fig. 25.1), this article is intended to describe the evidence for sex differences and the role of estrogen in the β-AR signaling system in an experimental model of HF due to volume overload. We have earlier examined the echocardiographic and hemodynamic changes in male and female rats in heart failure stage at 16 weeks due to volume overload induced by an arteriovenous (AV) shunt [19]. It was observed that the increase in heart muscle mass was higher in females than males and unlike the females, rate of pressure development/contraction (+dP/dt) and rate of pressure decay/relaxation (−dP/dt) were depressed whereas left ventricular end-diastolic pressure (LVEDP) was increased in male rats. In addition, while an increase in cardiac output occurred in both sexes, this was more marked in males [19]. Furthermore, increases in circulating catecholamine levels was only seen in male rats, whereas a decrease or no change was observed in female failing hearts [19].

Interestingly, ovariectomy resulted in depressed +dP/dt, −dP/dt, and fractional shortening, whereas a marked increase in cardiac output as well as increased LVEDP and LV internal diameters were observed at 16 weeks post-AV shunt. Although treatment with 17-βestradiol normalized ±dP/dt, LVEDP remained elevated [19]. Taking these observations together, it is thus evident that sex differences in cardiac

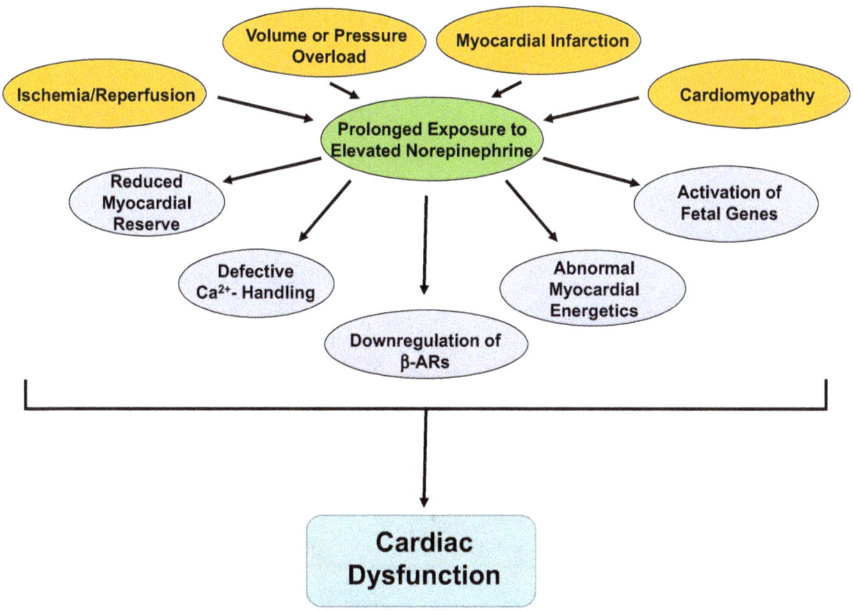


**Fig. 25.1** Different cardiac pathologies leading to the development of congestive heart failure

function and remodeling subsequent to volume overload exist, which may also be related to the sex differences in plasma catecholamine concentration. Moreover, estrogen may have a protective role and preserve cardiac contractile function, but the mechanisms involved in defining these functional sex differences remain to be fully elucidated.

The β-AR signaling system is known to regulate the activities of adenylyl cyclase, which generates cAMP with the subsequent activation of protein kinase A (PKA) and regulation of diverse functions including sarcolemmal L-type $Ca^{2+}$-channel activity, sarcoplasmic reticulum $Ca^{2+}$-release and myofibrillar sensitivity and subsequent cardiac contractile function (Fig. 25.2). It is interesting to note that PKA has also been reported to influence the sex-dependent differences in cardiac contractile function as well as the predisposition to heart disease [20]. Previously, we have reported sex differences in the various components of the β-AR system, including β-ARs and adenylyl cyclase, as well as Gi and Gs proteins, β-arrestin 1, β-arrestin 2 content, and GRK2 and GRK3 activities in cardiac hypertrophy due to volume overload [21]. In this chapter, we describe the available evidence that demonstrates sex differences in the β-AR signaling system, and associated components in heart failure due to volume overload in addition to indicating some of the mechanisms of the protective actions of estrogen that may be related to prevention of the effects of over-activation of the β-AR system and cardiac remodeling.

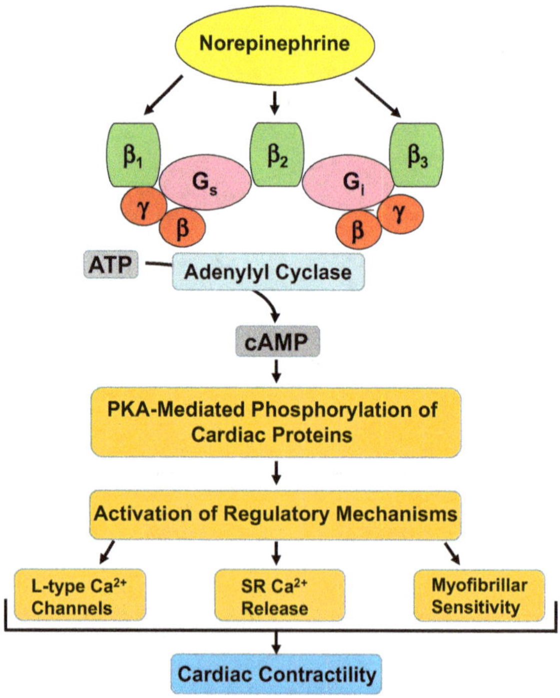


**Fig. 25.2 Mechanism of action of norepinephrine in the heart.** $G_s$ and $G_i$, stimulatory and inhibitory guanine nucleotide binding proteins; cAMP, cyclic AMP; PKA, protein kinase A; SR, sarcoplasmic reticulum

## Role of β-AR-Signal Transduction in Heart Function

Stimulation of the sympathetic nervous system (SNS) is known to release norepinephrine (NE), which exerts positive inotropic effect in the heart through the activation of β-ARs [22]. The β-AR signaling system not only regulates the cardiac contractile function, but is believed to provide essential support for the maintenance of cardiac function during the development of HF [23, 24]. The binding of NE to β-ARs initiates the activation of $G_s$ proteins, and stimulation of adenylyl cyclase. The consequent generation of cAMP from ATP, activates PKA and results in the increase of intracellular $Ca^{2+}$ levels and augmentation of contraction of the heart [25]. Furthermore, GRKs, a family of serine/threonine kinases, regulate the β-AR signal transduction system through phosphorylation of the agonist-occupied receptors [26, 27]. GRK2 (β-ARK1), GRK3 (β-ARK2), and GRK5 are predominantly expressed in the heart [28]. Upon NE-induced stimulation of β-ARs, GRK2 translocates from the cytosol to the membrane, phosphorylates β-ARs and uncoupling of $G_s$ proteins from β-ARs [29].

It is well established that β-AR-mediated signal transduction is altered in HF and the most evident change involves the β-ARs, where $\beta_1$-AR is markedly downregulated. Alterations in other components of the β-AR signaling system including uncoupling of $\beta_2$-ARs and increased activity of $G_i$ protein have also been shown to occur [30]. These changes appear to be related to increased activity of the SNS and increased exposure of the heart to NE. However, it should be mentioned that chronic activation of the SNS exerts a net long-term adverse effect on the heart and the myocardial adrenergic desensitization is partially maladaptive in the setting of LV dysfunction [30]. Data from human end-stage heart failure [22, 23] and several experimental models of HF such as the spontaneously hypertensive rat [31], pressure overload induced HF [32], myocardial infarction (MI) [33, 34] and volume overload induced HF have been reported to reveal different degrees of changes in β-AR signal transduction, including downregulation of $\beta_1$-AR. Specifically, in patients with volume overload due to left heart valvular disease and mitral regurgitation, a decrease in β-AR density has been observed [35, 36]. A decrease in myocardial β-AR density, adenylyl cyclase activity and response to isoproterenol in pigs subsequent to an AV shunt has also been reported [37]. On the other hand, while no changes in the maximal binding ($B_{max}$) and dissociation constant ($K_d$) characteristics of β-AR were found in the rat subsequent to AV shunt, a reduction in the maximal adenylyl cyclase response to NE was observed [38]. It is interesting to note that an increase in β-AR density was reported in the hypertrophic stage, whereas a decrease was seen at HF stage without having any effect on $K_d$ following AV shunt in rats [39].

## Sex Differences in Cardiovascular Disease

With the aging population, the incidence and prevalence of HF continue to rise, which remains a major health hazard [40]. Both clinical and experimental lines of evidence have demonstrated the existence of sex differences in the prevalence of heart disease. It is generally believed that pre-menopausal women have a lower risk for heart disease, but markedly increases after menopause [41]. Women tend to develop heart disease some 10–15 years later than men, which has led to the hypothesis that estrogen is cardioprotective in women [42]. In this regard, experimental studies have demonstrated that females are protected from developing HF in several different animal models. For example, a tenfold less mortality in female rats as compared to their male counterparts subsequent to volume overload has been reported; in addition, the female rats showed no signs of HF [43]. Female hearts have also been shown to be less susceptible to ischemia–reperfusion (I/R) injury and better recovery of cardiac function as well as cardiomyocyte survival following I/R [44, 45]. The intravenous administration of epinephrine in both male and female rats resulted in a higher frequency and occurrence of epinephrine-induced premature ventricular contractions, missed beats, and conduction blocks in the male rats as compared to female rat hearts; loss of cardioprotection was also observed upon ovariectomy [46].

Human observations have shown that the incidence of HF is higher for men in all age groups [47]; however, the lifetime risk is similar in men and women because women tend to live longer [48]. Indeed, a Framingham study reported, after adjustment for age, that the incidence of HF in women is approximately 30% lower than in men [47]. Women with heart disease tend to be sicker than men by the time HF is diagnosed; furthermore, women appear to benefit less following by- pass surgery and tend to have more severe symptoms in HF. These differences could be accounted for the fact that women are older and frailer when they develop HF.

In contrast to men, hypertension and diabetes represent the major risk factors for the development of heart failure in women and the clinical course of heart failure is generally more benign and more frequently characterized by HF with preserved systolic function and diastolic dysfunction [47]. Women tend to be older and more hypertensive but are less likely to demonstrate any clinical signs of coronary heart disease (CHD) and more often have preserved ventricular function [48]. The combination of obesity and hormonal disturbances in postmenopausal women is of serious concern [49]; in fact, it has been shown that diabetic women have a three-to six-fold risk of MI whereas diabetic men are at an increased risk of two to four-fold [49]. Following the induction of a mechanical load, women have a lower tendency to develop myocardial hypertrophy than men [50] and seem to be better protected with later onset of cardiac decompensation than men with heart failure [51].

It has become very apparent that there is also a sex bias in the diagnosis or treatment of HF, [52]. In fact, it has been reported that women are more frequently prescribed diuretics and digitalis, and are much less often prescribed a combination therapy of the two [52, 53]. It should be mentioned that digoxin was associated with a significantly higher risk of death in women when compared to placebo [54] and the administration of diuretics may actually result in more adverse effects in women [55]. Interestingly, older women with HF have better survival rates than men [56]. In women, an increase in LV mass or the extent of cardiac hypertrophy is a stronger predictor of mortality than traditional measures of LV size and function [57]. Taken together, it is clear that there is a need for sex specific clinical studies or trials which should be designed to include sufficient numbers of women [49]; this will assist to not only understand the influence of sex in the response to treatment, but also help to further understand the role of sex in HF outcomes.

Although the role of estrogen has drawn much attention for scientific investigation regarding the mechanisms for the cardioprotection observed in pre-menopausal women, this aspect remains to be completely understood. ERs are known to be present in the human heart, and are considered to have a role in cardiac hypertrophy and subsequent progression to HF [49, 58, 59]. In fact, HRT has been reported to attenuate cardiac hypertrophy in women [11]. There are two main types of ERs (ER-α and ER-β) ER-α is predominantly expressed in the adult rat heart whereas ER-βexpression has been reported to be the predominant ER expressed in the neonatal heart, but decreases later in life [60]. Clinical examination has revealed a marked upregulation of ER-α mRNA and protein levels as well as increased ER-β mRNA in cardiac hypertrophy [49, 61]. 17β-estradiol has been reported to modulate the expression of natriuretic peptides in the atria and the LV via the mitogen-activated protein kinase (MAPK) pathway [58, 62]; these alterations may reduce the activity of the p38 MAPK pathway, by stimulation of its inhibitor, MAPK phosphatase 1 [63], and thus attenuate cardiac hypertrophic response.

In many experimental models, the more severe cardiovascular phenotype in male transgenic animals or ovariectomized females can be rescued by the administration of estrogen; suggesting that estrogen prevents or at least slows down the development of cardiac hypertrophy and HF. In this regard, estrogen has been shown to attenuate cardiac hypertrophy induced by pressure overload in vivo [64], and antagonize cardiomyocyte hypertrophy in vitro involving ER-dependent mechanisms [49, 65]. Sex differences in LV remodeling and fibrosis in hypertrophic cardiomyopathy have also been investigated [66]. In this study it was observed that a disproportionate number of females had a more pronounced degree of remodeling in different forms of hypertrophic cardiomyopathy, but no sex differences were seen in the extent of fibrosis [65].

Another factor that may explain female cardioprotection is that there is a lower β-AR density in female cardiomyocytes and they are less responsive to isoproterenol, resulting in a reduced calcium influx after β-AR stimulation [67]. It has also been shown that there is a decreased β-AR density and β-AR inotropic response in the aging female heart [49]. The most convincing line of evidence for the protective effect for estrogen originates from large cohort studies that have compared CHD risk in postmenopausal women currently using estrogen versus never-users. These studies have shown consistently that CHD risk is 35–50% lower in estrogen users after adjusting for other risk factors [68, 69].

It has been suggested that HRT may only be beneficial when its administration is initiated early during a narrow window of opportunity, which is around the time of menopause and before women develop an excessive buildup of atherosclerotic plaque. In this regard, the Women's Health Initiative (WHI) study, which discounted any benefit of HRT was criticized because it included women much older than the average hormone user, who typically initiate HRT around the time of menopause. The average age of participants in this study was 64 years. The average age of menopause has been suggested to be 51.4 years of age, and some studies suggest that women who initiate HRT later may miss the chance to benefit from this treatment [70, 71]. Interestingly, in a study with healthy women aged between 45 and 58 and exhibited postmenopausal or perimenopausal symptoms, it was reported that after 10 years of randomized HRT early after menopause has significantly less risk of HF, without any increase in the risk of cancer, venous thromboembolism or stroke [72]. Also, HRT does not independently predict mortality or thromboembolism or bleeding in women with atrial fibrillation [73].

## Sex Differences in the β-AR Mechanisms in Heart Failure Due to Volume Overload

The β-AR signaling system is a key regulator of heart rate, systolic and diastolic functions, and myocardial metabolism [74]; however, its responses in HF are markedly altered [75]. Indeed, adrenergic over-activity is a major pathogenic event for the development of HF and is often associated with poor prognosis [75]. Initially, there is an increase in β-AR signaling allowing the heart to adapt quickly to increased workloads; this allows the heart to increase its output within a matter of seconds by increasing pacemaker frequency and myocardial contractility [74]. Although some compensatory mechanisms maintain cardiac performance at an acceptable level under pathologic situations, the prolonged activation of the SNS and increased levels of NE have an adverse action on the heart [75]. It is interesting to note that a decreased capacity to respond to β-adrenergic stimulation, particularly under conditions of increased demand (i.e. faster pacing rates and treatment with isoproterenol) has been reported in isolated female rabbit hearts; furthermore, this reduced responsiveness was also observed to be associated with diminished arrhythmic activity [76]. Since $\beta_1$-AR signaling accounts for the majority of pathologic hypertrophy in HF, reverse remodeling has been shown to occur by using the β-AR blocking agents in the failing human heart [77]. Alterations in the β-AR signaling system have also been shown to be involved in the development of HF due to volume overload in the rat, which was linked to a hypersensitivity of the myocardium to β-AR stimulation [78]. Furthermore, upregulation of β-AR system as well as changes in the subcellular distribution of regulatory proteins, GRK isoforms and β-arrestins have been observed in this experimental model [79].

It is interesting to note that no sex differences have been reported with respect to basic mechanical ability and functioning of healthy myocardium [80], and thus under pathophysiological conditions, it is the response to stressors that may be a major factor in eliciting sex differences in cardiac remodeling and contractile function [80]. In accordance to this suggestion, Chu et al. [81] found that there were also no functional sex differences in the heart as well as no difference in expression of β-ARs in the rat. On the other hand, sex differences have been shown in the inotropic response to isoproterenol where male atrial and ventricular cells exert a greater increase in cell shortening; in fact, the twofold higher β-AR density and larger cAMP production in male rat cardiomyocytes as compared to female cardiomyocytes suggested an elevated β-adrenergic signaling in males [82]. In human studies, it has been reported that sex affects age-related β-AR downregulation in normal human ventricles, with females having a greater extent of decrease in aging [83]. It was found that this curvilinear relationship between age and receptor density, which plateaus at approximately 40 years of age in women was suggested to be due to an effect of sex hormones on $\beta_1$-AR expression in the human heart [83]. Females have also demonstrated a lower sympathetic activation and parasympathetic withdrawal in nonischemic HF as compared to males that may be a contributory factor for cardioprotection in females [84]. Interestingly, males demonstrated a greater decline in the inotropic and chronotropic response to β-adrenergic stimulation with age as compared to females [84]. Since the β-adrenergic responsiveness is reduced in males with HF as compared to females, it has been suggested that sex-specific differences in calcium handling may be partially responsible for the improved survival of females in HF [85]. Indeed, sex-specific differences in calcium handling proteins have also been reported in rat heart ventricle [81] and in failing pig cardiomyocytes [85] that may be related to reduced β-AR responsiveness in males as compared to females [81, 85].

Overall, it is evident that HF is associated with abnormalities in the β-AR signaling system in different models of HF and that sex differences in outcomes of different cardiac pathologies may be related to sex-specific alterations or preservation of the β-AR and its signaling components. The following section describes the evidence demonstrating sex differences in the modifications in the β-AR signaling system in HF due to volume overload induced by AV shunt. Furthermore, some mechanisms associated with estrogen-induced cardioprotection in the female heart have been highlighted.

We have observed that at 16 weeks post AV shunt in male rats there was a significant decrease in both $\beta_1$-AR and $\beta_2$-AR mRNA levels in the failing hearts. In contrast, a small, but significant increase in the expression levels of both $\beta_1$-AR and $\beta_2$-AR gene expressions at 16 weeks post AV shunt was seen in the female heart [86] (Table 25.1). The changes in the gene expression of both receptor types were concurrent to the changes in protein contents of both $\beta_1$-AR and $\beta_2$-AR at 16 weeks in male and female hearts (Table 25.1). It should be mentioned that stimulation of the $\beta_2$-AR attenuates cardiac remodeling and infarct expansion in rats and that a combination therapy of $\beta_2$-AR stimulation with $\beta_1$-AR blockade exceeded the therapeutic effectiveness of sole $\beta_1$-AR blockade [87]. From these reports it is apparent that $\beta_2$-AR mediated signal transduction is cardioprotective [87, 88]. Therefore, our observations of elevated $\beta_2$-AR gene and protein expression at 16 weeks post AV shunt, may have an important contributory role in inducing appropriate remodeling and preservation of

**Table 25.1** Gene and protein expression of $\beta_1$-AR and $\beta_2$-AR in male and female heart failure due to volume overload induced by arteriovenous shunt for 16 weeks

| | Male | | Female | |
|---|---|---|---|---|
| | Control | AV shunt | Control | AV shunt |
| *A: mRNA (% of control)* | | | | |
| $\beta_1$-AR (680 b.p.) | 109 ± 11 | 83 ± 6* | 102 ± 3 | 123 ± 8* |
| $\beta_2$-AR (525 b.p.) | 102 ± 5 | 79 ± 4* | 106 ± 4 | 127 ± 8* |
| *B: Protein (Densitometric units)* | | | | |
| $\beta_1$-AR (65 kDa) | 1.6 ± 0.1 | 0.7 ± 0.1* | 2.2 ± 0.2# | 3.3 ± 0.4* |
| $\beta_2$-AR (46 kDa) | 2.2 ± 0.2 | 1.4 ± 0.1* | 2.6 ± 0.2 | 3.7 ± 0.3* |

Quantified data showing β-AR mRNA levels (A) and protein contents (B) in male and female rats at 16 weeks volume overload induced by AV shunt. Values are means ± SE of 3–5 different RNA or membrane preparations in each group. * $P < 0.05$ versus control (sham) values; # $P < 0.05$ versus corresponding male control (sham) values. Data presented are based on results from our publication—Dent et al. [86]

cardiac contractile function of the female heart subsequent to volume overload [86]. In this regard, overstimulation of the β-adrenergic system during the hypertrophic phase leads to downregulation of the β-ARs at the failing stage and concomitant decrease in cardiac function, which is the case in male rats subjected to volume overload; however; the female heart appears to be less sensitive to the stimulation of the adrenergic system under stress and a downregulation in the β-AR at 16 weeks post AV shunt was not observed; this may be one explanation for why cardiac contractile function is sustained by the female heart subjected to volume overload.

We have also examined different components of the β-AR signaling system for possible sex differences [86]. Table 25.2 shows the status of adenylyl cyclase V/VI gene expression and protein contents in male and female hearts subsequent to volume overload induced by an AV shunt. While no changes were seen in the expression of adenylyl cyclase V/VI in the female heart, adenylyl cyclase V/VI mRNA levels were depressed in male rats at 16 weeks post AV shunt; however, protein amounts for adenylyl cyclase V/VI were increased in females at this time point, while no change was detected in the male heart (Table 25.2). No alterations in the mRNA levels of both $G_i$ and Gs were observed in the heart of either male or female hearts at 16 weeks post-AV (Table 25.3). Further examination of the other components of the β-AR signaling system revealed an increase in GRK 2 gene expression in the female hearts, while β-arrestin 2 mRNA level was decreased in the male heart following the induction of volume overload at 16 weeks post-AV shunt. In this regard, it is pointed out that GRK-arrestin system is considered to play a critical role in desensitization and downregulation of GPCRs [35] as well as desensitization of β-ARs are important contributors to contractile dysfunction in HF [35]. The observed increase in GRK2 mRNA in females and the decrease in β-arrestin 2 mRNA level in the male heart in response to AV-shunt, is suggestive of a compensatory role of these regulatory proteins [25, 89]. Since the upregulation of GRK2 has been linked

**Table 25.2** Gene and protein expression of adenylyl cyclase (AC) in failing heart stage to volume overload induced by arteriovenous shunt for 16 weeks

| | Male | | Female | |
|---|---|---|---|---|
| | Control | AV shunt | Control | AV shunt |
| *A: mRNA (% of control)* | | | | |
| AC V (680 b.p.) | 101 ± 4 | 87 ± 6* | 104 ± 6 | 111 ± 6 |
| AC VI (525 b.p.) | 102 ± 3 | 33 ± 7* | 105 ± 7 | 102 ± 5 |
| *B: Protein (Densitometric units)* | | | | |
| AC (65 kDa) | 1.6 ± 0.1 | 1.8 ± 0.1 | 1.7 ± 0.2 | 2.7 ± 0.3* |

Quantified data showing AC subtype mRNA levels (A) and AC protein contents (B) in male and female rats at 16 weeks volume overload induced by AV shunt. Values are means ± SE of 3–5 different RNA or membrane preparations in each group. * $P < 0.05$ versus control (sham) values. Data presented are based on results from our publication—Dent et al. [86]

**Table 25.3** mRNA Levels of G-proteins, GRKs and arrestins in heart failure due to volume overload induced by arteriovenous shunt for 16 weeks

| | Male | | Female | |
|---|---|---|---|---|
| | Control | AV shunt | Control | AV shunt |
| *mRNA (% of control)* | | | | |
| Gi (680 b.p.) | 103 ± 5 | 112 ± 6 | 106 ± 7 | 110 ± 4 |
| Gs (525 b.p.) | 100 ± 4 | 101 ± 6 | 102 ± 3 | 101 ± 6 |
| GRK2 (680 b.p.) | 101 ± 3 | 103 ± 5 | 102 ± 4 | 123 ± 4* |
| GRK3 (525 b.p.) | 100 ± 4 | 108 ± 6 | 102 ± 2 | 110 ± 6 |
| β-arrestin 1 (65 kDa) | 104 ± 5 | 105 ± 4 | 103 ± 11 | 118 ± 5 |
| β-arrestin 2 (46 kDa) | 102 ± 3 | 88 ± 6* | 104 ± 5 | 109 ± 7 |

Quantified data showing mRNA levels of Gi, Gs, GRK2, GRK3, β-arrestin 1 and β-arrestin 2 in male rats at 16 weeks volume overload induced by AV shunt. Values are means ± SE of 3–5 different RNA preparations in each group. * $P < 0.05$ versus control (sham) values; # $P < 0.05$ versus corresponding male control (sham) values. Data presented are based on results from our publication—Dent et al. [86]

to depressed contractile function in HF [32], the observed increase of GRK2 gene expression levels in the female heart may attenuate the augmented function of β-ARs in the female heart. It is well recognized that β-arrestin plays a role in β-AR desensitization, and β-arrestin signaling and activation of EGFR/ERK by the $\beta_1$-AR has been suggested to be cardioprotective [87]. We observed a specific decrease in the mRNA expression level of β-arrestin 2 in the male hearts (Table 25.3), which may be seen as a compensatory mechanism to prevent the downregulation of β-AR function in males post AV shunt.

Some investigators have reported that 17β-estradiol reduced the contraction of the female rat tail artery in response to adrenergic stimulation indicating that females may be less sensitive to adrenergic stimulation and that estrogen may have a modulatory role on the vascular response to adrenergic stimulation [90]. In addition, ovariectomy has been reported to result in an upregulation of $\beta_1$-AR expression and downregulation of $\beta_2$-AR expression during I-R; these alterations were restored by treatment with estrogen, suggesting that estrogen might play a cardioprotective role in female rat hearts subjected to I-R [91]. From the aforementioned, it is evident that unlike males, the ability of females to maintain cardiac function following AV-shunt in the rat can be linked to female sex hormones. This view is based on our observations [86] that showed increased protein content for both $\beta_1$-AR and $\beta_2$-AR and adenylyl cyclase as well as basal adenylyl cyclase activity in the intact AV-shunt female, which were depressed by ovariectomy (Table 25.4). Furthermore, it was observed that the augmented epinephrine-stimulated adenylyl activity in the intact 16 weeks post AV shunt female heart was also attenuated by ovariectomy. It should be noted that β-AR proteins were increased in the ovariectomized control female rats. Treatment of the 16 week AV-shunt ovariectomized females with 17β-estradiol resulted in a reversal of β-ARs and basal adenylyl cyclase activity, whereas a partial normalization of the increased epinephrine-stimulated adenylyl cyclase activity was observed (Table 25.4). Interestingly, estrogen has been reported to facilitate coupling of the β-AR to adenylyl cyclase [92]. Overall, from these data, it is evident that 17β-estradiol treatment of ovariectomized female rats, at least partially, can reverse the effects of ovariectomy on some components of the β-AR signal transduction system, which correlate to cardioprotective effects when the female heart is subjected to volume overload by AV-shunt [86].

## Conclusion

Heart failure continues to be a major factor in morbidity and mortality in both men and women, but how the predisposition, etiology and treatment of HF differs between the sexes is underexplored [93]. Nonetheless, estrogen has been shown to play an essential role in the response of the β-AR signal transduction system to AV-shunt in the female heart, whereby upregulation in the female heart can be seen as an important mechanism for the maintenance of cardiac function. On the other hand, the downregulation of the β-AR signal transduction system in the male heart in response to AV-shunt can be seen as a contributory factor for the occurrence of cardiac dysfunction and HF. It should be noted that since the changes in some of the components of the β-AR signal transduction system were only partially corrected in the female heart upon treatment with 17β-estradiol is suggestive of involvement of other female sex hormones (e.g. progesterone) that may also contribute to the observed sex differences in the β-AR signaling system. In addition, the timing of instituting HRT in perimenopausal and/or postmenopausal women for cardioprotection needs to be further examined.

It should, however, be mentioned that a large multivariable analysis has demonstrated that there are no effects of sex (and age) on the β-AR-mediated inotropy or catecholamine sensitivity in human atrial trabeculae [94], which may have some

**Table 25.4** $\beta_1$-AR, $\beta_2$-AR and adenylyl cyclase protein contents in heart failure due to volume overload induced by arteriovenous shunt for 16 weeks in ovariectomized rats with or without 17-β-estradiol treatment

| A: Protein (Densitometric units) | Control | | | AV shunt | | |
|---|---|---|---|---|---|---|
| | Pre-OVX | OVX | OVX + 17-β-E | Pre-OVX | OVX | OVX + 17-β-E |
| $\beta_1$-AR (65 kDa) | 2.0 ± 0.2 | 5.6 ± 1.2$^{\#}$ 1.8 ± 0.2$^{**}$ | 1.8 ± 0.2$^{**}$ | 3.8 ± 0.4$^{*}$ | 1.1 ± 0.1$^{\#}$ | 5.1 ± 1.6$^{**}$ |
| $\beta_2$-AR (46 kDa) | 1.1 ± 0.1 | 3.5 ± 0.2$^{\#}$ | 0.9 ± 0.3$^{**}$ | 1.7 ± 0.3$^{*}$ | 1.8 ± 0.2 | 3.4 ± 0.2$^{**}$ |
| AC (139 kDa) | 3.7 ± 0.2 | 7.3 ± 0.3$^{\#}$ | 3.1 ± 0.3$^{**}$ | 4.6 ± 0.2$^{*}$ | 2.1 ± 0.1$^{\#}$ | 4.3 ± 0.3$^{**}$ |
| *B: AC Activities (pmol cAMP/mg/10 min)* | | | | | | |
| Basal | 100 ± 12 | 95 ± 8 | 102 ± 8 | 212 ± 18$^{*}$ | 43 ± 8$^{\#}$ | 205 ± 9$^{**}$ |
| Epi | 320 ± 20 | 200 ± 8$^{\#}$ | 180 ± 15 | 1060 ± 105$^{*}$ | 310 ± 8$^{\#}$ | 920 ± 57$^{**}$ |

Quantified data showing β-AR and AC protein contents (A) and AC activities (B) in ovariectomized (OVX) female rats, treated with or without 17- β-estradiol (17-β-E) at 16 weeks volume overload induced by AV shunt. Values are means ± SE of 3–5 different membrane preparations in each group. * $P < 0.05$ versus control (Pre-OVX) values; # $P < 0.05$ versus corresponding value of intact female (Pre-OVX); ** $P < 0.05$ versus corresponding OVX values. Data presented are based on results from our publication—Dent et al. [86]. AC = adenylyl cyclase; Epi = epinephrine

repercussions in the clinical use of β-blockers in HF. In addition, sex differences in the cardiac response to β-blockers have also been reported [95]. In this regard, while evidence-based therapies for HF are available, there is a paucity of information on sex-specific efficacy [4]. Thus, such information should be considered when using β-blockers in clinical practice. It is pointed out that testosterone has been reported to modulate AR gene expression in orchidectomized male rats subjected to ischemic insults [96]. Furthermore, an upregulation of $\beta_2$-AR, Gs and Gi2, but downregulation of $\beta_3$-AR, GRKK2 and Gi3 protein levels has been observed following testosterone treatment of castrated rats subjected to doxorubicin-induced HF [97]. Taken together, further investigation in understanding the role of this male sex hormone in cardiac function under the setting of AV-shunt as well as HF due to other etiologies is warranted. Nonetheless, significant progress has been made recently, in understanding sex differences in the pathophysiology of HF, which may ultimately assist in individualized sex-specific therapeutic strategies and in the approach for prevention that may also take sex of the individual into consideration.

**Acknowledgements** We thank the St. Boniface Hospital Albrechtsen Research Centre for infrastructural support.

## References

1. Shuaishuai D, Jingyi L, Zhiqiang Z, Guanwei F (2023) Sex differences and related estrogenic effects in heart failure with preserved ejection fraction. Heart Fail Rev 28(4):937–948
2. Teramoto K, Teng TK, Chandramouli C et al (2022) Epidemiology and clinical features of heart failure with preserved ejection fraction. Card Fail Rev 8:e27
3. Arcopinto M, Valente V, Giardino F et al (2022) What have we learned so far from the sex/gender issue in heart failure? An overview of current evidence. Intern Emerg Med 17:1589–1598
4. Khan SS, Beach LB, Yancy CW (2022) Sex-based differences in heart failure. J Am Coll Cardiol Focus Seminar 7/7 J Am Coll Cardiol 79:1530–1541
5. Bello N, Mosca L (1995) Epidemiology of coronary heart disease in women. Prog Cardiovasc Dis 46:287–295
6. Hoppe BL, Hermann DD (2003) Sex differences in the causes and natural history of heart failure. Curr Cardiol Rep 5:193–199
7. Pĩna IL, Buchter C (2003) Heart failure in women. Cardiol Rev 11:337–344
8. Shah RU, Winkleby MA, Van Horn L et al (2011) Education, income, and incident heart failure in post-menopausal women: the Women's Health Initiative Hormone Therapy Trials. J Am Coll Cardiol 58:1457–1464
9. Cleland JG, Goode K, Erhardt L et al (2004) Carvedilol or metoprolol European trial investigators. A description of the clinical characteristics at baseline of patients recruited into the Carvedilol or Metoprolol European Trial (COMET). Cardiovasc Drugs Ther 18:139–152
10. Modena MG, Molinari R, Muia N Jr et al (1999) Double-blind randomized placebo-controlled study of transdermal estrogen replacement therapy on hypertensive postmenopausal women. Am J Hypertens 12:1000–1008
11. Arosio B, Corbi G, Davinelli S et al (2022) Sex differences in cardiovascular diseases: a matter of estrogens, ceramides, and sphingosine 1-phosphate. Int J Mol Sci 23:4009
12. Swaraj S, Kozor R, Arnott C et al (2021) Heart failure with reduced ejection fraction-does sex matter? Curr Heart Fail Rep 18:345–432

13. Miličić D, Bergami M, Pavasović S (2022) Sex differences in therapies for heart failure. Curr Pharm Des 28:1295–1303
14. Parry M, Van Spall HGC, Mullen KA et al (2022) The Canadian women's heart health alliance atlas on the epidemiology, diagnosis, and management of cardiovascular disease in women: sex- and gender-specific diagnosis and treatment. Can J Cardiol Open 4:589–608
15. Kim I, Field TS, Wan D et al (2022) Sex and gender bias as a mechanistic determinant of cardiovascular disease outcomes. Can J Cardiol S0828–282X(22)00858-3
16. Ueda K, Fukuma N, Adachi Y et al (2021) Sex differences and regulatory actions of estrogen in cardiovascular system. Front Physiol 12:738218
17. Wang N, Evans J, Sawant S, Sindone J, Lal S (2023) Sex-specific differences in the efficacy of heart failure therapies: a meta-analysis of 84,818 patients. Heart Fail Rev 28(4):949–959
18. Pacheco C, Mullen KA, Coutinho T (2021) The Canadian women's heart health alliance atlas on the epidemiology, diagnosis, and management of cardiovascular disease in women: sex- and gender-unique manifestations of cardiovascular disease. Can J Cardiol Open 4:243–262
19. Dent MR, Tappia PS, Dhalla NS (2010) Gender differences in cardiac dysfunction and remodeling due to volume overload. J Card Fail 16:439–449
20. Liu Y, Chen J, Fontes SK et al (2022) Physiological and pathological roles of protein kinase A in the heart. Cardiovasc Res 118:386–398
21. Dent MR, Tappia PS, Dhalla NS (2011) Gender differences in β-adrenoceptor system in cardiac hypertrophy due to arteriovenous fistula. J Cell Physiol 226:181–186
22. Bristow MR, Hershberger RE, Port JD et al (1990) β-adrenergic pathways in nonfailing and failing human ventricular myocardium. Circ 82:112–125
23. Bohm M (1995) Alterations of β-adrenoceptor-G-protein-regulated adenylyl cyclase in heart failure. Mol Cell Biochem 147:147–160
24. Chakraborti S, Chakraborti T, Shaw G (2000) β-adrenergic mechanisms in cardiac diseases: a perspective. Cell Signal 12:499–513
25. Dhalla NS, Wang X, Sethi R et al (1997) β-adrenergic linked signal transduction mechanisms in failing hearts. Heart Fail Rev 2:55–65
26. Schumacher SM, Koch WJ (2017) Noncanonical roles of G protein-coupled receptor kinases in cardiovascular signaling. J Cardiovasc Pharmacol 70:129–141
27. Dzimiri N, Muiya P, Andres E, Al-Halees Z (2004) Differential functional expression of human myocardial G protein receptor kinases in left ventricular cardiac diseases. Eur J Pharmacol 489:167–177
28. Penn RB, Pronin AN, Benovic JL (2000) Regulation of G protein coupled receptor kinases. Trends Cardiovasc Med 10:81–89
29. Inoko M, Kihara Y, Sasayama S (1995) Neurohumoral factors during transition from left ventricular hypertrophy to failure in Dahl salt-sensitive rats. Biochem Biophys Res Commun 206:814–820
30. Weil J, Schunkert H (2006) Pathophysiology of chronic heart failure. Clin Res Cardiol 95:16–17
31. Anderson KM, Eckhart AD, Willette RN, Koch WJ (1999) The myocardial β-adrenergic system in spontaneously hypertensive heart failure (SHHF) rats. Hypertension 33:402–407
32. Choi DJ, Kock WJ, Hunter JJ, Rockman HA (1997) Mechanism of β-adrenergic receptor desensitization in cardiac hypertrophy is increased β-adrenergic receptor kinase. J Biol Chem 272:17223–17229
33. Ishigai Y, Mori T, Moriyama S, Shibano T (1999) Induction of cardiac β-adrenergic receptor kinase 1 in rat heart failure caused by coronary ligation. J Mol Cell Cardiol 31:1261–1268
34. Vinge LE, Oie E, Andersson Y et al (2001) Myocardial distribution and regulation of GRK and β-arrestin isoforms in congestive heart failure in rats. Am J Physiol Heart Circ Physiol 281:H2490–H2499
35. Dzimiri N, Moorji A (1996) Relationship between alterations in lymphocyte and myocardial β-adrenoceptor density in patients with left heart valvular disease. Clin Exp Pharmacol Physiol 23:498–502
36. Sakagoshi N, Nakano S, Taniguchi K et al (1991) Relation between myocardial β-adrenergic receptor and left ventricular function in patients with left ventricular volume overload due to chronic mitral regurgitation with or without aortic regurgitation. Am J Cardiol 68:81–84

37. Hammond HK, Roth DA, Insel PA et al (1992) Myocardial β-adrenergic receptor expression and signal transduction after chronic volume-overload hypertrophy and circulatory congestion. Circ 85:269–280
38. Sullegarger JT, D'Ambra PM, Clark LC et al (1998) Effect of digoxin on ventricular remodeling and responsiveness of β-adrenoceptors in chronic volume overload. J Cardiovasc Pharmacol Ther 3:281–290
39. Cartagena G, Sapag-Hagar M, Jalil J et al (1993) Changes in β-adrenergic receptors of rat heart and adipocytes during volume-overload induced cardiac hypertrophy. Int J Clin Pharmacol Ther Toxicol 31:198–203
40. Jaarsma T (2002) Are women different than men? Aspects of heart failure in special populations: elderly women. Eur J Cardiovasc Nurs 1:29–31
41. Schonfelder G (2005) The biological impact of estrogens on gender differences in congestive heart failure. Cardiovasc Res 67:573–574
42. Rossouw JE (2002) Hormones, genetic factors, and gender differences in cardiovascular disease Cardiovasc Res 53: 550–557
43. Gardner JD, Brower GL, Janicki JS (2002) Gender differences in cardiac remodeling secondary to chronic volume overload. J Card Fail 8:101–107
44. Bae S, Zhang L (2005) Gender differences in cardioprotection against ischemia/reperfusion injury in adult rat hearts: focus on Akt and protein kinase C signaling. J Pharmacol Exp Ther 315:1125–1135
45. Wang M, Wang Y, Weil B et al (2009) Estrogen receptor β mediates increased activation of PI3K/Akt signaling and improved myocardial function in female hearts following acute ischemia. Am J Physiol Regul Integr Comp Physiol 296:R972–R978
46. Teplitz L, Igic R, Berbaum ML, Schwertz DW (2005) Sex differences in susceptibility to epinephrine-induced arrhythmias. J Cardiovasc Pharmacol 46:548–555
47. Stromberg A, Martensson J (2003) Gender differences in patients with heart failure. Eur J Cardiovasc Nurs 2:7–18
48. Rosengren A, Hauptman P (2008) Women, men and heart failure: a review. Heart Fail Monit 6:34–40
49. Regitz-Zagrosek V, Lehmkuhl E (2005) Heart failure and its treatment in women: role of hypertension, diabetes and estrogen. Herz 30:356–367
50. Carroll JD, Carroll EP, Feldman T et al (1992) Sex-associated differences in left ventricular function in aortic stenosis of the elderly. Circ 86:1099–1107
51. Guerra S, Leri A, Wang X et al (1999) Myocyte death in the failing human heart is gender dependent. Circ Res 85:856–866
52. Agvall B, Dahlstrom U (2001) Patients in primary health care diagnosed and treated as heart failure, with special reference to gender differences. Scand J Prim Health Care 19:14–19
53. Harjai KJ, Nunez E, Stewart Humphrey J et al (2000) Does gender bias exist in the medical management of heart failure? Int J Cardiol 75:65–69
54. Rathore SS, Wang Y, Krumholz HM (2002) Sex-based differences in the effect of digoxin for the treatment of heart failure. N Engl J Med 347:1403–1411
55. Gasse C, Hense HW, Stieber J et al (2002) Factors associated with differences in antihypertensive drug treatment: results from the MONICA Augsburg Population Surveys 1989/90 and 1994/95. Soz Präventivmed 47:128–142
56. Parashar S, Katz R, Smith NL et al (2009) Race, gender, and mortality in adults > or = 65 years of age with incident heart failure (from the Cardiovascular Health Study). Am J Cardiol 103:1120–1127
57. Mejhert M, Kahan T, Edner M, Persson HE (2008) Sex differences in systolic heart failure in the elderly: the prognostic importance of left ventricular mass in women. J Womens Health 17:373–381
58. Grohe C, Kahlert S, Lobbert K et al (1997) Cardiac myocytes and fibroblasts contain functional estrogen receptors. FEBS Lett 416:107–112
59. Taylor AH, Al-Azzawi F (2000) Immunolocalisation of oestrogen receptor β in human tissues. J Mol Endocrinol 24:145–155

60. Jankowski M, Rachelska G, Donghao W et al (2001) Estrogen receptors activate atrial natriuretic peptide in the rat heart. Proc Nat Acad Sci USA 98:11765–11770
61. Nordmeyer J, Eder S, Mahmoodzadeh S et al (2004) Upregulation of myocardial estrogen receptors in human aortic stenosis. Circ 110:3270–3275
62. Nuedling S, Kahlert S, Loebbert K et al (1999) 17 β-estradiol stimulates expression of endothelial and inducible NO synthase in rat myocardium in-vitro and in-vivo. Cardiovasc Res 43:666–674
63. Dash R, Schmidt AG, Pathak A et al (2003) Differential regulation of p38 mitogen-activated protein kinase mediates gender-dependent catecholamine-induced hypertrophy. Cardiovasc Res 57:704–714
64. Van Eickels M, Grohe C, Cleutjens JP et al (2001) 17 β-estradiol attenuates the development of pressure-overload hypertrophy. Circ 104:1419–1423
65. Babiker FA, De Windt LJ, van Eickels M et al (2004) 17 β-estradiol antagonizes cardiomyocyte hypertrophy by autocrine/paracrine stimulation of a guanylyl cyclase A receptor-cyclic guanosine monophosphate-dependent protein kinase pathway. Circ 109:269–276
66. Schulz-Menger J, Abdel-Aty H, Rudolph A et al (2008) Gender differences in left ventricular remodeling and fibrosis in hypertrophic cardiomyopathy: insights from cardiovascular magnetic resonance. Eur J Heart Fail 10:850–854
67. Vizgirda VM, Wahler GM, Sondgeroth KL et al (2002) Mechanisms of sex differences in rat cardiac myocyte response to β-adrenergic stimulation. Am J Physiol Heart Circ Physiol 282:H256–H263
68. Johnstone D, Limacher M, Rousseau M et al (1992) Clinical characteristics of patients in studies of left ventricular dysfunction (SOLVD). Am J Cardiol 70:894–900
69. Gasse C, Stieber J, Doring A et al (1999) Population trends in antihypertensive drug use: results from the MONICA Augsburg Project 1984 to 1995. J Clin Epidemiol 52:695–703
70. Barrett-Connor E (2007) Hormones and heart disease in women: the timing hypothesis. Am J Epidemiol 166:506–510
71. Liu L, Klein L, Eaton C et al (2020) Menopausal hormone therapy and risks of first hospitalized heart failure and its subtypes during the intervention and extended postintervention follow-up of the Women's Health Initiative Randomized Trials. J Card Fail 26:2–12
72. Schierbeck LL, Rejnmark L, Tofteng CL et al (2012) Effect of hormone replacement therapy on cardiovascular events in recently postmenopausal women: randomised trial. Br Med J 345:e6409
73. Apostolakis S, Sullivan RM, Olshansky B, Lip GY (2014) Hormone replacement therapy and adverse outcomes in women with atrial fibrillation: an analysis from the atrial fibrillation follow-up investigation of rhythm management trial. Stroke 45:3076–3079
74. Lamba S, Abraham WT (2000) Alterations in adrenergic receptor signaling in heart failure. Heart Fail Rev 5:7–16
75. Clark AL, Cleland JGF (2000) The control of adrenergic function in heart failure: therapeutic intervention. Heart Fail Rev 5:101–114
76. Hoeker GS, Hood AR, Katra RP et al (2014) Sex differences in β-adrenergic responsiveness of action potentials and intracellular calcium handling in isolated rabbit hearts. PLoS ONE 9:e111411
77. Frigerio M, Roubina E (2005) Drugs for left ventricular remodeling in heart failure. Am J Cardiol 96:10L-18L
78. Wang X, Ren B, Liu S et al (2003) Characterization of cardiac hypertrophy and heart failure due to volume overload in the rat. J Appl Physiol 94:752–763
79. Wang X, Sentex E, Saini HK et al (2005) Upregulation of β-adrenergic receptors in heart failure due to volume overload. Am J Physiol Heart Circ Physiol 289:H151–H159
80. Monasky MM, Varian KD, Janssen PM (2008) Gender comparison of contractile performance and β-adrenergic response in isolated rat cardiac trabeculae. J Comp Physiol B 178:307–313
81. Chu SH, Sutherland K, Beck J et al (2005) Sex differences in expression of calcium-handling proteins and β-adrenergic receptors in rat heart ventricle. Life Sci 76:2735–2749

82. Vizgirda VM, Wahler GM, Sonderoth KL et al (2002) Mechanisms of sex differences in rat cardiac myocyte response to β-adrenergic stimulation. Am J Physiol Heart Circ Physiol 282:H256–H263
83. Lindenfeld J, Cleveland JC Jr, Kao DP et al (2016) Sex-related differences in age-associated downregulation of human ventricular myocardial β1-adrenergic receptors. J Heart Lung Transplant 35:352–361
84. Turner MJ, Mier CM, Spina RJ et al (1999) Effects of age and gender on the cardiovascular responses to isoproterenol. J Gerontol A Biol Sci Med Sci 54:B393–B400
85. Wei SK, McCurley JM, Hanlon SU, Haigney MC (2007) Gender differences in Na/Ca exchanger current and β-adrenergic responsiveness in heart failure in pig myocytes. Ann N Y Acad Sci 1099:183–189
86. Dent MR, Tappia PS, Dhalla NS (2012) Gender related alterations of β-adrenoceptor mechanisms in heart failure due to arteriovenous fistula. J Cell Physiol 227:3080–3087
87. Ahmet I, Krawczyk M, Zhu W et al (2008) Cardioprotective and survival benefits of long-term combined therapy with β2 adrenoreceptor (AR) agonist and β1 AR blocker in dilated cardiomyopathy postmyocardial infarction. J Pharmacol Exp Ther 325:491–499
88. Mieno S, Watanabe F, Sawa Y, Horimoto H (2005) Gene transfer of β2 adrenergic receptor enhances cardioprotective effects of ischemic preconditioning in rat hearts after myocardial infarction. Interact Cardiovasc Thorac Surg 4:163–167
89. Brodde OE, Bruck H, Leineweber K (2006) Cardiac adrenoceptors: physiological and pathophysiological relevance. J Pharmacol Sci 100:323–337
90. Garcia-Villalon AL, Buchholz JN, Krause DN, Duckles SP (1996) Sex differences in the effects of 17 β-estradiol on vascular adrenergic responses. Eur J Pharmacol 314:339–345
91. Wu Q, Zhao Z, Sun H et al (2008) Oestrogen changed cardiomyocyte contraction and β-adrenoceptor expression in rat hearts subjected to ischaemia-reperfusion. Exp Physiol 93:1034–1043
92. Shima S, Okeyama N, Akamatu N (1989) Effects of oestrogen on adenylate cyclase system and glucose output in rat liver. Biochem J 257:407–411
93. Lala A, Tayal U, Hamo CE et al (2022) Sex differences in heart failure. J Card Fail 28:477–498
94. Pecha S, Geelhoed B, Kempe R et al (2021) No impact of sex and age on β-adrenoceptor-mediated inotropy in human right atrial trabeculae. Acta Physiol 231:e13564
95. Tuncay E, Seymen AA, Sam P et al (2009) Effects of β-adrenergic receptor blockers on cardiac function: a comparative study in male versus female rats. Can J Physiol Pharmacol 87:310–317
96. Tsang S, Wu S, Liu J, Wong TM (2008) Testosterone protects rat hearts against ischaemic insults by enhancing the effects of $\alpha_1$-adrenoceptor stimulation. Br J Pharmacol 153:693–709
97. Sun J, Fu L, Tang X et al (2011) Testosterone modulation of cardiac β-adrenergic signals in a rat model of heart failure. Gen Comp Endocrinol 172:518–525

# Chapter 26
# Cardiovascular Consequences of Metabolic Disturbances in Women

**Belma Turan**

**Abstract** Metabolic disorders in humans are characterized by hyperglycemia and hyperinsulinemia, insulin resistance, impaired glucose intolerance, dyslipidemia, hypercholesterolemia, and hypertriglyceridemia affecting main systemic parameters as well as cells and organs such as cardiometabolic disturbances. These create high risks for the prevalence of various diseases, including cardiovascular diseases (CVDs). Metabolism is defined as a process in cells to provide energy and remove waste products via catabolism and anabolism. Metabolic syndrome (MetS) is present in about 5% of the population with normal body weight, 20% who are overweight, and 60% of those who are considered obese, while it increases with age in a sex-specific manner. For instance, MetS are slightly higher in men below 50 years, with a marked reversal after 50 years (https://www.webmd.com/heart-disease/guide/metabolic-syndrome). Sex differences in CVDs have been reported in human and animal studies, how they are involved in women's diseases, effectiveness of therapies, and clinical outcomes compared to men, however, there are various complex molecular mechanisms underlining these differences which needs to be clarified. According to current knowledge on sex-dependent differences, electrophysiological parameters, contractility, several intracellular signaling mechanisms concentrated on cellular metabolism, gene and protein expressions, and posttranslational protein modification in the heart. Taking into consideration, the pleiotropic effects of estrogen, it exerts a protective effect on the cardiovascular system throughout the premenopausal period of women. The cardioprotective action of estrogen on the cardiovascular system largely depends on its critical role in the prevention and/or regulation of oxidative stress in the heart.

**Keywords** Sex hormones · Electrophysiology · Mitochondria · Oxidative stress · Metabolic syndrome · Obesity · Diabetes

---

B. Turan (✉)
Department of Biophysics, Faculty of Medicine, Lokman Hekim University, Ankara, Turkey
e-mail: belma.turan@medicine.ankara.edu.tr; belma.turan@lokmanhekim.edu.tr

© The Author(s), under exclusive license to Springer Nature Switzerland AG 2023 
L. Kirshenbaum and I. Rabinovich-Nikitin (eds.), *Biology of Women's Heart Health*, Advances in Biochemistry in Health and Disease 26,
https://doi.org/10.1007/978-3-031-39928-2_26

## Introduction

The various ongoing biochemical processes in mammalians' bodies such as catabolism and anabolism are in the concept of metabolism, which has a role to maintain the mammalian life with the physiological condition. Metabolism includes two main components such as catabolism and anabolism, ongoing with many complex processes in organs, tissues, and cells. A metabolic disorder can occur under the impairment of an abnormal chemical reaction raised in a living cell, which in turn affects energy production and regulation. So, metabolic disorders are abnormal conditions, including a range of conditions further affecting negatively the mammalian body and causing different symptoms and complications [1, 2]. Various factors such as genetic factors, lifestyle and diet habits, and sexdifferences play important roles in the induction of metabolic disorders [3, 4]. The metabolic disorders affect negatively the mammalian body function altering the distribution of proteins, fats, and carbohydrates, which are associated with the induction of metabolic syndrome, diabetes, or both in humans.

Metabolic syndrome (MetS) or Syndrome X, characterized by insulin resistance [5], is a complex syndrome including a combination of several risk factors such as glucose intolerance, abdominal obesity, dyslipidemia, and high blood pressure. The combination of these changes leads to the development of type 2 diabetes mellitus (T2DM) as well as the induction of cardiovascular disorders [6–9]. In this concept, there are various data in the literature to demonstrate the association between insulin resistance and the development of cardiovascular disease [10].

In the MetS, not only one risk factor but also combinations of various risk factors can lead to further additional risks for the development of cardiovascular dysfunction [11–15]. From a clinical perspective, a special heart disease detected in MetS individuals seems to be associated with mostly insulin resistance. Indeed, cardiovascular disease (CVD), whether or not associated with metabolic disturbances, is the main cause leading to death at an increasing raise-rate worldwide [16].

Various external factors including sedentary lifestyle and changes in nutritional habits have led to an increase in the prevalence of overweight body and obesity, thereby the development of MetS among humans, not only in adults but also in childhood ages [17–19]. Recently, the notable level increases in the prevalence of MetS, and also the associated risks for CVDs have raised concerns about well-understand their risk factors in humans. More importantly, these concerns are also focused on well-understand whether or not sexdifferences can play roles in the prevalence of CVDs via metabolic disturbances. Generally, it is accepted that sex differences can arise from both reversible hormonal effects and irreversible organizational processes, including body fat distribution, glucose homeostasis, insulin signaling, ectopic fat accumulation, and lipid metabolism during normal growth as well as differences in responses to hormonal or nutritional factors. Moreover, there is a great effort to identify novel gender-difference associated factors to contribute to both the induction and/or susceptibility to MetS and also to CVDs. In these regards, there are studies focused on how the tissue-specific gene regulation can be different between men and women, contributing to differential metabolism and thereby bringing novel insights to open doors for personalized therapeutic approaches in cardiometabolic diseases [20–22]. This chapter will particularly focus on the role of metabolic disturbances in the development of CVDs in women. Insights into novel therapeutic interventions will be also discussed as well.

## Sex Differences in Lifestyle and Risk Factors of Metabolic Syndrome

Metabolic syndrome (MetS) is mainly associated with an increased risk not only for CVDs but also for T2DM. The MetS prevalence is high and increases with age in a gender-specific manner. For instance, the MetS prevalence is slightly higher in men below 50 years old, but it reverses after 50 years old [19, 23, 24]. Epidemiological studies also emphasize the expectation of the rising prevalence of the MetS and associated CVDs with the global obesity epidemic and suggest early weight management to reduce risk in pre-symptomatic individuals with large waists. In this context, documented data have shown the marked role of differences between women and men, at most, through differences in gene expression from the sex chromosomes, mostly, under the light of epidemiological and pathophysiological findings, clinical manifestations, and outcomes of clinical therapies [24]. These factors are leading to subsequent differences in sexual hormones further associated with different gene expressions and functions in the CV system. These differences can raise from biological differences between women and men, named sexdifferences. However, it is truly not possible to distinguish between the effects of sex (S) and sex(G) in the medical field. For instance, some clinicians suggested getting under consideration both of them by using the term S&G for all medical relevant differences between women and men [24, 25]. Supporting, Beigh and Jain [26] performed a comparative study among MetS males and females aged 30 years. A high prevalence of MetS in women than men is existing via comparative findings of some parameters such as the body mass index, waist circumference, blood HDL and cholesterol levels, and fasting blood glucose levels among humans. In addition, besides these parameters, age status is also effective in gender-based differences and risk factors for MetS-associated syndromes in adults [27].

Several factors can underline gender-dependent differences in the risk factors associated with MetS-induced CVDs. The considerable differences on this topic seem to be associated with the different responses to various pathological stimuli such as inflammatory factors. In this regard, interestingly, it has been shown that metabolic impairment in humans leads to different inflammatory profiles across sex with high pro-inflammatory cytokine productions in males higher than in females [28, 29]. It has been also observed varying cardiovascular profiles secondary to metabolic deterioration among male and female samples [30]. Furthermore, epidemiological data also demonstrated clinically relevant sex- and gender-dependent differences in the rates of T2DM and CVD outcomes across the life span associated with age-related data including not only midlife but also youth [31–33].

Stress has well-documented adverse effects on a range of health outcomes, including obesity, and metabolic disturbances, which are important contributors to the development of CVDs, as well [34]. During every period of life span, an important percentage of populations is exposed to a wide range of emotional stress, which is an independent risk factor for chronic diseases such as metabolic syndromes and CVDs [35], while chronic stress is also a risk factor for different pathological states in individuals [36]. Furthermore, several experimental and clinical studies mentioned the involvement of oxidative stress in the multifactorial pathological stimuli in the induction of MetS [37]. Due to our general knowledge, various sources including inducible nitric oxide synthase (iNOS) and mitochondria can contribute to the enhancement of reactive oxygen species (ROS) generation in cells. In these regards, it has been shown that MetS is characterized by an inappropriate rise of plasma-free fatty acids (FFAs) that is related to and activates ROS production [38, 39]. Although this mechanism is not known well, studies showed triggering of ROS augmentation as well as the augmentations in NAPDH) oxidase (NOX), NOS, and oxygenase besides reactive nitrogen species (RNS) in adipose tissue [39, 40], likely by stimulating NOX and lowering antioxidative enzymes [41] are among the factors to contribute increasing oxidative stress in cells. Obesity has an increasing prevalence worldwide and is generally related to the dysregulation of caloric intake and energy expenditure, which can promote oxidative stress being likely a mechanistic connexion between obesity and related complications, including insulin resistance [41]. For instance, FFAs in obese mammalians can prevent ROS production in the mitochondria, whereas, oxidative stress can increase fat accumulation by stimulating pre-adipocyte proliferation and adipocyte differentiation [41, 42].

As mentioned in previous paragraphs, various hormones play an important role in function of the heart. Supporting this statement, it has been shown an important protective effect of estrogen on cardiovascular disease is mediated by oxidative stress. Studies have shown that estrogen has important and critical physiological effects, including antioxidant effect via reversing the decreases in oxidized/reduced glutathione ratio, in the levels of superoxide dismutase and glutathione peroxidase, and preventing the cardiovascular risk in postmenopausal women, which have been attributed to part to lowering the increased ROS level [25, 43–46].

In this rergard, there is growing awarenss mitochondrial funcitn is critical for maintiaing normla cardaic function. Furthermore, there are both clinical and experimental data demonstraing the relationship between heart failure in humans and mitochondrial disruption. For instance, a close relationship exists between mitochondrial defects such as alterations in mitochondrial phosphorylation system, complex I respiration, fatty acid oxidation, and decreases in the mitochondrial content in heart failure [47–49]. Therefore, it can be concluded that significant metabolic defects, associated with mitochondrial dysfunctin are an underlying cause of CVDs.

## Insulin Resistance and Association with Cardiovascular Diseases

Insulin resistance has been shown to play important role in the development of oxidative stress in MetS besides other roles in the glucose pathway [10, 50–52]. In a general aspect, insulin is a key hormone to regulate cellular metabolism in many tissues in the human body, while tissue response to insulin stimulation is decreased under insulin resistance. In general aspects, there are significant defects in both uptake and oxidation of glucose and a decrease in glycogen synthesis, thereby, inducing the development of cardiovascular disease via several metabolic alterations in mammalians under insulin resistance. For instance, insulin resistance can induce an imbalance in glucose metabolism that generates chronic hyperglycemia, which in turn triggers oxidative stress and causes an inflammatory response that leads to cell damage. Insulin resistance, therefore, is closely associated with the stimulation of oxidative stress at systemic and cellular levels in mammalians.

The maintenance of normal circulating carbohydrate and lipid levels and their metabolism are under control with a normal insulin sensitivity of cells. Under physiological conditions, blood glucose levels can modulate the release of insulin from the pancreas, and therefore, induce stimulation of glucose uptake and metabolism in muscles, adipose tissue, and several other insulin-sensitive organs, including the heart. So, in the concept of muscle-associated high metabolic demand, insulin resistance has significant effects on heart muscle besides other tissues as main targets of intracellular glucose transport as well as glucose and lipid metabolisms [53–55]. These abnormal regulations of cellular glucose are further leading to impaired responses to any external pathological stimuli in mammalian cells.

Insulin resistance in the heart tissue can directly affect intracellular insulin signaling with a reduced glucose oxidation rate in cardiomyocytes. As demonstrated by several studies, an acute insulin application to myocytes can provide a significantly different effect than that of its chronic effect. For instance, as shown by some studies in the liver of obese mice, insulin inhibits fatty acid synthase delivered acutely, while chronic hyperinsulinemia may induce fatty acid synthase activity and increase fatty acid synthesis [56]. At the organ level, early and later studies mentioned that there exists an organ-dependent insulin resistance, including heart, as a manner of tissue-dependency [57, 58]. Correspondingly, it can be accepted that the effect of insulin resistance in different tissues is depending on either the physiological function or the metabolic function of the targeting tissue.

The effect of insulin resistance at the cellular level mainly includes disruptions in mitochondrial function and endoplasmic reticulum stress besides the alterations in other cellular signaling pathways associated with lipotoxicity. As summarized in Fig. 26.1, the increases in the mediators, such as ROS and RNS together with impairments in both glucose and fatty acid oxidations and also other metabolic and unfolded protein response disturbances, are all responsible for the induction of mitochondrial dysfunction [59–63]. In these regards, as mentioned in the previous section, there are close relations between increases in the production of ROS and mitochondrial

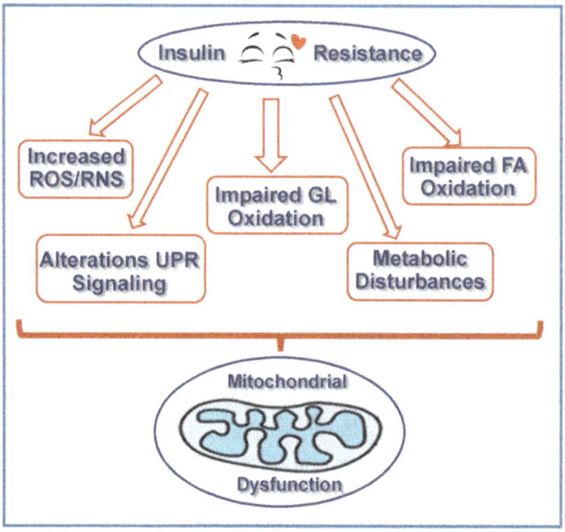


**Fig. 26.1** A schematic presentation of the relationship between insulin resistance and mitochondrial dysfunction. Here; ROS and RNS: Reactive oxygen and nitrogen species, GL and FA oxidation: Glucose and fatty acid oxidation, UPR: Unfolded protein response

dysfunction in various cell types, under either hyperglycemia, hyperinsulinemia, or both. In this regard, one can summarize that abnormal regulation of the redox-sensitive signaling pathway together with increased production of not only ROS but also RNS promote oxidative stress in the cells. Therefore, all these abnormalities in redox-regulation of the cells further lead to an intersection with the development of cellular insulin resistance.

Furthermore, mitochondria are targets of cellular ROS. A high cellular ROS level severely affects mitochondrial function, leading to defects in mitochondrial electron transport, complex enzyme activities, ATP diminution, caspase 3 liberations, and mitochondrial DNA lessening [64–66], thereby contributing to insulin resistance and its related syndromes. In turn, mitochondrial dysfunction can put effect on intermediary metabolism [65, 67], cellular redox signaling [14, 64], $Ca^{2+}$- and $Zn^{2+}$-homeostasis [68–70], and cell death [71]. Therefore, it is the truly correct definition for the insulin working ways: Insulin in cells has multiple processes basically to allow the correct balance between nutrient supply and demand [72]. However, the target cells fail to respond to ordinary levels of circulating insulin in insulin resistance conditions [73]. Briefly, it can be summarized that insulin exerts a lower biological effect under insulin resistance in comparison to the physiological level, leading to defective insulin-stimulated glucose uptake.

## Sex-Related Differences in Cardiac Function

There is a great effort to determine the role of sex differences in cardiovascular function in not only patients but also healthy populations. Indeed, studies demonstrated that the hearts of men and women respond differently to physiologic stresses and pathological stimuli [74–76]. Sex-dependent differences in myocardial contractile performance, resting heart rate, cardiac growth, stroke volume, response to exercise, and arterial blood pressure have been observed [77–80]. Taking into consideration the stress-associated adverse effects of some pathological stimuli (i.e. hypertension, obesity, insulin resistance, and other metabolic disturbances), sexdifferences, particularly observed in postmenopausal women, play an important role in comorbidities during the lifespan induction and the overcoming the CVDs [81, 82]. A recent clinical study, performed in Northeast China characterized by low levels of income and education, demonstrated that the risk factors for MetS have both similarities and differences depending on gender, particularly in the prevalence of CVDs [27]. There are supporting early studies that demonstrated why sexdifferences could be described by the biology-linked differences between women and men. These are basically due to differences in sex chromosomes, sex-specific gene expression of autosomes, sex hormones, and their effects on organ systems, including the CV system. In this regard, it has been shown that women show more dramatic changes in hormones during their lifetime than men [24]. In these regards, it is a fact that sex hormones have a great impact on energy metabolism, body composition, vascular function, and inflammatory responses. Thus, endocrine imbalances relate to unfavorable cardiometabolic traits, observable in women with androgen excess. Also, both biological and psychosocial factors are responsible for sex and sexdifferences in diabetes risk and outcome [83, 84]. In a supporting study, the authors demonstrated that estrogen, having a wide range of critical physiological effects including actions on both CV and metabolic systems, has marked cardioprotection on CVD mediated by oxidative stress [43].

The increasing prevalence of obesity, MetS, and comorbidities among women, particularly following the beginning of the menopause period brings a high risk for CVDs. So, it is clear evidence that metabolic risk factors are more strongly associated with CVDs in women than men and need more data (or metabolic predictors) yet to overcome the metabolic disturbances in humans starting from childhood age [85]. Sex and sexdifferences in the prevalence of MetS can translate into different CV risks associated with metabolic disturbances. As discussed by many review articles, however, there are differences related to the role of sexdifferences in the induction of T2DM and CV risks across the life course of men and women. In the presence of T2DM, the difference in the absolute rates of CVD between sexes seems to get lessen, with a higher level in men than women [86–88]. Large-scale observational studies suggest that T2DM women have a significantly high risk for the incidence of CVD than T2DM men [89]. Overall, as documented in a recent clinical study, the authors mentioned the striking role of sex differences in cardiometabolic health and disease and how sexdifferences have a role in cardiometabolic risk factors. Among them, the prevalence of the MetS is strongly related to different body fat distribution patterns, particularly abdominal adiposity, adiponectin, and related body and blood biomarkers [90].

Although cardiovascular disease remains essentially an age-related condition in both men and women and despite similarities for CV disorders, the number of CV events is higher in elderly women because of the higher prevalence of MetS and/or obesity compared to men in this age group [25, 86, 88, 91]. Clinical outcomes strongly support not only the sex-dependent effects but also age-dependent effects on the development of CVDs in humans. However, as a general understanding of how and why CVD inductions differ between the sexes remains limited, there are great amounts of data available to provide the differences between the sex in patterns of age-related cardiac remodeling. It can be considered that both hormonal and non-hormonal factors underlie sexdifferences in CV aging and the development of age-related CVDs.

There are various mechanistic pathways to demonstrate the role of sexdifferences in CVDs. Figure 26.2 can summarize the underline factors associated with obesity, MetS, T2DM, and aging to induce the main risk factors leading to gender-dependent induction of CVDs. Major differences between men and women exist in epidemiology, manifestation, pathophysiology, treatment, and outcomes of CVDs, such as coronary artery disease, pressure overload, hypertension, cardiomyopathy, and heart failure [92]. In terms of biological contributors, there are three main topics to describe the mechanisms of sexdifferences in CVDs such as genetic mechanisms, epigenetic mechanisms, and sex hormones. In these topics, genetic mechanisms mainly include sex-specific heterochromatin effects of sex chromosomes as genetic mechanisms [93], sex-specific effects of autosomal genetic variants [94], and genome-wide expression profiling and proteomics [95–98]. In the context of epigenetic mechanisms, histone and DNA modifications and non-coding RNA are the main factors related to the gender-dependent induction of CVDs. Both sex hormones and their receptors have an important impact on risk factors for the prevalence of gender-dependent CVDs [99, 100]. There are gender-dependent differences in the synthesis and metabolism of sex hormones, in subtypes and regulation of sex hormone receptors in the CV system, and genomic actions of sex hormone receptors [101–103]. There are also nongenomic effects of sex hormone receptor activation on the induction of gender-dependent CVDs. Studies mentioned that sex hormones can exert rapid nongenomic effects mediated either by the classical receptors [104, 105]. Interestingly, the nongenomic actions of sex hormones are not dependent on changes in gene expression for their actions.

## Metabolic Disturbances in the Heart Induce Cardiovascular Dysfunction in Women

There are several differences between the size and function of men's and women's hearts such as the smaller ratio of heart chambers with thinner walls and finer veins in a woman's heart than the man's heart. A woman's heart also pumps faster than a man's heart whereas a man's heart ejects more blood with each pump than the woman's heart. The differences between heart functions of women and men have been documented by both cardiovascular physiology and the incidence, severity, and comorbidities of several pathological conditions such as obesity, diabetes, and

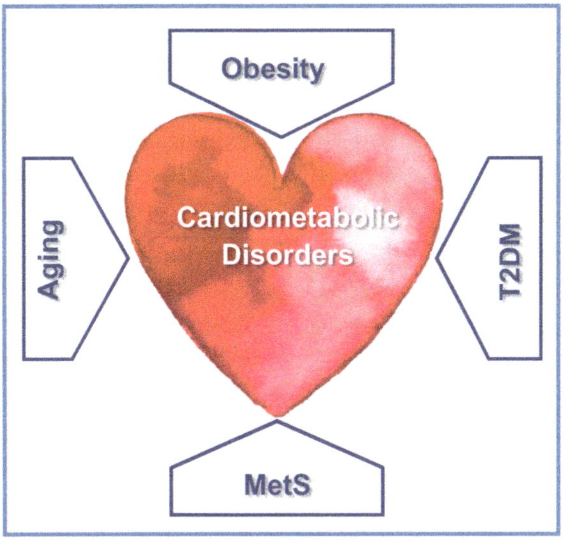


**Fig. 26.2** Cardiometabolic risk factors leading to cardiovascular diseases. Type 2 diabetes (T2DM), metabolic syndrome (MetS), aging, and obesity are the main responsible mechanisms and contributors to cardiometabolic disturbances which are further leading to cardiovascular diseases

CVDs [77] Consequently, the hearts of women and men do not only differ in physical parameters but also display several functional, structural, genetic, and hormonal differences which either individually or as a manner of their combination can lead to different responses to the similar pathological stimuli. Taking into consideration the documents in the literature and also presented in the previous sections, the overall gender-related mechanisms in the induction of gender-dependent MetS-associated cardiac remodeling and dysfunction can be summarized in Fig. 26.3 [21, 43, 85, 87, 106–108].

There are important differences in myocardial metabolism of women's hearts and men's hearts, such as decreased glucose utilization and oxygen consumption by the myocardium at most via the effect of estrogen [109, 110]. Besides, early studies documented the impact of the inherent sexinfluence on the myocardial contractile performance of healthy and diseased mammalians, mostly dependent on gender-dependent alterations in the regulation of diastolic and systolic $Ca^{2+}$ levels in cardiomyocytes as one of the predominant mechanisms underlying the gender-based distinctions [111–114]. Recent studies on this topic are also supporting the early ones. In these regards, Sorrentino and co-workers examined the effects of hyperglycemia on cardiac function and myocyte physiology in experimental diabetic mice [70]. Their data demonstrated that hyperglycemia-induced defective cardiac function is associated with increased chamber dilation, thinning of the LV, and significant loss in the myocyte population together with marked alterations in cellular level $Ca^{2+}$-homeostasis which are further associated with induction of diabetic cardiomyopathy.

Sex differences in the heart have been examined and presented by using electrophysiology, contractility, signaling metabolism, and cardioprotection at cellular levels. These observed differences are associated, in part, with differences due to gene, protein expression, and posttranslational protein modifications via primarily

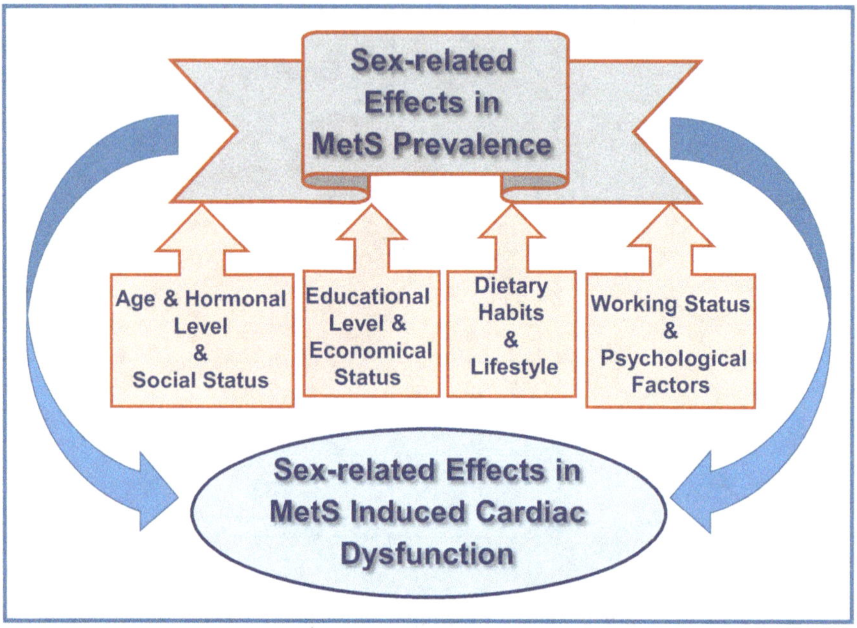


**Fig. 26.3** Schematic possible pathways demonstrating the association between the gender-related effects on the prevalence of metabolic syndrome (MetS) and sex-dependent effects on MetS-induced cardiac dysfunction

estrogen-mediated mechanisms [115]. Estrogen can regulate mRNA, miRNA, and protein expression levels, particularly rapid signaling pathways in cells [116–119]. In line with these statements, the proteomic differences in mitochondria between female rats and male rats are found to be due to differences in the phosphorylation levels of some proteins such as high levels of aldehyde dehydrogenase 2, pyruvate dehydrogenase, and alpha-ketoglutarate dehydrogenase in females than males, via the estrogen-mediated increase in PI3-kinase signaling pathway [120]. They also demonstrated that the levels of some metabolic enzymes are significantly different between female and male rat hearts and these differences could underline the important differences in ROS and RNS levels, as well [121, 122].

The early animal studies on the gender-dependent differences in CV function and CVDs are focused on electrophysiological changes in cellular $Ca^{2+}$-homeostasis-related parameters [123, 124]. The sex differences in the expression levels of some inwardly rectifying $K^{+}$-channels and their regulation with estrogen have been shown in animal heart cardiomyocytes [125] as well as in late $Na^{+}$-channels and L-type $Ca^{2+}$-channels [126, 127]. Supporting, Johnson and co-workers, in their early study, demonstrated the effects of estrogen on the heart and found an increased expression of L-type $Ca^{2+}$-channels in the hearts of estrogen receptor knockout mice [128]. Therefore, these results infer a modulation of cardiomyocyte $Ca^{2+}$-regulation by estrogen to limit the level of $Ca^{2+}$ in the ventricular myocytes. Estrogen can also exert a direct effect on the heart and be involved in the fast regulation of transcription. For instance, an acute application of 17β-estradiol has a negative inotropic effect [129] and inhibits L-type $Ca^{2+}$-current [130] in the guinea-pig heart. Furthermore, several lines of evidence have suggested the gender-related differences in SR function as the main culprit of this dimorphism under hyperglycemia, such as the observation of the smaller amplitude of transient $Ca^{2+}$ increase under electrical stimulation of female cardiomyocytes than those of males [131, 132]. Importantly, it has been mentioned a reduction in SR $Ca^{2+}$-ATPase activity can be the most obvious reason for these results in the above statements, whereas the possible contribution of $Na^{+}/Ca^{2+}$ exchanger to the above pathways cannot be fully excluded [131]. With further studies, it has been documented less thiol oxidation in diabetic female rat cardiomyocytes compared to those of males accounting for lesser variation in the voltage-gated $K^{+}$-channels currents [133]. The authors also reported that the activation of autocrine/paracrine mechanisms is absent or less pronounced in cardiomyocytes from those diabetic female rats via a protective action of estrogen with a lower level of oxidative stress expressed as lower superoxide ion generation [133, 134].

Previously, the role of sex differences in CV function under hyperglycemia has been studied in T1DM mimicking rat cardiomyocytes, focusing on cardiomyocyte level $Ca^{2+}$ regulating mechanisms [135]. In that study, the authors demonstrated defective intracellular $Ca^{2+}$-signaling was more prominent in males than females. Furthermore, additional examinations also showed these differences mainly arise due to differences in the protein and phosphorylation levels of cardiac RyR2 and its regulator protein FKBP12.6. More importantly, that study also reported the apparent less hyperphosphorylation of RyR2 with less thiol oxidation in diabetic female cardiomyocytes providing a lower risk of cardiac abnormalities in females compared to males. These marked gender-associated differences in the intracellular $Ca^{2+}$-signaling can be in part related to either oxidative stress status, antioxidant defense status, or both. Indeed, it is accepted that the differences in ovarian sex hormones, especially the potent antioxidant capacity associated with estrogen, have been documented to play a major role in this "sex bias" of myocardial contractile function [114, 136]. Overall, already published data provide excessive information about gender-dependent differences in hyperglycemia-induced electrophysiological changes in the hearts, in part, associated with differences in a sex-hormone-dependent generation of oxidative stress in cardiomyocytes. Consequently, it can be concluded that although both sex and sex are interpenetrated in humans, alterations in energy metabolism are essential features of CVDs, and sexual dimorphism of energy metabolism and more specifically mitochondria occupies an important part of these processes.

Metabolic predictors of incident coronary heart disease in women have been measured by determination of 371 metabolites in a discovery set of postmenopausal women with coronary heart disease (CHD) and demonstrated their importance in the relationship between lipid oxidation and subsequent CHD as novel markers [85]. Moreover, as mentioned in previous sections, Murphy and co-workers emphasized the role of sex differences in the development of metabolic cardiomyopathy [87] comparing the different CV dysfunction under different pathological conditions. In these regards, in contrast to ischemic cardiomyopathies, diabetic cardiomyopathies are more prominent in women than in men. Although diabetes is a risk factor for CVD, however, there is sexual dimorphism in this risk factor with a fivefold high common in diabetic women compared to that only twofold in diabetic men. Interestingly, the authors in the above study demonstrated that heart failure with preserved ejection fraction is more prevalent in MetS women than age-matched men. Their data presented that the parameters such as $Ca^{2+}$-handling, ROS- and NO-signaling, and cardiac structural proteins have sex-dependent cardiometabolic differences. So, the disturbances in the heart, and metabolic cardiomyopathies (at most observed in MetS, obese, and/or diabetic individuals), are strongly arising as a manner of sexual dimorphism.

Supporting the above statements, it has been observed gender-dependent differential responses to a 28-week high-carbohydrate diet-induced MetS male and female rats. The MetS male rats had increased blood pressure, prolonged Q–T interval in surface electrocardiograms, low heart rate with depressed left ventricular function in hypertrophic hearts, and relaxation of the aorta [15]. A parallel study with the same experimental protocol in aged-matched MetS female rats has presented no significant changes in their heart rate and heart weight, and no significant prolongation in Q-T interval and heart rate (Fig. 26.4). In addition, the light microscopy analysis of the heart tissue from MetS male rats demonstrated high-level fibrosis than those of MetS females (Fig. 26.5). The further biochemical analysis also supported these gender-dependent differences. Although it has been determined significantly a low ratio of phospho-Akt to Akt and a high ratio of phospho-PKA to PKA in the ventricular cardiomyocytes from male MetS rats [137], the phospho-Akt to Akt ratio was significantly increased and phospho-PKA to PKA ratio was significantly decreased in the ventricular cardiomyocytes from female MetS rats compared to the age-matched normal rats (Fig. 26.6a, b). Furthermore, the ROS and RNS levels in the same cells were significantly high in male MetS rats compared to the aged-matched male normal rats [15, 137]. However, the ROS level was significantly high with no significant change in the RNS level determined in the ventricular cells from the female MetS (Fig. 26.6d). There were no significant differences in the ratios of Bax/Bcl-2 between the male and female groups (Fig. 26.6c).

In line with the previously demonstrated data, the authors evaluated the SR $Ca^{2+}$-leak to sex differences in humans either with afterload-induced cardiac hypertrophy with preserved LV function or with end-stage heart failure [138]. Although cardiac function did not differ between sex in both cardiac pathologies, there was a significantly lower $Ca^{2+}$-spark frequency in ventricular cardiomyocytes in females with hypertrophy than those in males. Their further studies strongly pointed out the role of low-level SR $Ca^{2+}$-leak in the low risk of ventricular arrhythmias in female individuals than the males. These findings are also in line with the others, particularly associated with the different levels of oxidation levels among female and male hearts [135]. In addition, as discussed by authors [42, 106], estrogen has a definite protective effect on the cardiovascular system, at most, through oxidative stress levels associated with oxidative-stress induced pathogenesis of atherosclerosis, myocardial dysfunction, cardiac hypertrophy, heart failure, and myocardial ischemia.

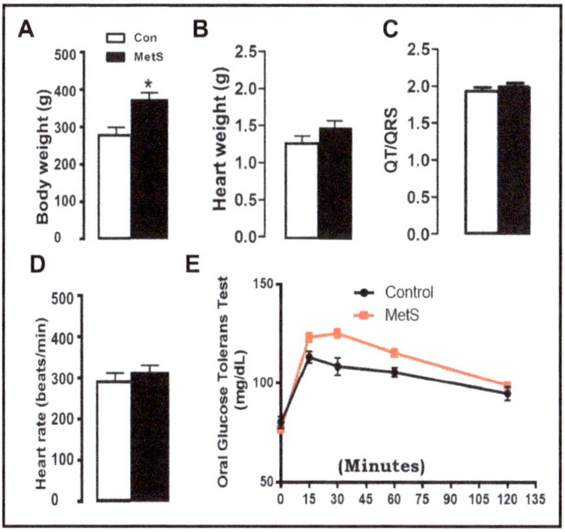


**Fig. 26.4** General parameters female Wistar rats with the age of 2 months old and treated with 32% sucrose in their drinking water for 28 weeks to induce metabolic syndrome (MetS) compared to aged-matched female rats (controls; Con group) the similar animal model used for MetS male rat groups as described, previously. Their body and heart weights (**a**, **b**), the QT-interval calculated from surface electrocardiograms (**c**), and heart rate (**d**). Validation of the development of insulin resistance is determined by the HOMA-IR index (**e**). Data presenting the mean (± SEM) values. The total number of rats/group; n = 7–8. Significance level at *$p$ < 0.05 versus Con group with student-t test

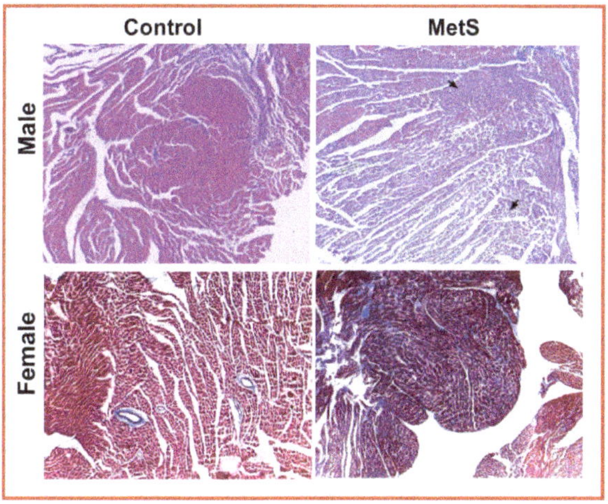


**Fig. 26.5** Light microscopy examinations represent Masson trichrome-stained heart sections from the male (upper) and female (lower) rats in either control (left) or metabolic syndrome, MetS, (right) groups. The abnormal arrangements and loss of integrity in myofibres are more prominent in the MetS male heart compared to the MetS female heart. The fibrosis level was also high in the male heart than in the female heart. The total number of rats/group; n = 4–5. Magnification: 50 μM

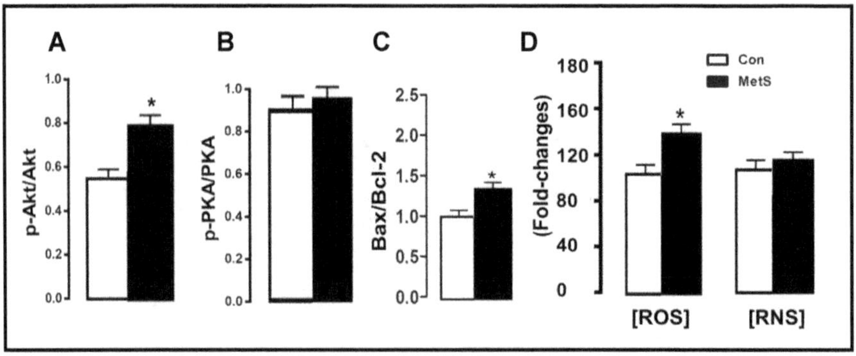


**Fig. 26.6** The biochemical analysis of some heart parameters from female MetS rats compared to those of aged-matched controls and oxidative stress status of isolated cardiomyocytes. The phosphorylation and protein expression levels of insulin-regulated protein Akt (p-Akt and Akt) (**a**) and protein kinase A (p-PKA and PKA) (**b**), the ratio of apoptosis markers Bax/Bcl-2 (**c**), and (**d**) the cardiomyocyte levels of reactive oxygen species (ROS) (left) and reactive nitrogen species (RNS) (right). The intracellular level of ROS ([ROS]) and RNS ([RNS]) were imagined with confocal microscopy in the loaded cells with either specific dyes DCFDA or DAF, respectively. Maximal fluorescence intensity was achieved by a HEPES-buffered solution supplemented with $H_2O_2$ (100 μM) and with NO donor ZipNONO (100 μM). Data are presented as mean (±SEM) values. The total number of rats/group; n = 6–8. Significance levels at $*p < 0.05$vs Con group with student-t test

## Conclusions

There are marked sex-related differences in mammalian cardiac function not only under physiological conditions but also in responses to pathological stimuli. Although metabolomic profiling offers promise for the prediction of CVDs, metabolic risk factors are more strongly associated with CVD in female than males. There is also a strong association between the levels of oxidations and CVDs in women and men, however, it is needed to have an additional investigation of the mechanism and potential clinical use of oxidative risks. More importantly, since alterations in energy metabolism are essential features of CVDs, sex-related dimorphism, particularly in energy metabolism and more specifically mitochondria, plays a vital role in the induction and progress of CVDs with different levels and different health problems. Despite these facts, sex differences are still not under serious consideration in the interest of both experimental and clinical research. Indeed, the mechanisms underlying the sex-dependent differences in CVDs are still poorly understood.

**Acknowledgements** Not applicable.

**Conflict of Interest** The author declares that there is no competing interests.

**Ethical Declaration** The experimental protocol with rats was by the standards of the European Community guidelines on the care and use of laboratory animals and they have been approved by the local ethics committee of Ankara University (No: 2015–12-137).

## References


1. Yoon C et al (2019) Problematic eating behaviors and attitudes predict long-term incident metabolic syndrome and diabetes: the coronary artery risk development in young adults study. Int J Eat Disord 52(3):304–308
2. Jeong E-M et al (2012) Metabolic stress, reactive oxygen species, and arrhythmia. J Mol Cell Cardiol 52(2):454–463
3. Suhre K et al (2011) Human metabolic individuality in biomedical and pharmaceutical research. Nature 477(7362):54–60
4. Boyer SW, Barclay LJ, Burrage LC (2015) Inherited metabolic disorders: Aspects of chronic nutrition management. Nutr Clin Pract 30(4):502–510
5. Reaven GM (1988) Role of insulin resistance in human disease. Diabetes 37(12):1595–1607
6. Ferrannini E et al (1997) On behalf of the European Group for the study of insulin resistance (EGIR). Insulin resistance and hypersecretion in obesity. J Clin Invest, 1997. **100**(5): p. 1166–1173.
7. Zhang Y, Sowers JR, Ren J (2012). Pathophysiological insights into cardiovascular health in metabolic syndrome. Exp Diabetes Res 2012:320534
8. Borghetti G et al (2018) Diabetic cardiomyopathy: current and future therapies. Beyond glycemic control. Front Physiol 9:1514
9. Reaven GM (1995) Pathophysiology of insulin resistance in human disease. Physiol Rev 75(3):473–486
10. Ormazabal V et al (2018) Association between insulin resistance and the development of cardiovascular disease. Cardiovasc Diabetol 17(1):1–14
11. Chinali M et al (2008) Cardiac markers of pre-clinical disease in adolescents with the metabolic syndrome: the strong heart study. J Am Coll Cardiol 52(11):932–938
12. Voulgari C et al (2010) The impact of metabolic syndrome on left ventricular myocardial performance. Diabetes Metab Res Rev 26(2):121–127
13. Galassetti P (2012) Inflammation and oxidative stress in obesity, metabolic syndrome, and diabetes. Exp Diabetes Res 2012:943706
14. Ilkun O, Boudina S (2013) Cardiac dysfunction and oxidative stress in the metabolic syndrome: an update on antioxidant therapies. Curr Pharm Des 19(27):4806–4817
15. Durak A et al (2018) A SGLT2 inhibitor dapagliflozin suppresses prolonged ventricular-repolarization through augmentation of mitochondrial function in insulin-resistant metabolic syndrome rats. Cardiovasc Diabetol 17(1):1–17
16. Gaziano TA et al (2010) Growing epidemic of coronary heart disease in low- and middle-income countries. Curr Probl Cardiol 35(2):72–115
17. Bakhtiyari M et al (2022) Contribution of obesity and cardiometabolic risk factors in developing cardiovascular disease: a population-based cohort study. Sci Rep 12(1):1544
18. Hruby A, Hu FB (2015) The Epidemiology of obesity: a big picture. Pharmacoeconomics 33(7):673–689
19. Han TS, Lean ME (2016) A clinical perspective of obesity, metabolic syndrome and cardiovascular disease. JRSM Cardiovasc Dis 5:2048004016633371
20. Krishnan KC, Mehrabian M, Lusis AJ (2018) Sex differences in metabolism and cardiometabolic disorders. Curr Opin Lipidol 29(5):404
21. Bentley-Lewis R, Koruda K, Seely EW (2007) The metabolic syndrome in women. Nat Clin Pract Endocrinol Metab 3(10):696–704

22. Dwaib HS, AlZaim I, Ajouz G, Eid AH, El-Yazbi A (2022). Sex differences in cardiovascular impact of early metabolic impairment: interplay between dysbiosis and adipose inflammation. Mol Pharmacol 102(1):481–500
23. Murphy MO, Loria AS (2017) Sex-specific effects of stress on metabolic and cardiovascular disease: are women at higher risk? Am J Physiol Regul Integr Comp Physiol 313(1):R1-r9
24. Regitz-Zagrosek V et al (2016) Sexin cardiovascular diseases: impact on clinical manifestations, management, and outcomes. Eur Heart J 37(1):24–34
25. Hegner P et al (2021) The effect of sex and sex hormones on cardiovascular disease, heart failure, diabetes, and atrial fibrillation in sleep apnea. Front Physiol 12:741896
26. Beigh SH, Jain S (2012) Prevalence of metabolic syndrome and sexdifferences. Bioinformation 8(13):613–616
27. Li F-E et al (2021) Sex-based differences in and risk factors for metabolic syndrome in adults aged 40 years and above in Northeast China: results from the cross-sectional China national stroke screening survey. BMJ Open 11(3):e038671
28. Ter Horst R et al (2020) Sex-specific regulation of inflammation and metabolic syndrome in obesity. Arterioscler Thromb Vasc Biol 40(7):1787–1800
29. Mosca L, Barrett-Connor E, Wenger NK (2011) Sex/sexdifferences in cardiovascular disease prevention: what a difference a decade makes. Circulation 124(19):2145–2154
30. Gerdts E, Regitz-Zagrosek V (2019) Sex differences in cardiometabolic disorders. Nat Med 25(11):1657–1666
31. Mayer-Davis EJ et al (2017) Incidence trends of Type 1 and Type 2 diabetes among youths, 2002–2012. N Engl J Med 376(15):1419–1429
32. Urakami T et al (2005) Annual incidence and clinical characteristics of type 2 diabetes in children as detected by urine glucose screening in the Tokyo metropolitan area. Diabetes Care 28(8):1876–1881
33. Sattar N (2013) Sexaspects in type 2 diabetes mellitus and cardiometabolic risk. Best Pract Res Clin Endocrinol Metab 27(4):501–507
34. Brent DA, Silverstein M (2013) Shedding light on the long shadow of childhood adversity. JAMA 309(17):1777–1778
35. Felitti VJ et al (1998) Relationship of childhood abuse and household dysfunction to many of the leading causes of death in adults. The adverse childhood experiences (ACE) study. Am J Prev Med 14(4):245–58
36. Danese A et al (2009) Adverse childhood experiences and adult risk factors for age-related disease: depression, inflammation, and clustering of metabolic risk markers. Arch Pediatr Adolesc Med 163(12):1135–1143
37. Spahis S, Borys J-M, Levy E (2017) Metabolic syndrome as a multifaceted risk factor for oxidative stress. Antioxid Redox Signal 26(9):445–461
38. Haffner S, Taegtmeyer H (2003) Epidemic obesity and the metabolic syndrome. Circulation 108(13):1541–1545
39. Inoguchi T et al (2000) High glucose level and free fatty acid stimulate reactive oxygen species production through protein kinase C–dependent activation of NAD(P)H oxidase in cultured vascular cells. Diabetes 49(11):1939–1945
40. Jacob C, Jamier V, Ba LA (2011) Redox active secondary metabolites. Curr Opin Chem Biol 15(1):149–155
41. Furukawa S et al (2004) Increased oxidative stress in obesity and its impact on metabolic syndrome. J Clin Invest 114(12):1752–1761
42. Lee H et al (2009) Reactive oxygen species facilitate adipocyte differentiation by accelerating mitotic clonal expansion. J Biol Chem 284(16):10601–10609
43. Xiang D, Liu Y, Zhou S, Zhou E, Wang Y (2021) Protective effects of estrogen on cardiovascular disease mediated by oxidative stress. Oxid Med Cell Longev 2021:5523516
44. Bellanti F et al (2013) Sex hormones modulate circulating antioxidant enzymes: impact of estrogen therapy. Redox Biol 1:340–346
45. Strehlow K et al (2003) Modulation of antioxidant enzyme expression and function by estrogen. Circ Res 93(2):170–177

46. Wenger NK, Speroff L, Packard B (1993) Cardiovascular health and disease in women. N Engl J Med 329(4):247–256
47. Park S-Y et al (2016) Mitochondrial function in heart failure: The impact of ischemic and non-ischemic etiology. Int J Cardiol 220:711–717
48. Lemieux H et al (2011) Mitochondrial respiratory control and early defects of oxidative phosphorylation in the failing human heart. Int J Biochem Cell Biol 43(12):1729–1738
49. Olgar Y et al (2018) Increased free $Zn^{(2+)}$ correlates induction of sarco(endo)plasmic reticulum stress via altered expression levels of $Zn^{(2+)}$-transporters in heart failure. J Cell Mol Med 22(3):1944–1956
50. Anderson EJ et al (2009) Mitochondrial $H_2O_2$ emission and cellular redox state link excess fat intake to insulin resistance in both rodents and humans. J Clin Invest 119(3):573–581
51. Reaven G (2012) Insulin resistance and coronary heart disease in nondiabetic individuals. Arterioscler Thromb Vasc Biol 32(8):1754–1759
52. Gast KB et al (2012) Insulin resistance and risk of incident cardiovascular events in adults without diabetes: meta-analysis. PLoS ONE 7(12):e52036
53. Dimitriadis G et al (2011) Insulin effects in muscle and adipose tissue. Diabetes Res Clin Pract 93(Suppl 1):S52–S59
54. Wang CC, Gurevich I, Draznin B (2003) Insulin affects vascular smooth muscle cell phenotype and migration via distinct signaling pathways. Diabetes 52(10):2562–2569
55. Bonora E (2005) Insulin resistance as an independent risk factor for cardiovascular disease: clinical assessment and therapy approaches. Av Diabetol 21(4):255–261
56. Najjar SM et al (2005) Insulin acutely decreases hepatic fatty acid synthase activity. Cell Metab 2(1):43–53
57. Zhang H, Zhang C (2010) Adipose "talks" to distant organs to regulate insulin sensitivity and vascular function. Obesity (Silver Spring, Md) 18(11):2071
58. Yokoyama I et al (1998) Organ-specific insulin resistance in patients with noninsulin-dependent diabetes mellitus and hypertension. J Nucl Med 39(5):884–889
59. Tuncay E et al (2019) $Zn^{2+}$-transporters ZIP7 and ZnT7 play important role in progression of cardiac dysfunction via affecting sarco (endo) plasmic reticulum-mitochondria coupling in hyperglycemic cardiomyocytes. Mitochondrion 44:41–52
60. Wang CCL, Goalstone ML, Draznin B (2004) Molecular mechanisms of insulin resistance that impact cardiovascular biology. Diabetes 53(11):2735–2740
61. Cho H et al (2001) Insulin resistance and a diabetes mellitus-like syndrome in mice lacking the protein kinase Akt2 (PKB beta). Science 292(5522):1728–1731
62. Wei MC et al (2001) Proapoptotic BAX and BAK: a requisite gateway to mitochondrial dysfunction and death. Science 292(5517):727–730
63. Dresner A et al (1999) Effects of free fatty acids on glucose transport and IRS-1-associated phosphatidylinositol 3-kinase activity. J Clin Invest 103(2):253–259
64. Daiber A (2010) Redox signaling (cross-talk) from and to mitochondria involves mitochondrial pores and reactive oxygen species. Biochim Biophys Acta 1797(6–7):897–906
65. Herrmann JM et al (2012) Biogenesis of mitochondrial proteins. Adv Exp Med Biol 748:41–64
66. Montgomery MK, Turner N (2015) Mitochondrial dysfunction and insulin resistance: an update. Endocr Connect 4(1):R1-r15
67. Malhotra JD, Kaufman RJ (2011) ER stress and its functional link to mitochondria: role in cell survival and death. Cold Spring Harb Perspect Biol 3(9):a004424
68. Calì T, Ottolini D, Brini M (2012) Mitochondrial $Ca^{(2+)}$ as a key regulator of mitochondrial activities. Adv Exp Med Biol 942:53–73
69. Tuncay E et al (2019) β3-adrenergic receptor activation plays an important role in the depressed myocardial contractility via both elevated levels of cellular free $Zn^{2+}$ and reactive nitrogen species. J Cell Physiol 234(8):13370–13386
70. Sorrentino A et al (2017) Hyperglycemia induces defective $Ca^{2+}$ homeostasis in cardiomyocytes. Am J Physiol Heart Circ Physiol 312(1):H150-h161
71. Giorgi C et al (2012) Mitochondrial $Ca^{(2+)}$ and apoptosis. Cell Calcium 52(1):36–43

72. Choi CS et al (2008) Paradoxical effects of increased expression of PGC-1alpha on muscle mitochondrial function and insulin-stimulated muscle glucose metabolism. Proc Natl Acad Sci U S A 105(50):19926–19931
73. Christian P, Su Q (2014) MicroRNA regulation of mitochondrial and ER stress signaling pathways: implications for lipoprotein metabolism in metabolic syndrome. Am J Physiol Endocrinol Metab 307(9):E729–E737
74. Kolar F, Ostadal B (2013) Sex differences in cardiovascular function. Acta Physiol (Oxf) 207(4):584–587
75. Humphries KH et al (2017) Sex differences in cardiovascular disease - Impact on care and outcomes. Front Neuroendocrinol 46:46–70
76. Pianosi PT et al (2018) Sex differences in fitness and cardiac function during exercise in adolescents with chronic fatigue. Scand J Med Sci Sports 28(2):524–531
77. Huxley VH (2007) Sex and the cardiovascular system: the intriguing tale of how women and men regulate cardiovascular function differently. Adv Physiol Educ 31(1):17–22
78. Celentano A et al (2003) Sexdifferences in left ventricular chamber and midwall systolic function in normotensive and hypertensive adults. J Hypertens 21(7):1415–1423
79. McKenna DS et al (2006) Gender-related differences in fetal heart rate during first trimester. Fetal Diagn Ther 21(1):144–147
80. Papanek PE et al (1998) Gender-specific protection from microvessel rarefaction in female hypertensive rats. Am J Hypertens 11(8 Pt 1):998–1005
81. Chopra KK et al (2009) Sex differences in hormonal responses to a social stressor in chronic major depression. Psychoneuroendocrinology 34(8):1235–1241
82. Möller-Leimkühler AM (2010) Higher comorbidity of depression and cardiovascular disease in women: a biopsychosocial perspective. World J Biol Psychiatry 11(8):922–933
83. Kautzky-Willer A, Harreiter J, Pacini G (2016) Sex and sex differences in risk, pathophysiology and complications of Type 2 diabetes mellitus. Endocr Rev 37(3):278–316
84. Fourny N et al (2021) Sex differences of the diabetic heart. Front Physiol 12:661297
85. Paynter NP et al (2018) Metabolic predictors of incident coronary heart disease in women. Circulation 137(8):841–853
86. Pucci G et al (2017) Sex- and gender-related prevalence, cardiovascular risk and therapeutic approach in metabolic syndrome: a review of the literature. Pharmacol Res 120:34–42
87. Murphy E et al (2017) Sex differences in metabolic cardiomyopathy. Cardiovasc Res 113(4):370–377
88. Sergi G et al (2020) Sexdifferences in the impact of metabolic syndrome components on mortality in older people: a systematic review and meta-analysis. Nutr Metab Cardiovasc Dis 30(9):1452–1464
89. Huebschmann AG et al (2019) Sex differences in the burden of type 2 diabetes and cardiovascular risk across the life course. Diabetologia 62(10):1761–1772
90. Strack C et al (2022) Sexdifferences in cardiometabolic health and disease in a cross-sectional observational obesity study. Biol Sex Differ 13(1):8
91. Merz AA, Cheng S (2016) Sex differences in cardiovascular ageing. Heart 102(11):825–831
92. Regitz-Zagrosek V, Kararigas G (2017) Mechanistic pathways of sex differences in cardiovascular disease. Physiol Rev 97(1):1–37
93. Charchar FJ et al (2003) Y is there a risk to being male? Trends Endocrinol Metab 14(4):163–168
94. Silkaitis K, Lemos B (2014) Sex-biased chromatin and regulatory cross-talk between sex chromosomes, autosomes, and mitochondria. Biol Sex Differ 5(1):2
95. Haddad GE et al (2008) Human cardiac-specific cDNA array for idiopathic dilated cardiomyopathy: sex-related differences. Physiol Genomics 33(2):267–277
96. Heidecker B et al (2010) The gene expression profile of patients with new-onset heart failure reveals important gender-specific differences. Eur Heart J 31(10):1188–1196
97. Kararigas G et al (2014) Comparative proteomic analysis reveals sex and estrogen receptor β effects in the pressure overloaded heart. J Proteome Res 13(12):5829–5836

98. Kararigas G et al (2014) Genetic background defines the regulation of postnatal cardiac growth by 17β-estradiol through a β-catenin mechanism. Endocrinology 155(7):2667–2676
99. Grohé C et al (1997) Cardiac myocytes and fibroblasts contain functional estrogen receptors. FEBS Lett 416(1):107–112
100. Tobi EW et al (2009) DNA methylation differences after exposure to prenatal famine are common and timing- and sex-specific. Hum Mol Genet 18(21):4046–4053
101. Tatsuguchi M et al (2007) Expression of microRNAs is dynamically regulated during cardiomyocyte hypertrophy. J Mol Cell Cardiol 42(6):1137–1141
102. Toischer K et al (2010) Differential cardiac remodeling in preload versus afterload. Circulation 122(10):993–1003
103. van Rooij E et al (2006) A signature pattern of stress-responsive microRNAs that can evoke cardiac hypertrophy and heart failure. Proc Natl Acad Sci U S A 103(48):18255–18260
104. de Jager T et al (2001) Mechanisms of estrogen receptor action in the myocardium. Rapid gene activation via the ERK1/2 pathway and serum response elements. J Biol Chem 276(30): 27873–27880
105. Deroo BJ, Korach KS (2006) Estrogen receptors and human disease. J Clin Invest 116(3):561–570
106. Murphy E (2011) Estrogen signaling and cardiovascular disease. Circ Res 109(6):687–696
107. St Pierre SR, Peirlinck M, Kuhl E (2022) Sex matters: a comprehensive comparison of female and male hearts. Front Physiol 13:831179
108. Ventura-Clapier R et al. (2020) Sexissues in cardiovascular diseases. Focus on energy metabolism. Biochim Biophys Acta Mol Basis Dis 1866(6):165722
109. Wittnich C et al (2013) Sex differences in myocardial metabolism and cardiac function: an emerging concept. Pflugers Arch 465(5):719–729
110. Regitz-Zagrosek V (2020) Sex and sexdifferences in heart failure. Int J Heart Failure 2(3):157–181
111. Brown RA et al (1996) Influence of sex, diabetes and ethanol on intrinsic contractile performance of isolated rat myocardium. Basic Res Cardiol 91(5):353–360
112. Curl CL, Wendt IR, Kotsanas G (2001) Effects of sexon intracellular. Pflugers Arch 441(5):709–716
113. Schwertz DW et al (2004) Sex differences in the response of rat heart ventricle to calcium. Biol Res Nurs 5(4):286–298
114. Ren J, Ceylan-Isik AF (2004) Diabetic cardiomyopathy: do women differ from men? Endocrine 25(2):73–83
115. Murphy E, Steenbergen C (2014) Estrogen regulation of protein expression and signaling pathways in the heart. Biol Sex Differ 5(1):6
116. Liu D et al (2008) Estrogen-enhanced gene expression of lipoprotein lipase in heart is antagonized by progesterone. Endocrinology 149(2):711–716
117. Hsieh YC et al (2005) PGC-1 upregulation via estrogen receptors: a common mechanism of salutary effects of estrogen and flutamide on heart function after trauma-hemorrhage. Am J Physiol Heart Circ Physiol 289(6):H2665–H2672
118. Ambrosi CM et al (2013) Sexdifferences in electrophysiological gene expression in failing and non-failing human hearts. PLoS ONE 8(1):e54635
119. Di Leva G et al (2013) Estrogen mediated-activation of miR-191/425 cluster modulates tumorigenicity of breast cancer cells depending on estrogen receptor status. PLoS Genet 9(3):e1003311
120. Lagranha CJ et al (2010) Sex differences in the phosphorylation of mitochondrial proteins result in reduced production of reactive oxygen species and cardioprotection in females. Circ Res 106(11):1681–1691
121. McKee LA et al (2013) Sexually dimorphic myofilament function and cardiac troponin I phosphospecies distribution in hypertrophic cardiomyopathy mice. Arch Biochem Biophys 535(1):39–48
122. Lin J et al (2009) Estrogen receptor-beta activation results in S-nitrosylation of proteins involved in cardioprotection. Circulation 120(3):245–254

123. Capasso JM et al (1983) Sex differences in myocardial contractility in the rat. Basic Res Cardiol 78(2):156–171
124. Schwertz DW et al (1999) Sexual dimorphism in rat left atrial function and response to adrenergic stimulation. Mol Cell Biochem 200(1–2):143–153
125. Saito T et al (2009) Estrogen contributes to sexdifferences in mouse ventricular repolarization. Circ Res 105(4):343–352
126. Lowe JS et al (2012) Increased late sodium current contributes to long QT-related arrhythmia susceptibility in female mice. Cardiovasc Res 95(3):300–307
127. Sims C et al (2008) Sex, age, and regional differences in L-type calcium current are important determinants of arrhythmia phenotype in rabbit hearts with drug-induced long QT type 2. Circ Res 102(9):e86-100
128. Johnson BD et al (1997) Increased expression of the cardiac L-type calcium channel in estrogen receptor-deficient mice. J Gen Physiol 110(2):135–140
129. Jiang C et al (1992) Effect of 17 beta-oestradiol on contraction, $Ca^{2+}$ current and intracellular free $Ca^{2+}$ in guinea-pig isolated cardiac myocytes. Br J Pharmacol 106(3):739–745
130. Meyer R et al (1998) Rapid modulation of L-type calcium current by acutely applied oestrogens in isolated cardiac myocytes from human, guinea-pig and rat. Exp Physiol 83(3):305–321
131. Belke DD, Swanson EA, Dillmann WH (2004) Decreased sarcoplasmic reticulum activity and contractility in diabetic db/db mouse heart. Diabetes 53(12):3201–3208
132. Leblanc N et al (1998) Age and sex differences in excitation-contraction coupling of the rat ventricle. J Physiol 511(Pt 2): 533–548
133. Shimoni Y, Liu XF (2003) Sex differences in the modulation of $K^+$ currents in diabetic rat cardiac myocytes. J Physiol 550(Pt 2):401–412
134. Shimoni Y, Liu XF (2004) Sexdifferences in ANG II levels and action on multiple $K^+$ current modulation pathways in diabetic rats. Am J Physiol Heart Circ Physiol 287(1):H311–H319
135. Yaras N et al (2007) Sex-related effects on diabetes-induced alterations in calcium release in the rat heart. Am J Physiol Heart Circ Physiol 293(6):H3584–H3592
136. Brown RA, Walsh MF, Ren J (2001) Influence of sex and diabetes on vascular and myocardial contractile function. Endocr Res 27(4):399–408
137. Durak A, Bitirim CV, Turan B (2020) Titin and CK2α are new intracellular targets in acute insulin application-associated benefits on electrophysiological parameters of left ventricular cardiomyocytes from insulin-resistant metabolic syndrome rats. Cardiovasc Drugs Ther 34(4):487–501
138. Fischer TH et al (2016) Sex-dependent alterations of $Ca^{2+}$ cycling in human cardiac hypertrophy and heart failure. Europace 18(9):1440–1448

Printed in the USA
CPSIA information can be obtained
at www.ICGtesting.com
LVHW021126140924
790985LV00003B/247